ENCYCLOPEDIA OF
Microbiology

Volume 4 S–Z, Index

Editorial Advisory Board

Editor-in-Chief
Joshua Lederberg
Rockefeller University
New York, New York

ENCYCLOPEDIA OF
Microbiology

Volume 4 S–Z, Index

ACADEMIC PRESS, INC.
A Division of Harcourt Brace & Company
San Diego   New York Boston London Sydney Tokyo Toronto

Academic Press, Inc.
1250 Sixth Avenue, San Diego, California 92101-4311

United Kingdom Edition published by
Academic Press Limited
24–28 Oval Road, London NW1 7DX

Library of Congress Cataloging-in-Publication Data

Encyclopedia of microbiology / edited by Joshua Lederberg
      p.   cm.
   Includes bibliographical references and indexes.
   ISBN 0-12-226891-1 (v. 1). -- ISBN 0-12-226892-X (v. 2). -- ISBN
0-12-226893-8 (v. 3). -- ISBN 0-12-226894-6 (v. 4)
   1. Microbiology--Encyclopedias.   I. Lederberg, Joshua.
QR9E53   1992
576'.03--dc20   92-4429
                                                              CIP

PRINTED IN THE UNITED STATES OF AMERICA
   93 94 95 96 97   EB   9 8 7 6 5 4 3 2

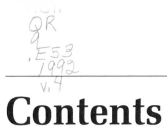

Contents

Preface

For the purposes of this encyclopedia, microbiology has been understood to embrace the study of "microorganisms," including the basic science and the roles of these organisms in practical arts (agriculture and technology) and in disease (public health and medicine). Microorganisms do not constitute a well-defined taxonomic group; they include the two kingdoms of Archaebacteria and Eubacteria, as well as protozoa and those fungi and algae that are predominantly unicellular in their habit. Viruses are also an important constituent, albeit they are not quite "organisms." Whether to include the mitochondria and chloroplasts of higher eukaryotes is a matter of choice, since these organelles are believed to be descended from free-living bacteria. Cell biology is practiced extensively with tissue cells in culture, where the cells are manipulated very much as though they were autonomous microbes; however, we shall exclude this branch of research. Microbiology also is enmeshed thoroughly with biotechnology, biochemistry, and genetics, since microbes are the canonical substrates for many investigations of genes, enzymes, and metabolic pathways, as well as the technical vehicles for discovery and manufacture of new biological products, for example, recombinant human insulin.

Within these arbitrarily designated limits, let us consider the overall volume of published literature in microbiology, where to find its core, and strategies for searching for current information on particular topics. Most of the data for this preface are derived from the 1988 Journal Citation Reports Current Contents (T) of the Institute for Scientific Information (ISI). Table I lists the 53 most consequential journals in microbiology, assessed by citation impact factor, the average number of literature citations per article published in a given journal. Table II presents that list sorted by the total number of articles printed in each journal in 1988. Table III shows the distribution of journals citing the *Journal of Bacteriology* and the distribution of journals cited in it.

Obviously, the publications of the American Society for Microbiology (indicated by AMS in the tables) play a commanding role. The society is now making its journals available in electronically searchable form (on optical disks), which will greatly facilitate locating and retrieving the most up-to-date information on any given subject. In addition, interdisciplinary journals such as *Nature (London), Science,* and the *Proceedings of the National Academy of Sciences, U.S.A.* are important sources of prompt news of scientific developments in microbiology. It is difficult to assess how much of their total publication addresses microbiology. As seen in Table III, the bibliographies in the *Journal of Bacteriology* cite half as many articles from the *Proceedings* (2348) as from the *Journal of Bacteriology* itself (5708). The 7038 articles indicated in Table II probably reach some 10,000 per year when these interdisciplinary and other dispersed sources are taken into account. An equal number might be added from overlapping aspects of molecular biology and genetics. To find and read all these titles would tax any scholar, although it could be done as a near full-time occupation with the help of the weekly Current Contents (T) of the ISI. To start afresh, with perhaps a decade's accumulation of timely background, would be beyond reasonable human competence. No one person would intelligently peruse more than a small fraction of the total texts.

The "Encyclopedia of Microbiology" is intended to survey the entire field coherently, complementing material that would be included in an advanced undergraduate and graduate major course of university study. Particular topics should be accessible to talented high school and college students, as well as graduates involved in teaching, research, and technical practice of microbiology.

Even these hefty volumes cannot embrace all current knowledge in the field. Each article does provide key references to the literature available at the time of writing. Acquisition of more detailed and up-to-date knowledge depends on (1) exploiting the review and monographic literature and (2) bibliographic retrieval of the preceding and current research literature. To make greatest use of review literature and monographs, the journals listed in Table II are invaluable. Titles such as *Annual Reviews* should not be misunderstood: these journals appear at annual intervals, but 5 or 10 years of accumulated research is necessary for the inclusion of a focused treatment of a given subject.

To access bibliographic materials in microbiology, the main retrieval resources are Medline, sponsored by the U.S. National Library of Medicine, and the Science Citation Index of the ISI. With governmental subsidy, Medline is widely available at modest cost: terminals are available at every medical school and at many other academic centers. Medline provides searches of the recent literature by author, title, and key word, and offers on-line displays of the relevant bibliographies and abstracts. Medical aspects of microbiology are covered exhaustively; general microbiology is covered in reasonable depth. The Science Citation Index must recover its costs from user fees, but is widely available at major research centers. It offers additional search capabilities, especially by citation linkage. Therefore, starting with the bibliography of a given encyclopedia article, one can quickly find (1) all articles more recently published that have cited those bibliographic reference starting points and (2) all other recent articles that share bibliographic information with the others. With luck, one of these articles may be identified as another comprehensive review that has digested more recent or broader primary material.

On a weekly basis, services such as Current Contents on Diskette (ISI) and Reference Update offer still more timely access to current literature as well as abstracts with a variety of useful features. Under the impetus of intense competition, these services are evolving rapidly, to the great benefit of a user community desperate for electronic assistance in coping with the rapidly growing and intertwined networks of discovery. The bibliographic services of Chemical Abstracts and Biological Abstracts would also be potentially invaluable; however, their coverage of microbiology is rather limited.

In addition, major monographs have appeared from time to time—"The Bacteria," "The Pro-

karyotes," and many others. Your local reference library should be consulted for these volumes.

Valuable collections of reviews also include *Critical Reviews for Microbiology, Symposia of the Society for General Microbiology, Monographs of the ASM,* and *Proceedings of the International Congresses of Microbiology.*

The articles in this encyclopedia are intended to be accessible to a broader audience, not to take the place of review articles with comprehensive bibliographies. Citations should be sufficient to give the reader access to the latter, as may be required. We do apologize to many individuals whose contributions to the growth of microbiology could not be adequately embraced by the secondary bibliographies included here.

The organization of encyclopedic knowledge is a daunting task in any discipline; it is all the more complex in such a diversified and rapidly moving domain as microbiology. The best way to anticipate the rapid further growth that we can expect in the near future is unclear. Perhaps more specialized series in subfields of microbiology would be more appropriate. The publishers and editors would welcome readers' comments on these points, as well as on any deficiencies that may be perceived in the current effort.

My personal thanks are extended to Kathryn Linenger at Academic Press for her diligent, patient, and professional work in overseeing this series; to my coeditors, Martin Alexander, David A. Hopwood, Barbara H. Iglewski, and Allan I. Laskin; above all, to the many very busy scientists who took time to draft and review each of these articles.

Joshua Lederberg

Table I   The Top Journals in Microbiology Listed by Impact Factor

Citation impact rank	Journal title	Number of articles published in 1988	Citation impact rank	Journal title	Number of articles published in 1988
1	Microbiol. Rev.	28	28	FEMS Microbiol. Lett.	365
2	Adv. Microb. Ecol.	10	29	Am. J. Reprod. Immunol.	50
3	Annu. Rev. Microbiol.	29	30	Infection	103
4	FEMS Microbiol. Rev.	13	31	Can. J. Microbiol.	236
5	Yeast	NA	32	Curr. Microbiol.	87
6	J. Bacteriol.	915	33	J. Appl. Bacteriol.	125
7	Mol. Microbiol.	94	34	J. Microbiol. Meth.	34
8	Antimicrob. Agents Ch.	408	35	B. I. Pasteur	20
9	Rev. Infect. Dis.	213	36	ZBL Bakt. Mikr. Hyg. A	164
10	CRC Crit. Rev. Microbiol.	12	37	Ann. Inst. Pasteur Mic.	58
11	Syst. Appl. Microbiol.	52	38	Vet. Microbiol.	104
12	Int. J. Syst. Bacteriol.	83	39	Acta Path. Micro. Im. B	NA
13	J. Antimicrob. Chemoth.	352	40	Protistologica	NA
14	Appl. Environ. Microb.	588	41	Med. Microbiol. Immun.	37
15	J. Clin. Microbiol.	619	42	Diagn. Micr. Infec. Dis.	60
16	Adv. Appl. Microbiol.	8	43	Int. J. Food Microbiol.	66
17	Curr. Top. Microbiol.	53	44	J. Gen. Appl. Microbiol.	27
18	Arch. Microbiol.	173	45	Microbiol. Immunol.	122
19	J. Gen. Microbiol.	367	46	Lett. Appl. Microbiol.	81
20	Enzyme Microb. Tech.	108	47	Gen. Physiol. Biophys.	57
21	Eur. J. Clin. Microbiol.	161	48	A. Van Leeuw. J. Microb.	51
22	FEMS Microbiol. Ecol.	42	49	Symbiosis	14
23	J. Med. Microbiol.	124	50	Comp. Immunol. Microb.	27
24	J. Infection	68	51	Microbios.	61
25	Eur. J. Protistol.	37	52	ZBL Bakt. Mikr. Hyg. B	76
26	Microbiol. Sci.	70	53	J. Basic Microb.	69
27	Appl. Microbiol. Biot.	270			

NA, Not available.

Table II   Microbiology Journals Listed by Total Number of Articles Published per Year (1988)

Journal title	Number of articles published in 1988	Journal title	Number of articles published in 1988
J. Bacteriol.	915	Int. J. Food Microbiol.	66
J. Clin. Microbiol.	619	Microbios.	61
Appl. Environ. Microb.	588	Diagn. Micr. Infec. Dis.	60
Antimicrob. Agents Ch.	408	Ann. Inst. Pasteur Mic.	58
J. Gen. Microbiol.	367	Gen. Physiol. Biophys.	57
FEMS Microbiol. Lett.	365	Curr. Top. Microbiol.	53
J. Antimicrob. Chemoth.	352	Syst. Appl. Microbiol.	52
Appl. Microbiol. Biot.	270	A. Van Leeuw. J. Microb.	51
ZBL Bakt. Mikr. Hyg. A	240	Am. J. Reprod. Immunol.	50
Can. J. Microbiol.	236	FEMS Microbiol. Ecol.	42
Rev. Infect. Dis.	213	Med. Microbiol. Immun.	37
Arch. Microbiol.	173	Eur. J. Protistol.	37
Eur. J. Clin. Microbiol.	161	J. Microbiol. Meth.	34
J. Appl. Bacteriol.	125	Eur. J. Protistology	29
J. Med. Microbiol.	124	Annu. Rev. Microbiol.	29
Microbiol Immunol.	122	Microbiol. Rev.	28
Enzyme Microb. Tech.	108	J. Gen. Appl. Microbiol.	27
Vet. Microbiol.	104	Comp. Immunol. Microb.	27
Infection	103	B. I. Pasteur	20
Mol. Microbiol.	94	Acta Path. Micro. Im.	18
Curr. Microbiol.	87	Symbiosis	14
Int. J. Syst. Bacteriol.	83	FEMS Microbiol. Rev.	13
Lett. Appl. Microbiol.	81	CRC Crit. R. Microbiol.	12
Microbiol. Sci.	70	Adv. Microb. Ecol.	10
J. Basic Microb.	69	Adv. Appl. Microbiol.	8
		Total	7038

Table III.A   Distribution of Journals Cited in *Journal of Bacteriology*, 1979–1988

Journal cited	Number of citations	Journal cited	Number of citations
J. Bacteriol.	5708	Genetics	183
P. Natl. Acad. Sci. U.S.A.	2348	Can. J. Microbiol.	139
J. Biol. Chem.	1698	Arch. Biochem. Biophys.	127
Mol. Gen. Genet.	1157	Virology	123
J. Mol. Biol.	1148	Bacteriol. Rev.	118
Gene	902	Cold Spring Harb. Sym.	110
Nature (London)	820	Antimicrob. Agents Ch.	109
Nucleic Acids Res.	804	Escherichia Coli Sal.	95
Cell	802	Plant Physiol.	80
J. Gen. Microbiol.	701	J. Biochem.-Tokyo	78
Infect. Immun.	478	J. Virol.	78
Methods Enzymol.	434	Mol. Cell. Biol.	68
Anal. Biochem.	411	J. Infect. Dis.	67
Biochim. Biophys. Acta	401	Bio-Technol.	61
Eur. J. Biochem.	376	Exp. Gene Fusions	60
Mol. Cloning Laboratory	363	Trends Biochem. Sci.	60
Microbiol. Rev.	361	Mutat. Res.	59
Arch. Microbiol.	347	Syst. Appl. Microbiol.	55
Embo J.	327	Phytopathology	51
Biochemistry-U.S.	310	Adv. Bacterial Genet.	50
Science	301	Photochem. Photobiol.	50
Appl. Environ. Microb.	294	Biochimie	49
FEMS Microbiol. Lett.	257	J. Exp. Med.	48
Exp. Mol. Genetics	234	Agr. Biol. Chem. Tokyo	47
Plasmid	234	Int. J. Syst. Bacteriol.	44
Biochem. Bioph. Res. Commun.	224	FEMS Microbiol. Rev.	43
FEBS Lett.	213	J. Clin. Microbiol.	42
Biochem. J.	207	Curr. Microbiol.	41
Annu. Rev. Microbiol.	194	J. Cell Biol.	41
Annu. Rev. Biochem.	188		
Annu. Rev. Genet.	187	All other (1301)	4311

(continues)

Table III.B (*continued*) Distribution of Journals Citing *Journal of Bacteriology*, 1979–1988

Journal citing	Number of citations	Journal citing	Number of citations
J. Bacteriol.	5708	Curr. Genet.	117
J. Biol. Chem.	1119	FEMS Microbiol. Rev.	115
J. Gen. Microbiol.	963	J. Basic Microb.	115
Mol. Gen. Genet.	896	J. Antimicrob. Chemoth.	112
Appl. Environ. Microb.	890	Microb. Pathogenesis	110
Microbiol. Rev.	759	Science	104
Infect. Immun.	663	Ann. Inst. Pasteur Mic.	101
FEMS Microbiol. Lett.	648	Methods Enzymol.	99
Gene	599	ZBL Bakt. Mikr. Hyg. A	98
P. Natl. Acad. Sci. U.S.A.	588	A. Van Leeuw. J. Microb.	95
Can. J. Microbiol.	579	Annu. Rev. Biochem.	94
Arch. Microbiol.	484	Plant Physiol.	88
Mol. Microbiol.	452	J. Infect. Dis.	86
J. Mol. Biol.	434	J. Med. Microbiol.	85
Nucleic Acids Res.	431	Folia Microbiol.	79
Biochim. Biophys. Acta	378	Genetika	79
Eur. J. Biochem.	350	Gene Dev.	78
Antimicrob. Agents Ch.	340	Microbios.	77
Annu. Rev. Microbiol.	316	Arch. Biochem. Biophys.	75
Cell	246	Biotechnol. Bioeng.	73
Biochimie	238	Nature (London)	69
Biochemistry-U.S.	236	Syst. Appl. Microbiol.	69
Plasmid	236	Zh. Mikrob. Epid. Immun.	67
Embo J.	234	J. Antibiot.	66
J. Clin. Microbiol.	214	Annu. Rev. Genet.	65
Genetics	201	Microbiol. Immunol.	65
Adv. Microb. Physiol.	199	J. Biochem.-Tokyo	64
Agr. Biol. Chem. Tokyo	198	Microbial Ecol.	60
Mol. Cell. Biol.	197	Plant Soil	58
CRC Crit. R. Microbiol.	194	Anal. Biochem	56
Curr. Microbiol	193	Annu. Rev. Cell Biol.	55
Appl. Microbiol. Biot.	183	Biotechnol. Lett.	54
J. Appl. Bacteriol.	169	Adv. Microb. Ecol.	53
Mutat. Res.	160	Enzyme Microb. Tech.	53
Biochem. Bioph. Res. Commun.	152	Curr. Sci. India	52
Rev. Infect. Dis.	141	Eur. J. Clin. Microbiol.	51
Biochem. J.	137	J. Theor. Biol.	51
Microbiol. Sci.	135	Bot. Acta	50
Int. J. Syst. Bacteriol.	128	Photochem. Photobiol.	50
FEBS Lett	125		

How to Use the Encyclopedia

This encyclopedia is organized in a manner that we believe will be the most useful to you, and we would like to acquaint you with some of its features.

The volumes are organized alphabetically as you would expect to find them in, for example, magazine articles. Thus, "Foodborne Illness" is listed as such and would not be found under "Illness, Foodborne." If the first words in a title are not the primary subject matter contained in an article, the main subject of the title is listed first (e.g., "Heavy Metals, Bacterial Resistances," "Marine Habitats, Bacteria," "Method, Philosophy," "Transcription, Viral"). This is also true if the primary word of a title is too general (e.g.,"Bacteriocins, Molecular Biology"). Here, the word "bacteriocins" is listed first because "molecular biology" is a very broad topic. Titles are alphabetized letter-by-letter so that "Cell Membrane: Structure and Function" is followed by "Cellulases" and then by "Cell Walls of Bacteria."

Each article contains a brief introductory Glossary wherein terms that may be unfamiliar to you are defined *in the context of their use in the article*. Thus, a term may appear in another article defined in a slightly different manner or with a subtle pedagogic nuance that is specific to that particular article. For clarity, we have allowed these differences in definition to remain so that the terms are defined relative to the context of each article.

Articles about closely related subjects are identified in the Index of Related Titles at the end of the last volume (Volume 4). The article titles that are cross-referenced within each article may be found in this index, along with other articles on related topics.

The Subject Index contains specific, detailed information about any subject discussed in the *Encyclopedia*. Entries appear with the source volume number in boldface followed by a colon and the page number in that volume where the information occurs (e.g., "DNA repair by bacterial cells, 2:9"). Each article is also indexed by its title (or a shortened version thereof), and the page ranges of the article appear in boldface (e.g. "Lyme disease, **2:639–646**" means that the primary coverage of the topic of Lyme disease occurs on pages 639–646 of Volume 2).

If a topic is covered primarily under one heading but additional related information may be found elsewhere, a cross-reference is given to the related material. For example, "Biodegradation" would contain all the page numbers where relevant information occurs, followed by "*See also* Bioremediation; Pesticide biodegradation" for different but related information. Similarly, a "*See*" reference refers the reader from a less-used synonym (or acronym) to a more specific or descriptive subject heading. For example, "Immunogens, synthetic. *See* Vaccines, synthetic." A *See under* cross-reference guides the reader to a specific subheading under a term. For example, "Mixis. *See under* Genome rearrangement."

An additional feature of the Subject Index is the identification of Glossary terms. These appear in the index where the word "defined" (or the words "definition of") follows an entry. As we noted earlier, there may be more than one definition for a particular term, and as when using a dictionary, you will be able to choose among several different usages to find the particular meaning that is specifically of interest to you.

S

Sexually Transmitted Diseases

Stephen A. Morse
Centers for Disease Control

Glossary

Epidemiology Science of epidemics and epidemic diseases

Gonorrhea Contagious sexually transmitted infection of the genital mucosa caused by the bacterium *Neisseria gonorrhoeae*

Incidence Cases per population accumulated over a period of observation

Prevalence Number of cases ascertained over a fixed (usually short) period of time

Syndrome The aggregate of signs and symptoms associated with any morbid process, together constituting the picture of the disease

Syphilis Acute and chronic infectious disease caused by *Treponema pallidum* and transmitted by direct contact, usually through sexual intercourse

SEXUALLY TRANSMITTED DISEASES (STDs) are infections that are transmitted through sexual contact. Sexually transmitted infections may also be transmitted from pregnant women to their fetuses or during the birth process. Sexually transmitted agents include bacteria, viruses, protozoa, fungi, and ectoparasites. Sexually transmitted agents can cause diseases that range from mildly symptomatic to fatal. STDs are very common infections in both developing and developed countries.

I. Introduction

A. Scope

STDs are a significant medical and public health problem. STDs represent a large proportion of ambulatory patient visits and a growing proportion of hospitalized patients seen by many primary care physicians. In addition, STDs are also implicated in a wide spectrum of acute inflammatory conditions and in a variety of preneoplastic, neoplastic, and postinflammatory complications seen by many dermatologists, urologists, obstetricians, and gynecologists. Pediatricians more frequently diagnose and manage sexually transmitted infections in the neonate and infant, the abused older child, and the adolescent. Internists and infectious disease specialists must deal with the increasing number of patients infected by the human immunodeficiency virus (HIV-1 and HIV-2) who subsequently develop acquired immunodeficiency syndrome (AIDS) and related opportunistic infections. [*See* ACQUIRED IMMUNODEFICIENCY SYNDROME (AIDS).]

Until recently, the scope of STD was limited to five classical infections—gonorrhea, syphilis, chancroid, lymphogranuloma venereum (LGV), and granuloma inguinale (Donovanosis). A major shift in thinking has occurred over the past two decades as clinicians, laboratory scientists, and public health workers became increasingly aware of the broader spectrum of STDs. Our concepts of these infections have changed considerably. The classical "venereal diseases," such as gonorrhea and syphilis, have been overshadowed by other sexually transmitted infections that are actually more common. Syndromes associated with *Chlamydia trachomatis,* human papillomavirus (HPV), and herpes simplex virus type 2 (HSV-2), have become prominent, in part because of a wider array of laboratory diagnostic techniques. [*See* HERPESVIRUSES; CHLAMYDIAL INFECTIONS, HUMAN.]

With the exception of HIV, the "newly" described sexually transmitted infections long have been recognized, but have achieved recent prominence as STDs for several reasons. The development of new diagnostic tests and the improvement of existing tests have facilitated investigations of the etiology, prevalence, mode of transmission, and clinical consequences of STDs. Changes in sexual behavior have resulted in a higher rate of sexual transmission of infectious agents that have multiple modes of transmission [e.g., hepatitis A, B, and C viruses, cytomegalovirus (CMV), and some enteric pathogens]. In addition, the maturing of the "baby boom" generation in the 1960s and 1970s resulted in an increase in the proportion of young adults in the population that was faster than the increase of the total population; these individuals were also more sexually active than their predecessors. The association of STDs such as AIDS, HPV-associated genital cancers, and chronic recurrent genital herpes with incurable and fatal infections has captured the attention of the media and the public. It has also been recognized that women and children suffer from a disproportionate share of the complications of STDs, for example, pelvic inflammatory disease (PID), infertility, ectopic pregnancy, adverse pregnancy outcome, infant pneumonia, and developmental disability. PID and its sequelae (ectopic pregnancy and infertility) are responsible for the majority of the cost associated with STDs (excluding AIDS). For example, it has been estimated that the economic cost of pelvic inflammatory disease alone exceeds $2 billion per year in the United States.

Even the longest-known STDs are undergoing a resurgence. The number of cases of syphilis and chancroid has increased in recent years to the highest in decades. Much of this increase has been associated with casual sex and the exchange of sex for drugs or money. Although reported gonorrhea cases continue to decline, both the number and variety of antibiotic-resistant strains have increased dramatically, resulting in the use of more expensive antimicrobial therapy regimens.

B. Epidemiology

Currently, at least 25 organisms and 25 syndromes are recognized as sexually transmitted. Many of these have been recognized recently, partly as a result of improved laboratory techniques. Common sexually transmitted agents and the diseases and syndromes they cause are listed in Table I. An in-depth discussion of the biology and epidemiology of all these sexually transmitted agents is beyond the scope of this article. The interested reader is referred to the bibliography for additional references. [*See* EPIDEMIOLOGIC CONCEPTS.]

Common behavioral and biological features characterize the epidemiology of STDs. These infections typically have a long latent or incubation period, so transmission to another individual often occurs before symptoms are apparent. With one exception (hepatitis B), vaccination is ineffective in preventing these infections. Many of the agents listed in Table I are highly adapted to humans and reside on the genital mucosa. Thus, contact between genital mucosal surfaces—sexual intercourse, in most instances—is the major mode of transmission. Behavioral and environmental factors determine whether sexual transmission is the primary route of spread for these agents. For example, in many developing countries hepatitis B and CMV infections are acquired in childhood because of low standards of hygiene or poor living conditions. However, in industrialized countries the same infections are acquired in adulthood through sexual contact. In general, infectious agents are more often spread by sexual contact in persons with an increasing standard of living, since opportunities for person-to-person transmission during childhood are reduced.

Factors that affect the spread of STDs (risk factors) are directly related to patterns of sexual behavior. Risk factors include multiple sexual partners, a history of STDs, urban residence, single marital status, and young age. Thus, the highest rates of STDs are found in urban men and women in their sexually most active years, that is, between 15 and 35 years of age.

Prostitutes are an important reservoir of STDs and play an important role in the spread of these infections in many parts of the world. It is difficult to make generalizations concerning the role of prostitutes because major differences in sexual behavior patterns are known to exist within continents and even within countries. For example, prostitutes as a

Table I Major Sexually Transmitted Agents and the Diseases They Cause

Agents	Disease or syndrome
Bacteria	
Neisseria gonorrhoeae	Urethritis; epididymitis; proctitis; bartholinitis; cervicitis; endometritis; salpingitis and related sequelae (infertility, ectopic pregnancy); perihepatitis; complications of pregnancy (e.g., chorioamnionitis, premature rupture of membranes, premature delivery, postpartum endometritis); conjunctivitis; disseminated gonococcal infection (DGI)
Chlamydia trachomatis	Same as *N. gonorrhoeae*, except for DGI; also, lymphogranuloma venereum,[a] Reiter's syndrome; infant pneumonia
Treponema pallidum	Syphilis[a]
Haemophilus ducreyi	Chancroid[a]
Calymmatobacterium granulomatis	Donovanosis[a]
Mycoplasma hominis	Postpartum fever; salpingitis
Ureaplasma urealyticum	Urethritis; low birth weight (?); chorioamnionitis (?)
?Gardnerella vaginalis, Mobiluncus sp., *?Bacteroides* sp.	Bacterial vaginosis
?Group B β-hemolytic streptococci	Neonatal sepsis; neonatal meningitis
Virus	
Herpes simplex virus (HSV-2, HSV-1)	Primary and recurrent genital herpes[a]
Hepatitis B virus (HBV)	Acute, chronic, and fulminant hepatitis B, with associated immune complex phenomena and sequelae including cirrhosis and hepatocellular carcinoma
Cytomegalovirus (CMV)	Congenital infection: gross birth defects and infant mortality, cognitive impairment (e.g., mental retardation, 8th nerve deafness), heterophile-negative infectious mononucleosis, protean manifestations in the immunosuppressed host
Human papilloma virus (HPV)	Condyloma acuminata; laryngeal papilloma in infants; squamous epithelial neoplasias of the cervix, anus, vagina, vulva, penis
Molluscum contagiosum virus (MCV)	Genital molluscum contagiosum
Human immunodeficiency virus (HIV-1, HIV-2)	AIDS and related conditions
Human T-lymphotropic virus type 1	T-cell leukemia/lymphoma; tropical spastic paraparesis
Protozoa	
Trichomonas vaginalis	Vaginitis; urethritis (?); balanitis (?)
Fungus	
Candida albicans	Vulvovaginitis; balanitis; balanoposthitis
Ectoparasite	
Phthirus pubis	Pubic lice infestation
Sarcoptes scabiei	Scabies

[a] These infections are responsible for the syndrome known as genital ulcer disease.

source of disease are named by up to 80% of male patients in some parts of the developing world, compared with less than 20% in Europe and North America.

From a global perspective, societal problems are also important in the epidemiology of STDs. Increasing urbanization with disruption of traditional social structures, increased mobility for economic or political reasons, poor medical facilities, a population with a high proportion of teenagers and young adults, poverty, and high unemployment rates all

contribute to the high incidence of STDs and their sequelae.

It is difficult to estimate the global incidence of STDs because many developing countries do not conduct surveillance for these infections or lack the resources for the laboratory confirmation. In developing countries such as those in Africa, the prevalence of gonorrhea, syphilis, and chlamydial infections has been estimated by screening pregnant women in urban settings. Using this approach, the prevalence of gonorrhea has been reported to vary between 2 and 15%, whereas the prevalence of chlamydial infections and syphilis has been reported to be as high as 13% and 18%, respectively. These figures may actually underestimate the prevalence of these infections, since infertile women are excluded from these surveys. Nevertheless, these figures provide evidence that STDs are a significant component of the morbidity and mortality due to infectious diseases in the developing world.

It is not uncommon for individuals to be infected with more than one sexually transmitted agent at a time. Some studies have found individuals infected with as many as six different STDs. Co-infection is a significant problem. For example, several studies have indicated that approximately 25% of individuals with gonorrhea are co-infected with *C. trachomatis*. Co-infection with *N. gonorrhoeae* and *C. trachomatis* has influenced the treatment recommendations for gonorrhea. Treatment recommendations now include antibiotics that will be effective against both agents. Reinfections are common and probably represent continued high risk behaviors as well as infections that do not generate a protective immune response.

In the United States, gonorrhea, syphilis, chancroid, LGV, and granuloma inguinale are reportable diseases. The Centers for Disease Control compile the statistics of the case reports that are collected at the state and local level. The incidences of these STDs in 1990 are shown in Table II. The figures for the other STDs are based on estimates from various surveys. More than 17 million infections were transmitted sexually in the United States in 1989–1990. Thus, STDs remain major medical and public health problem in developed countries.

II. Selected Sexually Transmitted Diseases

The STDs described in this section have been selected because they reflect differences in our understanding of the pathogenesis, epidemiology, diagnosis, and control of sexually transmitted agents. Gonorrhea represents a STD with a decreasing reported incidence; however, increasing antimicrobial resistance has presented problems in the effective control of this infection. Syphilis represents a STD with an incidence that has increased dramatically in recent years. Our inability to culture the causative agent has severely restricted our understanding of this infection.

Table II Incidence of Various STDs in the United States

Disease/syndrome	Estimated no. of cases[a]	Reported no. of cases[b]
Gonorrhea	1,400,000	690,323
Syphilis		134,255
C. trachomatis infections	3,000,000	
Human papilloma virus	425,000	
Genital herpes	200,000–500,000	
Trichomoniasis	1,000,000	
Other vaginitis[c]	9,500,000	
Hepatitis[b]	200,000	
Urethritis[d]	1,200,000	
Mucopurulent cervicitis[d]	1,000,000	
Pelvic inflammatory disease	750,000	
Chancroid		4215
Lymphogranuloma venereum		277
Granuloma inguinale		97

[a] Annual report, Division of STD/HIV Prevention, National Center for Prevention Services, Centers for Disease Control. Estimates are for fiscal year 1989.

[b] Number of cases reported in 1990.

[c] The evidence concerning the role of sexual transmission in these infections is not clear.

[d] Infections not due to N. gonorrhoeae or C. trachomatis.

A. Gonorrhea

1. Clinical Manifestations

Gonorrhea is an ancient disease. Descriptions of an illness resembling gonorrhea can be found in writings as early as 2637 B.C. Infections caused by *Neisseria gonorrhoeae* are generally limited to superficial mucosal surfaces lined with columnar epithelial cells. Anatomic sites most frequently infected are the cervix, urethra, rectum, pharynx, and conjunctiva. Squamous epithelium, which lines the adult vagina, is not susceptible to infection by the gonococcus. However, the prepubertal vaginal epithelium, which has not been keratinized under the influence of estrogen, may be infected (gonorrhea in young girls may present as vulvovaginitis). Mucosal infections are usually characterized by a marked local inflammatory response characterized by a purulent discharge.

The most common symptom of uncomplicated gonorrhea in men is a discharge that may range from a scanty, clear, or cloudy fluid to one that is copious and purulent. Pain upon urination (dysuria) is often present. Men with asymptomatic urethritis are an important reservoir for transmission. In addition, asymptomatic men and those who ignore their symptoms are at increased risk for developing complications.

Endocervical infection, the most common form of uncomplicated gonorrhea in women, is usually characterized by vaginal discharge and sometimes by dysuria (because of coexistent urethritis). About 50% of women with cervical infections are asymptomatic. Local complications include abcesses in Bartholin's and Skene's glands.

Rectal infections with *N. gonorrhoeae* occur in up to one-third of women with cervical infection. These infections most probably result from autoinoculation with cervical discharge and are rarely symptomatic. Rectal infections in homosexual men usually result from anal intercourse and are more often symptomatic (proctitis).

Pharyngeal infections with *N. gonorrhoeae* are diagnosed most often in women and homosexual men with a history of fellatio. Pharyngeal gonococcal infections may be a focal source of gonococcemia (i.e., bacteremia due to *N. gonorrhoeae*).

Ocular infections can have serious consequences such as corneal scarring or perforation of the cornea; thus, prompt diagnosis and treatment are important. Ocular infections occur most commonly in newborns who are exposed to infected secretions in the birth canal (opthalmia neonatorum). The instillation of silver nitrate drops into the eyes of newborn infants is recommended to prevent gonococcal opthalmia neonatorum. Erythromycin or tetracycline opthalmic ointments have been used, but have been associated with occasional treatment failures due to antibiotic resistance. Gonococcal keratoconjunctivitis is occasionally seen in adults as a result of autoinoculation.

Disseminated gonococcal infections (DGI) result from bacteremic spread from a primary mucosal infection. Asymptomatic infections of the pharynx, urethra, or cervix often serve as focal sources for bacteremia. The most common form of DGI is the dermatitis—arthritis syndrome. It is characterized by fever, chills, skin lesions, and pain in the joints (arthralgia) of the hands, feet, and elbows that is due to inflammation of the tendon sheaths surrounding the joint. Rarely, DGI causes endocarditis or meningitis.

Gonococci may also spread by a nonbacteremic

route that involves ascension from the endocervical canal through the endometrium to the fallopian tubes and ultimately to the peritoneal cavity, resulting in endometritis, salpingitis, parametritis, and peritonitis. This is commonly referred to as PID. PID may also be caused by other sexually transmitted organisms (e.g., *C. trachomatis*), as well as by nonsexually transmitted bacteria (e.g., anaerobes) that are part of the normal vaginal flora. As many as 15% of women with uncomplicated cervical infections develop PID. Sequelae of PID may be serious, including increased probability of infertility and ectopic pregnancy.

2. Diagnosis

The Gram stain is an inexpensive and highly sensitive and specific method for the presumptive diagnosis of gonorrhea in men with symptomatic urethritis. Isolation of *N. gonorrhoeae* on selective medium is often used for cervical, rectal, and urethral specimens. For medicolegal reasons, it is important to confirm the identity of the isolate as *N. gonorrhoeae* in order to avoid confusion with other *Neisseria* spp. that inhabit these sites. Newer technologies include antigen detection tests and nucleic acid probe tests.

3. Epidemiology and Control

Gonorrhea is the most frequently reported communicable disease in the United States. The incidence of gonorrhea has decreased steadily since 1978 (Fig. 1). There were 690,323 cases of gonorrhea reported to the Centers for Disease Control in 1990, a 6% decrease over the 1989 figure. Reported cases represent only about half the true number of cases because of underreporting by private physicians, nonspecific treatment without culture diagnosis, and self-treatment. Incidence and prevalence rates from developing countries are scarce and less reliable than those from the United States because of limited surveillance and variable diagnostic criteria. Various estimates suggest that prevalence rates of gonococcal infections in developing countries may be 5–50 times those in the United States.

The peak incidence and prevalence of gonorrhea in the United States occurs in adolescents and young adults. The peak rate is at 20–24 years in men and at 15–19 years in women. In 1990, 70% of gonorrhea in women occurred in women 24 years of age or younger. Gonorrhea rates are substantially higher among minorities. In 1990, the gonorrhea rate among hispanics and blacks was 2.8 and 39 times greater, respectively, than the rate for whites.

Epidemiologic data suggest that continued transmission occurs, in part, to a core group of highly effective transmitters who disproportionately contribute to endemic levels of disease. The core group includes individuals residing in areas of high population density and low socioeconomic status.

Several measures have been useful in controlling the spread of gonorrhea as well as other STDs. Condoms are effective in preventing the transmission of gonorrhea. Screening has been useful in identifying asymptomatic individuals. Contact tracing has been used also to identify contacts who were exposed to the index patient and who may have become infected or those with ignored symptoms.

The evolution of antimicrobial resistance in *N. gonorrhoeae* has had an important effect on the control of gonorrhea. Treatment costs have increased substantially as relatively ineffective and inexpensive antibiotics such as penicillin have been replaced by more expensive broad-spectrum third-generation cephalosporins. Strains with multiple chromosomal resistance to antibiotics such as penicillin, tetracycline, erythromycin, and cefoxitin have been identified in the United States and in most other parts of the world. Sporadic high-level resistance to the aminocyclitol spectinomycin also has been reported.

Penicillinase-producing strains of *N. gonorrhoeae* were first described in 1976. Five related beta-lactamase plasmids of different sizes have been identified in these organisms. Penicillinase-producing strains cause more than one-half of all gonococcal infections in parts of Africa and Asia. Their prevalence has increased dramatically in the United States since 1984 (Fig. 1) and has affected nearly every major metropolitan area. Plasmid-mediated high-level resistance of *N. gonorrhoeae* to tetracycline is a more recent phenomenon. This resistance is due to the presence of the streptococcal *tetM* determinant on a gonococcal conjugative plasmid. Strains with plasmid-mediated tetracycline resistance have been isolated in many areas of the world.

4. Organism and Pathogenesis

The gonococcus is an obligate human pathogen. It is one of two *Neisseria* species that cause significant human infections. Gonococci are gram-negative cocci, 0.6–1.0 μm in diameter. The organisms are usually seen in pairs with the adjacent sides flattened. Pili—proteinaceous hairlike appendages—extend several micrometers from the cell surface and play a role in adherence. The outer membrane is composed of proteins, phospholipids, and lipopolysaccharide (LPS). Features that distinguish gonococcal LPS from enteric LPS are the highly branched basal oligosaccharide structure and the absence of repeating O-antigen subunits. For these reasons, gonococcal LPS is often referred to as lipo-oligosaccharide (LOS).

Our knowledge of the molecular basis of gonococcal pathogenesis is incomplete. Mucosal surfaces susceptible to gonococcal infection are lined by ciliated and nonciliated columnar epithelial cells. Attachment of gonococci to the microvilli of nonciliated columnar epithelial cells is mediated in part by pili and an outer membrane protein called protein II, although nonspecific factors such as surface charge and hydrophobicity may be important. Gonococci do not appear to attach to ciliated cells. Following attachment, gonococci invade the host cell. Much of our knowledge of gonococcal invasion comes from studies with tissue culture cells and human fallopian tube organ culture. After gonococci attach to the nonciliated cells of the fallopian tube, they are surrounded by the microvilli, which draw them to the surface of the mucosal cell. The gonococci appear to enter the epithelial cells by a process called parasite-directed endocytosis. The major porin protein of the gonococcal outer membrane, protein I, is thought to stimulate this process. During endocytosis, the membrane of the mucosal cell retracts, pinching off a membrane-bound vacuole that contains gonococci; this vacuole is rapidly transported to the base of the cell, where gonococci are released by exocytosis into the subepithelial tissue. Gonococci are not

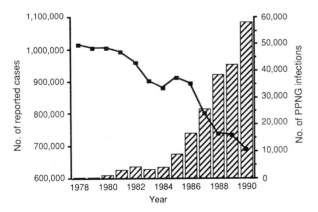

Figure 1 Reported gonorrhea cases and infections caused by penicillinase-producing *N. gonorrhoeae* in the United States, 1978–1990.

destroyed in the phagocytic vacuole; it is not clear whether they replicate in the vacuoles.

The presence of an antiphagocytic capsule has not been demonstrated in *N. gonorrhoeae*. Gonococci do not elaborate extracellular toxins that would function as virulence factors. However, gonococci are highly autolytic and release LOS and peptidoglycan fragments during growth. The LOS appears to be important in the pathogenesis of gonococcal infections. Gonococcal LOS produces mucosal damage in fallopian tube organ cultures and brings about the release of enzymes, such as proteases and phospholipases, that may be important in pathogenesis. More recent evidence suggests that gonococcal LOS stimulates the production of tumor necrosis factor (TNF) in fallopian tube organ cultures; inhibition of TNF with specific antiserum prevents tissue damage. Thus, LOS may play an indirect role in mediating tissue damage. Peptidoglycan fragments are also toxic for fallopian tube mucosa and may contribute to the intense inflammatory reactions characteristic of gonococcal disease.

Gonococci have evolved numerous strategies for evading the immune responses of the host. Pili and protein II undergo both phase and antigenic variation. Protein III, a highly conserved outer membrane protein, shares partial homology with the ompA protein of *Escherichia coli*. Antibodies to this shared region of protein III, induced either by a neisserial infection or by colonization with *E. coli*, block bactericidal antibodies directed against protein I and LOS. Gonococci can also express several antigenic types of LOS; the type of LOS expressed can be altered by an as yet unknown mechanism. Recently, it has been shown that gonococci can use host-derived cytidine monophospho-*N*-acetylneuraminic acid *in vivo* to sialylate the oligosaccharide component of one of its LOSs, converting a serum-sensitive organism (i.e., killed by normal human serum) to a serum-resistant one (i.e., not killed by normal human serum). When such organisms are grown *in vitro*, their resistance to normal human serum is rapidly lost. Gonococci may also evade the hosts immune response by molecular mimicry. Neisserial LOS and antigens present on human erythrocytes share antigenic similarity. This similarity to self may preclude an effective immune response to these antigenic determinants. Gonococci also produce an extracellular enzyme (IgA_1 protease) that inactivates human IgA_1 by cleaving the heavy chain of this immunoglobulin in the hinge region. Human IgA_2 is not susceptible to cleavage, since this isotype lacks the hinge region. The role of IgA_1 protease in pathogenesis remains unclear.

5. Host Defenses

Not everyone exposed to *N. gonorrhoeae* acquires the disease. This may be due to variations in the size or virulence of the inoculum, to nonspecific resistance, or to specific immunity. A 50%- infective dose (ID_{50}) of about 1000 organisms has been established, based on the experimental urethral inoculation of male volunteers. There is no reliable ID_{50} for women, although it is assumed to be similar.

Most uninfected individuals have serum antibodies that react with gonococcal antigens. These antibodies probably result from colonization or infection with various gram-negative bacteria that possess cross-reactive antigens. These "natural" antibodies are both bactericidal and opsonic and may be important in an individual's natural resistance to infection. Infection with *N. gonorrhoeae* stimulates both mucosal and systemic antibodies to a variety of gonococcal antigens. Mucosal antibodies are primarily IgA and IgG. In general, the IgA response is brief and declines rapidly after treatment; IgG levels decline more slowly. In spite of this humoral immune response, there is no conclusive evidence that antibodies protect against reinfection.

The complement system is important in host defense. Individuals with inherited complement deficiencies have a markedly increased risk of acquiring systemic neisserial infections and are subject to recurring episodes of systemic gonococcal infections.

There is no effective vaccine to prevent gonorrhea. Candidate vaccines consisting of pilus protein or protein I are of little benefit. The development of an effective vaccine has been hampered by the lack of a suitable animal model and the fact that an effective immune response has never been demonstrated.

B. Syphilis

1. Clinical Manifestations

Untreated syphilis is a chronic disease caused by the bacterium *Treponema pallidum*. This microorganism spreads throughout the body via the blood stream and can infect virtually every organ system. Because of this ability to spread, syphilis exhibits diverse clinical manifestations that mimic many other bacterial or viral infections. Transmission by sexual contact requires contact with moist mucosal or cutaneous lesions. Although the disease becomes systemic within a matter of hours after infection, multiplication of *T. pallidum* at the site of entry produces the primary stage. After an incubation period of 10–90 days, a papule forms at the site of entry that eventually progresses to a superficial ulcer with a firm base (chancre). Numerous treponemes are present in this highly contagious open lesion. Regional lymph nodes enlarge (i.e., regional lymphadenopathy). After 2–6 wk of symptoms, the primary lesion heals. Dissemination of the treponemes to other tissues and their subsequent multiplication result in the secondary stage. The secondary stage begins after an asymptomatic period of 2–24 wk. Clinical manifestations include slight fever, generalized lymphadenopathy, malaise, and a mucocutaneous rash. The rash initially appears on the palms and soles and eventually spreads to other areas. White mucoid patches of moist papules (condylomata) occur on mucous membranes of the mouth, vagina, or rectum. These lesions team with treponemes and are highly contagious. After 2–6 wk of secondary syphilis, host defenses bring about healing. However, about 25% of patients will experience up to three relapses of this secondary stage. About two-thirds of primary and secondary syphilis infections will resolve without treatment due to an effective immune response by the host. Those individuals who remain infected enter the latent stage for which there are no clinical signs or symptoms despite the fact that *T. pallidum* can still be demonstrated in some tissues and reactive serologic test results are still obtained. Latency can last 3–30 yr; early latency refers to the first 4 yr and late latency refers to the period beyond 4 yr. Individuals are infectious for their partners during the primary, secondary, and early latent stages.

A significant proportion of individuals with untreated syphilis will progress and develop tertiary syphilis, which may involve the skin, bones, central nervous system, heart, and arteries. Approximately 80% of fatalities result from cardiovascular involvement; most of the remaining 20% are due to neurologic involvement. Cardiovascular problems are attributed to multiplication of the treponemes in the aorta. The subsequent aortitis may result in complications such as stenosis, angina, myocardial insufficiency, and aneurysms. Neurologic syphilis may involve the membrane surrounding the brain (meningovascular form) or the brain and spinal cord (parenchymatous form). Syphilis involving the brain

is called general paresis; infection involving the spinal cord is called tabes dorsalis.

Gummas are highly destructive granulomatous lesions that are characteristic of tertiary syphilis. They usually occur in skin and bones but may also occur in other tissues. The pathogenesis of gumma formation is unclear; however, delayed hypersensitivity similar to that found in tuberculosis may be responsible.

T. pallidum also damages fetuses. Pregnant women may transmit the infection to the fetus, resulting in an infant with congenital syphilis. The organism passes through the placenta and infects most organs and tissues of the fetus. Congenital syphilis can have devastating and costly consequences for the fetus. Approximately 50% of fetuses are aborted or stillborn; the rest exhibit diverse manifestations of syphilis. In early congenital syphilis, symptoms are apparent before 2 yr of age. They include cutaneous lesions, mucous membrane lesions, inflammation of the bones, anemia, and enlargement of the liver and spleen (hepatosplenomegaly). In untreated late congenital syphilis, an infected child may be asymptomatic until after 2 years of age and then exhibit syphilitic manifestations such as interstitial keratitis and blindness, tooth deformation, eighth-nerve deafness, neurosyphilis, cardiovascular lesions, and bone deformation.

2. Diagnosis

Treponema pallidum cannot be grown on convential laboratory medium. The diagnosis of syphilis is based on clinical symptoms, results of serological tests, or the visualization of treponemes by dark-field microscopy. Newer technologies include fluorescent antibody staining with labeled monoclonal antibodies and the polymerase chain reaction; however, these newer techniques have not been adequately evaluated.

3. Epidemiology

Syphilis was a major public health problem in the preantibiotic era; the Public Health Service estimated that about 2.5% of the total United States population was infected with some stage of syphilis at the beginning of World War II. However, after the introduction of penicillin in the 1940s, the number of cases of primary and secondary syphilis in the United States declined by about 93%. Resistance to penicillin has not been observed in *T. pallidum*.

However, in spite of effective therapy, this STD has not been effectively controlled. The incidence of primary and secondary syphilis in the United States has increased from a low of 3.9 cases per 100,000 population in 1957 to 22.1 cases per 100,000 population in 1990. Reported primary and secondary syphilis cases totaled 55,132 in 1990, a 20.2% increase over the 45,854 cases reported in 1989. This is the highest number of reported cases since 1949, and reflects a trend in increased numbers of primary and secondary syphilis that began in 1985 (Fig. 2).

Before people were aware of the sexual practices that facilitated the spread of HIV, syphilis was a common STD among homosexual men. As recently as 1982, more than 40% of infected men named other men as sex partners. Among white males, the number of reported cases of primary and secondary syphilis has been influenced by homosexual behavior, increasing during the pre-AIDS era of the 1970s but decreasing substantially since 1982. It is presumed that this decline reflects changes in behavior in response to the threat of AIDS.

Primary and secondary syphilis rates are higher in large urban areas, especially in low-income minority populations, than in less-populated areas. The increase in syphilis since 1985 has occurred largely in minority heterosexual populations. The factors responsible for this increase in syphilis have not been ascertained with certainty. However, during the past two decades, resources for the public health programs involved with the control of syphilis have decreased progressively. In addition, there has been

a dramatic increase in cocaine use in United States cities during the second half of the 1980s. Available information suggests that the exchange of sexual services for drugs, especially crack cocaine, is an important factor in the resurgence of syphilis and in the spread of other STDs.

Trends in congenital syphilis reflect both the recent rise in the rate of heterosexual syphilis and varying definitions of the condition. Congenital syphilis cases steadily declined throughout the early 1980s (Fig. 2); however, substantial increases have been reported in recent years. The rates of congenital syphilis are substantially higher for minorities than for whites. Congenital syphilis is a preventable disease if pregnant women obtain adequate prenatal care. Several studies have suggested that the increased rates of congenital syphilis may be related in part to either underutilization of or inadequate prenatal care. It is suspected that substance abuse (i.e., crack cocaine, marijuana, or intravenous drugs) may play a significant role in the lack of motivation to obtain prenatal care.

Epidemiological studies carried out during the early years of the AIDS epidemic in the United States, and later in developing countries, revealed that most patients with AIDS gave a past history of STDs. STDs may be nothing more than a marker for "risky behavior." However, epidemiological studies controlling for sexual activity have implicated sexually transmitted infections characterized by genital ulceration (i.e., syphilis, genital herpes, chancroid, LGV, and donovanosis) as a co-factor in the transmission of HIV. Two biologically plausible mechanisms have been proposed to explain how genital ulcers increase the risk of acquiring HIV. First, genital ulcers can increase the infectivity of the index partner as a consequence of the excretion of HIV or HIV-infected lymphocytes through the ulcer. Second, the presence of an ulcer could enhance the susceptibility of the contact following exposure to semen or cervicovaginal fluid from an infected partner by disrupting the integrity of the mucosal barrier or recruitment of HIV-susceptible cells into the ulcer.

4. Organism and Pathogenesis

Treponema pallidum is a helical corkscrew-shaped gram-negative bacterium 6–15 μm long and 0.1–0.2 μm wide. These organisms are almost too slender to be visualized by light microscopy. They can be visualized by several staining techniques or by

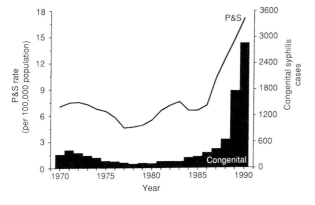

Figure 2 Rates of primary and secondary syphilis among women and reported cases of congenital syphilis in infants <1 year of age in the United States, 1970–1990. Note that the surveillance case definition for congenital syphilis changed in 1989.

dark-field microscopy. *Treponema pallidum* exhibits a characteristic motility that consists of rapid rotation about its longitudinal axis and bending and flexing about its full length. [*See* SPIROCHETES.]

Treponema pallidum differs in structure from other bacteria. The outer membrane is almost devoid of proteins, a characteristic that may be responsible for the limited antibody response to this pathogen. The flagella (axial filaments) arise at either end of the organism and are located between the outer membrane and cytoplasmic membrane. Multiplication occurs through binary transverse fission. The *in vivo* generation time is relatively long and has been estimated to be approximately 30 hrs.

Treponema pallidum is a fastidious organism that exhibits narrow optimum ranges of pH, E_h, and temperature. Despite intense efforts, *T. pallidum* has not been successfully cultured *in vitro*. However, viable organisms can be maintained several weeks in tissue cell culture. The inability to grow *T. pallidum in vitro* has hindered studies of the antigenic composition of this organism. Recent evidence indicates that this microorganism is microaerophilic and requires low concentrations of oxygen. *Treponema pallidum* is very fragile; it is rapidly inactivated by mild heat, cold, desiccation, and most disinfectants.

Our understanding of the pathogenesis of *T. pallidum* is incomplete. *In vitro* studies have demonstrated that *T. pallidum* can specifically attach to different tissues. This probably accounts for it's ability to infect most tissues and organs of the body. It has been suggested that *T. pallidum* synthesizes hyaluronidase, an enzyme that degrades the intercellular hyaluronic acid in the ground substance to facilitate the spread of the organism.

Treponema pallidum appears to be very adept at avoiding the immune response of the host. The organism is apparently covered by a layer of glycosaminoglycans similar to hyaluronic acid. It is unclear whether this material is synthesized by the organism or is derived from the host. Nevertheless, this material appears to have anticomplementary activity and to interfere with treponemal killing by the classical (antibody-dependent) complement pathway. *Treponema pallidum* may also have surface-associated sialic acid. Sialic acid inhibits activation and killing by the alternative (antibody-independent) complement pathway. A recent report suggests that *T. pallidum* synthesizes the immunodulator prostaglandin E_2. Prostaglandin E_2 is thought to modulate the immune response by stimulating macrophage supressor activity.

5. Host Defenses

Data obtained prior to the development of effective antibiotic therapy about the natural history of syphilis suggests that, in many cases, the body can effectively mount an immune response and eliminate the treponemes.

Syphilis generates a wide variety of antibodies that are the basis for the serologic tests that are so important in making an accurate clinical diagnosis. The contribution of humoral antibodies to host defenses against syphilis is unclear. Cellular mechanisms may be more important in immunity to syphilis. Macrophages appear to play a prominent role in the clearance of *T. pallidum* from infected tissues. In some instances, most but not all treponemes are killed; the residual treponemes multiply and induce the next clinical stage. The mechanism(s) responsible for the failure of the immune system to kill all the treponemes is (are) not known.

Bibliography

Barnes, R. C. (1989). *Clin. Microbiol. Rev.* **2,** 119–136.
Holmes, K. K., Mardh, P.-A., Sparling, P. F., Wiesner, P. J., Cates, W., Lemon, S. M., and Stamm, W. E. (eds.) (1990). "Sexually Transmitted Diseases." McGraw-Hill, New York.
Morse, S. A. (1989). *Clin. Microbiol. Rev.* **2,** 137–157.
Morse, S. A., Moreland, A. A., and Thompson, S. E. (eds.) (1990). "Atlas of Sexually Transmitted Diseases." Lippincott, Philadelphia.
Wentworth, B. B., Judson, F. N., and Gilchrist, M. J. R. (eds.) (1991). "Laboratory Methods for the Diagnosis of Sexually Transmitted Diseases." American Public Health Association, Washington, D.C.

Single-Cell Proteins

John H. Litchfield
Battelle Memorial Institute

I. Photosynthetic Microorganisms
II. Nonphotosynthetic Microorganisms
III. Product Quality and Safety
IV. Future Trends

Glossary

Bioreactor Any vessel or reaction device used for microbiological or enzymatic processes

Cell protein concentrates Protein preparations made from microbial cells with cell walls remaining, with or without reduction of nucleic acid contents

Cell protein isolates Protein preparations made from disrupted microbial cells in which the cell walls have been removed, with or without reduction of nucleic acid contents

Fermenter Production vessel in which a suspension of microbial cells is grown in a liquid nutrient medium

Single-cell proteins Dried cells of microorganisms for use as protein sources in human foods or animal feeds

SINGLE-CELL PROTEINS (SCPs), also called microbial proteins, are the dried cells of microorganisms such as algae, bacteria, yeasts, molds, and higher fungi grown in large-scale culture systems for use as protein sources in human foods or animal feeds. Also included in this general product class are SCP products including SCP concentrates and SCP isolates.

In this article, we will consider the microorganisms and substrates used in SCP production including growth conditions, product recovery, product quality, and safety for consumption by humans or domestic livestock. We will emphasize commercial-scale processes but will cite experimental-scale processes where they illustrate an important point.

I. Photosynthetic Microorganisms

Photosynthetic processes for SCP production utilize either algae or bacteria. Typical processes are discussed in the following sections.

A. Algae

Algae have been consumed by humans since ancient times. The ancient Aztecs harvested *Spirulina* from alkaline ponds in Mexico. At the present time, people in the Lake Chad region of Africa also harvest *Spirulina* from alkaline waters for food.

Modern controlled large-scale cultivation of algae for human consumption was initiated in Japan and Taiwan from 1950–1960. Also, during the 1960s, research was conducted on growing *Chlorella* sp. in closed-cycle systems for life support of astronauts in long-term space exploration missions (>4 mo duration) under the sponsorship of the U.S. Air Force and National Aeronautics and Space Administration.

More recently, controlled culture systems for producing algal protein for human food or animal feed applications have been developed in the United States in California and Hawaii and in India, Israel, Japan, Mexico, Taiwan, and Thailand. The organisms used in these systems are blue-green algae including *Scenedesmus* sp. and *Spirulina* sp. (cyanobacteria) and green algae in the genus *Chlorella*.

1. Substrates and Growth Conditions

Algae are grown photosynthetically in the presence of light as the energy source with carbon dioxide, bicarbonate, or carbonate as carbon sources, or in the dark under heterotrophic conditions with organic compounds as carbon and energy sources. For cell growth and protein synthesis, algae require a nitrogen source such as ammonia, nitrate, or urea and sources of phosphorus, sulfur, and mineral nutrients including calcium, iron, magnesium, manganese, potassium, sodium, and various micronutrients.

The following equation represents photosynthetic cell protein synthesis by *Chlorella* sp.:

$$6.14\,CO_2 + 3.65\,H_2O + NH_3 \xrightarrow{\text{light}} C_{6.14}H_{10.3}O_{2.24}N + 6.85\,O_2$$

The carbon dioxide concentration normally present in air, about 0.03%, is inadequate to support economic production of algae. Supplemental carbon can be provided by CO_2 or as bicarbonate or carbonate ions.

Spirulina maxima grows optimally in the pH range of 9–11, which is provided by bicarbonate in naturally alkaline waters such as Lake Texcoco in Mexico. *Scenedesmus acutus* is grown with CO_2 as the carbon source at pH 7–8, or heterotrophically in the dark with acetate, or sucrose in the form of molasses as substrates. Pure CO_2 is too costly for use in large-scale processes. This substrate can be supplied from combustion gases or from industrial sources. *Chlorella* sp. are grown in Japan or Taiwan either photosynthetically or heterotrophically in the dark using sucrose syrups or molasses as substrates at pH 6–7.

Although numerous experimental studies have been conducted on algal cultivation using artificial illumination, this source of light energy is too costly for use in commercial production systems. Consequently, large-scale photosynthetic algal production is conducted in outdoor pond systems using sunlight as the energy source.

Typical production systems include batch ponds, or semicontinuous fill and draw systems including circular ponds or raceways provided with suitable agitation to circulate algal cells to the upper 2–10-cm layer of the culture system, because a major portion of the incident light is attenuated in this upper layer.

The remaining algal cells below this level are essentially metabolizing organic substrates under heterotrophic conditions in the dark. Also, agitation prevents depletion of nutrients at the surface.

Tropical, semitropical, and arid regions between 30° north and south latitudes, where skies at the culture site are clear and temperatures are 20°C or above for most of the year, are best suited for outdoor algal production. Also, variations in light intensity throughout the year should be minimal.

Table I presents a summary of the characteristics of selected algal SCP production processes. Algal productivities as high as 30–40 g (dry weight)/m^2/day may be possible under optimal conditions, but these productivities have not been achieved in actual operating systems. Generally, cell densities are in the range of 1–2 g (dry weight)/liter so large pond areas are required.

Algal production systems involving open ponds operate under nonsterile conditions. Preventing contamination by undesired algae may be a problem, particularly in *Spirulina* production where *Chlorella* sp. may compete with the desired culture. Research in Israel has shown that *Chlorella* can be prevented by maintaining high bicarbonate concentrations in the range of 0.2 M, minimizing the dissolved organic load and the dissolved oxygen concentration, treating the culture system with 1 mM ammonia, and increasing the winter temperature by heating the medium.

Also, possible contamination of the product by pathogenic bacteria may be a problem if algae are grown in sewage oxidation ponds. Furthermore, contamination can occur from bird droppings and insects. Algae produced in sewage or animal manure oxidation ponds are suitable only for animal feeds

Table I Selected SCP Processes Based on Photosynthetic Microorganisms

Organism	Carbon and energy source	Output	Crude protein %	Location
Algae				
Chlorella sp.	CO_2, sunlight sucrose or molasses	2 metric tons/d	50–60	Taiwan
Scenedesmus acutus	CO_2, sunlight	20 g/m^2/d	55	Thailand
Spirulina maxima	CO_2, HCO_3, CO_3^{2-}, sunlight	15 g/m^2/d	62	United States Mexico Israel
Bacteria				
Rhodopseudomonas capsulata	Industrial wastes, sunlight	1.2–2.0 g/liter	61	Japan

after thermal drying at times and temperatures sufficient to destroy pathogens.

Considerable losses in yield of algal cell mass may occur from respiration during the dark cycle at night. Strains with low respiration rates in the dark are desirable. Maintaining low temperatures during the dark cycle may also decrease respiration.

2. Product Recovery

The relatively low concentrations of algal biomass in open pond and raceway systems requires handling of large volumes of water during harvesting. Centrifugation is a very effective method for separating *Chlorella* sp. and *Scenesdemus* sp., which do not settle readily. However, this method is too costly to use from an energy-use standpoint, except for very high-value algal specialty food products produced in Taiwan.

For animal feed grade products, flocculation with inorganic chemical agents has been employed, but a nontoxic flocculent must be used to prevent contamination of the product. Chitosan, derived from shellfish processing wastes is a possibility, if large quantities could be available at a low enough cost. Cells of *Spirulina* sp. are larger than those of *Chlorella* and *Scenedesmus,* form clumps, and can be separated by passing through stainless steel screens or by filtration.

In the United States, food grade *Spirulina* is prepared by spray-drying. The temperature of the product reaches 60°C for a few seconds and damage to heat-sensitive nutrients and pigments is avoided.

Chlorella is produced and sold in either a dry powder or pellet form in Japan and Taiwan. *Chlorella* cell extracts are prepared by mechanically disrupting the cells, extracting with alkali or organic solvents concentrating under vacuum, and freeze-drying.

B. Bacteria

Photosynthetic bacteria have been used to improve waste treatment processes such as activated sludge in Japan and yield a protein-enriched product. *Rhodopseudomonas capsulata* is a photoorganotrophic bacterium that can be grown photosynthetically in pure culture with an organic carbon source or in mixed culture with algae or other bacteria such as the aerobic heterotroph *Bacillus megaterium* or the nitrogen fixer *Azotobacter agilis*. Food fermentation and chemical industry wates have been used as substrates. The preferred temperature and pH ranges for *R. capsulata* are 25–30°C and 5.5–8.5, respectively. The cell product has been used as a protein source for egg-laying hens in Japan with acceptable results.

II. Nonphotosynthetic Microorganisms

Nonphotosynthetic microorganisms useful for SCP production include bacteria, yeasts, molds, and higher fungi. These organisms utilize organic carbon and energy sources together with nitrogen, phosphorus, sulfur, and mineral element sources under aerobic growth conditions.

To be suitable for commercial processes, these organisms must not be pathogenic to plants, animals, or humans and must not contain toxins that would give adverse effects in human and animal diets.

Processes for SCP production by nonphotosynthetic microorganisms are aerobic. Achieving oxygen transfer to the cells to meet the growth requirements of microorganisms on various substrates is an important factor in obtaining economically satisfactory growth rates and yields.

Various bioreactor or fermenter designs have been used to achieve suitable aeration. The bioreactor configurations used most frequently are the baffled, stirred tank and airlift types. [*See* BIO-REACTORS.]

Sterile air must be supplied to the bioreactor, and the nutrients should be sterilized by passing through a heat exchanger before addition to the presterilized bioreactor.

Large amounts of heat are released during aerobic growth of microorganisms. The amount of heat generated decreases with increasing yield of cells (calculated as weight of cells per unit weight of substrate utilized) and is approximately 0.46 kJ/millimole of oxygen consumed.

SCP processes are operated in either batch, continuous, or fed-batch modes. In conventional batch systems, the nutrients are supplied to the bioreactor initially and the cells are harvested after they consume the nutrients and stop growing. In continuous processes, the nutrients are supplied continuously to the bioreactor in concentrations required to support growth and the cells are harvested continuously when the cell population reaches the desired concentration.

In fed-batch processes, which are now being used in high cell density production systems, the nutrients are added in an incremental manner to meet growth requirements and yet maintain very low nutrient concentrations in the production medium at any time. The cells are harvested in the same manner as that in batch systems.

A. Bacteria

1. Substrates and Growth Conditions

During the 1960s and 1970s, there was considerable interest in growing selected bacteria for feed grade SCPs due to their high growth rates of up to 2 hr^{-1} (generation time, 30 min) and ability to utilize a wide range of carbon and energy sources including carbohydrates in pure form or in agricultural or forest product wastes, hydrocarbons, and petrochemicals (e.g., methanol, ethanol). Some actinomycetes are similar to bacteria in their substrate utilization patterns and growth rates and also have been considered for use in SCP production. Table II presents a summary of bacterial-based SCP processes.

Bacteria that are useful for SCP production should have temperature tolerance in the range of 35–45°C to minimize cooling costs. Most bacterial strains of interest grow best in the pH 5–7 range. pH can be adjusted by the rate of addition of ammonia and phosphoric acid to the medium as nitrogen and phosphorus sources.

Because bacterial processes operate at a pH range that is favorable for growth of a wide range of microorganisms, bacterial SCP processes must be operated under aseptic conditions using sterile air and a sterilized medium and bioreactor.

In batch processes, substrate concentrations are in the 1–5% range. Continuous and fed-batch processes are operated so that the rate of feed of nutrients is adjusted to meet the requirements of the growing organism.

Pilot plant-scale investigations on the production of bacterial SCPs utilizing methane as the carbon and energy sources were conducted in the United Kingdom using the methanotroph *Methylococcus capsulatus* or a mixed culture of *Pseudomonas* sp., *Hyphomicrobium* sp., *Acinetobacter* sp., and *Flavobacterium* sp. Factors affecting the productivity of this process included methane and oxygen transfer from the gas phase to the cells, single or double gaseous substrate limitation, heat production requiring cooling of the bioreactor, and possible formation of inhibitory products. Also the system's capital costs were higher than those with other substrates due to equipment and operational requirements for avoiding explosion hazards.

Imperial Chemical Industries (ICI) developed the Pruteen process for producing SCPs from methanol using *Methylophilus methylotrophus*. A novel "pressure cycle" airlift bioreactor was developed and used in this process to provide high oxygen transfer rates without oxygen limitation, removal of the heat liberated during growth at high productivity, a homogeneous liquid phase, and contamination control by avoiding shafts and seals used in conventional baffled, agitated–aerated fermenter vessels. Methanol conversion by *M. methylotrophus* to SCPs is represented by the following equation:

$$1.72\ CH_3OH + 0.23\ NH_3 + 1.51\ O_2 \longrightarrow$$

$$1.0\ CH_{1.68}O_{0.36}N_{0.22} + 0.72\ CO_2 + 2.94\ H_2O$$

Typical yields obtained by ICI were about 0.50 g (dry weight)/g of methanol utilized, at cell densities in the range of 30 g (dry weight)/liter.

ICI operated a commercial-scale plant at Billingham, United Kingdom, producing 6000 metric tons/mo in the early 1980s; however, production was discontinued because the product was not economically competitive with conventional feed stuffs such as soybean meal in West European markets.

Other bacterial SCP processes that have been

Table II   Selected Bacterial SCP Processes

Organism	Carbon and energy source	Specific growth rate (hr^{-1})	Yield (dry weight basis) g/g substrate used	Crude protein content %
Methylococcus capsulatus	Methane	0.14	1.00–1.03	N.D.
Methylophilus methylotrophus	Methanol	0.38–0.50	0.50	72
Methylomonas clara	Methanol	0.50	0.50	70–72

investigated on a small pilot plant scale include the Hoechst-Uhde process (Frankfurt, Germany), based on methanol utilization by *Methylomonas clara;* the Exxon–Nestle process, based on ethanol utilization by *Acinetobacter calcoaceticus;* and the Chinese Petroleum Company process (Taiwan), based on diesel oil or fuel oil utilization by *Achromobacter delvacvate* or *Pseudomonas* strain 5401. Also, during the 1970s, General Electric Company conducted studies on feed-lot waste utilization for feed grade SCP production using thermophilic actinomycetes on a small pilot plant at Casa Grande, Arizona. However, none of these processes were successful from an economic standpoint.

During the 1960s, the National Aeronautics and Space Administration sponsored research on the use of the hydrogen-fixing bacterium *Alcaligenes eutrophus* for recycling expired carbon dioxide and urinary excretion products during extended space missions, respectively.

Breathable oxygen and hydrogen could be produced by the electrolysis of water. *Alcaligenes eutrophus* utilized the hydrogen, a portion of the oxygen, and the carbon dioxide in a 6:2:1 ratio with urea as the nitrogen source. Productivities in continuous cultures were about 2 g (dry weight)/liter/hr.

This organism accumulates poly-β-hydroxybutyrate (PHB) readily, which may reach 20% or greater of cell dry weight when nitrogen is limiting. ICI has developed a process using this organism for manufacturing PHB and other polmers such as PHB-valerate on a commercial scale since these materials have the advantage of being biodegradable in the environment. [*See* BIOPOLYMERS.]

2. Product Recovery

Bacterial SCP is difficult to recover from the growth medium by conventional centrifugation because bacterial cells are small (1–2 μm) and have densities close to that of water (1.003 g/cc). Also, because cell concentrations are in the range of 30 g (dry weight)/liter, large volumes of water must be handled.

Cells of *M. methylotrophus* were recovered by a novel agglomeration method providing a higher solids slurry for feeding to the centrifuges than could be practiced in a conventional process. Hoechst-Uhde used electrochemical coagulation to separate *M. clara* from the production medium followed by centrifuging the concentrated slurry and spray-drying.

A process for bacterial cell separation developed by the Phillips Petroleum Company utilized a foam breaker inside of the bioreactor to concentrate the cells, which are then withdrawn or centrifuged to separate the cells from the production medium. Phillips also developed a bacterial cell separation process in which the fermentation broth containing bacterial cells is mixed with a broth containing yeast cells and then centrifuged to recover the mixed cell product.

In any event, the spent cell production medium and wash waters will have a sufficiently high biochemical oxygen demand (BOD) to require purification, sterilization by passing through a heat exchanger, and recycling. The residual aqueous stream should be treated by appropriate waste treatment methods before discharge.

B. Yeasts

Yeast-based processes for food or feed production originated during World War I based on technology for aerobic growth of baker's yeast developed during the nineteenth century. In Germany, baker's yeast (*Saccharomyces cerevisiae*) was produced as food and fodder yeast during World War I. During World War II, Germany produced torula yeast (*Candida utilis*) from wood hydrolysates containing pentoses for food and fodder uses. Modern food and fodder yeast SCP processes are still largely based on these earlier technologies.

1. Substrates and Growth Conditions

Various yeasts have been investigated for use for SCP production utilizing a variety of substrates as carbon and energy sources, ranging from carbohydrates such as glucose, xylose, sucrose, or lactose, or agricultural and forest products or food-processing wastes containing these substrates, to hydrocarbons, methanol, and ethanol.

Important factors in selecting a yeast strain for SCP production are the same as those for all SCP processes and include specific growth rate, yield and productivity with a given substrate, pH and temperature tolerance, aeration requirements, genetic stability, and nonpathogenicity or nontoxigenicity.

Yeasts that have been used in SCP production generally have both fermentation and respiratory metabolic patterns. For example, in *S. cerevisiae,* under aerobic conditions, glucose metabolism is regulated by both the Pasteur effect (glucose uptake and glycolysis inhibited by oxygen) and the Crabtree effect (biosynthesis of respiratory enzymes inhibited by glucose). Initially, under aeration, respira-

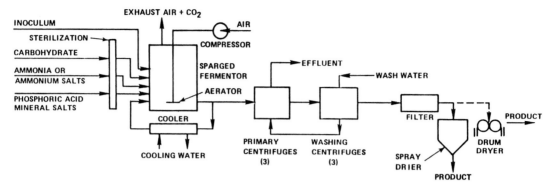

Figure 1   General process for producing SCP from carbohydrates. [From Litchfield, J. H. (1989). Single-cell protein. *In* "A Revolution in Biotechnology," (J. L. Marx, ed.) pp. 71–81. Cambridge University Press, Cambridge.]

tory enzyme synthesis is repressed, and biomass and ethanol accumulates. Then, after glucose is consumed, and an adaptation period, respiratory enzymes are synthesized and ethanol is utilized.

In contrast, *C. utilis* exhibits a Pasteur effect but is Crabtree effect negative and glucose is converted to biomass mostly by respiration under the aeration conditions employed in SCP production.

Typical yeast SCP processes are shown in Fig. 1 and Table III. Generally, most yeasts used in SCP processes have the highest specific growth rates in the range of 30–34°C. An exception is *Hansenula polymorpha*, which utilizes methanol as carbon and energy sources and has an optimum growth temperature range of 37–42°C. Again, as with bacteria, operation at the highest possible temperature minimizes cooling costs.

Typical reactions for yeasts grown on carbohydrates or hydrocarbons are the following:

Carbohydrates: $6.67\ CH_2O + 2.1O_2 \longrightarrow$
$C_{3.92}H_{6.5}O_{1.94} + 2.75\ CO_2 + 3.42\ H_2O$
Hydrocarbons: $7.14\ CH_2 + 6.135\ O_2 \longrightarrow$
$C_{3.92}H_{6.5}O_{1.94} + 3.22\ CO_2 + 3.89\ H_2O$

Yeast processes may operate under clean but non-sterile conditions where waste substrates and an animal feed grade product is produced. The pH is adjusted to the range of 3.0–4.5, depending on the particular strain of yeast to minimize the growth of bacteria. However, molds will grow readily under aerobic conditions in this pH range also. Food grade yeast must be produced under the Food and Drug Administration's (FDA's) Good Manufacturing Practice (GMP) conditions to minimize the possibility of contamination by pathogenic organisms.

Oxygen transfer is critical in maintaining high growth rates and yields in yeast processes, as was

discussed earlier for bacterial processes. Oxygen requirements for growth of *S. cerevisiae* on carbohydrates are 1 kg/kg (dry weight) of cells, whereas the requirements for growth of *Candida* sp. on hydrocarbon stubstrates such as n-alkanes is about 2 kg of oxygen/kg (dry weight) of cells.

Food grade yeast products include *S. cerevisiae*, grown on molasses or other sugar substrates; *C. utilis*, grown on glucose, ethanol, or papermill wastes; and *Kluyveromyces marxianus* var. *fragilis*, grown on cheese whey and cheese whey permeates. Either batch or fed-batch modes of operation are employed in most present-day commercial yeast SCP operations.

Saccharomyces cerevisiae (baker's yeast) has been produced in aerated fermentation vessels for over 100 years. Typical batch processes produce about 10–15 g (dry weight) of yeast/liter of medium, whereas conventional fed-batch processes achieve concentrations of about 35–45 g (dry weight) of yeast/liter. Generally, baffled, aerated stirred tank-type bioreactors are employed with air supply through a perforated screen, or perforated pipes in the bottom of the vessel, or by a rotating aeration wheel, as in the Waldhof-type fermenter.

At the present time, primary-grown *S. cerevisiae* for food use is produced by the fed-batch process in large aerated bioreactor vessels with a usual capacity of 150 m^3 and a working volume of 112.5 m^3 to allow for expansion during aeration and for foam formation. Internal coils are provided for cooling because heat in the range of 3.5–4.4 kcal/g (dry weight) of yeast is generated during growth. A blend of beet and cane molasses containing at least 20% cane molasses is used as the substrate because cane molasses supplies biotin required for baker's yeast growth. Ammonia or ammonium sulfate or phos-

Table III Selected Yeast SCP Processes

Organism	Carbon and energy source	Specific growth rate[a], or dilution rate[b] (hr^{-1})	Yield (dry weight) g/g substrate used	Organization
Candida lipolytica	n-Alkanes	0.16[b]	0.88	British Petroleum
Candida utilis	Paper Mill Waste	0.30[b]	0.50	Rhinelander Paper Corporation
	Ethanol	0.20[b]	0.80	Phillips Petroleum Company (Provesta)
	Sucrose	0.16[b]	0.52	Phillips Petroleum Company (Provesta)
	Molasses	0.20[b]	0.51	Phillips Petroleum Company (Provesta)
Hansenula jadinii	Sugars			Phillips Petroleum Company (Provesta)
Kluyveromyces marxianus var. *fragilis*	Cheese whey (lactose)	0.66[a]	0.55	Universal Foods Corporation
	Cheese whey (lactose)	0.20[a]	0.45	Phillips Petroleum Company (Provesta)
	Cheese whey (lactose) Cheese whey permeate (lactose)		0.56	Fromagerie Bel
Pichia pastoris	Methanol		0.40	Phillips Petroleum Company (Provesta)
Saccharomyces cerevisiae	Molasses	0.20[a]	0.50	Universal Foods Corporation

[a] Specific growth rate.
[b] Dilution rate.

phate are suitable nitrogen sources. Phosphorus is furnished as phosphoric acid or as ammonium phosphate to provide a P_2O_5 concentration of 33% of the nitrogen concentration. Magnesium salts may also be added to compensate for deficiencies in molasses.

The process is operated under GMP conditions with the substrate and nutrients being sterilized in a heat exchanger. The fermenter vessel is cleaned and sanitized using clean-in-place systems. The temperature is controlled at 30°C and the initial pH is 4.5. (See Table III for growth rate and yield data.)

Torula yeast (*C. utilis*) is produced as both a food and a feed grade product. This yeast is grown in a continuous process using stirred-tank bioreactors with total capacities up to 300 m^3. The usual substrate is papermill waste, particularly sulfite waste liquor, although wood hydrolysates can also be used because *C. utilis* metabolizes both hexoses and pentoses. The process is operated at 30–32°C and pH 4.5.

Ethanol, sucrose, and molasses can also be used as substrates for producing *C. utilis*. Phillips Petroleum Company's Provesteen process uses high cell density continuous culture with a specially designed bioreactor to optimize oxygen and heat transfer. The composition of the medium and operating conditions are adjusted to give maximum yields of cells and protein contents. Generally, the carbon and energy source is limiting in concentration, and dissolved oxygen concentrations are maintained at low residual levels. The Provesteen process uses a specially designed 25,000-liter bioreactor to produce *Hansenula jejunii*, the sporulating form of *C. utilis* from sucrose. This bioreactor system optimizes oxygen and heat transfer and allows continuous operation to maintain cell concentrations of 140–150 g/liter with productivities of 15–22 g/liter/hr at oxygen transfer rates of 800–1000 mmoles O_2/liter/hr. Typical temperature and pH ranges are 30–38°C and 3.5–4.5, respectively. (See Table III for typical values for growth rates and yields of *C. utilis* from various substrates.)

Large quantities of cheese whey are produced from cheese manufacturing operations. It contains about 4% lactose. The yeast *Kluyveromyces marxianus* var. *fragilis* (fragilis yeast), formerly classified as *K. fragilis*, utilizes lactose readily as a substrate and is used to make a food grade product from cheese whey collected from cheese plants.

Lactalbumin must be removed from the whey by acid precipitation, heat coagulation, or ultrafiltration to minimize foaming during aerobic growth.

Cheese whey permeate from ultrafiltration processes is also a suitable substrate.

Operating conditions for fragilis yeast production are similar to those used for *C. utilis*. Either fed-batch or continuous fermentation is employed with temperatures maintained at 30–38°C and pH at 3.0–4.5.

Kluyveromyces marxianus var. *fragilis* has been produced at high cell densities of 112 g/liter at a productivity of 22 g/liter/hr with the Phillips Petroleum Provesteen process. Table III also presents typical values for yields of *K. marxianus* var. *fragilis*.

In the 1960s, British Petroleum Company, Ltd., developed a feed grade SCP process based on *Candida* sp. using straight-chain paraffin hydrocarbons (n-alkanes) as substrates. The required substrate purity of 95.7–97.0% was achieved by molecular sieve preparation. Kanegafuchi Chemical Industry Company, Ltd., in Japan also developed a similar process.

British Petroleum constructed a 100,000 metric ton/yr plant in Sardinia designed to operate continuously using an agitated–aerated fermenter vessel. In contrast, Kanegafuchi's process employed a modified airlift type of bioreactor.

The British Petroleum plant was never operated due to a dispute with the Italian government over the product quality, including the presence of hydrocarbons. The All-Union Research Institute of Protein Biosynthesis in Russia also constructed and operated a yeast SCP plant based on hydrocarbon feedstocks for producing an animal feed grade protein product. Typical growth rate and yield values for the British Petroleum process are shown in Table III.

2. Product Recovery

Yeasts are harvested by centrifuging the production medium since the density of yeast cells of 1.13 g/cm^3 is significantly greater than that of water. *Saccharomyces cerevisiae* is separated by two passes through centrifuges to obtain maximum recovery and reduce the amount of soluble solids other than those of yeast. After washing, the yeast is spray-dried to obtain a product with superior functional properties and lighter color over as compared with drum-drying. In the case of *C. utilis*, the extent of centrifuging and washing determines whether the product is suitable for animal feed or human food. For a food grade product, the yeast must be centrifuged twice following three washing stages to remove lignosulfonates and other impurities. The high cell density

processes mentioned previously allow the production medium to be spray-dried directly without centrifuging due to the high solids content in the feed to the spray-drier. As in bacterial SCP processes, the waste liquids from separating and washing yeast give a high BOD effluent, which must be treated before discharge.

C. Molds and Higher Fungi

People have eaten molds in oriental fermented foods and certain higher fungi such as mushrooms since ancient times. During World War II, in Germany, fungal mycelium was produced in submerged culture for human food. A large number of mold species have been investigated for food or feed use in recent years.

1. Substrates and Growth Conditions

SCP processes based on molds and higher fungi are similar to those for producing yeasts in terms of strain selection for high specific growth rates, yields, pH and temperature tolerance, aeration requirements, genetic stability, and lack of pathogenicity or toxigenicity.

Various carbohydrates including simple sugars (glucose, maltose, sucrose, lactose) and agricultural, food-processing, and forest product wastes have been evaluated for mold and higher fungi-based processes. Carbohydrate concentrations generally are in the 1–10% range with carbon : nitrogen (C:N) ratios of 5:1 to 20:1; C:N ratios below 15:1 are preferable to give mycelia with high protein and low fat contents.

Suitable nitrogen sources include anhydrous ammonia and ammonium sulfate, phosphate, or nitrate. Phosphoric acid or ammonium phosphate serve as phosphate sources. Mineral nutrient requirements include calcium, copper, cobalt, iron, magnesium, manganese, potassium, sodium, and zinc and may be met in part by waste materials used as substrates, but they may have to be added as supplements to the production medium to compensate for deficiencies. It is important to limit the amounts of added mineral nutrients to the concentrations actually required to prevent the development of bitter flavors in the mycelium.

Some higher fungi such as *Coprinus comatus* have vitamin requirements (thiamin in this case). Temperature and pH values for fungal SCP processes range from 25° to 36°C and 3–7, respectively. For organisms with optimal growth rates above pH 5.0, such

Table IV SCP Processes Based on Molds and Higher Fungi

Organism	Carbon and energy source	Specific growth rate (hr^{-1})	Yield g/g substrate utilized	Location
Fusarium graminearum	Glucose	0.18	0.50	Ranks Hovis McDougall United Kingdom
Morchella hortensis	Glucose	—	0.48	Mid-America Dairymen United States
Paecilomyces varioti	Sulfite waste liquor	0.20	0.55	United Paper Mills Finland
Penicillium cyclopium	Cheese whey (lactose)	0.20	0.68	Heurtey France

as *Morchella* sp. (morel mushrooms), aseptic conditions must be employed to prevent the growth of bacteria, yeast, and mold contaminants.

During aerobic growth, fungal mycelium takes either a filamentous or pellet form, depending on agitation and aeration conditions. An aeration system that provides agitation is preferable because the mycelium may wrap around the impeller blades and shaft in mechanically agitated systems.

Several mold processes have been investigated for producing animal feeds from waste substrates. Examples include *Aspergillus niger* on carob bean extract in Cypress and Greece, *Gliocladium deliquescers* on corn wet milling wastes in the United States *Trichoderma harzianum* on coffee processing wastes in Guatemala, and *Penicillium cyclopium* on cheese whey in France.

During the 1970s, the Pekilo process was developed for producing *Paecilomyces varioti* from sulfite waste liquor at the United Paper Mills in Finland. A 10,000-metric ton/yr capacity plant with a 360-m^3 continuous fermenter was used to produce mycelium for animal feed and human food. Although animal feeding and limited human feeding studies were conducted with satisfactory results, the product was not manufactured on a commercial scale for economic reasons. Morel mushroom mycelium (*Morchella* sp.) was produced for sale as a food-flavoring ingredient during the 1960s by Special Products Division of Mid-America Dairymen, Springfield, Missouri.

Morchella sp. were grown with glucose as the substrate supplemented with corn steep liquor in an aerated tank without mechanical agitation to favor pellet formation. Lower-cost imported dried mushrooms led to the process being abandoned.

Ranks Hovis McDougall developed a process for producing a food-grade microbial protein called myco-protein from glucose using *Fusarium graminearum*. The nitrogen source is gaseous ammonia, which is also used for pH control along with mineral salts. The fermenter is operated under aseptic conditions. The process is operated at 30°C and pH 6 in batch culture. Production capacity is 50–100 tons/yr. Table IV presents growth rates and yields for selected fungal processes.

2. Product Recovery

Fungal mycelium can be recovered readily on drum filters, rotary vacuum filters, basket filters, or inclined screens, depending on the form of the mycelium (pellet or filamentous). The resulting filter cake is then dried in a rotary tray or belt driers. It is important to control drying temperatures to prevent excessive darkening of the product and damage to heat-sensitive flavor constituents.

III. Product Quality and Safety

A. Product Quality

SCP product quality includes nutritional value for humans and animals and functional quality for use as a food ingredient. In addition to their protein contents, microbial cells also contain carbohydrates, lipids, vitamins, and minerals that also contribute to nutritional value and to functional properties such as flavor, water and fat binding, emulsion stability, whipability and foam stability, dispersibility, gel formation, and thickening. [*See* FOOD BIO-TECHNOLOGY.]

1. Nutritional Value

A large body of information has been published on the composition and nutritional value of SCP prod-

Table V Protein and Amino Acid Contents of Selected Microorganisms Used in SCP Processes

Organism and substrate	Protein content (%)	Amino acid content (g/16 g N)[a]								
		HIS	ILE	LEU	LYS	MET	PHE	THR	TRY	VAL
Algae										
Spirulina maxima (photosynthetic, CO_2)	62	1.7	5.8	7.8	4.8	1.5	4.6	4.6	1.3	6.3
Bacteria										
Methylophilus methylotrophus (methanol)	72	1.8	4.3	6.6	5.9	2.4	3.4	4.6	0.9	5.2
Yeasts										
Candida utilis (sulfite waste liquor)	55	1.2	3.8	7.6	4.8	1.1	8.6	5.4	1.3	3.8
Kluyveromyces marxianus (cheese whey)	54	2.1	4.0	6.1	6.9	1.9	2.8	5.8	1.4	5.4
Saccharomyces cerevisiae	53	4.0	5.5	7.9	8.2	2.5	4.5	4.8	1.2	5.5
Molds and Higher Fungi										
Fusarium[b] *graminearum* (glucose)	47–50	4.8	5.1	8.3	8.1	2.2	4.8	5.5	1.7	5.9
Morchella hortensis (glucose)	34	1.9	2.4	5.0	3.0	0.7	2.3	2.7	1.0	2.9
Paecilomyces varioti (sulfite waste liquor)	55	—	4.3	6.9	6.4	1.5	3.7	4.6	1.2	5.1
FAO/WHO Reference	—	—	4.2	4.8	4.2	2.2	2.8	2.8	—	4.2

[a] Key to amino acid abbreviations: HIS, histidine; ILE, isoleucine; LEU, leucine; LYS, lysine; MET, methionine; PHE, phenylalanine; THR, threonine; TRY, tryptophan; VAL, valine.

[b] Amino acid values are expressed in g/100 g amino acid.

ucts related to human food and animal feeding uses. Protein contents and amino acid compositions of various microorganisms vary widely even within the same species, depending on growth conditions and the composition of the production medium. The effects of the C:N ratio in the medium cell protein and lipid formation was discussed earlier. Also, true protein contents based on amino acid contents are a better index of nutritional value of a SCP product than crude protein values calculated by multiplying Kjeldahl nitrogen values by the factor 6.25, which includes nonprotein nitrogen sources. It is important to determine nucleic acid contents in SCP products to be used as a significant source of protein in human foods to avoid exceeding intakes >2 g/day, which leads to kidney stone formation and gout.

Table V presents protein and amino acid contents of some algae bacteria, yeasts, and molds used in SCP processes. A comparison of the amino acid compositions of microorganisms with the Food and Agricultural Organization/World Health Organization of the United Nations reference values reveals that algae, yeasts, molds, and higher fungi tend to have lower sulfur amino acid contents (methionine) than is desirable, whereas bacterial proteins have better balances of these amino acids. Either synthetic methionine or food protein sources of sulfur amino acids (cystine plus methionine) must be used to supplement these SCP products.

Many animal-feeding studies have been conducted with SCP products using domestic livestock including broiler chickens, egg-laying hens, and pigs. Protein digestibility and feed conversion ratio (weight of ration consumed : weight gain) are the usual measures of feedng performance. In general, SCP in rations must be supplemented with DL-methionine to give equivalent performance to a conventional corn soybean meal diet.

Protein efficiency ratio (PER) as compared with casein, biological value (BV), and net protein utilization in rats are used to measure nutritional value of SCP products for human foods. Typical PERs and BVs for unsupplemented dried *C. utilis* range from 0.9 to 1.4 and 32 to 48, respectively, which increase to 2.0–2.3 (compared with 2.5 for casein) and 88, respectively, when supplemented with 0.5% DL-methionine.

One must interpret PER values with caution, because the rat assay may underestimate the protein quality of SCP products that are deficient in sulfur amino acid contents due to the higher requirements of rats for these amino acids as compared with those of humans.

Human feeding studies have been conducted with algae, yeasts, and molds. A major problem is human acceptance from a flavor standpoint when algae and yeasts are added to foods in nutritionally significant quantities. Also, digestibility and amino acid nitrogen limit SCP utilization in human diets. *Candida utilis*, *F. graminearum* (myco-protein), and *S. maxima* gave acceptable performance in human feeding studies.

2. Functional Ingredient Applications

SCP products can be used as functional food ingredients as mentioned earlier. In addition to dried microbial cells, cell protein concentrates and isolates have been evaluated for various functional characteristics of importance in food formulations.

Published values indicate that freeze-dried *C. utilis* had better fat binding and at least equivalent water binding but lower emulsifying capacity and foam volume than soy flour. Furthermore, a cell protein isolate prepared from *K. marxianus* var. *frasilis* by cell disruption, water extraction, and acid precipitation had higher emulsifying capacity and nearly equivalent foam volume to those of soy isolate.

Yeast proteins have been succinylated and phosphorylated in attempts to modify or improve functional properties. However, these products have not been developed commercially due to the necessity of obtaining FDA clearance in the United States.

Yeast products have been used primarily as flavoring ingredients in foods. In the United States, the FDA has approved *C. utilis*, *K. marxiamus* var. *fragilis* and *S. cerevisiae* for food use. Baker's yeast protein is also approved in the United States.

Fusarium graminearum (myco-protein) has been formed into textured fiber products. The mycelium is formed into meal and fish analogues with appropriate flavoring and coloring additives. The product has been test-marketed in the United Kingdom under the tradename Quorn.

B. Safety

Extensive safety evaluations have been conducted with SCP products including algae, bacteria, yeast, and molds. These evaluations generally follow the guidelines of the Protein Advisory Group of the Food and Agricultural Organization of the United Nations for production and preclinical and clinical testing of novel protein products for human consumption. [*See* FOODS, QUALITY CONTROL.]

In human studies, nausea, vomiting and diarrhea were observed with bacterial proteins such as *A. calcoaceticus* and *A. eutrophus*. *Caudida utilis* grown on glucose exhibited no adverse effects in human subjects whereas a rash was observed in those subjects fed with this organism grown on sulfite waste liquor. *Fusarium graminearum* (myco-protein) has been fed to human subjects without adverse effects.

The need to reduce nucleic acid contents of SCP products to prevent kidney stones and gout was mentioned earlier. Several processes have been developed for reducing nucleic acid contents including chemical treatment with acids or alkalis, enzymatic treatment by inducing endogenous ribonucleases or adding exogenous ribonucleases, or chemical modification of the protein by succinylation or phosphorylation. Extraction of *F. graminearum* with isopropanol followed by treatment with NH_4OH/NH_4Cl at pH 8.5 reduced nucleic acid contents from 9.10 to 0.59–1.7%.

IV. Future Trends

It is clear that a variety of SCP products have been produced that meet established regulatory agency requirements for nutritional value and safety. *Spirulina*, *C. utilis*, *K. marxianus accharomyces S. carlsbergensis* (brewer's yeast), *S. cerevisiae*, and *F. graminearum* fall into this category and will be the major products in this class in the future. It is not clear that new specialty protein products prepared from microorganisms will be developed in the near future due to the extensive testing for safety required by regulatory agencies such as the FDA. The cost of this testing apparently cannot be justified by the size of future markets.

A genetically modified baker's yeast with more efficient genes for elaborating maltose permease has been developed for use in bread by Gist Brocades, Delft, The Netherlands. This "new" form of *S. cerevisiae* has been approved for use in bread manufacture in the United Kingdom.

The safety of genetic modifications of existing microorganisms for use in SCP production will apparently have to be demonstrated to the satisfaction of regulatory agencies. Price–performance benefit, nutritional value, functional performance, and consumer acceptability will be key factors in the success of any new SCP product once regulatory requirements are met.

Bibliography

Benemann, J. R. (1990). Microalgal products and production. An overview. *In* "Developments in Industrial Microbiology" (G. E. Pierce, ed.) Vol. 31. pp. 247–256. Elsevier, Amsterdam.

Boze, H., Moulin, G., and Galzy, P. (1992). *"Crit. Rev. Biotechnol.* **12,** 65–86.

Goldberg, I. (1985). "Single-cell Protein," pp. 260 Springer Verlag, Berlin.

Litchfield, J. H. (1979). Production of single cell protein for use in food or feed. *In* "Microbial Technology," 2nd ed. (H. J. Peppler and D. Perlman, eds.) Vol. 1, pp. 93–155. Academic Press, New York.

Litchfield, J. H. (1986). Bacterial biomass. *In* "Comprehensive Biotechnology" (H. N. Blanch, S. Drew, and D.I.C. Wang, eds.) Vol. 3, pp. 463–481. Pergamon Press, Oxford.

Litchfield, J. H. (1989). Single-cell protein. *In* "A Revolution in Biotechnology," (J. L. Marx, ed.) pp. 71–81. Cambridge University Press, Cambridge.

Litchfield, J. H. (1991). Food supplements from microbial protein. *In* "Biotechnology and Food Ingredients" (I. Goldberg and R. Williams, eds.) pp. 65–109. Van Nostrand Reinhold, New York.

Moo Young, M., and Gregory, K. F. eds. "Microbial Biomass Proteins," pp. 1–10; 19–26; 27–32. Elsevier Applied Science, London.

De la Noue, J., and de Pauw, N. (1988). *Biotechnol. Adv.* **6,** 725–770.

Reed, G., and Nagodawithana, T. W. (1991). "Yeast Technology," 2nd ed, pp. 413–440. Van Nostrand Reinhold, New York.

Sharp, D. (1989). "Bioprotein Manufacture; a Critical Assessment," pp. 140 Ellis Horwood, Chichester, United Kingdom.

Solomons, G. L. (1985). Production of biomass by filamentous fungi. *In* "Comprehensive Biotechnology" (H. W. Blanch, S. Drew, and D.I.C. Wang, eds.) Vol. 3 pp. 483–505. Pergamon Press, Oxford.

Skin Microbiology

William D. James
Walter Reed Army Medical Center

Rudolf R. Roth
Travis Air Force Base

I. Resident Flora Organisms
II. Ecologic Considerations
III. Protective Mechanisms Limiting
   Colonization and Infection

Glossary

Atopic dermatitis Itchy inflammatory skin condition characterized by oozing and cracked skin
Blastospore form Stage of growth of a fungus characterized by budding yeast forms
Commensals Organisms that live on the skin surface and cause no harm to the individual
Hyphal stage Stage of growth of a fungus characterized by filamentous elements
Immunocompromised patients Patients whose defense system for fighting infection is not effective
Intertriginous areas Body folds or flexural areas such as the groin and axillae
Microflora Microorganisms that are characteristically present in a specific location

SKIN MICROBIOLOGY is the study of microorganisms that may be cultured from the surface of the skin. The microflora is relatively limited in the number of types of bacteria and fungi that can colonize and remain present on the skin over long periods of time. The skin possesses protective mechanisms to limit colonization, and the survival of organisms on the skin surface lies in part in the ability of the organisms to resist these mechanisms. While the composition of the resident flora is of limited complexity, the ecologic factors that influence which organisms may survive in each body area are relatively complex. These factors, such as humidity, lipid content, exogenous climatic or occupational environment, and endogenous bacterial properties all must be considered when discussing which organisms will be found routinely on the skin surface of different body surface sites.

I. Resident Flora Organisms

Normal skin is colonized by large numbers of organisms that live harmlessly as commensals on its surface. Those organisms that grow on the skin in a relatively stable number and composition are called resident flora. They live as small microcolonies on the surface of the stratum corneum and within the outermost layers of the epidermis. Flora from any one area of the skin cannot be taken as representative of the entire flora, and samples must be taken from multiple sites to determine composition.

Another group of organisms live on the skin surface and will not be discussed in this article. These are transient or temporary flora that are mainly derived from exogenous environmental sources. They can adhere and multiply on mainly exposed surfaces for only limited periods of time. They include pathogenic organisms such as the streptococci.

A. Micrococcaceae

Staphylcocci and micrococci are gram-positive, catalase-positive cocci in the family Micrococcaceae. Most microbiologists subclassify these organisms based on the ability of staphylococci to produce acid anaerobically from glycerol in the presence of erythromycin (0.4 mg/ml) and on their susceptibility to lysis by lysostaphin and nitrofuran. [See Gram-Positive Cocci.]

1. Staphylococci

Staphylococci are divided into the coagulase-positive *Staphylococcus aureus* and the coagulase-negative species. *Staphylococcus aureus* is of particular importance because it is one of the primary causes of clinical infection. If breaks in the skin surface occur, such as abrasions or burns, and *S. aureas* is present in the resident flora, overt infection in the form of impetigo or abscesses may result. Humans have a high degree of natural resistance to skin colonization by *S. aureus*. In most body sites, the skin is usually not colonized with *S. aureus*, but it can be found in intertriginous areas, particularly the perineum, of up to 20% of persons. In addition, persistent nasal carriage of the organism is present in 20–40% of normal adults from whence it can cause persistent skin colonization and recurrent infection. Some persons are more susceptible to *S. aureus* colonization of the skin, including hospital workers, diabetic patients, intravenous drug abusers, and patients undergoing hemodialysis. Patients with certain skin diseases tend to harbor *S. aureus*. In psoriasis and atopic dermatitis, *S. aureus* may be found widely in both diseased and normal skin, often constituting up to 80% of the normal flora.

Coagulase-negative staphylococci are the most frequently found organisms of the normal flora. This type of staphylococci rarely cause clinical infection in the normal individual. It may cause disease in immunocompromised states such as in patients with leukemia. At least 18 different species have been isolated from normal skin, the primary residents are *Staphylococcus epidermidis*, *S. hominis*, *S. haemolyticus*, *S. capitis*, *S. warneri*, *S. saprophyticus*, *S. cohnii*, *S. xylosus*, and *S. simulans*. *Staphylococcus epidermidis* and *S. hominis* are the species recovered most frequently. *Staphylococcus epidermidis* colonizes the upper part of the body preferentially and constitutes >50% of the resident staphylococci. After these in prevalence are several species with almost identical DNA base pairings: *S. haemolyticus*, *S. capitis*, and *S. warneri*. *Staphylococcus saprophyticus* is often found as a resident in the perineum.

Peptococcus saccharolyticus, also called *Staphylococcus saccharolyticus*, is a strict anaerobic staphylococcus and a member of the normal flora in 20% of persons. It may be present in large numbers on the forehead and antecubital fossa.

2. Micrococci

Although less frequently present than staphylococci, at least eight different *Micrococcus* species have been identified from human skin. By order of prevalence, these are *M. luteus*, *M. varian*, *M. lylae*, *M. nishinomiyacnsis*, *M. kristinae*, *M. roseus*, *M. sedentarius*, and *M. agieis*. *Micrococcus varians* represents the dominant species. *Micrococcus lylae* and *M. kristinae* may be more common in children, and *M. lylae* is more frequently present during the colder months of the year.

B. Coryneform Organisms

Coryneform organisms are gram-positive pleomorphic rods, which were originally all thought to be of the *Corynebacterium* genus. The term *diphtheroids*, given to these organisms in much of the medical literature, is best avoided because they have very little resemblance to the pathogen *Corynebacterium diphtherae*. The coryneforms have been poorly classified in the past and were simply divided into lipophilic organisms, which require lipid supplements for growth in artificial media, and nonlipophilic organisms. More recently, attempts have been made to classify the organisms on the basis of cell wall analysis. On the basis of the presence and type of amino acids, mycolic acid, and neutral sugars, only 60% of cutaneous coryneform organisms were found to belong to the classical *Corynebacterium* genus. Another 20% were classified as *Brevibacterium* species, organisms that had previously been identified in milk. The final 20% of organisms fell into various other groups and were thought to be transient flora because they are found primarily on exposed areas. [*See* GRAM-POSITIVE RODS.]

1. Classical Species

Classical *Corynebacterium* species compose a significant portion of the normal flora, particularly in moist intertriginous areas. Most of these organisms are lipophilic; the name *Corynebacterium lipophilicus* has been suggested for them. Similar organisms were found in >50% of the toe web spaces of Danish military recruits, with the incidence greater in those with hyperhidrosis. *Corynebacterium minutissimum*, once thought to be a single organism distinguished by the ability to produce porphyrin, is actually a complex of as many as eight different species.

2. Group JK Organisms

Group JK coryneforms are organisms that have acquired resistance to nearly all antibiotics except vancomycin. Studies of their cell walls have shown that they are nearly identical to normal cutaneous lipophilic coryneforms in all aspects except antibiotic resistance. They are found on the skin of normal persons but are found more frequently in immunosuppressed hosts. They are thought to colonize the intertriginous areas preferentially and can be found in up to 35% of patients in hospitals. There is an inverse correlation between the number of antibiotic-sensitive lipophilic coryneforms that colonize the skin and the number of antibiotic-resistant group JK bacteria. Although the latter are sensitive to vancomycin *in vitro*, systemic use of vancomycin neither prevents nor controls their growth on the skin.

3. *Brevibacterium*

Brevibacterium species, also known as large-colony coryneforms, produce proteolytic enzymes, are penicillin-resistant, and are probably the most rapidly growing of the coryneforms. They are frequently isolated from the toe webs, especially in patients with fungal infections of the feet, and are implicated in foot odor.

Corynebacteria may cause several superficial skin diseases. These entities, known as erythrasma, trichomycosis, and pitted keratolysis are all seen in humid intertinginous areas or in areas partially occluded by clothing. They do not cause invasive serious infection except in immunocompromised patients or, as in the case of the group JK organisms, in hospitalized patients.

C. Propionibacteria

Propionibacterium species are non-spore-forming, anaerobic, gram-positive bacteria that are normal inhabitants of hair follicles and sebaceous glands. They are the most prevalent anaerobes of the normal flora and are also known as anaerobic coryneforms. On the basis of colony morphology and susceptibility to lysis by bacteriophages, they have been divided into three species.

Propionibacterium acnes is most numerous on the skin of the scalp, forehead, and back and is the predominant species by far, present in almost 100% of adults. A direct correlation has been shown between the total number of *P. acnes* organisms and the amount of sebum present. The follicular density of *P. acnes* reaches its peak around puberty, when sebaceous glands are enlarging because of androgen stimulation, and remains relatively constant through adult life. *Propionibacterium granulosum* is next in prevalence, composing almost 20% of this genus,

and is found in small numbers at all tested sites. *Propionibacterium avidum* strains are found most often in moist intertriginous areas, especially the axilla. *Propionibacterium acnes* is the predominant organism colonizing acne lesions, and adolescent patients with acne have more surface *P. acnes* than matched controls. They play particular importance in the inflammatory, pustular lesions. The use of antibiotics in the treatment of acne has focused on decreasing the density of these organisms, with resultant clinical improvement.

D. Gram-Negative Rods

Gram-negative rods are uncommon components of the resident flora of the skin, probably because of the skin's desiccation. Transient organisms are frquently found as contaminants from the gastrointestinal system. These organisms will occasionally become resident flora in moist intertriginous areas and mucosal surfaces, such as the perineum, axilla, toe webs, or nasal mucosa of some persons.

Acinetobacter species are nonfermentive aerobic gram-negative rods that are widely dispersed in nature and are found in up to 25% of persons as normal flora. They include species that were formerly referred to as members of the genera *Mima* and *Herellea*. Male persons are more frequently colonized than female persons, and colonization significantly increases during the summer. These trends are probably secondary to an increase in perspiration because a high moisture content increases the growth of these organisms.

Acinitobacteria infections occur in hospitalized and immunosuppressed patients and may include bacteremia, endocarditis, meningitis, infection of the genitourinary tract, and, most commonly, the respiratory tract. Skin infections caused by gram-negative organisms include postoperative wound infections and burn wound infections with *Acinetobacter* and gram-negative folliculitis and gram-negative toe web infections, both secondary to a wide variety of organisms.

E. Mycoflora

Fungi are regularly present in the normal human flora. Surveys of the resident cutaneous mycoflora have revealed a predominance of yeast organisms, both in temperate and tropical environments.

1. *Pityrosporum*

Pityrosporum species organisms are lipophilic yeasts that require lipids for growth. Their *in vitro* growth must be accomplished on olive oil-enriched media. *Pityrosporum ovale* and *Pityrosporum orbiculare* are probably identical organisms that are prominent in sebaceous areas. In the normal flora, they exist in the blastospore form. In quantitative testing, the organisms are most numerous on the back and chest; the highest numbers parallel areas of highest sebum excretion.

Under proper environmental conditions, *P. orbiculare* may change from the resident blastospore stage to the pathogenic hyphal stage. Factors that predispose to this pathogenic transformation include increased heat and humidity, hyperhydrosis, immunosuppression, and exogenous or endogenous steroids.

Malasessezia furfur is the name given to the hyphal stage. The extremely common superficial infection tinea versicolor, which causes increased and decreased pigmentation of the skin of the trunk, is caused by this organism as is the less common condition pityrosporum folliculitis.

2. *Candida*

Candida species, normally found on up to 40% of the oral mucous membranes, seldom colonize the normal skin. When present, *Candida albicans* is the most common species found, existing in the blastospore form. Increased colonization of skin by *Candida* species is seen in immunosuppressed and diabetic patients and in patients with psoriasis or atopic dermatitis.

Infection secondary to *Candida* species is usually caused by *Candida abicans* and, less commonly, by other *Candida* species such as *Candida tropicalis* and *Candida parapsilosis*. *Candida* can rapidly colonize damaged skin, although active invasion of intact stratum corneum can occur when the organism transforms to the filamentous or hyphal form. Just as environmental factors such as heat and humidity may predispose a person to infection with *Pityrosporum* species, these same factors can predispose the person to disease with candidal organisms. In addition to the factors mentioned for *Pityrosporum*, other factors that may increase susceptibility to *Candida* include suppression of the normal flora with oral or topical antibiotics and systemic factors that might alter the immune response such as pregnancy or cancer. Although candidal organisms only rarely colonize normal skin, their prevalence on mucous membrane surfaces results in an endogenous source of infection when cutaneous host defense mechanisms break down. Localized infection occurs most commonly on mucous membranes but can also occur cutaneously, especially on moist macerated skin or skin in close contact with the mucous membranes.

II. Ecologic Considerations

Although the resident flora remains relatively constant, several factors can change the quantity of organisms and the relative percentages of each organism in the normal flora. These factors can be either endogenous to the individual or secondary to environmental or bacterial influences.

A. Patient Specific Factors

1. Effects of Age

The age of the person has a profound influence on the microflora. The flora is most varied in young children, who carry micrococci, coryneform bacteria, and gram-negative organisms more frequently and in larger numbers than older children and adults. Infants also carry a higher proportion of pathogens or potential pathogens on their skin.

On the other hand, *Pityrosporum* and *Propionibacterium* species are present at much lower levels before puberty. These organisms require higher skin lipid levels, and their appearance parallels age-related changes in sebum production. *Pityrosporum orbiculare,* for example, is rare in children <5 yr of age and becomes increasingly established in persons during the next 10 yr, achieving adult levels by 15 yr. *Propionibacterium acnes* population levels are also directly proportional to the amount of secreted sebum and free fatty acids in surface lipid. Elderly patients have a decrease in sebum production, and infections with these lipophilic organisms are rare.

Colonization of the skin begins at birth. The skin of babies born by cesarean section is sterile and becomes colonized on first contact with the outside world, whereas the skin of babies born by vaginal delivery is already colonized by organisms encountered in the birth canal before delivery. The organisms that are found at birth are usually present in small numbers, except for *S. epidermidis,* which is the predominant organism of the vaginal flora just before birth. Coryneform bacteria are also prominent in the resident flora of the newborn, but, unlike *S. epidermidis,* they take several hours after delivery to establish themselves.

Before these organisms become fully established, however, other organisms may readily colonize the skin surface, such as gram-negative rods and streptococcal organisms. *Staphylococcus aureus* infection is much more common in the newborn and colonizes the nasopharynx and umbilicus of many infants. Despite the readiness of these pathogenic organisms to colonize the skin, the infant's flora begins to resemble that of the adult after the first few weeks of life.

2. Effects of Sex

Evidence indicates that men carry higher absolute numbers of organisms and more biotypes. This may be due to higher production of sweat in men as well as to the tendency of men to wear more occlusive clothing. Other possible factors include increased sebum production in men and hormonal differences between the sexes.

3. Effects of Race

Various reports have distinguished differences among the races in some areas, such as a higher nasal carriage rate of *S. aureus* in Whites, an earlier appearance of *P. acnes* in Blacks, and fewer cutaneous streptococcal infections or neonatal streptococcal infections in Blacks. Differences in carriage rates of organisms may be due to differences in human lymphocyte antigen (HLA) expression, differences in adhesion, or different environmental conditions.

4. Effects of Body Location

The composition of the normal flora varies depending on body location. The face, neck, and hands represent exposed areas of the body and, as a result, may have a higher proportion of transient organisms and a higher bacterial density. The head and upper trunk have more sebaceous glands and a greater number of lipophilic organisms, with *Propionibacterium* the dominant organism. The axilla, perineum, and toe webs are areas under partial occlusion, with an increased temperature and moisture level. These areas are colonized more heavily with all organisms, but particularly with gram-negative rods or coryneforms that need moisture for survival. Coryneform organisms are frequently reported to be the predominant organism in these areas. The upper portions of arms and legs are relatively dry and often

have lower bacterial counts. When the ecologic conditions of an area are changed, as with occusion of a dry skin surface, the flora will change to adapt to the new environment.

5. Effects of Disease

The presence of systemic disease may predispose to colonization or infection with different organisms, which may be due to an associated immunologic abnormality or to changes in bacterial adherence (see later section). In diabetes mellitus, another possible factor affecting colonization is an increase in skin glucose concentration. An increase in the nasal carriage of *S. aureus* occurs in diabetic children and in insulin-dependent diabetic adults compared with either non-insulin-dependent diabetic persons or control subjects. Whether or not this leads to a statistically significant increase in infections is still not resolved. However, diabetic persons do have an increased prevalence of candidal infections, particularly in females. The morbidity of these infections is often proportional to the degree of diabetic control.

B. Exogenous Factors

1. Effects of Climate

The resident flora can be influenced by their microenvironment. Increased temperature and humidity increase the density of bacterial colonization and alter the relative ratios of organisms. Artificially applied organisms survive longer on wet skin than on dry skin. These environmental changes have been experimentally reproduced by application of occlusive materials on forearm skin. After 24 hr of occlusion, bacterial counts are increased 10,000-fold and the relative numbers of gram-negative rods and coryneform bacteria are increased over coccal forms. Once the occlusive material is removed, the numbers of bacteria decrease toward normal but only very slowly. The effect of experimental occlusion is similar to localized proliferation of the resident flora in areas of the body normally under partial occlusion from body surface-to-surface contact, such as the toe webs. Studies have shown that increases in both temperature and humidity are necessary to duplicate these results and that an increase in either alone will not cause as significant a change in the normal flora.

The increased heat and humidity provided by occlusion are environmental factors that favor the growth of fungal organisms as well. Cutaneous candidiasis, tinea versicolor, or *Trichophyton rubrum* infections can be experimentally reproduced on human skin only if applied under occlusion. The raised carbon dioxide tension produced by occlusion may also favor the conversion of yeasts and dermatophytes to a more infectious stage.

Any decrease in ambient temperature or humidity would be expected to cause decreased colonization. Persons tend to wear more clothing as it gets colder, however, thereby keeping the microenvironment warm and humid and negating any inhibitory effect of the external environment. A definite ecologic advantage has been found to favor the survival of nonpathogenic over pathogenic staphylococci when the temperature decreases, decreasing virulent infections when the microenvironmental temperature decreases.

Minor differences have been reported in the resident organisms in different geographic locations. The species of coagulase-negative staphylococci in patients in North Carolina differed from those in London and in New Jersey. Whether these differences are due solely to variables in climate or reflect other local differences is unknown.

2. Effects of Ultraviolet Light

Although no statistically significant change was seen in the normal flora after psoralen ultra violet light A therapy for psoriasis, ultra violet light B has been shown to inhibit growth of certain organisms. *Pityrosporum* and *Candida* species were more sensitive than staphylococci.

3. Effects of Occupation

Just as other factors in the environment can influence the pattern of the resident flora, occupation can also have an effect. Those who work in environments that have high temperature and humidity, for example, might develop microflora favored by these factors, such as *Candida*, gram-negative organisms, and coryneforms. Hospital workers have also been shown to harbor more pathogenic organisms as transient organisms, which may become established as resident flora if these persons are exposed continually.

4. Effects of Soaps and Disinfectants

Repeated washing with soap makes the skin more alkaline than washing with medicated disinfectants. Neither of these products significantly altered the count of coagulase-negative staphylococci, but propionibacteria were markedly increased when soap was used and depressed with a medicated disinfec-

tant. Preoperative disinfection by washing with chlorhexidine soap decreased the incidence of postoperative infections caused by *S. aureus* from 8 to 2%. Quantitative studies showed a significant reduction of aerobic bacteria, whereas control washing with a nonmedicated soap gave an increase in measured organisms. In a related observation, it has also been found that scrubbing the hands with water alone decreased the numbers of bacteria faster than when nonmedicated soap was used.

5. Effects of Hospitalization

Various studies have shown significant differences between the flora of healthy persons and those of hospitalized patients. The results of one study show that hospitalized patients have increased colonization with pathogenic and antibiotic-resistant organisms, such as group JK coryneforms. As the numbers of group JK organisms increased in hospitalized patients, there was a corresponding decrease in the number of normal antibiotic-sensitive coryneforms, adding to the number of normal antibiotic-sensitive coryneforms, adding to the evidence that group JK organisms are simply normal lipophilic coryneforms that have developed antibiotic resistance. Other changes of resident flora in hospitalized patients include an increase in total gram-negative organisms, favoring especially *Proteus* and *Pseudomonas* species, and in *C. albicans*. Another study showed that the frequency of nasal carriage of *S. aureus* in neonates or in patients with eczema increases with the number of days hospitalized. The changes in the resident flora of these persons help to explain the patients propensity to have nosocomial infections with more aggressive organisms.

6. Effects of Medications

Of all exogenous influences, drugs are capable of making the most rapid and radical changes in the normal flora. Antibiotics may suppress the normal flora and increase colonization by other organisms. The major effect of antibiotic treatment on the nasal flora is to depress the coryneforms with a corresponding increase in coagulase-negative micrococci and gram-negative organisms. In addition to inhibiting the growth of certain organisms in the normal resident flora, antibiotics may impair bacterial adherence to epithelial cells and allow the natural selection of other organisms, such as gram-negative rods, *Candida,* or *Pityrosporum*.

Oral retinoids, such as isotretinoin, cause a decrease in sebum excretion, and a significant decrease in the numbers of *P. acnes* persists even after treatment is discontinued and sebum excretion returns to pretreatment levels. Retinoids cause drying of the mucous membranes, which results in a decrease in the total number of organisms. The number of gram-negative rods decreases significantly, although *S. aureus* recovery from the anterior nares and the skin is increased. This increase in *S. aureus* colonization can be significantly minimized if the anterior nares are treated with a topical antibiotic ointment during isotretinoin therapy. The increase in staphylococcal colonization, well documented with isotretinoin, has also been suggested to occur with etretinate.

Oral steroids and hormones are also associated with changes in the normal flora. Corticosteroids suppress the immune system and can increase susceptibility to various bacterial, fungal, viral, and parasitic infections. Susceptibility to infection is related to the dosage, duration of therapy, and strength of the steroid used. Patients with renal transplants showed a statistically significant correlation between bacterial infection and prednisone dosage, and patients with lupus erythematosus showed an increased incidence of infection when treated with >20 mg of prednisone daily. Patients given alternate-day steroid therapy have fewer infectious complications than those taking steroids daily. The duration of therapy is also important because long-term steroid therapy is more often associated with infection, whereas a short course is not. Women on estrogen therapy may experience an increase in vaginal as well as cutaneous candidiasis.

Not only do systemic drugs change the composition of the normal flora, but topical medications can as well. Treatment of the axilla with topical neomycin resulted in an initial decrease in the total number of microorganisms present. After continued treatment, the number of gram-negative organisms increased and became the dominant flora; the coryneform population also markedly decreased. Even after the topical antibiotic was discontinued and the resident flora returned, the gram-negative organisms were slightly more numerous than before treatment. The use of topical antibiotics may also induce the appearance resistant strains. In acne vulgaris studies, the staphylococcal and enterococcal isolates developed resistance in 40% of patients treated with topical tetracycline, and topical clindamycin resulted in 60% of patients having resistant organisms. The number of resistant strains normalized after treatment was discontinued.

However, topical steroids have little effect on the numbers or types of cutaneous microflora, although local candidiasis is not uncommon after the application of topical steroids to the flexures. In one study, no significant change was found when 0.1% triamcinolone acetonide in petrolatum was applied to normal skin. On eczematous skin, however, topical steroids actually decreased bacterial counts, probably by healing the dermatitis.

C. Bacterial Adherence Factors

To colonize the skin, organisms must first become attached to the epithelial surface. The capacity to colonize is proportional to the ability of the organism to adhere to the surface. Adherence involves the attraction of a specific molecular structure on the cell walls of the bacterium, called an adhesin, to a specific receptor on the host cell surface. Adhesins are microbial surface antigens, or lectins, often present in the form of filamentous projections. The adhesin functions as a bridge between the microbe and the host cell. The binding between adhesin and receptor is virtually irreversible. Epithelial cells from different anatomic sites have variability in their receptors for bacteria, which helps to explain the differences in resident microflora found in different locations on the body. Group A streptococci isolated from skin adhere much better to skin epithelium than to buccal mucosa, whereas streptococci isolated from the oral cavity bind better to the buccal mucosa. The inability of certain organisms, such as viridans streptococci and *C. albicans*, to attach to normal unbroken skin may explain why they are so rarely found as resident organisms. [*See* ADHESION, BACTERIAL.]

Teichoic acids are cell wall components of staphylococci and streptococci that are adhesins for those organisms. When epithelial cells are treated with teichoic acid, it competitively binds with the epithelial receptors and the binding of *S. aureus* is inhibited. The receptor on the epithelial cell for teichoic acid is fibronectin, which binds the adhesin in at least two separate sites. Soluble fibronectin inhibits the adherence of streptococci to epithelial cells. The binding of *S. aureus* to fibronectin is time-dependent and irreversible, and occurs on both live and heat-killed cells. In contrast, fibronectin seems to have a barrier effect against adhesion of gram-negative organisms. None of 147 strains of gram-negative organisms was able to bind to fibronectin.

The inability of streptococci and *S. aureus* to colonize the unbroken epithelium may be due to the absence of available fibronectin on the skin surface. Patients with dermatitis may, through microscopic breaks of the skin surface, uncover dermal fibronectin receptors and increase the adherence of *S. aureus* to skin. Another possibility is that some atopic patients may colonize *S. aureus* because of an innate increase in adherence for the organism.

Pathogenic bacteria have increased potential for adherence to the host, giving them greater virulence. The enhanced potential may be due to the presence of multiple adhesins, which would increase the propensity for adherence and give the organism a selective advantage over other organisms. Studies with *S. aureus* have found another adhesin in the cell wall—protein A. Additional adhesins might allow the organism to bind to additional receptors. *Staphylococcus aureus* has been found to bind to fibrinogen as well as to fibronectin. Pathogenic bacteria, such as *S. aureus,* may also increase adherence by having the ability to degrade fibronectin.

In addition to these factors determined by the organism, the host can also influence adherence. Patients with increased tendencies to colonize pathogens might have a greater proclivity for adherence to those organisms. For example, *S. aureus* adheres more readily to skin cells from atopic patients than to skin cells from either normal persons or psoriatic patients. The propensity for bacterial adhesion in atopic patients may be due to additional receptors for *S. aureus* or to more numerous primary receptors available. A genetic susceptibility toward nasal carriage of *S. aureus* has been demonstrated, and evidence indicates an association between HLA expression and colonization. Adherence of *S. aureus* to nasal mucosal cells, while markedly low for the first 4 days of life, apparently approaches adult levels on day 5. The period of reduced adherence may be due to immature receptor sites.

Pityrosporum species have been found to adhere selectively to stratum corneum cells. The number of organisms that can attach per cell is lower for these yeasts than for bacteria because the yeast cells are larger. Propulsive forces between yeast cells on the stratum corneum surface and the availability of binding sites may also influence adherence. Adherence of these organisms increases with increased binding time and elevated skin temperature and is not influenced by the presence of *S. aureus. Corynebacterium albicans* can also adhere to oral epithelial cells; this adherence is most likely the initial step in the onset of clinical infection. Adherence is enhanced by increased concentration of organisms and in conditions where germ tube formation, or the filamentous form of the organism, is favored. Saliva increases adherence, perhaps because of its inhibition of bacterial attachment. Adherence of pathogenic dermatophytes to human keratinocytes has also been shown, and adherence is inhibited by antifungal agents.

Virally infected cells have been found to cause changes on the cell surface, altering existing cell receptors and forming new potential receptors for bacteria. Acutely ill patients have increased adherence of unusual organisms such as *Pseudomonas,* perhaps because of changes on the epithelial surface. Infected cells lose their fibronectin surface coating and presumably uncover cell surface receptors for other organisms.

Knowledge of the importance of bacterial adherence might affect the treatment of bacterial infection. Application of purified bacterial adhesin (such as teichoic acid) or host receptor (such as exogenous fibronectin) might selectively block the adherence of specific organisms by competitive inhibition. Vaccines may also be developed against specific microbial adhesins to prevent adherence.

III. Protective Mechanisms Limiting Colonization and Infection

A. Host Defense

Normal skin is resistant to colonization and invasion by most bacteria. The presence of bacteria on the surface does not make infection inevitable. Many factors prevent colonization and invasion by pathogenic organisms.

1. Intact Stratum Corneum

The most important feature offering protection of the skin from invasion is an intact stratum corneum. Overlapping cells, joined by modified desmosomes, function as an armor against organisms. In addition, the relative dryness of intact skin limits the growth of organisms that require moisture, such as *Candida* species. This protective armor is strong but not impenetrable because appendageal structures may offer a route for infection by some organisms, particularly *S. aureus.* These pathogenic staphylococci have also been observed with electron microscopy to intercalate between corneocytes *via* the intercellular species in the stratum corneum.

Experimental production of localized infection with pathogenic organisms is difficult if the skin is intact. A break in the stratum corneum is an absolute requirement for the induction of a streptococcal infection. *Staphylococcus aureus* can infrequently cause infection on intact skin, but the presence of a single silk suture through the skin increases the infectiousness of *S. aureus* by a factor of 10,000. In addition, *S. aureus* colonization of patients with dermatitis is directly related to the degree of epidermal change.

2. Rapid Cell Turnover

Another feature of the stratum corneum that offers protection from invasion is its rapid turnover; the transit time is only 14 days. If pathogenic organisms adhere, they have a limited time to invade. The resident flora are better suited for reattachment after desquammation than the weakly adhering transient organisms.

3. Lipid Layer

Many components of the resident microflora have lipase activity and liberate fatty acids, such as oleic, stearic, or palmitic acid, from the triglycerides of sebum. These free fatty acids create an acid mantle on the surface of the skin, which has a potent antimicrobial effect on *S. aureus* and streptococcal organisms and a stimulating effect on some other organisms such as *Propionibacterium* species. Propionibacteria also produce proprionic acid, which has an antimicrobial effect on many organisms. At the mean pH value of normal skin (5.5), this antimicrobial effect is much more selective against transient organisms than against resident flora. Occluded areas of the body, such as intertriginous areas, have a neutral or slightly alkaline pH because of the diluting effect of the skin's secretions and become more densely populated with microorganisms.

Not only do sebum-derived lipids play a role in host defense, but epidermal lipids do as well. Free fatty acids, polar lipids, and glycosphingolipids derived exclusively from the stratum corneum have significant antistaphylococcal activity. These lipids exist in the intercellular spaces of the stratum corneum where they provide a significant antimicrobial line of defense.

Factors that change the composition of these lipids or the pH of skin may affect the normal flora.

Occlusion of the skin of the forearm results in two concomitant and probably interrelated events: growth of resident organisms and increased pH (from 4.9 to a relatively alkaline 7.1). When the skin surface layer is stripped with acetone, pathogenic organisms can colonize. Antimicrobial activity is restored when the skin surface lipids are replaced. *Pityrosporum* organisms possess lipoxygenases that are capable of oxidizing oleic acid to azelaic acid. This product has antimicrobial activity against both propionibacteria and staphylococci and has been reported to be of therapeutic value in the treatment of acne.

4. Immune System

The skin has humoral and cellular immune systems that can influence the composition of the microbial flora. In normal skin, the humoral mechanisms involve the secretion of primarily IgA and IgG antibodies, brought to the surface through the eccrine system. The secretory IgA in sweat may prevent infection by several possible mechanisms, and its presence may explain the absence of colonization in the eccrine duct.

Elements of cellular immunity in the epidermis include antigen presentation by Langerhans cells and T-cell activation by epidermal thymocyte-activating factor, a product of epidermal cells. The skin also plays a role in T-cell differentiation. These interactions are vital in preventing cutaneous infection. Patients with defects in cellular immunity are more susceptible to infection. These infections may include diseases with a narrow, limited defect such as patients with chronic mucocutaneous candidiasis or with more widespread defects, such as patients with acquired immunodeficiency syndrome.

B. Organism-Specific Factors

1. Antibacterial Substances

Many microorganisms produce protein or protein-complex antibiotics that have an antagonistic effect on other organisms, but not on the producer bacterium. Substances produced by gram-negative bacteria generally have a wide range of antibacterial activity, whereas those produced by gram-positive organisms are usually effective only against strains of the same or closely related species. These latter substances, those with a narrow spectrum of activity, are called bacteriocins.

In the resident flora, cyclic peptide bacteriocins

are produced by the coagulase-negative staphylococci and, to a lesser extent, by coryneform bacteria. Their bactericidal activity exerts a lethal effect on closely related organisms, and this effect is initiated by adsorption to specific outer membrane receptors. Bacteriocin producers may be found in up to 20–25% of persons, but bacteriocins, if present, are usually only produced by a small (<5%) number of organisms. When skin becomes diseased, however, these bacteriocin producers increase in number and become the predominant members of the flora. Patients with dermatitis and organisms producing bacteriocins have a significantly decreased incidence of secondary infection. [*See* BACTERIOCINS: ACTIVITIES AND APPLICATIONS.]

Dermatophytes on the skin can also produce antibiotics, including penicillin and other substances with antibacterial and antifungal activity. Penicillin production can suppress the bacterial flora in fungal lesions but also tends to select a penicillin-resistant flora, such as *Brevibacterium* or penicillin-resistant staphylococci.

2. Bacterial Interference

A resident bacterium sometimes prevents colonization by another strain of a similar species, which most likely occurs by competitive inhibition of binding sites. The best example of this is the inhibition of colonization by other strains of *S. aureus* when a nonvirulent strain, *S. aureus* 502A, colonizes the skin or anterior nares. This strain has been implanted iatrogenically to prevent recurrent furunculosis, to eradicate persistent nasal carriage, and to prevent life-threatening nursery epidemics. Similarly, commensal staphylococci inhibit colonization with *S. aureus,* and only when they are removed can *S. aureus* colonize forearm skin. The protective role of the resident staphylococcal flora is also demonstrated by its absence in the newborn, when during the first week of life pathogenic organisms such as *S. aureus* frequently colonize the skin because of the absence of an established resident microflora.

3. Other Interactions

Certain resident organisms also exercise antagonistic inhibition over other residents that are of dissimilar species. Gram-positive organisms in the axilla, for example, exert a restraining influence on the number of gram-negative organisms there. Suppression of the gram-positive organisms by topical antibiotics causes a corresponding proliferation of gram-negative bacteria. A reciprocal relationship between carriage of *S. aureus* and gram-negative organisms in the nose has also been seen.

Large numbers of *S. aureus* in patients with atopic dermatitis seem to eliminate the lipophilic coryneform bacteria from the flora. Similarly, large numbers of staphylococci also displace coryneform bacteria on the plaques of psoriasis.

The development of candidiasis after suppression of the normal flora suggests that the inhibitory function of the normal flora has an influence on fungal organisms as well. Normal oral microflora were found to block the adherence of *C. albicans* to host epithelium in normal mice, whereas germ-free mice had a significant increase in adherence of *C. albicans* to buccal cells. Similarly, when the normal flora is suppressed with oral antibiotics, *C. albicans* quickly colonizes.

In summary, the microbiology of human skin is of limited complexity when considering the composition of the resident organisms. Ecologic factors that influence the type of flora that is present in diverse body areas and under varying external conditions is essential to the understanding of skin microbiology.

Bibliography

Feingold, D. S. (1986). *Arch. Dermatol.* **122,** 161–163.
Hartmann, A. A. (1983). *Arch. Dermatol. Res.* **275,** 251–254.
Leyden, J. J., McGinley, K. J., Nordstrom, K. M., and Webster, G. F. (1987). *J. Invest. Dermatol.* **88,** 655–705.
Mok, W. Y., and Barrett de Silva, M. S. (1984). *Can. J. Microbiol.* **30,** 1205–1209.
Noble, W. C. (1984). *J. Med. Microbiol.* **17,** 1–12.
Roth, R. R., and James, W. D. (1989). *J. Am. Acad. Dermatol.* **20,** 367–390.
Spriggs, D. R. (1986). *J. Infect. Dis.* **153,** 809–810.

Smallpox

Abbas M. Behbehani

The University of Kansas School of Medicine

Glossary

Conjunctivitis Inflammation of the mucous membrane covering the eyelids and eyeball

Disseminated intravascular coagulation Altered blood coagulation caused by depletion of clotting factors and platelets, causing rapid formation of multiple small intravascular clots

Endothelium Cells lining blood and lymphatic vessels, heart, and other body cavities

Eosinophilic Stainable rosy-red by the eosin dye

Epidermis; dermis; prickle cells Epidermis: outer layer of skin; dermis: layer under epidermis; prickle cells: those with rod-shaped processes for connection to adjoining cells

Macule, papule, vesicle, and pustule Sequential developing stages of skin eruptions

Malpighian epithelium Innermost layer of epidermis

Pathogenesis Origination and development of a disease

Plasmid Extrachromosomal selfreplicating structure carrying genes for functions not essential for bacterial growth

Reticuloendothelial system Body cells that ingest particulate matter such as bacteria

SMALLPOX is a disease of antiquity. It killed more people than any other infectious disease. The ancient Chinese protected themselves against this infection by inhaling dried powder of smallpox skin crusts. Edward Jenner of England introduced the protective measure of vaccination in which cowpox virus is inoculated into the skin. The World Health Organization (WHO) started the intensified program of global eradication of smallpox in 1967 and the disease was eradicated from the entire world in 1977. However, there was one fatal laboratory-acquired case of smallpox in 1978 at Birmingham University, England.

Smallpox is caused by a member of poxviruses, which are the largest of DNA vertebrate viruses. The disease is strictly a human one and is generally acquired by close personal contact, usually through the respiratory tract. After systemic multiplication, the virus invades the skin, producing typical centrifugally distributed lesions, which, when healed, leave permanent scars (pockmarks). A mild form of this disease (variola minor), which caused less than 1% mortality, replaced the classical form (variola major), which caused greater than 15% mortality, in Africa, Europe, and the Americas in late 1800s.

Jenner's vaccine virus is called vaccinia virus, which is speculated to be a hybrid arising from the inadvertent mixing of smallpox, cowpox, and horsepox viruses during the early years of vaccination. Vaccinia virus is now being genetically engineered to produce a variety of recombinant vaccinia virus vaccines in which genes of other organisms are expressed.

I. Historical Notes

Smallpox is an ancient disease; its description appears in the earliest Egyptian and Chinese writings. The mummy of Pharaoh Ramses V (at Cairo Museum), who reportedly died of an acute illness in his early thirties in 1157 B.C., shows numerous pustular lesions that closely resemble those of smallpox (Fig. 1). Smallpox is mentioned neither in the Old and New Testaments nor in the classical Greek (including Hippocratic writings) and Roman literatures. Although smallpox was prevalent in China and India since about 1000 B.C., the first recorded smallpox epidemic farther west was initiated in Arabia by an

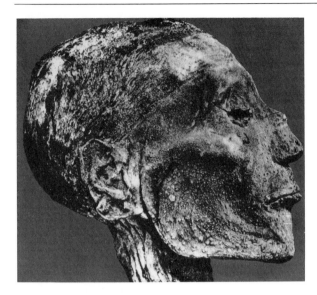

Figure 1   Mummy of Pharaoh Ramses V (died 1157 B.C.) showing smallpox lesions on the lower face and neck. (Courtesy of the WHO, Geneva, Switzerland.)

invading Ethiopian army in A.D. 570. This disease was then disseminated to North Africa and Spain (A.D. 710) by the Arab invaders. The celebrated Persian physician, Rhazes (A.D. 865–935), wrote the first differential and graphic description of symptoms of smallpox and measles in his classic monograph "A Treatise on Smallpox and Measles." The returning Crusaders (1096–1291) reintroduced smallpox more extensively throughout Europe. Early in the sixteenth century, smallpox appeared in Britain, while the Spaniards and slave ships from Africa spread the disease to the West Indies and Central America. The disease was called smallpox to distinguish it from pox or great pox, which were the common names for the then prevalent syphilis. The army of Conquistador Hernando Cortes introduced smallpox to Mexico in 1520 and subsequently $3\frac{1}{2}$ million Aztec Indians died of smallpox within 2 years. During the seventeenth and eighteenth centuries, smallpox caused deadly endemics and epidemics throughout Europe killing some 400,000 people annually. Five European reigning monarchs died of smallpox during the eighteenth century. Indeed, smallpox was so prevalent in Europe that a woman who had no pockmarks on her face was considered a beauty.

In the United States, smallpox epidemics, involving occasionally one third of the population, occurred frequently during the eighteenth century, and in contrast to Britain where 90% of cases occurred in children younger than 10 years, Americans of all ages were afflicted. For example, during a smallpox epidemic in Boston in 1721, 5759 cases with 842 deaths were recorded among 10,700 citizens of that city. All in all, smallpox killed more people of all ages, all classes, and all races than any other infectious disease.

The ancient Chinese initiated a protective measure against this disease by the practice of variolation (i.e., dried powder of smallpox skin crusts was inhaled through the nose in a manner similar to taking the snuff). In the Near East, a modified method of variolation, namely, removing some of the thick liquid from a smallpox pustule and rubbing it into a small scratch on the arm of a child, made with a needle, was widely practiced.

Although a considerable number of European physicians became aware of variolation around the seventeenth century and two scientific communications by Emanuel Timoni and Jacob Pylarini, who practiced variolation in Turkey, appeared in the *Philosophical Transactions of the Royal Society of London* (in 1714 and 1717, respectively), introduction of this practice into Britain apparently needed a third element. This was very effectively provided by Lady Mary Wortley Montagu (Fig. 2), the wife of the British Ambassador to the Ottoman Empire during 1717 to 1718, who became enthusiastically interested in the practice of variolation among the Turks as a protective measure against smallpox.

She had her 6-year-old son variolated on March 18, 1718, with the supervision and participation of Embassy surgeon, Charles Maitland. The immunization was successful and after her return to Britain, she informed her friend the Princess of Wales (Caroline of Ansbach) of her experience with this practice.

In the spring of 1721, a severe smallpox epidemic, which involved both children and adults, broke out in Britain. Consequently, in April 1721, Lady Mary requested Maitland to variolate her 3-year-old daughter. This was of signal importance as variolation had been hitherto considered as a virtuoso-amusement until Lady Mary sponsored it by having her own daughter variolated. Moreover, Lady Mary gained support of the Princess of Wales who, after the successful variolation of a number of volunteer condemned criminals and orphan children, had two of her daughters variolated. Subsequently, variolation became popular in Britain.

In Colonial America, the Reverend Cotton Mather of Boston had some knowledge of the prac-

Figure 2 Lady Mary Wortley Montagu (1689–1762) of England who introduced variolation to Britain in 1721. (Courtesy of Wellcome Institute Library, London.)

Revolutionary War was that, in contrast to the British soldiers, only a small percentage of the American soldiers had been variolated.

Edward Jenner (1749–1823), a country physician of Berkeley, England (Fig. 3), practiced variolation with increasing uneasiness, as this practice had a mortality rate of about 2%. He had been variolated as a boy in 1756 and nearly died from it. Thus the search for a safe and effective procedure against smallpox became of immense importance to him.

The notion of cowpox protecting against smallpox was presented to Jenner while an apprentice to Daniel Ludlow in Sodbury by a dairymaid with a pustular skin infection who told Ludlow that she could not have smallpox because she had had cowpox. This notion was well established among some local farmers; it is reported that in 1774, farmer Benjamin Jesty, who had had cowpox as a young man, vaccinated his wife and his two sons with material taken directly from the udder lesion of a cow with cowpox. Most probably, other laymen performed

tice of variolation in Africa through his black slave and had read Timoni's above-mentioned communication. Thus, when an outbreak of smallpox occurred in Boston in 1721 (as described above), he actively urged Boston physicians to practice variolation. Initially only Zabdiel Boylston was persuaded, and on June 26, 1721, he successfully variolated his son Thomas and two slaves. Boylston, who lacked the M.D. degree, was promptly opposed by all his medical colleagues. However, six of the most prominent clergymen and some prominent Boston citizens defended Boylston, and subsequently variolation became popular.

Variolation was widely practiced in the United States both before and during the Revolutionary War. However, concern about the possibility of variolated individuals transmitting the disease to others caused several of the 13 colonies, at one time or another, to pass laws against variolation or at least against its practice outside very strictly controlled variolation hospitals. Thus, it has been suggested that the reason for the severe outbreaks of smallpox among the colonial troops early during the

Figure 3 Edward Jenner (1749–1823) of England who introduced vaccination to Britain in 1796. (Courtesy of Wellcome Institute Library, London.)

similar prophylactic measures. The dairymaid's remark left an indelible impression on Jenner. Later, the smallpox epidemic of 1778 induced him to investigate this notion for more than 10 years, and he consequently became convinced of the protective effect of cowpox against smallpox. Thus, on May 14, 1796, he immunized the 8-year-old James Phipps of Berkeley with material obtained from a typical cowpox lesion on the hand of a milkmaid named Sarah Nelmes. On July 1, 1796, Jenner challenged Phipps (variolated on both arms) with material from a real smallpox lesion; he observed a solid immunity. In June 1798, he published a 64-page monograph entitled "An Inquiry into the Causes and Effects of the Variolae Vaccinae, etc." Jenner's proposal, which was based on conclusive evidence, met with strong opposition of his medical colleagues, who characterized his immunization procedure as unnatural and dangerous. Meanwhile, in London a surgeon, Henry Cline, used some of Jenner's dried lymph vaccine, which was inadvertently left with him by Jenner, as a counter irritant in a boy with an inflamed hip joint. Cline was subsequently informed by his medical colleague, Lister, that his patient became resistant to variolation. This observation, publicized by Cline and Lister, produced a blaze of publicity and gave Jenner's vaccination its needed impetus, and thus the practice of vaccination became popular in Britain. Elsewhere (e.g., France, Germany, and Russia) the Jennerian vaccination became even more popular. In the United States, Benjamin Waterhouse of Boston, who vaccinated members of his household with Jenner's vaccine in 1800, is credited with the introduction of this practice. Subsequently, President Thomas Jefferson played a valiant role in the popularization of vaccination. The Jennerian innovation was adopted by almost the whole world within 10 years. Jean de Carro of Switzerland and Francis Xavier de Balmis of Spain introduced vaccination to the East (Middle East, India, and Ceylon) and to the West (South America, The Philippines, and China), respectively.

Jenner's vaccine was maintained by arm-to-arm passage for many years. In 1845, Negri of Naples propagated the vaccine virus in cows. The technique of scarifying the entire flank of a cow and producing a large quantity of vaccine was eventually introduced.

The vaccine virus is called vaccinia, which is, in its present form, different from both the original cowpox virus and the smallpox virus. It is believed to be a hybrid that arose from the inadvertent mixing of smallpox, cowpox, and horsepox viruses during the early vaccination practices. The vaccines last used in the United States were mainly glycerinated or freeze-dried lymph from infected skins of calves or sheep. It produces good immunity for up to 10 years. However, there are low risks (>1 in million) of life threatening adverse reactions to this vaccination and the incidence of other various adverse reactions is about 1 in 50,000.

The last cases of smallpox in the United States occurred in 1949 when eight cases (one fatal) were diagnosed in lower Rio Grand Valley in Texas. In Europe, however, many importations of smallpox have occurred since 1950 (1113 cases with 107 deaths). The last importation was from Iraq to Yugoslavia in 1972, which involved 175 cases with 35 deaths.

As the risk of acquiring smallpox in the United States became essentially zero in late 1960s, routine vaccination of U.S. children was discontinued in 1971, and by 1976, routine smallpox vaccination of U.S. hospital employees was likewise discontinued. In most European countries, routine vaccination of children was also discontinued in 1970s.

II. The Disease

A. Etiology

The causative viral agent is a member of subfamily Chordopoxvirinae of family Poxviridae, which are the largest and most complex of DNA viruses measuring 230×400 nm. It is brick-shaped or ellipsoid, with a lipoprotein outer membrane (envelope), which contains lateral bodies of unknown function and a biconcave core enclosing a single linear double-stranded DNA. It contains several enzymes and about 100 polypeptides and has at least 20 different antigens. However, all poxviruses have a common nucleoprotein (NP) antigen located in the inner core, and all replicate in the infected cell's cytoplasm. The virus can survive in the dry state for months, as evidenced by an outbreak of smallpox in England that was traced to infected cotton shipped from Egypt. [See DNA REPLICATION.]

B. Pathogenesis and Pathology

Smallpox was classified as two related diseases, namely, variola major (classical or Asian smallpox) and variola minor (alastrim or African smallpox) caused by two viruses with distinctive characteristics. In contrast to variola minor, which had a mortality rate of 1%, variola major had an overall mortality rate of 15% to 45%. The virus of variola minor seemingly originated during the last decades of eighteenth century in South Africa and spread to the rest of Africa, Europe, and the Americas, commonly replacing the virus of variola major. The variola minor virus was imported from the West Indies to Pensacola, Florida, in 1896. However, in the rest of the world, variola major virus remained dominant, and death and disfiguration from this disease continued to prevail. The WHO recognized six types of variola major with different mortality rates: (1) ordinary discrete, <10%; (2) ordinary semiconfluent, 25% to 50%; (3) ordinary confluent, 50% to 75%; (4) flat, >90%; (5) hemorrhagic, almost 100%; and (6) modified (altered by previous vaccination), <10%.

As regards the pathogenesis of smallpox, the virus enters the susceptible host through the mucous membrane of the oropharynx or upper respiratory tract and, after local multiplication, is drained by the lymphatics to the regional lymph nodes. Here, further multiplication occurs, and the virus then enters the bloodstream, causing the primary viremia. This is followed by the invasion of the reticuloendothelial system and extensive multiplication of the virus leading to the secondary viremia. The virus then invades the epidermis, causing the skin eruptions. The patient is not infectious during the incubation period and the first 1 to 2 days of the pre-eruptive phase. As the rash appears, which usually coincides with the development of oropharyngeal lesions, the patient becomes infectious, especially during the first week of this phase. Oropharyngeal secretions were the main source for contaminating the face, the body, the clothes, and the beddings of the patient. Much lower transmissibility was associated with skin scab material. Direct face-to-face contact with a patient via infected droplets and physical contact with a patient or the contaminated articles were usually responsible for the transmission of the disease. Indirect transmission, such as that which may occur with laundry workers and hotel chambermaids through infected bed clothes, was also observed.

Clinically, after an incubation period of about 12 days, the pre-eruptive symptoms of headache, fever, malaise, prostration, pain in the back and limbs, and vomiting may appear suddenly or gradually. The skin eruptions usually develop after 3 to 5 days; they appear as one crop of macules that successively change to papules, vesicles, and pustules. The pustules start to scab toward the end of the second week, and the crusts fall off in about 1 week, leaving pink scars (pockmarks) that gradually fade in color. The distribution of the skin lesions is typically centrifugal, showing the greatest concentration on the face, forearms, wrists, and palms and soles of the feet, as well as in the mouth and throat. The chest, abdomen, thighs, and upper arms are relatively spared (Fig. 4). Conjunctivitis was manifested in some patients during the first 8 days as part of the acute illness. Bacteria, especially *Staphylococci*, may contaminate the pustules and desquamated skin, causing cutaneous and systemic infections

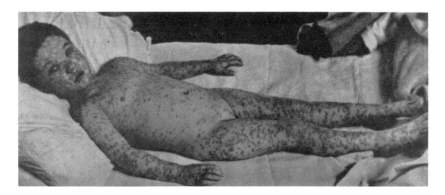

Figure 4 Classical smallpox (pustular stage) in an unvaccinated child. [Reprinted from Horsefall and Tamm. (1965). "Viral and Rickettsial Infections of Man," 4th ed. J. B. Lippincott Co., Philadelphia, Pennsylvania.]

leading to a variety of complications such as abscesses, pneumonia, septic joints, osteomyelitis, and corneal ulcers, which may cause blindness; this was often observed before the advent of antibiotics and where hygiene was poor. The cause of death was not well understood; general toxemia, septic shock, and disseminated intravascular coagulation (in the hemorrhagic type) have been suggested. The virus did not persist in the patient after recovery (no carrier state).

Variola minor (alastrim) showed a pathogenesis similar to that of variola major, but the clinical features were much milder. The skin lesions generally were fewer, smaller, and not as deep as in smallpox; they remained discrete and evolved quickly. The general condition of the patient was usually good and convalescence was rapid. Bacterial contamination and complications were very rare.

Histopathologically the skin lesions start with the proliferation of the prickle cells of epidermis caused by viral invasion. Dilation of capillaries in the dermis with swelling of the endothelial lining and infiltration of cells especially around the vessels are also present. The malpighian epithelium then becomes edematous and undergoes ballooning degeneration. The walls of the affected cells break down to form vesicles between the horny layer (as roof) and dermis (as floor). Small vesicles coalesce with their neighboring ones forming larger vesicles that become filled with tissue debris and white cells. Similar lesions also occur at the same time in the mouth and esophagus. The latter mucosal vesicles rupture early in the disease and shed virus into the secretion before the skin lesions become infectious. The infected epithelial cells of the skin and mucous membrane lesions show the typical intracytoplasmic eosinophilic Guarnieri inclusions. Permanent facial pockmarks were observed in about 75% and 7% of patients recovering from variola major and variola minor, respectively (Fig. 5).

C. Treatment, Immunity, and Laboratory Diagnosis

Vaccinia immune globulin was used for the treatment of smallpox and also for complications arising from the use of smallpox vaccine. Methisazone (Marboran) was also used as an effective prophylactic drug.

Immunity after clinical smallpox is believed to be permanent. The disease may occur in persons vaccinated many years earlier; however, it is milder, with

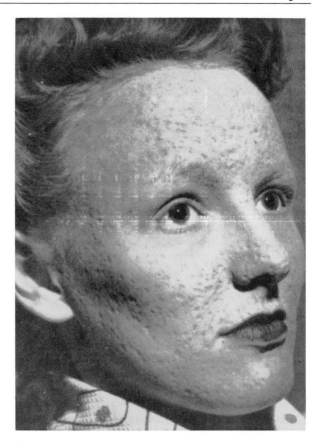

Figure 5 Residual scarring (pockmarks) 6 months after recovery from smallpox. [Reprinted from Dixon, C. W. (1962). "Smallpox." Churchill, Livingstone, Edinburgh, United Kingdom.]

less virus shedding and less transmission efficiency. Previously uninfected or unvaccinated persons of all ages were equally susceptible; however, as the ratio of the vaccinated to the unvaccinated persons in a population increased with age, the disease was most commonly observed in children.

Laboratory diagnosis of smallpox was made mainly by electron microscopy, using scrapings from macular and pustular lesions or suspensions prepared from crusts. Typical poxvirus particles are readily distinguished from all other human pathogenic viruses because of their characteristic shape and size as described above. Moreover, the virus can be grown on the chorioallantoic membranes of 10- to 12-day-old chick embryos; it produces characteristic lesions (pocks) on the membrane at a specific ceiling (maximum) temperature.

Primary monkey kidney and human embryonic lung cell cultures can also be used for the isolation of this virus. Serological tests (complement fixation, hemagglutination inhibition, radioimmunoassay, and serum neutralization), as well as the fluorescent antibody technique and certain enzymatic procedures, were also used for the diagnosis of smallpox. When no specimen for virus isolation was available, serological tests could not provide a definitive diagnosis as all members of orthopoxviruses are immunologically related. In such cases clinical history played an important role.

III. World Health Organization's Global Eradication Program

The concept of eradicating smallpox by vaccination was originally conceived by Jenner in 1801. Subsequently, data collected in recent years from extensive worldwide investigations indicated that the global eradication of this disease was achievable. This was based on the findings that, first, smallpox is a specifically human disease caused by one stable virus with no known animal or insect reservoir; second, the disease occurs only as an acute infection in which the infected person either dies or recovers with lifelong immunity and without relapse; and third, there exists a very effective, easily administered and fast-acting vaccine for protection against smallpox. This concept became a reality in 1950 when Fred L. Soper, Director of the Pan American Sanitary Bureau, proposed at the Third World Health Assembly of the WHO a program for smallpox eradication in the Americas. The proposal was

approved, and under Soper's direction, smallpox was eradicated from the Caribbean, Central America, and five of the South American countries within 8 years. Consequently, in 1958 at the Eleventh World Health Assembly, Victor M. Zhdanov of U.S.S.R. boldly proposed the adoption of the principle of global smallpox eradication as a policy. This was adopted by the assembly. However, preoccupation of many member countries with malaria eradication, unavailability of sufficient suitable vaccine doses, insufficient budget, and certain doubts about the feasibility of such a world-wide program hindered the progress. In 1965, however, the member countries took a much stronger position, and a more realistic budget ($2.5 million) was recommended. Subsequently, three essential elements of the eradication program, namely, availability of a stable vaccine (freeze-dried) of good quality, a more efficient vaccination method (using the bifurcated needle), and a more effective strategic approach (active case hunting and vigorous containment of outbreaks), were successfully worked out and implemented by the WHO experts.

In January 1967, when the intensified global smallpox eradication program was started, the disease was reported from some 46 countries, of which 33 were considered endemic areas (the latter in Asia, Africa, and South America involving a population of more than 1.2 billion). It is believed that the number of cases in 1967 was as many as 10 to 15 millions, with some 2 million deaths (Fig. 6). The aim of the program was to eradicate smallpox from the entire

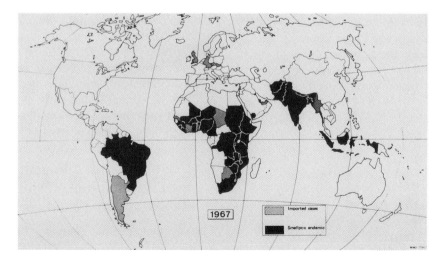

Figure 6 Countries reporting smallpox cases in 1967. (Courtesy of the WHO, Geneva, Switzerland.)

world in 10 years. This aim was achieved within the allotted time, and the last naturally occurring small-pox (variola minor) was in a 23-year-old male Soma-lian hospital cook in April 1977. The WHO, however, required 2 years of effective worldwide surveillance with negative results after the last rec-ognized case before certifying the global eradication of smallpox. During 1978 to 1979, a special WHO multilingual poster, which was widely distributed, announced a $1000 reward for the first person who reported a confirmed case of smallpox acquired by person-to-person transmission. Thus, on October 26, 1979, the WHO declared that smallpox had been eradicated from the entire world. Later, victory over smallpox was officially marked by solemn cere-monies at the Palais des Nations in Geneva on May 8, 1980. The global eradication of this most impor-tant infectious disease is indeed a unique event in the history of humankind and represents a sig-nal achievement by the WHO. The credit for this phenomenal feat indeed belongs to the thoughtful and dedicated field workers, both national and inter-national, who worked out remarkable solutions to many different and very difficult problems with due understanding and consideration of the local customs and circumstances. The successful global eradication of smallpox provides a model for the eradication of other similar viral diseases such as poliomyelitis and measles.

Since October 1977, many smallpox scares have been rumored throughout the world. Special WHO teams investigated all of these scares; all proved false (mostly chickenpox, herpes, or monkeypox).

Currently, smallpox virus stocks are stored safely at only two locations, namely, the Centers for Dis-ease Control (CDC) in Atlanta, Georgia, and the Research Institute for Viral Preparations in Mos-cow, U.S.S.R. Because the genetic content of vari-ous strains of smallpox virus are now being se-quenced by genetic engineering in recombinant bacterial plasmids, most WHO experts recommend that the remaining stocks of this virus be destroyed by the end of 1993.

IV. Post-Global Eradication Period

A. Laboratory-Acquired Smallpox

After the last naturally occurring case of smallpox, in Somalia, one accidental laboratory-acquired in-fection occurred in 1978 at Birmingham University

Medical School in Britain. A 40-year-old medical photographer, Janet Parker, of the Department of Anatomy worked on the floor above a research labo-ratory where a comparative study of smallpox and whitepox viruses were performed by the 48-year-old Henry S. Bedson of the Department of Medical Mi-crobiology. The service duct that connected Park-er's office to Bedson's research laboratory had ill-fitting inspection panels at both ends. The smallpox virus apparently escaped from Bedson's research laboratory, entering the service duct through the ill-fitting inspection panel and subsequently reached Parker in her office (here again coming out of com-mon service duct through the ill-fitting inspection panel).

Parker, last vaccinated in 1966, became ill on Au-gust 11 and died of smallpox on September 11, 1978. Bedson, who became depressed and blamed himself for the escape of the virus, had already died of self-inflicted throat wounds, on September 6, 1978. Park-er's mother and close contact was prophylactically vaccinated and administered vaccinia immune glob-ulin and methisazone. However, she developed a modified smallpox but recovered uneventfully.

B. Unjustified Use of Smallpox Vaccine

After the global eradication of smallpox, the vaccine was needlessly administered to international travel-ers. It was also occasionally used, without any dem-onstrable benefit, for the treatment of warts and herpes, which are caused by unrelated viruses. Many serious and even fatal reactions to such small-pox vaccinations have been reported by CDC; one of the last fatal cases was reported on March 14, 1980; it involved a 7-month-old child who was vacci-nated because of a 3-month history of recurrent mouth ulcers suspected of being caused by herpes simplex virus. The use of vaccine (provided in the United States only by CDC) became restricted to military personnel in the early 1980s. This was be-cause of the potential of biological warfare with the smallpox virus, which is a poor choice.

C. Use of Smallpox Vaccine Virus for Production of Various Recombinant Vaccines

The vaccinia virus has been recently used to prepare a number of recombinant vaccinia virus vaccines. As this virus has a large, double-stranded DNA mol-ecule of 190 kilobase pairs, it is possible to insert and express foreign genes coding for immunogenic proteins of other viruses at nonessential sites of the DNA of this virus. Such vaccines for rabies, AIDS, hepatitis B, influenza, herpes simplex, and many other viral diseases have been prepared and are being tested mainly in animals but also in humans. The advantages are that such recombinants effectively prime the immune system, are easily administered, produce effective immunity against one or more antigens, and can be used to produce diagnostic polyclonal antisera in animals. However, there are certain associated and potential problems. If such a vaccine is inadvertently administered to a person with active human immunodeficiency virus (HIV) infection, a fatal disseminated vaccinia might occur. Moreover, the use of such vaccines in animals (e.g., rabies vaccine in domestic and wild animals), theoretically may cause *in vivo* hybridization of the engineered vaccinia virus with naturally occurring active or latent animal orthopoxviruses in these animals with the possible emergence of a new human pathogenic poxvirus.

Bibliography

Behbehani, A. M. (1988). ''The Smallpox Story: In Words and Pictures.'' The University of Kansas Medical Center, Kansas City, Kansas.

Buller, R. M., and Palumbo, G. J. (1991). Poxvirus pathogenesis. *Microbiol. Rev.* **55,** 80–122.

Fenner, F., Henderson, D. A., Arita, I., Jezek, Z., and Ladnyi, I. D. (1988). ''Smallpox and Its Eradication.'' WHO, Geneva.

Moss, B. (1991). Vaccinia virus: A tool for research and vaccine development. *Science* **252,** 1662–1667.

Sorbic Acid, Mode of Action

John N. Sofos
Colorado State University

I. Chemistry
II. Antimicrobial Activity
III. Applications
IV. Mechanisms of Action

Glossary

Antimicrobial activity Inhibitory effect of a process, condition, or compound (e.g., sorbic acid) against a microorganism, which may be detected or expressed as a delay, reduction, long-term inhibition, or failure to metabolize and/or reproduce

Bacterial spore germination Degradative process during which the dormant bacterial spore is converted into a sensitive entity, which exhibits metabolic function

Preservation Lengthening of the time during which a food product is of acceptable eating quality and is safe through inhibition of the proliferation of microorganisms that are associated with this product

Proton-motive force Difference in electrical potential and in pH across the cell membrane, which provides the energy that acts as the driving force for uptake of nutrients such as amino acids by microbial cells

Safety Food product is not involved in outbreaks of microbial foodborne illness because pathogenic bacteria that may have been present are inhibited by preservatives such as sorbic acid

Shelf life Period of time during which a product is acceptable, fit, and safe for use or consumption because preservatives such as sorbic acid have inhibited microbial proliferation

Sorbates Sorbic acid and its derivatives that are approved and used for inhibition of microbial proliferation in foods and other substrates

Substrate Material, including foods or other products, that has been treated with sorbic acid or its derivatives for inhibition of microbial proliferation

SORBIC ACID is an unsaturated fatty acid. Together, sorbic acid and some of its derivatives are collectively known as sorbates and are used widely in the preservation of various foods and other items. The value of the sorbates as preservatives is based on their inhibitory activity against growth of yeasts, molds, and bacteria. Inhibition of microorganisms with sorbic acid, however, is variable among yeasts, molds, and bacteria among different species and strains and under changing environmental conditions. This variability can range from complete inhibition of microbial proliferation to resistance or metabolism of the compound by certain strains and under certain conditions. Although not fully defined, inhibition of microbial metabolism by sorbates may include inhibition of enzymes and transport functions in the microbial cell or alterations in cell membranes. Germination of bacterial spores may be inhibited through a potential action on spore membranes or protease enzymes involved in the initial stages of the germination process.

I. Chemistry

A. Properties

Sorbic acid is a naturally occurring *trans-trans* unsaturated fatty acid (2,4-hexadienoic; $CH_3—CH=CH—CH=CH—COOH$), with a highly reactive carboxylic group that forms various salts and esters. The compound is also reactive through its conjugated double bonds. The reactivity of sorbates in food systems may influence their activity as antimicrobial agents and may influence the quality and safety of the products. Commercially important salts of sorbic acid include calcium, sodium, and potassium sorbates. The volatility of the compound in steam is useful in its isolation for quantitative detection in foods or other materials.

Solubility of sorbic acid is higher in lipids than in water (0.15% w/v at 20°C), whereas its salts, especially potassium sorbate, are more soluble in water. Solubility increases with increasing temperature and with increasing concentration of ethanol, propylene glycol, and acetic acid, whereas it decreases with increasing concentrations of sucrose and sodium chloride. Partitioning of sorbic acid between the lipid and aqueous phases of foods depends on the pH of the food, the amount and type of lipid, other ingredients present such as sugar and ethanol, and temperature. The high solubility of potassium sorbate in water (58.2% w/v at 20°C) makes it the compound of choice for use in the form of solutions. In water, potassium sorbate hydrolyzes to yield 74% of its weight as sorbic acid. Sodium sorbate is soluble in water at the rate of approximately 30%, whereas calcium sorbate is soluble at 1.2% in water and insoluble in fats, which makes it valuable as a delayed-release form of sorbic acid.

Depending on the type of food or other material to be preserved by sorbic acid, its activity may be necessary either only on the surface or throughout the product. Diffusion of the compound into the product is desirable when preservation of the core of the product is necessary. When there is no need for microbial inhibition in the core of the product, then it is desirable that sorbate does not diffuse into the product but that it remains on the surface where its activity is needed. Uptake and diffusion of sorbic acid into model systems or foods have been affected by properties such as the moisture content and water activity of the substrate, the structure of the diffusion medium, and its hydration properties. Diffusivity appears to decrease with increasing concentrations of gelling agents, decreasing concentrations of sorbic acid, increasing concentration and molecular weight of solutes, and decreasing temperature.

Direct mixing of sorbic acid or its salts into a food product during processing is the ideal application procedure for uniform distribution and control of the concentration applied. In addition to direct mixing, however, sorbates may be applied by treating food wrappers and packaging materials that come in contact with the food or by immersing, dipping, spraying, or dusting of foods with solutions or the dry form of the compound. These latter methods of application may result in variable residual levels of sorbic acid present in different parts of the food. The residual levels will vary not only with method of application, but also with concentration, time/length of exposure, type of substrate, and handling of the substrate after exposure to sorbate. Because of such variations, before a given application, testing should be performed to determine appropriate solutions, exposure times, and handling for the desirable levels of residual sorbate in the product.

In the pure and dry form, sorbic acid and its derivatives are remarkably stable to oxidation, but in aqueous solutions they are relatively unstable and degrade by first-order reaction kinetics. The olefinic bonds of sorbate are attacked by oxidizing agents, resulting in formation of peroxides, followed by their degradation or polymerization. Products of sorbic acid oxidation include carbonyl compounds such as crotonaldehyde, malonaldehyde, acrolein, formic acid, and malonic acid. Factors influencing the rate of oxidation include pH, sugars, salt, amino acids, antioxidants, metal ions, light, temperature, time, types of food materials present, and properties of the packaging materials. Low or no oxygen permeability is desirable in packaging materials to avoid oxidation of sorbic acid.

Oxidation is more rapid at decreasing pH values and in the presence of certain acids such as acetic, sulfuric, and hydrochloric, whereas citric and malic acids appear to have no major effect on degradation of sorbic acid. Oxidation may also be accelerated in the presence of certain concentrations of copper, iron, cobalt, and manganese. Chloride salts appear to inhibit oxidation, whereas sulfates and phosphates have pro-oxidant activity. Oxidation is also inhibited by antioxidants and alcohols. In general, oxidative degradation of sorbic acid in foods is influenced by the nature of the food and its components and by processing, handling, and storage conditions, but it is generally less extensive than in aqueous solutions.

Naturally, sorbic acid is present in the berries of the mountain ash tree (*Sorbus aucuparia*, L., *Rocaceae*), in which it was first detected. The scientific literature contains numerous publications on original procedures and modifications for formation of sorbic acid and its derivatives. Commercial production of the compound, however, has involved either oxidation of 2,4-hexadienal or the condensation reaction of ketene and crotonaldehyde, with the second procedure being most widely used in practice. The crude sorbic acid produced by decomposition of the ketene–crotonaldehyde reaction polymer is purified and stabilized for storage, distribution, and application as an antimicrobial agent. Purification may involve treatment with sodium hydroxide, hydrochloric acid, and activated carbon; distillation and crystalization; or washing, distillation, and vacuum drying. Granulation is desirable for uniform mixing in foods and for improved solubility, and it is accomplished by extrusion and pelletization of a slurry.

In addition to alkaline salts formed by neutralization of sorbic acid, other derivatives of the compound manufactured and tested for antimicrobial activity include esters, alcohols, aldehydes, amine salts, and amide derivatives. The derivative with the widest commercial application as an antimicrobial agent is potassium sorbate. Its high solubility allows preparation of concentrated stock solutions to be used in dipping or spraying applications.

B. Toxicity

Sorbic acid and its salts are considered as generally safe food additives, which makes them popular for use in the preservation of many foods throughout the world. Extensive testing and feeding to test-animal species have indicated that they are among the safest of all food preservatives. The oral lethal dose (LD_{50}) of sorbic acid in rats is 7.4–10.5 g/kg body weight and of potassium sorbate is 4.2–6.2 g/kg body weight compared with the LD_{50} for sodium chloride, which is 5 g/kg body weight.

Increases in liver or other organ weights associated with long-term feeding studies involving diets high in sorbic acid have been attributed to functional hypertrophy through caloric utilization of sorbic acid, while the histopathological appearance of the liver has been normal. As an aliphatic carboxylic acid, sorbate can be used in the animal body as are other fatty acids, serving as a source of calories.

Skin irritation risks or allergenic reactions from sorbate used as a preservative in cosmetics and pharmaceuticals are low, and no such effects are expected when sorbates are used as indicated in the regulations for preservation of foods. Furthermore, no carcinogenic or mutagenic activity has been attributed to sorbates. A committee of the Food and Agriculture Organization and the World Health Organization of the United Nations has indicated that the acceptable daily intake (ADI) for sorbic acid is 25 mg/kg body weight.

C. Analysis

Analytical methods developed for qualitative and quantitative determination of sorbic acid and its salts in food products are acidimetric, bromometric,

colorimetric, spectrophotometric, polarographic, and chromatographic. The most commonly used methods are colorimetric and spectrophotometric, with chromatographic procedures gaining acceptance in recent years.

Before detection, there is a need for extraction and quantitative separation of sorbic acid from the food material. Extraction procedures include dialysis, acid-steam distillation, selective gas diffusion, and extraction with solvents such as ether, dichloromethane, and isooctane. Proper extraction is necessary to avoid interference by food constituents or other additives present in the product.

Analytical methods approved by the Association of Official Analytical Chemists (AOAC) include a colorimetric procedure for cheese and wine, whereas the ultraviolet absorption (spectrophotometric) method has been used in many foods and is approved by AOAC for use in dairy products, wines, and ground beef. A gas chromatographic method is recommended by AOAC as a "first action" procedure for various foods.

II. Antimicrobial Activity

A. Inhibition of Microorganisms

Inhibition of yeasts and molds by sorbic acid and its derivatives appears to be comprehensive, whereas inhibition of bacteria appears to be selective. Concentrations of the compound used as an antimicrobial agent in foods and other substrates are in the range of 0.001% to 0.30%. These amounts are generally static in activity, whereas higher levels may also be cidal to microorganisms.

Extensive testing has demonstrated inhibition of many yeast species associated with spoilage of various foods, including fermented vegetable products, wines, fruit juices, meat, fish and dairy products, salad dressings, carbonated beverages, jams, jellies, candy, and chocolate products. Depending on species, strains, and pH, concentrations of sorbate inhibitory against yeasts are in the range of 0.0025% to 0.2%.

Sorbic acid is also effective in inhibiting growth of molds, including mycotoxin-producing strains, at concentrations of 0.001–0.3%, depending on species and pH. Effective mold inhibition occurs in cheeses, butter, sausages, fruit products, bakery items, and fish products. Mold inhibition involves action on spore germination and mycelial growth.

Sorbates are also known to inhibit formation of mycotoxins such as aflatoxins, patulin, sterigmatocystin, citrinin, and ochratoxin by various molds and to various degrees. Mycotoxin inhibition is greater in richer substrates, in the presence of limited mold contamination, at lower pH values, and at higher sorbate concentrations.

Under certain conditions (e.g., higher temperature of storage), however, low concentrations of sorbic acid may stimulate production of mycotoxins by certain strains of molds. Although inhibition of mycotoxin synthesis is believed to be caused by inhibition of uptake of substrates by the cell, stimulation of mycotoxin formation may be due to a reduction of the activity of the tricarboxylic acid cycle by sorbate, which may lead to accumulation of acetyl coenzyme A, which is an intermediate in formation of certain mycotoxins.

Sorbic acid also inhibits many species of spoilage and pathogenic, gram-positive and gram-negative, catalase-positive and catalase-negative, aerobic and anaerobic, mesophilic and psychrotrophic bacteria, at concentrations of 0.001–1.0%. Foods in which sorbate has delayed bacterial growth include cheese, yogurt, meat, and seafood products. In addition to vegetative cell forms, sorbic acid is inhibitory against spore-forming species during their germination, outgrowth, or cell division. Pathogenic bacteria reported as inhibited by sorbate include *Clostridium botulinum, Salmonella* spp., *Staphylococcus aureus, Vibrio parahaemolyticus, Bacillus cereus,* and *Clostridium perfringens.*

Overall, sorbates are useful in the preservation of products with low levels of contamination. Sorbic acid is also considered a more potent antimicrobial agent than other common preservatives such as benzoate, propionate, diacetate, and parabens. Substrates in which sorbate has exhibited higher activity than the other compounds include pickles, cheeses, margarine, juices, fish, and pharmaceutical preparations. The higher antimicrobial activity of sorbic acid may be due to its higher dissociation constant (pH 4.76), which yields a higher amount of the effective undissociated form than either benzoic or propionic acid, which have lower pKa values. In high acid foods, however, use of benzoate may be more common than sorbate.

In certain applications, combinations of specific preservatives may be more beneficial for extending product shelf life than are single additives. Combinations are beneficial because the spectrum of microorganisms inhibited may be increased and the amounts of individual compounds used in the combination can be decreased, thereby reducing any undesirable effects on product flavor.

B. Interactions of Factors

The antimicrobial activity of sorbic acid can be affected by compositional, processing, and environmental variables. These influential factors include type of microorganisms, extent of initial microbial contamination, concentration of sorbate, product composition, pH, water activity, other chemical additives, processing and storage temperature, type of packaging, and gas atmosphere. Depending on individual factors, levels, and multifactor combinations, their effects on microorganisms can be synergistic or antagonistic to sorbate. The intensity and importance of these factors usually influence the extent of microbial inhibition by sorbate and, consequently, the length of shelf life or the safety of the product. The complexity of multifactor interactions in nonhomogeneous substrates (e.g., food), however, does not always allow accurate prediction of the extent of microbial inhibition and length of product shelf life. Although sorbate inhibits all types of microorganisms, including yeasts, molds, and bacteria, it is not effective against all species and strains. In addition, its antimicrobial activity is reduced in highly contaminated environments, which indicates the need for efficient sanitary practices even when chemical preservatives are used.

The antimicrobial activity of sorbate differs with types of food components because they affect solubility, uptake, distribution, partitioning, and reactivity of the compound in the substrate. Inhibition of microorganisms by sorbate is enhanced in presence of increased concentrations of solutes such as sucrose and sodium chloride, which reduce the amount of unbound water (water activity) in a food. In certain instances, however, solutes have decreased the synergistic effect of heat and sorbate on thermal inactivation of microorganisms.

The antimicrobial activity of sorbic as well as other weak acid preservatives is enhanced as the pH of the substrate approaches its dissociation constant, which for sorbate is at pH 4.76. The maximum pH for microbial inhibition by sorbate levels commonly used in foods is in the range of 6.0 to 6.5, but there are reports of measurable activity even at pH values approaching 7.0.

It is believed that the higher antimicrobial activity at pH values approaching its dissociation constant is due to an increase in the undissociated form of the compound. This, however, cannot explain the antimicrobial activity recorded at pH values near neutrality, at which the amount of sorbate in the undissociated state is negligible. Thus, certain studies demonstrate that even the dissociated sorbic acid exerts antimicrobial activity, but it is low compared with that of the undissociated form of the compound. As the pH value increases above 6.0, however, as much as 50% of the antimicrobial activity may be due to the dissociated species. The exact pH value at which microbial inhibition by sorbate becomes significant depends on factors such as concentration of sorbate, properties of the substrate, microbial contamination, and temperature of storage.

Another important interaction of sorbate is with heat processing treatments. Such interactions can affect microbial destruction by heat, dormancy of heated bacterial spores, and recovery of heat-injured organisms. Low concentrations of sorbate can stimulate thermal activation of mold ascospores and may increase thermal inactivation of yeasts, molds, and bacteria or inhibit repair and recovery of thermally injured cells. Increased thermal destruction and inhibition of repair is observed with sorbic acid concentrations in the range of 0.0025% to 3.0%, depending on type of microorganisms and magnitude of the effect. [*See* THERMOPHILIC MICROORGANISMS.]

Reduction in thermal resistance in the presence of sorbate should be expected regardless of solute concentration, even though presence of solutes reduces water activity and increases thermal resistance of microorganisms. Use of sorbate in fruit products before thermal processing can allow reduction in the extent of heat treatment needed for preservation, which can result in products of improved sensory and nutritional qualities and in reduced energy costs.

The exact mechanisms of increased thermal destruction and reduced recovery after heat treatment in the presence of sorbate is not known. A potential mechanism for increased mold sensitivity to heat may be related to an increased permeability caused by expansion of the cell wall and cytoplasmic membranes occurring at higher temperatures. This would allow a physically easier uptake of the preservative by the cell at higher temperatures.

In addition to heat, combinations of treatments with sorbate and low-dose ionizing radiation may be useful in lengthening the shelf life of food and other products. The potential for application of such combination processes has been evaluated in fish, fruit juice, and cheese products. The results have been promising, and it is believed that application of sorbate sensitizes microorganisms to destruction by irradiation or inhibits growth of microorganisms exposed to irradiation. These interactions have been attributed to radiation-induced changes in cell wall permeablity or may be due to formation of new antimicrobial compounds produced through irradiation of sorbate-treated substrates. Potential chemical changes and formation of radiolytic products, however, would be undesirable relative to food safety and quality. Thus, there is a need for research to examine such potential problems. Treatment with sorbate and low-dose irradiation could be combined, however, to delay spoilage and inhibit toxin production by *Clostridium botulinum* type E spores in refrigerated seafoods.

Storage at temperatures outside the range of growth of a given microorganism should enhance its inhibition by sorbic acid. Because microbial inhibition by the compound is more pronounced at refrigeration temperatures, shelf life of sorbate-treated foods is longer when they are refrigerated. [*See* REFRIGERATED FOODS.]

The type of packaging and gas atmosphere surrounding the product are also influential on the antimicrobial activity of sorbate. Inhibition is usually enhanced under vacuum packaging or modified gas atmosphere storage (e.g., increased carbon dioxide levels). Extent of inhibition, however, differs with types and species of microorganisms, as well as other variables such as type of food and its pH. The reduced inhibitory activity of sorbic acid against yeasts (e.g., *Candida albicans*) in aerobic, compared with anaerobic, environments may be due to detoxification of the compound in presence of oxygen. Aerobic detoxification, however, may be reduced by certain compounds such as cysteine and glutathione.

The antimicrobial activity of sorbic acid may also be influenced by the type of acid present naturally or added to foods during processing, because in addition to reduced pH, specific anions can themselves influence microbial inhibition. Thus, acidification may allow reduction of the amount of sorbic acid needed for preservation, but the activity may be different in different substrates. Inorganic acids are generally less effective potentiators of microbial inhibition by sorbate than acetic and citric acid.

Combinations of sorbate with antioxidants such as butylated hydroxyanisole, butylated hydroxytoluene, tertiary butyl hydroxyquinone, and propyl gallate have exhibited greater antimicrobial activities than single-compound formulations. Also, inhibitory interactions of sorbate with phosphate salts have been detected in laboratory culture media and in foods. Microbial inhibition can often be enhanced when sorbic acid is combined with other chemical preservatives, such as benzoate, propionate, and nitrite as well as certain antibiotics. Combinations of sorbate with sodium nitrite are effective in inhibiting pathogens such as *C. botulinum* in cured meat products and sausages. Although not approved for use in the United States, this interaction is interesting because it allows reduction in nitrite levels used in meat curing without compromising the microbiological shelf life of the products. A reduction in the amount of added nitrite should result in a decreased potential for formation of carcinogenic nitrosamines during meat product processing and cooking. [*See* MEAT AND MEAT PRODUCTS.]

The extensive published scientific literature on sorbic acid and its derivatives identifies a large number of other potential interactions that may result in increased antimicrobial activity. Other compounds reported as interacting with sorbate include ascorbate, sulfur dioxide, sucrose fatty acid esters, propylene glycol, amino acids, fatty acids, glucose oxidase, diethyl ester of polycarbonic acid, 2-(4-thiazolyl) benzimidazole, acyloxyalkanoic acid, glutathione, carboxymethyl cellulose, lysozyme, gelatin, sodium bicarbonate, furylfuramide, *o*-phenylphenol, 1,5-pentanedial, and levulinic acid.

In addition to interactions of sorbate with one other common or uncommon chemical compound or processing treatment, the scientific literature includes a large number of publications and patents reporting on multifactor interactions. Such multiple factor interactions include compounds or processes that are commonly used in preservation, as well as chemicals that are not approved for use because of potentially undesirable effects on product quality and consumer safety. In most commercial applications, however, sorbate contributes to extension of product shelf life as one of a series of hurdles that inhibit or delay microbial growth.

C. Resistance and Metabolism

As indicated earlier, the antimicrobial activity of sorbic acid is variable because it is dependent on multifactor interactions that can lead to complete, partial, or no inhibition of growth. In general, microbial inhibition by sorbate will depend on individual systems and conditions, whereas variations in sensitivity to sorbate by different species and strains of microorganisms may lead to defective products when proliferation of microorganisms is permitted.

Certain species and strains of molds, yeasts, and

bacteria may possess or develop resistance to inhibition by sorbate, and under certain conditions they may even be able to metabolize the compound. Traditionally, sorbate has been suggested for use as a selective ingredient in isolation media for catalase-negative lactic acid bacteria and clostridia, because it was found to exert a greater inhibitory effect against catalase-positive microorganisms. Subsequent studies, however, have indicated that sorbate may also inhibit catalase-negative microorganisms such as clostridia, especially when used at higher concentrations (>0.1%) and in combination with other chemical additives such as sodium chloride and sodium nitrite. In general, however, different species of bacteria exhibit different sensitivities to inhibition by sorbate, which may lead to shifts in the microbial contamination during storage of sorbate-treated products. Psychrotrophic microorganisms are inhibited somewhat more extensively than mesophiles, which coincides with the higher antimicrobial activity of sorbate at lower storage temperatures.

Certain species of yeasts may become resistant to inhibition by sorbate. Yeasts that have exhibited such resistance include strains of the species *Saccharomyces uvarum, Saccharomyces cerevisiae, Candida utilis,* and *Brettanomyces bruxellensis,* as well as the osmotolerant species of *Zygosaccharomyces bailii, Saccharomyces rouxii, Saccharomyces bisporus,* and *Saccharomyces acidifaciens.* The extent of their resistance to inhibition by sorbate depends on concentration, pH, inoculum level, storage gas atmosphere, and previous exposure of the organism to sorbate. Osmotolerant yeasts appear to acquire their resistance to sorbate when they are predisposed or preconditioned in the presence of low concentrations (<0.1%) of the preservative.

One proposed mechanism of resistance includes potential presence of an inducible, energy-requiring system in osmotolerant yeasts, which may transport the preservative out of cells. This mechanism appears to be induced by previous growth of the organism in the presence of subinhibitory levels of the preservative. Another proposed mechanism has suggested that at lower water activities, the yeast cells shrink and their membrane pore size is reduced in size, which may retard the flow of sorbate and similar antimicrobials into the cell cytoplasm. When the water activity and sorbate levels are increased, uptake of the compound is too rapid and results in intracellular accumulation, which becomes inhibitory to the yeast cell. Resistance of osmophilic yeasts at lower water activity may also involve production and intracellular accumulation of polyhydric alcohols, which are involved in osmoregulation, acting as compatible solutes and thus protecting enzymes from inactivation.

Certain bacteria, especially species of the group producing lactic acid, in addition to being resistant to sorbic acid, have been associated with metabolic breakdown of sublethal concentrations in presence of high contamination. In addition to strains of lactics, other bacteria reported as metabolizing sorbate include *Acetobacter* spp. and *Pseudomonas fluorescens*.

Metabolism of sorbate by certain bacteria has been associated under certain conditions with loss of quality in fermented vegetables and wines, which are described as developing "geranium-like" or "flowery" odors and flavors. This off-odor is believed to be due to metabolic degradation of sorbate into products such as ethyl sorbate, 4-hexenoic acid, 1-ethoxy-hexa-2,4-diene, and 2-ethoxy-hexa-3,5-diene. The defect may develop in wines improperly treated with low concentrations ($\leq 0.03\%$) of sorbic acid and contaminated with high microbial loads. One postulation, which has been disputed, has indicated that strains of lactic acid–producing bacteria, capable of metabolizing sorbate and isolated from wines, possess an aldehyde dehydrogenase enzyme, which is not present in other strains. Other scientists have postulated that this reaction is carried out by all strains of lactic acid bacteria and that sorbate metabolism may require presence of two different enzymes.

Published information on the potential metabolic degradation of sorbic acid by yeasts is conflicting. If at all, metabolism by yeasts should occur only at low concentrations of sorbate. Published evidence on metabolic degradation of the compound by molds, however, is more convincing. As a fatty acid, sorbate can be metabolized through β-oxidation under favorable environmental conditions, with eventual release of energy, carbon dioxide, and water.

Metabolic degradation of sorbate by molds has been associated with disappearance of the compound from highly contaminated cheeses. Decomposition has also been detected in moldy fruit products, and it has depended on pH, level of contamination, strains of molds, and concentration of sorbic acid. Certain molds, especially of the genus *Penicillium,* isolated from sorbate-treated moldy cheeses, have been able to proliferate in the presence of concentrations of sorbate in the range of 0.1% to 1.2%. Under normal conditions, sensitive strains of molds are inhibited by concentrations of up to 0.1%. Isolation of molds able to degrade sorbate from moldy cheeses that had been treated with the compound indicates selection and adaptation of molds for resistance to the preservative.

Metabolism of sorbic acid by molds has resulted in formation of products such as 1,3-pentadiene, methyl ketones, ethyl sorbate, and *trans*-4-hexenol. Formation of 1,3-pentadiene through decarboxylation of sorbate by *Penicillium* molds in cheeses results in a "hydrocarbon-like," "plastic-paint," "kerosene," or methyl-methacrylate off-flavor, which may also occur in sorbate-treated carbonated beverages.

In general, sorbate is an effective inhibitor of spoilage and mycotoxin-producing molds, but some species and strains can metabolize the compound, especially when exposed to subinhibitory levels. Degradation, then, depends on type of substrate, level of sorbic acid, and extent of contamination. Thus, sorbate should be an adequate preservative for foods produced and processed under sanitary conditions, but it should not be used as a replacement for appropriate hygiene and good manufacturing practices.

III. Applications

A. Food Products

Since the first patent for use of sorbic acid as a fungistatic agent for foods was issued in 1945, the compound and its derivatives have been tested and used extensively throughout the world as preservatives of foods and other materials. Foods preserved with sorbates include cheeses and other dairy products; bakery products; fruit and vegetable products; and other foods. Dairy products preserved with sorbate may include various types of natural and processed cheeses, yogurt, sour cream, cheese spreads, cottage cheese, and dips. Common concentrations of sorbate used in cheeses to inhibit molds and yeasts during ripening and in consumer packages are 0.05–0.3%. [*See* DAIRY PRODUCTS.]

Fruit and vegetable products preserved with sorbates include juices, beverages, wines, jams, jellies, dried fruits, and salads, as well as fermented and acidified products. A common original use of sorbic acid was to inhibit undesirable film-forming yeasts during fermentation of cucumbers. Other fermented products that may be preserved with sorbate include sauerkraut, tomato products, olives, relishes, and oriental fermented vegetable products. The concentration needed to preserve such products depends on pH and sodium chloride concentration, and it is in the range 0.025% to 0.1%. The success of sorbate as a yeast inhibitor in fermented vegetables is based on its limited activity against lactic acid–producing bacteria, which are involved in the fermentation, especially at the lower concentrations, used in these products.

Sorbates are useful as preservatives of wines, where in combination with sulfites, they control secondary fermentations. In nonalcoholic beverages, however, the compound may be used to prevent fermentation during storage of raw materials to be used for manufacture of soft drinks and similar products. Several beverages may also be preserved with sorbates in addition to other common preservatives such as benzoate. Sorbic acid may also be beneficial as a preservative of packaged fresh vegetable products such as salads, celery, cucumbers, lettuce, and sugar beets. Dried fruits, pulp, fruit syrups, fruit salads, jams, jellies, marmalade, and preserves can also be preserved with sorbic acid compounds.

Because mold growth is favored in bakery products, sorbic acid may be valuable in their preservation, especially in nonyeast-leavened items such as cake mixes, pie fillings, doughnuts, fudges, icings, toppings, muffins, and tortillas. In yeast-leavened products, however, propionate is the preservative of choice because sorbate may delay or stop fermentation by the *Saccharomyces* yeasts. In yeast-leavened products, sorbate may be applied as a surface spraying solution after baking. In addition to spoilage and mycotoxin-forming molds, sorbic acid can inhibit rope-forming bacteria in bread and the pathogen *S. aureus* in cream pies and cakes.

Other food products that may be preserved by sorbic acid and its derivatives include food emulsions such as butter, margarine, mayonnaise, and salad dressings; sugar and confectionary products such as chocolate syrups, creamy fillings, and pralines; cereal grain products; legumes and low-salt formulations of products such as soy sauce and miso; fresh, dried, smoked fish and fish sausages; and meat products. In the United States, the only approved use of sorbate in meat products is for the inhibition of molds on the surface of uncooked and dried sausages, whereas in certain other countries, preservation of meat products with sorbic acid is more common. The proposed use of sorbate as an inhibitor of *C. botulinum* in meat products for partial reduction of nitrite levels and potential for nitrosamine formation was not approved in the United States. Miscellaneous products that have been evaluated for preservation by sorbate alone or in combination with other additives include calf rennet preparations, egg products, olive husks, vitamin preparations, blood, marshmallow candy, tangerine sherbet, hot sauces, low-calorie imitation foods, pancake batter, eggroll dough, garlic oil, potato chips, and shrimp cocktail.

When the concentrations of sorbate used in food preservation are limited to the amounts approved, which, depending on the food, are in the range of 0.005% to 0.3%, they should not have undesirable effects on the sensory quality of the products. Concentrations greater than 0.3% are undesirable from a flavor standpoint. The effect of sorbate on the flavor of specific products, however, depends not only on concentration, but also on type of food, processing, and other flavorings added.

Methods of application of sorbates to foods and other products include direct addition and mixing in formulated items; spraying or immersing the product in a solution of sorbate; dusting of the product surface with a sorbate powder; or application of sorbate to a product coating or packaging material that will come in contact with the food. In each of these procedures, sorbate may be applied in the form of its acid or salts, singly or as a part of a multi-ingredient mixture. Sometimes (e.g., when applied to coatings or packaging materials) sorbate is combined with an organic carrier such as ethanol, propylene glycol, or vegetable oil. The specific method of application selected is determined on the basis of factors such as type of food, processing procedures, and convenience. The form of sorbate selected for specific uses is often determined by the method of application. In spray or immersion applications, the potassium salt, which is highly water-soluble, should be selected. For direct mixing with the formulation, the less soluble acid form may be used.

B. Other Applications

Although preservation of various food products constitutes the major application of sorbic acid and its derivatives, other important uses are also recognized throughout the world. Sorbates alone or as components of multicompound formulations may be used to inhibit yeasts and especially mycotoxin-producing molds in animal feeds. Their low toxicity and the relatively high tolerance by the skin makes them also useful for the preservation of pharmaceutical and cosmetic products. In addition, sorbates have been tested or applied as inhibitors of microbial growth in numerous and diverse items such as tobacco products, adhesives, films, copying paper, paints, deodorizers, fertilizers, rubber, pesticides, growth hormone preparations, animal hides, and feeds for silkworm larvae.

IV. Mechanisms of Action

A. Bacterial Spores

Sorbic acid has been described as a competitive and reversible inhibitor of bacterial spore germination induced by amino acids. Others, however, have presented evidence indicating that inhibition of bacterial spore germination takes place after initiation (triggering) of spore germination and during the undefined connecting reactions, which are believed to occur before loss of spore refractility in the germination process. Therefore, according to this theory, sorbate does not compete with the amino acids that trigger germination for a common binding site on the spore, but it acts after initiation of the germination process. Because the nature of the connecting reactions of spore germination is largely unknown, however, it is difficult to present a mechanism of their inhibition by compounds such as sorbate. Based on indirect evidence, it has been suggested that inhibition could involve alteration of permeability of spore membranes or inhibition of spore lytic enzymes that may be involved in germination by causing cortex hydrolysis and loss of refractility.

B. Microbial Growth

The amounts of sorbic acid employed in food preservation (<0.3%) inhibit or delay proliferation of microorganisms without causing permanent or lethal damage to their cells. Higher concentrations, however, may inactivate microorganisms. Several modes and mechanisms of action have been proposed as being involved in inhibition or delay of microbial cell growth. Inhibition of metabolic function may result from alterations in cell membranes and cell transport functions, inhibition of enzymes involved in transport or metabolic activity, alteration of the morphological structure of the cell, or

changes in the genetic material. It is possible that each one of these modes of inhibition may be operational under specific conditions of microbial species, substrates, and environmental as well as processing parameters.

Although published information on the effect of sorbic acid on the genetic material of microbial cells is limited, the compound is believed not to possess mutagenic activity. It may form mutagenic products, however, when it reacts under certain conditions with compounds such as sulfur dioxide and sodium nitrite.

In certain cases, sorbates have been reported to cause morphological changes in microbial cells. Reported alterations in cell appearance include development of yeast cells with dense lipoprotein granules, irregular nuclei, numerous mitochondria of various sizes, and vacuoles. Spores of *C. botulinum* treated with sorbate have become long and bulbous in shape and have exhibited defective division and elongation of emerging cells. The mechanism causing such morphological changes is unknown, but it may involve incorporation of sorbate into specific cell structures and/or inhibition or enhancement of specific biosynthetic processes. Another observed influence of sorbic acid treatment on microbial cells involves alterations in the morphology, integrity, permeability, and function of cell membranes. Elongation and aberrant cell formation are also associated with changes in cell walls and membranes.

Numerous studies have demonstrated inhibition of cell metabolism in microorganisms treated with sorbic acid, which has been detected as a decrease in assimilation of carbon from numerous substrates including glucose, acetate, lactate, ethanol, succinate, pyruvate, and oxaloacetate. Inhibition of cell metabolism, however, may be the result of adverse effects on enzymes, or transport mechanisms.

Sorbate is known to inhibit the *in vitro* activity of many enzymes, especially sulfhydryl-containing dehydronases. Although exact mechanisms of inhibition and inactivation of specific enzymes are unknown, it has been postulated from indirect evidence that binding of sorbic acid with sulfhydryl groups inhibits their activity. Inhibition of the enzyme catalase, however, has been attributed to autoxidation of sorbic acid and formation of sorbyl peroxide. It has also been suggested that sorbate combines with coenzyme A forming sorbyl coenzyme A, which causes microbial inhibition through inhibition of oxygen uptake.

Inhibition of microbial growth by sorbate may

also be accomplished through interference with mechanisms of electron and substrate transport in the cell. In a manner similar to that of other lipophilic acid antimicrobials, it has been suggested that sorbate inhibits transport of substrate molecules such as amino acids and glucose into the cells by uncoupling it from the electron transport system, which leads to cell starvation and inhibition of growth. Inhibition of nutrient uptake may be due to neutralization of the proton-motive force needed for substrate transport, inhibition of transport enzyme systems, or inhibition of synthesis and/or depletion of ATP. Inhibition of amino acid uptake has been attributed to inhibition of amino acid permease enzymes or the metabolic energy utilization of amino acid transport mechanisms. This effect may be occurring through incorporation of unsaturated fatty acids such as sorbic into the cell membrane, where they may cause steric disorganization of active membrane transport proteins.

One proposed mode for inhibition of mold growth by sorbic acid has been a potential depletion of ATP levels in conidia. This could potentially take place because ATP levels may be depleted as the cell attempts to maintain ion balance when dissociation of sorbic acid in the cytoplasm increases the intercellular cation concentration and because the primary sodium/hydrogen pump is directly linked to hydrolysis of ATP. As the hydrogen influx exceeds the pumped efflux, a shift in charge may potentially take place and lead to a decrease in the net negative intercellular change. This could then discharge the pH gradient required for ATP formation according to the chemostatic theory of oxidative phosphorylation. Studies with bacteria, however, found no decreases in ATP in the presence of sorbic acid. [See ATPASES AND ION CURRENTS.]

Another proposed mechanism to explain inhibition of microbial growth by sorbate supports the idea that weak, lipophilic acid preservatives cause cell starvation because they neutralize the transport driving proton-motive force that exists across cell membranes. In lower pH environments, the difference across the cell membrane is large, and the amount of acid entering and dissociating in the cytoplasm is higher. This accumulation of hydrogen ions inside the cell acts as an inhibitor by interfering with metabolic processes and causing a dissipation of the transmembrane proton gradient, which is one of the components of the proton-motive force. This elimination of the transport driving pH difference across the cell membrane has been presented as a major mechanism of microbial inhibition by compounds such as sorbic acid. According to this, undissociated sorbic acid acts as a protonophore, which decreases the intracellular pH and dissipates the proton-motive force of the membrane that energizes transport of compounds such as amino and keto acids. Inhibition of uptake of such components is believed to induce a stringent-type regulatory response in the cells, resulting in inhibition of growth but in maintenance of cell viability. A stringent response involves readjustments occurring in bacteria when amino acids become limiting or their specific ratios are disturbed.

Several published reports, however, have indicated that neutralization of the pH difference across the cell membrane is not the only mechanism of microbial inhibition by sorbate. Reasons for this conclusion include (1) although dissociation of the acid inside the cell may eliminate the pH difference across the membrane, its effect on the other component of the proton-motive force, which is the difference in electrical potential, is smaller. The remaining difference in electrical potential, however, is believed to be adequate to energize the uptake of substances needed for cell maintenance and growth. (2) As indicated earlier in this article, sorbic acid has been shown to inhibit microbial growth, although less efficiently, even at pH values near neutrality. In such situations, uptake of the compound based on pH difference across the cell membrane would not be adequate to explain the observed antimicrobial activity. (3) Neutralization of the proton-motive force is probably not involved in growth inhibition in presence of carbohydrates, because they do not depend on this force for their transport. Therefore, mechanisms involved in inhibition of microbial metabolism and proliferation by sorbic and other similar lipophilic acid preservatives appear to be different, depending on microbial types, substrates, and environmental conditions.

Bibliography

Liewen, M. B., and Marth, E. H. (1985). *J. Food Prot.* **48,** 364–375.

Sofos, J. N. (1989). "Sorbate Food Preservatives." CRC Press, Boca Raton, Florida.

Sofos, J. N., and Busta, F. F. (1981). *J. Food Prot.* **44,** 614–622.

Sofos, J. N., and Busta, F. F. (1983). Sorbates. *In* "Antimicrobials in Foods" (A. L. Branen and P. M. Davidson, eds.), pp. 141–175. Marcel Dekker, New York.

Sofos, J. N., Pierson, M. D., Blocher, J. C., and Busta, F. F. (1986). *Int. J. Food Microbiol.* **3,** 1–17.

Space Flight, Effects on Microorganisms

S. K. Mishra
KRUG Life Sciences

D. L. Pierson
NASA/Johnson Space Center

Glossary

Commensalism Symbiotic relationship in which one species benefits and the other is unharmed

Endogenous Originating or produced within an organism or one of its parts

Exogenous Originating outside an organism; infections can be of exogenous or endogenous origin

Microgravity Used to describe an environment in which acceleration due to gravity is approximately zero; also termed weightlessness

Pedicel Slender stalk that supports the fruiting or spore-bearing organ in some fungi

Solar particle events Sudden eruption on the surface of the sun that results in an increased flux of high-energy particles, which in turn increases the exposure of spacecraft to ionizing radiation

Spacelab Manned laboratory developed by the European Space Agency for flights aboard the U.S. Space Shuttle; pressurized habitable modules and pallets adapted to specific missions are carried in the Shuttle's payload bay

THE EFFECT OF SPACE FLIGHT on microbial function has been of concern to microbiologists since humans first began to explore space. Because microorganisms will be present on board manned and unmanned spacecraft, the potential exists for colonization of the vehicle itself as well as its inhabitants. The combination of the closed nature of spacecraft and the stressful nature of space flight (e.g., acceleration, weightlessness, radiation) increases the possibility of microbially induced allergic reactions and infections among space crews. Space flight is also suspected of altering human immune function as well as bringing new environmental selection pressures to bear on endogenous and exogenous flora. The combined effect of these processes may render normally harmless commensal or environmental microbes pathogenic to humans. Furthermore, colonization of the vehicle itself may result in system fouling, biodegradation of sealants, and perhaps the production of toxic metabolites and environmental pollutants. As the number and duration of manned flights increases, it has become imperative to characterize the effects of space flight and related factors on microbial growth, physiology, virulence, and susceptibility to antibiotic agents to protect the health of the crews and the integrity of the spacecraft.

At present, very few research data are available to address these concerns. Although many microorganisms have been used as models to study the effects of cosmic radiation, microgravity, vibration, and hypervelocity on living systems during a number of missions over the past 30 years (Tables I and II), the severe constraints involved in performing experiments in space have largely precluded exhaustive studies. The absence of gravity, which mandates the development of specialized equipment and procedures, the restrictions imposed on power, weight, and volume, and intense competition for the crew's time during space flights require that experiments be simple and easily performed with little or no crew involvement. Thus, many basic questions

Table I U.S. and Soviet Missions Carrying
Microbiology Experiments

Flights	Country	Launched	Manned duration
Sputnik	USSR	1957–1961	Unmanned
Vostok	USSR	1961–1965	
Gemini	US	1964–1966	10 Manned flights
Cosmos 110	USSR	Feb 1966	
Apollo	US	1967–1972	
Biosatellite II	US	Sept 1967	2 Days
Zond 5	USSR	Sept 1968	Unmanned
Zond 7	USSR	Aug 1969	
Cosmos 368	USSR	Oct 1970	
Salyut 1	USSR	April 1971	23 Days
Skylab 1	US	May 1973	Unmanned
Skylab 2	US	May 1973	28 Days
Skylab 3	US	July 1973	59 Days
Soyuz 12	USSR	Sept 1973	
Skylab 4	US	Nov 1973	84 Days
Salyut 3	USSR	June 1974	14 Days
Salyut 4	USSR	Dec 1974	41 Days
Apollo-Soyuz	US-USSR	July 1975	9 Days
Salyut 5	USSR	June 1976	33 Days
Salyut 6	USSR	Sept 1977	1,192 Days
Cosmos 1129	USSR	Sept 1979	
Salyut 7	USSR	April 1982	1,805 Days
Spacelab 1	US	Nov 1983	10 Days
Spacelab 3	US	July 1985	7 Days
Spacelab 2	US	July 1985	8 Days
Spacelab D-1	US	Oct 1985	7 Days
Mir	USSR	Feb 1986	366 Days (max)

[Modified from Cioletti, L. A., Pierson, D. L., and Mishra, S. K. (1991). S.A.E. Tech. Paper #911512.]

concerning the effects of space on microbial structure and function have yet to be resolved. This article presents an overview of information collected to date on the effects of the space flight environment on microbial function.

I. Microbial Survival and Growth in Space

Studies on the effect of extreme conditions on microorganismal growth and survival began as early as 1935, when high-altitude balloons were used to investigate the effects of low temperatures, decreased

pressures, and increased radiation. From 1954 to 1960, experiments on nearly 30 high-altitude balloons and sounding-rocket flights revealed that *Neurospora* spores and vegetative bodies could survive direct exposure to the environment at 35–150 km above the earth. In the 1960s, viable organisms from cultures of *Penicillium roqueforti* and *Bacillus subtilis* carried on the Gemini 9A and 12 missions were recovered after nearly 17 hr of direct exposure to space. Parallel attempts to detect microorganisms in the extraterrestrial environment in analyses of micrometeorites collected during the Gemini missions, and lunar samples collected during the Apollo flights, revealed no evidence of viable microorganisms nor any identifiable biological compounds. Therefore, the potential for contaminating Earth with extraterrestrial life forms, an early concern, was judged to be extremely unlikely.

Studies of microbial behavior in space performed to date, although numerous, have produced inconclusive, occasionally contradictory results (Tables III and IV). The first experiments on the unmanned Sputnik orbital satellites (1957–1961) used microorganisms to identify the gross effects of galactic radiation, weightlessness, and other related factors on biological systems. Studies of bacteriophage induction have been many (see later) and have dated back to the second Soviet satellite mission in August 1960, which included flight experiments with *Clostridium butyricum*, *Streptomyces* spp., *Aerobacteria aerogenes* 1321 bacteriophage, and T-2 coliphage. Viability and gas production by *C. butyricum* was no different after flight than that of ground-based control cultures; growth of flight cultures of *Streptomyces*, however, was accelerated sixfold on return to earth. Growth of *Escherichia coli* cultures flown on later Soviet missions was not appreciably different than that of ground-based controls. The characteristics of an alga, *Chlorella* sp., were studied on many Soviet missions during the 1960s and 1970s but showed no change in growth characteristics, viability, proportion of photosynthetically active cells, or mutation rate; however, chloroplast volume tended to decrease.

Continuing chronologically, the first serious attempts by U.S. investigators to study the effect of space flight on microorganisms began with the launch of Biosatellite II in September 1967. The biological experiments included in this 45-hr earth orbital flight were investigations of the effects of microgravity and radiation on the growth and life cycle of *Salmonella typhimurium*, *E. coli*, *Neurospora crassa*, and *Pelomyxa carolinensis*. Although the

Table II Organisms Used in Space Flight Experiments with Cells

Prokaryotes	Eukaryotes
Bacteria	Protozoa
Actinomyces aureofaciens	*Euglena gracilis*
Actinomyces erythreus	*Paramecium aurelia*
Actinomyces levoris*	*Paramecium tetraurelia*
Actinomyces streptomycin	*Pelomyxa carolinensis*
Aerobacteria aerogenes	*Tetrahymena periformis*
Aeromonas proteolytica	*Tetrahymena pyriformis*
Bacillus brevis	Fungi
Bacillus subtilis	Molds
Bacillus thuringiensis	*Aspergillus niger*
Clostridium butyricum	*Chaetomium globosum*
Clostridium sporogenes	*Neurospora crassa*
Escherichia coli	*Penicillium roqueforti*
Hydrogenomonas eutropha	*Phycomyces blakeleanus*
Methylobacterium organophilum	*Trichoderma viride*
Methylomonas methanica	*Trichophyton terrestre*
Methylosinus sp.	Yeasts
Proteus vulgaris	*Candida tropicalis*
Pseudomonas aeruginosa	*Rhodotorula rubra*
Staphylococcus aureus	*Saccharomyces cerevisiae*
Streptomyces levoris*	*Saccharomyces vivi*
Bacteriophage	*Zygosaccharomyces bailii*
Aerobacteria aerogenes phage 1321	Slime mold
Escherichia coli phage T1, T2, T4, T7, and lambda	*Physarum polycephalum*
Salmonella typhimurium phage P-22	Algae
	Chlamydomonas reinhardii
	Chlorella ellipsoidea
	Chlorella pyrenoidosa
	Chlorella sorokiniana
	Chlorella vulgaris
	Scenedesmus obliquus

* synonym: *Actinomyces* in USSR, *Streptomyces* in U.S.
[Adapted from Dickson, K. J. (1991). *ASGSB Bull.* **4**, 151–260.]

total cell density in cultures of *S. typhimurium* flown as nonirradiated controls on this mission increased 15% during stationary-phase growth, mean viable cell density increased by 30%. Some portion of these increases, however, was attributable to vehicular vibration and acceleration.

On Apollo 16 in April 1972, four fungal species, two yeastlike fungi, *Rhodotorula rubra* and *Sacchraomyces cerevisiae,* and two filamentous fungi, *Trichophyton terrestre* and *Chaetomium globosum,* were exposed outside the space capsule within a specially designed Microbial Ecology Evaluation Device. Both dry inocula and aqueous suspensions of vegetative yeast cells, conidia, and ascospores were exposed to the sun at a 90° angle for about 10 min in cuvettes equipped with a quartz window and bandpass filters. These cultures were subjected to many tests on their return to earth. The two filamentous fungi showed changes in the colony perimeters, growth density, protoplasmic leakage of hyphal apices damaged by ultraviolet (UV) irradiation, abnormal growth at the hyphal apex, forked hyphal branches, and irregular hyphal walls. The infectivity and ability to degrade human hair of *T. terrestre* increased after flight. The cellulolytic fungus *C. globosum* lost its ability to produce pigment, developed fewer fruiting bodies, and demonstrated variations in amylase production. Cultures of *R. rubra* also showed altered phosphoglyceride content. *Saccharomyces cerevisiae* nearly doubled its rate of phosphate uptake after space exposure, as well as being more susceptible to UV radiation and better able to survive in intradermal lesions in artificially infected mice. Finally, exposing all four strains to UV irradi-

Table III   Effect of Microgravity on Microbial Growth and Sensivity to Antibiotics

Organism	Result	Mission	Duration of experiment
Escherichia coli	Cell density slightly increased	Biosatellite II	2 Days
Salmonella typhimurium	Cell density increased 20%	Biosatellite II	
Proteus vulgaris	Cell numbers increased sevenfold	Biosatellite II	
Bacillus subtilis	Biomass significantly increased	Soyuz 12	2 Days
	Sporulation reduced	Spacelab D-1	3 Days
	Autolysis increased in stationary phase		
Physarum polycephalum	Growth reduced	Cosmos 1129	
Chlorella sp.	Growth not affected	Soyuz 12	
		Cosmos 573	Up to 10 days
Chlamydomonas sp.	Proliferation increased 100%	Spacelab D-1	6 Days
Paramecium tetraurelia	Growth increased twofold	Salyut 6	4 Days
		Spacelab D-1	5 Days
Staphylococcus aureus	MIC of following increased twofold: Oxacillin Chloramphenicol Erythromycin	Salyut 7	1 Day
Escherichia coli	MIC of following increased two- to fourfold: Colistin Kanamycin	Salyut 7	1 Day

ation at 254, 280, and 300 nm and then maintaining them in total darkness increased their susceptibility to antibiotics.

Growth of *Proteus vulgaris* studied aboard Soyuz 12 in 1973 showed a sevenfold increase compared with ground controls; a significant increase was also observed in the growth of *B. subtilis*, with a corresponding decrease in spore production (5×10^4 spores/ml in space-grown cultures versus 8×10^5 spores/ml in ground controls). Cultures grown in space also showed increased cell lysis. Fruiting bodies of the fungus *Polyporus brumalis* grown on the satellite Cosmos 690 (Oct–Nov 1974) showed disorientation and occasional flattening, with pedicels twisted into spirals or balls; ground-control cultures, by contrast, displayed strong negative geotropism and long pedicels.

During the joint Apollo–Soyuz mission in 1975, *Streptomyces levoris* was used to study growth rate periodicity. One of the eight cultures studied grew more quickly in the spacecraft environment than on the ground, and three of the space cultures developed double spore rings during the immediate postflight period. Of the many microbial species studied aboard the Salyut-6 mission, space cultures of the methanogenic bacteria *Methylosinus* sp. (AB-21) and *Methylomonas methanica* (AB-3) grew no differently than did ground controls; however, *Methylobacterium organophilum* (MB-67) showed a greater tendency to grow anaerobically in space. Electron microscopic analysis after landing, however, revealed no changes in cytoskeletal ultrastructure. Cultures of the protozoan *Paramecium tetraurelia* aboard Salyut 6 (1977–1981) and on Spacelab

Table IV   Effect of Microgravity on Microbial Genetics and Sensitivity to Radiation

Organism	Result	Mission	Duration of experiment
Escherichia coli	No effect on transduction, transformation; 40% increase in conjugation	Spacelab D-1	3 hr
Salmonella typhimurium	Increased sensitivity to P-22 phage	Biosatellite II	2 Days
	Increased resistance to high doses of radiation	Biosatellite II	2 Days
Escherichia coli	Increased resistance to high doses of radiation	Biosatellite II	2 Days

D-1 in 1985 had larger, spherical cells during the early log phase, with smaller cells during later growth stages. The growth rate of this protozoan nearly doubled in microgravity; its yield increased nearly four times after 5 days in space, and its cell-protein, magnesium, and calcium concentrations had decreased relative to ground controls. By contrast, earlier cultures of another protozoan, *Pelomyxa carolinensis*, on Biosatellite II showed possibly increased rate of cell division but no distinct effects on overall growth rate or physiological, morphological, and cytochemical variables.

A serious drawback in these early studies, both Soviet and American, was the lack of on-board controls that could be used to rule out the effect of vibration and acceleration in flight. This drawback was overcome in 1985 with the use of the European Space Agency's Biorack facility, which included an incubator-centrifuge that could be used in flight to simulate gravity. This facility was used successfully on the German Spacelab D-1 mission in 1985, during which several microbiological studies were conducted using cultures of *B. subtilis*, the alga *Chlamydomonas*, and the protozoan *Paramecium tetraurelia*. The period and phase of the photo-accumulation behavior of the two strains of *Chlamydomonas* flown on Spacelab D-1 did not change in flight, although their proliferation rates increased significantly. By contrast, growth of the slime mold *Physarum polycephalum* on Spacelab D-1 and on the Cosmos 1129 mission in 1979 was reduced relative to ground controls and also showed reduced protoplasmic movement after flight.

In sum, although many species of both prokaryotes and eukaryotes have been flown in space, the inconsistencies with regard to use of analytical equipment, procedures, and culture conditions have led in some cases to conflicting results. In addition, the relatively infrequent flight opportunities, as well as limited crew time during flight, means that the results that are available tend to be fragmentary and often lack a classic, controlled experimental context with which to interpret them. The absence of an overall structure to the study of microbial function in space is reflected throughout this article.

II. Sensitivity to Antibiotics

Because microorganisms will be ubiquitous on space vehicles and in space-based habitats, the risk of infectious disease from environmental or commensal flora cannot be ruled out. Little information is available, however, as to whether exposure to microgravity affects the sensitivity of microorganisms to antibiotics, either *in vitro* or *in vivo*. Evidence suggests that some forms of bacteria isolated from astronauts after space flight may be more resistant to antibiotic agents than the same strains isolated before flight. To supplement these observations, a group of French and Soviet scientists in July 1982 studied the effects of several antibiotics on space-grown cultures of *Staphylococcus aureus* and *E. coli* as part of the Cytos 2 experiment on Salyut 7. The minimal inhibitory concentrations (MIC) of oxacillin, chloramphenicol, and erythromycin on actively growing cultures of *S. aureus* were found to be roughly twice the ground-control values of 0.16, 4.0, and 0.5 μg/ml, respectively. The MIC for colistin and kanamycin against *E. coli* were greater than 16 μg/ml, compared with 4 μg/ml in ground-control cultures. The Antibio experiment conducted 3 yr later on Spacelab D-1 supported these observations. The tentative conclusion at present is that space flight may increase the required MIC for these organisms, possibly by stimulating microbial growth or by inducing physiological or biochemical changes in cell-wall structure or permeability. More data are needed to explore these speculations. [*See* ANTIBIOTIC RESISTANCE.]

III. Radiation, Bacteriophage Induction, and Microbial Genetics

Another environmental factor prompting concern as to whether humans and other earth-based biological systems can survive in space is radiation, with its preponderance of high-intensity solar UV light, radionuclides, protons, electrons, decay products, and galactic cosmic rays. Cosmic rays normally reach the earth's surface at about 0.027 rem (roentgen-equivalent-man) per year. By comparison, inside the U.S. Space Station "Freedom," (Figs. 1 and 2), which will be located 270 nautical miles above the earth's surface, organisms will be subjected to annual cosmic radiation of about 43 rem; additional peaks up of 1400 rem can take place in free space during solar particle events such as the one that took place in August 1972. The yearly dose of cosmic radiation at the lunar surface is ex-

Figure 1   U.S. Space Station "Freedom," with a Space Shuttle to the upper left.

pected to be 30 rem; at the Martian surface, even with atmospheric shielding of 10 g/cm², average annual cosmic radiation has been calculated to be 12 rem, with peaks to 83 rem.

The response of *B. subtilis* to high atomic number–high energy (HZE) particles, a component of galactic cosmic radiation, was studied in the Biostack I and II experiments on board Apollo 16 and 17, and again on the Apollo-Soyuz Test Project. In the Apollo experiments, germination was not influenced by an HZE "hit," but spore outgrowth was reduced significantly. In the later project, *B. subtilis* spores were inactivated by HZE particles at distances up to 4 μm. Dose values at these distances were approximately 0.1 Gray, whereas the D37 value (the dose needed to reduce survival to 37%) for electron radiation is about 800 Gray. These investigations prompted the conclusion that the biological hazard of cosmic HZE particles, especially

Figure 2   A model of the space station facility that includes the microbiological components discussed in the text.

its high Z component, has been much underestimated.

Ionizing radiation is well known for its mutagenicity. Mutagenic doses of radiation vary among species and with the life stage of the organism. For example, in 1960 on the second Soviet satellite, the number of spores produced by *Actinomyces erythreus* strain 2577, which is known to be resistant to UV light, increased six times over the spores produced by the UV-sensitive strain 8594. Bacterial endospores are quite resistant to gamma radiation; a 90% kill rate can be achieved with 3000–4000 Gray and only 10% of this amount to reach the same mortality in vegetative bacteria. Indeed, experiments with *Neurospora* spp. on Biosatellite II and Gemini XI showed that rapidly metabolizing cells showed greater effects of radiation in the control samples on the ground than did those from the flight. For nondividing, inactive spores, there was no difference in genetic effects between control and flight samples. Similarly, *B. subtilis* and *B. thuringiensis* spores exposed to solar UV and the space vacuum on Apollo 16 survived as well as did ground controls. Later exposure of other cultures of *B. thuringiensis* to full sun in space produced a lower survival rate compared with either exposure to solar UV alone or ground controls. Also on Apollo 16, the space vacuum enhanced the lethal effect of solar UV at 254 and 280 nm in one strain of *B. subtilis* but not in a repair-defective strain. Exposure to the space vacuum alone did not alter the survival rate of dried spores. These observations suggest that other components of solar radiation in space may act as powerful biocides, either alone or in combination with UV.

Lysogenic bacteria (those with a viral prophage integrated into the host genome) have been used frequently as biological indicators in tandem with physical methods of measuring radiation. Bacteriophage production from lysogenic bacteria can be induced by X-rays, gamma rays, protons, or neutrons, is dependent on radiation dose, and may be resistant to other flight factors. Soviet investigators have used the K-12 lambda phage of *E. coli* since the second earth-orbiting satellite as a kind of dosimeter to monitor radiation in the spacecraft interior. Phage production by these bacteria can be stimulated by as little as 0.3 rad of gamma radiation, or still smaller doses of protons or rapid neutrons; thus, information about the mutagenicity of cosmic radiation and its associated risks to the genetic structure can also be ascertained. This phage system was carried aboard most of the Sputnik, the manned flights of the

Vostok series, the unmanned biosatellite Cosmos 110, and Zond 5 and 7, the latter of which orbited the moon before returning to earth. In the early Soviet studies, phage production in flight did not exceed that of controls; the relationship between length of flight and number of phages produced is still not clear. Another Soviet observation that vibration increased the sensitivity of lysogenic bacteria to subsequent gamma irradiation suggests that space flight factors may have cumulative effects. Phage induction in *E. coli* cultures on the Soviet missions Vostok 4 and 6, Zond 5 and 7, and the U.S. Biosatellite II mission showed slightly higher survival rates than ground controls, with small increases in viable cell density.

U.S. experiments with the lysogenic bacterium *Salmonella typhimurium* BS-5 (P-22)/P-22 on Biosatellite II found that weightlessness and gamma irradiation had significant effects on growth rate and induction of prophages: Although not different statistically, the flight cultures tended to have fewer induced bacteriophages per viable bacterium at all levels of radiation relative to ground cultures. Research with *B. subtilis* on Apollo 16 indicated that filtered solar radiation at 254 and 280 nm appeared to be less lethal to spores at these doses and induced lower mutation frequencies than in ground-control studies. The T-7 bacteriophage was inactivated to a greater degree by the 254-nm solar UV than in ground studies, although the shapes of the dose-response curves were comparable. [*See* BACTERIOPHAGES.]

Studies of microbial genetics with respect to radiation effects in space have been few; however, it was observed during the Apollo Microbial Ecology Evaluation Device experiments that *Trichophyton terrestre* and *Chaetomium globosom* showed greater numbers of nucleotide pairs per haploid nucleus after flight. On the Spacelab D-1 mission, space flight was found to have no effect on transduction and transformation in *E. coli;* however, conjugation was enhanced by up to 40%. Far more data are needed to separate and characterize the multiple effects of space flight factors on microbial genetics at the molecular level.

IV. Future Studies

Because gravity has been an omnipresent force during the evolution of all earth-based forms of life, it is logical to assume that exposing microorganisms to factors such as weightlessness and space radiation can lead to physiological or genetic adaptation. These adaptive changes in turn may affect the host-parasite relationship, leading to increased risk of infectious disease. Designers of space vehicles and habitats naturally are concerned with the effects of microorganisms on the closed spacecraft environment and its inhabitants; however, a clear understanding of how the space environment affects microorganisms themselves has yet to be achieved. With the relative brevity of most recent U.S. space flights and the "operational" focus of the U.S. Space Shuttle program, the complex research necessary for in-depth study of long-term interactions among spacecraft inhabitants has remained a relatively low priority in the highly competitive Space Shuttle program (Fig. 3). Moreover, as mentioned above, the contraints placed on on-board stowage

Figure 3 A Space Shuttle launch.

and the difficulties involved in performing even simple tests in the absence of gravity have hampered detailed study of microbial function in microgravity. However, with the recent emphasis on extending Shuttle flights to perhaps 90 days and lengthy tours of duty on Soviet and U.S. space stations, interest in characterizing microbial responses and adaptations to the space flight environment has been renewed, and new, more comprehensive studies are being planned. It is hoped that these new studies, which will take full advantage of advances in earth-based technologies and automation, will produce much-needed basic information about the specific effects of space on the microbial life cycle.

Acknowledgments

The authors appreciate the assistance of the following employees of KRUG Life Sciences in preparing this article: Christine Wogan, Louis Cioletti, David Koenig, and Nellie Clayton.

Bibliography

Cioletti, L. A., Pierson, D. L., and Mishra, S. K. (1991). Microbial growth and physiology in space: a review. Presented at the 21st International Conference on Environmental Systems, San Francisco, California, July 1991. Society for Automotive Engineers Technical Paper Series No. 911512, Warrendale, Pennsylvania.

Diskson, K. J. (1991). *ASGSB Bull* **4,** 151–260.

Gmunder, F. K., and Cogoli, A. (1988). *Appl. Microgravity Tech.* *1* **3,** 115–122.

Mattoni, R. H. T., Ebersold, W. T., Eiserling, F. A., Keller, E. C., and Romig, W. R. (1971). Induction of lysogenic bacteria in the space environment. *In* ''The Experiments of Biosatellite II [NASA SP 204.]'' (J. F. Sanders, ed.), pp. 309–324. National Aeronautics and Space Administration, Washington, D.C.

Pierson, D. L., McGinnis, M. R., Mishra, S. K., and Wogan, C. F., eds. (1989). ''Microbiology on Space Station Freedom [NASA Conference Publication 3108].'' National Aeronautics and Space Administration, Washington, D.C.

Rodgers, E. (1986). The ecology of microorganisms in a small closed system: potential benefits for space station [NASA Technical Memorandum 86563]. National Aeronautics and Space Administration, Marshall Space Flight Center, Huntsville, Alabama.

Taylor, G. R. (1974). *Annu. Rev. Microbiol.* **28,** 121–137.

Taylor, G. R. (1976). *Bioscience* **27,** 102–108.

Tixador, R., Richoilley, G., Gasset, G., Templier, J., and Bes, J. C. (1985). *Aviation Space Environmental Med.* **56,** 748–51.

Volz, P. A. (1990). *Mycopathologia* **109,** 89–98.

Zhukov-Vererzhnikov, N. N., Volkov, M. N., Maiskii, I. N., Guberniev, M. A., Rybakov, N. I., Antipov, V. V., Kozlov, V. A., Saksonov, P. P., Parfenov, G. P., Kolobov, A. V., Rybakova, K. D., and Aniskin, E. D. (1968). *Kosmicheskie Issledovaniya* **6,** 144–149. (In Russian; English translation: *Cosmic Res.* **6,** 121–125.)

Zhukov-Verezhnikov, N. N., Volkov, M. N., Rybakov, N. I., Maiskii, I. N., Saksonov, P. P., Guberniev, M. A., Podoplelov, I. I., Antipov, V. V., Kozlov, V. A., Kulagin, A. N., Aniskin, E. D., Rybakova, K. D., Sharyi, N. I., Voronkova, Z. P., Parfenov, G. P., Orlovskii, V. I., and Gumenyuk, V. A. (1971). *Kosmicheskie Issledovaniya* **9,** 292–299. (In Russian; English translation: *Cosmic Res.* **9,** 267–273.)

Specimen Collection and Transport

I. Jerome Abramson[1]
U.S. Food and Drug Administration

Glossary

Anaerobe Microorganism with the capacity to live without utilizing free oxygen
Aseptic Free from microorganisms
Bacteriostatic Inhibiting the growth or multiplication of bacteria
Polymicrobial Characterized by the presence of several species of minute living organisms
Reduced transport medium Fluid or solidified environment from which oxygen has been removed (low-oxidation-reduction potential)
Transport medium Device that preserves a specimen in a fluid or solidified environment from the time it is obtained until it is removed from that environment

---

COLLECTION AND TRANSPORT are physical procedures by which microbiological specimens are obtained, transported, and maintained temporarily until they can be processed and evaluated. Various devices and media provide a protective environment for a specimen to arrive in a laboratory quickly and safely under conditions similar to those from which it was obtained, provided that qualified personnel adhere to proper collection and transport procedures. Whether in a clinical or nonclinical setting, proper collection and transport procedures are critical to the correct identification of a microorganism.

---

[1] Opinions expressed in this article are those of the author and do not reflect, necessarily, those of the U.S. Food and Drug Administration.

Some scientific fields in which collection and transport of specimens are required include human and veterinary medicine, pharmaceutical, food, and environmental.

I. The Process of Collection

The most critical factor in collection of a specimen is that it be obtained by aseptic technique. Any contamination could result in incorrect identification of a microorganism and a misconception as to the cause of an infection. Additionally, many specimens are available only once or for a limited length of time (e.g., during surgery, biopsy) and, therefore, a contaminated specimen may preclude study. To reduce the risk of self-infection, qualified personnel should ensure a safe environment for themselves. This is usually accomplished by instituting "universal precautions."

Prior to collection, the individual obtaining the specimen should decide what type of transport medium and device is appropriate. He or she may select sterile dishes, tubes, or vials and should determine whether an aerobic or anaerobic environment will be effective for the purpose intended. Also to be determined is the quantity of specimen to obtain and the optimal time for collection. It is appropriate for the microbiologist to predict which microorganism(s) is (are) suspected so that precautions can be taken. The microbiologist should be gowned and/or gloved as needed during collection of a specimen. In a clinical setting, it should be verified that a specimen from a patient is obtained prior to drug treatment when possible and that any ingested substances, potential complications, and history are noted. Using aseptic techniques, the microbiologist (or other professional) should then obtain a specimen and place it immediately into a collection and transport device. Specimens may be obtained using various medical devices, including a needle and syringe for fluid cultures; contact plates for surfaces such as floors, walls, and ceilings; and slit-to-agar or exposed agar for air samples from sterile rooms in industrial environments, however, many times the most appropriate material to use for collection is a swab. The type of swab material as well as the type of shaft to support the swab is an important choice to consider to maintain the integrity and viability of a specimen (Tables I and II). [*See* CULTURE COLLECTION, FUNCTIONS; CULTURE COLLECTIONS: METHODS AND DISTRIBUTION.]

II. Transport Methods

After obtaining a specimen, a microbiologist should then identify that specimen on a label attached to a transport device by indicating clearly the source of the specimen, the name of the clinician ordering the specimen, identification number or name of the patient, the date and time of collection, and other information suitable for the circumstances. The person obtaining the sample should follow transport instructions peculiar to the specimen, device, and medium. These may include storage temperature, limited exposure to sunlight, or appropriate culture medium. In all cases, a specimen should be delivered to the laboratory for processing as expeditiously as possible because some organisms may be sensitive to environmental factors such as O_2, may be subject to drying, or, in case of polymicrobial infections, may result in one organism masking a slower-dividing microorganism.

In the laboratory, the specimen will be processed to isolate microorganisms and identify them. Antimicrobial agent susceptibility testing as well as other microbiological and/or biochemical procedures may be performed.

III. Devices

A collection and transport device can be a sterile, empty container or a container filled with a medium that preserves and protects a specimen and its contents. It is critical to select the most appropriate device and medium to maintain the integrity of a

Table I Types of Materials for Swabs

Swab material	Characteristics	Clinical purposes
Calcium alginate	Free of fatty acids that may inhibit some fastidious bacteria, nonirritating to tissue, absorbed by body fluids, inactivates HSV May be toxic to some chlamydia	Specimens from any source (e.g., infectious disease of the ear, nose, throat; sexually transmitted diseases)
Cotton	Absorbent	Culture any source General purpose, and used for specimens suspected of *Chlamydia trachomatis* Can be used for HSV
Dacron	Not recommended for open wounds Free of fatty acids	Recommended for viruses, chlamydia, gonococcal culture Can be used for HSV
Polyester	Nonirritating to patient tissue, free of fatty acids	Group A streptococci General use Can be used for HSV
Rayon	Capacity for high absorbency Not recommended for open wounds and eyes Does not dissolve in body fluids Fibers may detach	Specimens suspected of low concentrations of bacteria Cervical, vaginal, or vesicle cultures Can be used for HSV
Nylon brushes	Nylon bristle wound on stainless steel around a plastic shaft	Specimens from female patients: endocervix for Pap smear *Chlamydia trachomatis* Sexually transmitted diseases

HSV, Herpes simplex virus.

Table II Types and Purposes of Shaft Materials for Swabs[a]

Material	Clinical purposes
Wood (rigid)	Throat and various other clinical specimens Foods Industrial and other purposes (e.g., walls, floors)
Plastic (semiflexible)	Various clinical specimens (e.g., throat, nose, fluids) Industrial and other purposes (e.g., walls, floors)
Aluminum (flexible)	Small size, thin, wire follows the contour of an anatomical passage (e.g., nose, ear, urethra)

[a] Each type of shaft may contain one of the swab material in Table I.

specimen. Generally, it is accepted that microorganisms, whether aerobic or anaerobic, remain viable longer when transported anaerobically.

There are many microbiological collection and transport devices available from commercial sources (Table III). Some of these devices were developed for both generalized (aerobes and anaerobes) and for specific (e.g., chlamydia, virus) collection and transport purposes and may be quite similar in design and purpose. For example, Cultureswab™ Transport System products (Fig. 1) were developed to transport bacteria and fungi (yeasts and molds) in Amies, Cary-Blair, or Stuart's medium, either with or without charcoal. Charcoal inactivates the toxic effects of fatty acids in cotton and is useful in maintaining viability of some pathogens.

The aerobic Port-A-Cul® Transport System (Fig. 2) is similar to Cultureswab™ Transport System

Table III Commerical Products and Sources for Some Collection and Transport Devices

Product	Source
Port-A-Cul (aerobic and anaerobic) Culturette Systems (aerobic, anaerobic, and viruses)	Becton Dickinson Microbiology Systems, Cockeysville, Maryland
CultureSwab	DIFCO Laboratories, Detroit Michigan
Accu-CulShure (aerobes and anaerobes)	Technology for Medicine, Pleasantville, New York
Transette systems Calgiswab	Spectrum, Los Angeles, California
KultSure	NCS Diagnostics Corp., Buffalo, New York
Transwab (aerobes and anaerobes) Transtube (aerobes and anaerobes) Virocult (viruses)	Medical Wire and Equipment Co., USA, New York, New York
Culture-Pen (viruses) Culture-Eze (bacteria)	Medical Media Laboratory, Boring, Oregon
Bioport Culture CATS	Precision Dynamics Corp., Burbank, California
Vi-Pak (aerobes and anaerobes) Chlamydia Port Chlamydia Viral Transport Media	Adams Scientific, West Warwick, Rhode Island
Transporter (viruses)	Bartels Immuno-diagnostic, Bellevue, Washington
Mycotrans (mycoplasma)	Hana Media, Berkeley, California
Anatrans (anaerobes)	Carr Scarborough, Stone Mt., Georgia
Plastic tubes containing media and swabs	M&H Plastics, Red Bluff, California
Dyna-Trans (mailing containers)	Dyan-A-Med Products, Barrington, Illinois
Safe-Pak (mailing containers)	Med-Pak, Bridgeport, Pennsylvania
Lab Guard (specimen bags)	Minigrip, Inc., Orangeburg, New York
Specimen cups and jars	Monoject Scientific, Plainview, New York Stockwell Scientific, Walnut California Sigma Diagnostic, St. Louis, Missouri
Urine cups, fecal transport	Evergreen Scientific, Los Angeles, California

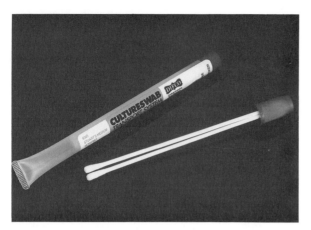

Figure 1 Transport system for collection and transport of clinical specimens (CultureSwab™). [Courtesy of Nancy Laskowski, DIFCO Laboratories, Detroit, Michigan.]

products. Other commercial products are modifications of these devices or more sophisticated versions, such as Accu-CulShure®.

Port-A-Cul® Tubes for swabs and Port-A-Cul® Vials for fluids (Figs. 3 and 4) contain reduced transport media with a low-oxidation-reduction potential that were developed to maintain a specimen in an anaerobic, moist, nonnutrient, buffered, solidified environment. Each tube or vial may be introduced into a sterile environment. These products were among the first to be developed that provide a me-

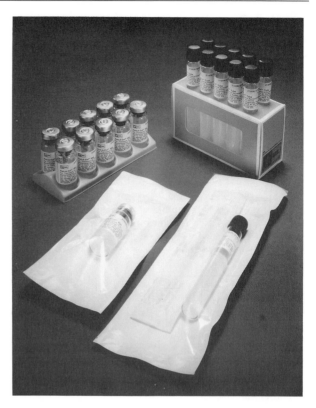

Figure 3 Tubes, vials, and sterile packs (BBL® Port-A-Cul®). [Courtesy of David Power, Becton Dickinson Microbiology Systems, Cockeysville, Maryland.]

dium to maintain viability of, while not being deleterious to, the many and various microorganisms in a specimen. Microorganisms may remain viable in these systems for varying lengths of time (24–72 hr or longer), depending on the particular physiological requirements of a microorganism.

Various types of media to maintain viability of bacteria and fungi may be incorporated in collection and transport devices, but three remain prevalent: Amies, Cary-Blair, and Stuart's. Each has a specific formulation that makes it unique (Table IV).

Originally, Stuart's medium was designed for transport of *Neisseria gonorrhoeae*, the bacterium that causes gonorrhea, but it was determined to be an effective medium to maintain the viability of numerous other micoorganisms, including enteric pathogens.

Amies transport media, both with and without charcoal, are modifications of Stuart's medium and are used to transport specimens on swabs suspected of containing the *N. gonorrhoeae*. Amies medium does not contain glycerophosphate as does Stuart's

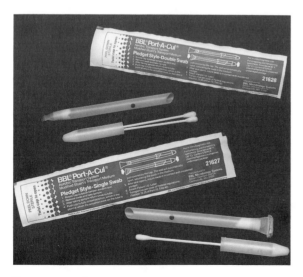

Figure 2 Aerobic transport system with modified Stuart's transport medium—single and double swabs (BBL® Port-A-Cul®). [Courtesy of David Power, Becton Dickinson Microbiology Systems, Cockeysville, Maryland.]

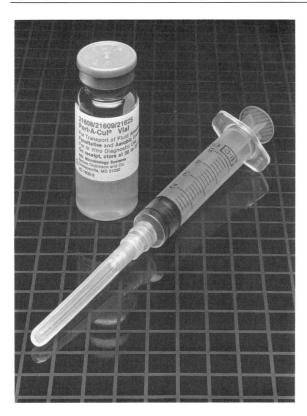

Figure 4 Vial for collection of fluid specimens (BBL® Port-A-Cul®). [Courtesy of David Power, Becton Dickinson Microbiology Systems, Cockeysville, Maryland.]

Table IV Formulations of Some Transport Media

Amies transport medium[a]	
Sodium chloride	8.0 g
Potassium chloride	0.2 g
Calcium chloride	0.1 g
Magnesium chloride	0.1 g
Monopotassium chloride	0.2 g
Disodium phosphate	1.15 g
Sodium thioglycollate	1.0 g
Agar	3.6 g
Cary-Blair transport medium[b]	
Sodium thioglycollate	1.5 g
Disodium phosphate	1.1 g
Sodium chloride	5.0 g
Agar	5.0 g
Stuart's transport medium	
Sodium glycerophosphate	10.0 g
Sodium thioglycollate	1.0 g
Calcium chloride dehydrate	0.1 g
Sodium chloride	5.0 g

[a] If charcoal is included in Amies trans port medium, 10 g charcoal is added/1000 ml of medium.
[b] Final pH 8.0 ± 0.5.

medium; instead, an inorganic phosphate buffer replaced sodium glycerophosphate and has been effective in transporting a wide spectrum of microorganisms.

Cary-Blair transport medium is formulated with a minimum of nutrients to maintain viability of microorganisms without promoting division of bacteria. Sodium thioglycollate is included in the formulation as a reducing agent. The medium is adjusted to approximately pH 8 to inhibit an acid pH (<7) to prevent destruction of bacterial cells.

IV. Summary

When specimens are obtained, the importance of using aseptic techniques and observing extreme precautions by personnel cannot be overstressed. These precautions must be observed to ensure both the integrity of a specimen and the safety of the individual.

It is critical to determine the appropriate collection and transport device and medium for each specimen, because individual specimens require special handling, and the microbiologist must be able to determine what is required. Speed in transport and processing is a crucial factor in identification of a microorganisms, although some commercial transport products may extend the viability of an organism when cultured properly.

Collection and transport of a specimen is a relatively simple process for the trained, knowledgable professional.

Acknowledgment

The author expresses his deep appreciation to Kim D. Abramson for her devotion in preparation of the manuscript.

Bibliography

Finegold, S. M., and Baron, E. J. (1986). Selection, collection, and transport of specimens for microbiological examination. *In* "Bailey and Scott's Diagnostic Microbiology," 7th ed., pp. 53–62. C. V. Mosby, St. Louis.

Isenberg, H. D., Washington, J. A. II, Doern, G., and Amsterdam, D. (1991). Specimen collection and handling. *In* "Manual of Clinical Microbiology," 5th ed. (A. Balows, W. J. Hausler, Jr., K. L. Herrmann, H. D. Isenberg, and H. J. Shadomy, eds.) pp.15–28, American Society for Microbiology, Washington, D.C.

Koneman, E. W., Allen, S. D., Dowell, V. R., Jr., Janda, W. M., Sommers, H. M., and Winn, W. C., Jr. (1988). Introduction to medical microbiology Part I: Presumptive laboratory diagnosis of infectious diseases. *In* "Color Atlas and Textbook of Diagnostic Microbiology," 3rd ed., pp. 6–37. J. B. Lippincott, Philadelphia.

Power, D. A., and McCuen, P. J. (1988). Collection and handling of clinical specimens. *In* "Manual of BBL Products and Laboratory Procedures" (D. A. Power, ed.), pp. 2–3. Becton Dickinson Microbiology Systems, Cockeysville, Maryland.

Ryan, K. J., and Ray, C. G. (1990). Laboratory diagnosis of infectious diseases. *In* "Medical Microbiology: An Introduction to Infectious Diseases," 2nd ed. (J. C. Sherris, ed.), pp. 233–235, 252–253, 842–845, 861–863. Elsevier, New York.

Smith, T. F. (1986). Specimen requirements: Selection, collection, transport and processing. *In* "Clinical Virology Manual" (S. Specter and G. J. Lancz, eds.), pp. 15–29. Elsevier, New York.

Spirochetes

Thomas W. Huber

Texas A&M University

Glossary

Outer sheath Lipoprotein envelope of spirochete that encases the periplasmic flagella and protoplasmic cylinder

Periplasmic flagella Internal flagella that arise near the cell terminus, also known as axial filaments, axial fibrils, and endoplasmic flagella

Protoplasmic cylinder Cell components contained by and including the cell wall–cell membrane complex

Reagin Syphilis antibody that cross-reacts with lipids extracted from mammalian organs

Spirochete Helical bacterium with a lipoprotein outer sheath and internal flagella belonging to the order Spirochaetales

THE SPIROCHETES represent an order of bacteria characterized by having a helical shape, an outer sheath, and internal flagella. Some are free-living, but many are parasitic or pathogenic for animals and humans. Spirochetes often fail to grow or grow only under special nutritional and/or environmental conditions. Organisms classified as spirochetes cause syphilis, Lyme disease, relapsing fever, and leptospirosis. Disease symptoms, arthropod vector, immunologic response, and microbial size, morphology, and motility must be used in taxonomy because cultivation of most spirochetes, if possible, is difficult.

I. Classification and Taxonomic Features of Spirochetes

Spirochetes belong to the order Spirochaetales, which has two families, Spirochaetaceae and Leptospiraceae. Genera in the family Spirochaetaceae are *Spirochaeta, Cristispira, Treponema, Serpula,* and *Borrelia. Leptospira* and *Leptonema* are genera in the family Leptospiraceae.

The important morphologic features of spirochetes are diagrammed in Fig. 1. Spirochetes are spiral-shaped (helical), motile bacteria surrounded by a multilaminate outer membrane or sheath. Inside the outer sheath lies the protoplasmic cylinder consisting of protoplasm limited by the cytoplasmic membrane–cell wall complex. Spirochetes differ from all other bacteria in that they possess internal periplasmic flagella. Data in Table I show that the number of periplasmic flagella per cell is genus-dependent. Periplasmic flagella arise near the poles of the helical organisms and extend toward the center of the cell. Periplasmic flagella, also known as axial filaments, axial fibrils, endoplasmic flagella, and periplasmic fibrils, are organelles of locomotion for spirochetes. The protoplasmic cylinder winds around the periplasmic flagella. Periplasmic flagella overlap in the center of the cell in *Borrelia* and *Treponema* but not in *Leptospira*. Spirochetes are highly motile as a result of their special periplasmic flagella. Motility is due to movement by creeping, rotation, and flexing. Rotation of cells about their periplasmic flagella results in a "corkscrew motion." Motility is maintained in viscous environments. Spirochetes penetrate cellulose acetate membrane filters with pore sizes as small as 0.22 μm. Spirochetes can move through 2% agar gels, and it is believed that pathogenic spirochetes enter unbroken skin and mucous membranes. A number of morphological types of spirochetelike organisms have been observed in the hindgut of ter-

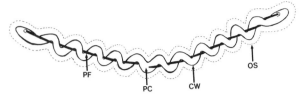

Figure 1 Morphologic features of spirochetes. CW, cell membrane–cell wall complex; OS, outer sheath; PC, protoplasmic cylinder; PF, periplasmic flagellum.

mites and roaches. Classification of insect hindgut organisms in existing or new genera awaits further characterization. Genera of other helical bacteria such as *Spirillum* and *Campylobacter* resemble spirochetes but have external flagella and no outer sheath and are classified in the order Spirillaceae. [*See* MOTILITY; FLAGELLA.]

II. Spirochaetales

A. Spirochaeta

The data in Table I show the major differential characteristics of genera classified as spirochetes. The genus *Spirochaeta* currently contains the species *plicatilis, stenostrepta, zuelzerae, litoralis, aurantia,* and *halophila. Spirochaeta* are found in H₂S containing mud from freshwater, brackish water, or saltwater. Substrate for energy metabolism appears to be carbohydrates, and end-products fail to indicate a role for *Spirochaeta* in the sulfur cycle. All

species are free-living, and no human or animal pathogens or parasites have been recognized.

B. Cristispira

The periplasmic flagella of *Cristispira* occur in bundles of 100 or more, which distend the outer sheath of the spirochete. The resultant ridge is called a crista. *Cristispira* are found in clams, mussels, oysters, and gastropods and are most likely commensals because they occur in greater numbers in healthy hosts. *Cristispira* decrease or are absent in mollusks with disturbed metabolism due to adverse living conditions. *Cristispira pectinis* is the only species known at this time. Biochemical and physiologic information on *Cristispira* is difficult to obtain because no artificial medium has been devised to support their growth in pure culture. No members of *Cristispira* have been discovered to be pathogens or parasites in humans or animals.

C. Treponema

Members of the genus *Treponema* are pathogenic or parasitic in humans and animals. *Treponema* are difficult if not impossible to cultivate in artificial culture media or tissue culture. Visualization of most treponemes requires dark-field microscopy, direct immunofluorescent staining, or silver impregnation staining such as the Warthin–Starry stain since Gram and Giemsa stains fail to sufficiently stain most species. Species that grow in artificial culture media require carbohydrates, amino acids, and long-

Table I Differential Features of Spirochaetales

Feature	Spirochaeta	Cristispira	Treponema	Serpula	Borrelia	Leptospira	Leptonema
				Genus			
Length (μm)	5–250	30–180	5–20	7–9	3–20	6–20	6–20
Width (μm)	0.2–0.75	0.5–3	0.1–0.4	0.4	0.2–0.5	0.1	0.1
No. periplasmic flagella	2[a]	100+	1–8	8–9	15–22	2	2
O₂ requirements[b]	An, F	?	An	An	An	Ae	Ae
Energy source[c]	CHO	?	CHO, AA	CHO	CHO	FA	FA
Pathogenic	no	no	yes	yes	yes	yes	no
Vector-borne	no	no	no	no	Tick, louse	no	no
DNA (G + C) (moles%)	50–66	?	36–53	25–26	28–31	35–41	51–53

[a] *Spirochaeta plicatilis* produces many periplasmic flagella.
[b] Ae, aerobic; An, anaerobic; F, facultative.
[c] AA, amino acids; CHO, carbohydrates; FA, fatty acids or alcohols.

or short-chain fatty acids. Cultivatable organisms are microaerophilic (require reduced oxygen tension) or are strict anaerobes.

Species within the genus *Treponema* include *pallidum* subspecies (ss) *pallidum*, ss *pertenue*, and ss *endemicum*. These subspecies are all pathogenic in humans, causing syphilis, yaws, and nonvenereal syphilis, respectively. Syphilis is a sexually transmitted systemic disease of humans. Yaws is characterized by cutaneous lesions and is spread through casual contact. Subspecies *endemicum* causes a cutaneous form of syphilis called bejel and is spread by casual contact. In humans, *Treponema carateum* causes cutaneous lesions known as pinta in tropical regions of the New World. Organisms involved in all of the diseases previously mentioned are similar to *T. pallidum* ss *pallidum*. Subspecies *pertenue* has been shown to have 100% DNA homology with ss *pallidum*. Persons with any treponematosis become reactive in serological tests for syphilis with reactivity indistinguishable from that of homologous antibody. Researchers have theorized that yaws, pinta, and bejel are caused by attenuated forms of the same syphilis spirochete. Pinta was believed to have been common among New World Indians. The epidemic of syphilis, which scourged Europe following Columbus' return, may have been due to a New World strain of *T. pallidum* ss *pallidum* or conversion of *T. carateum* to a new agent of venereal syphilis in a nonimmune population. Other evidence suggests that the epidemic was due to a mutation of *T. pallidum* ss *endemicum* of African origin or possibly European *T. pallidum* ss *pallidum* without any relationship to Columbus' voyages.

Infection with any subspecies of *T. pallidum* or *T. carateum* causes the production of a heterospecies-specific antibody for which the term reagin was coined. It is unfortunate that reagin now has two meanings. IgG and IgM immunoglobulins specifically elicited by treponemal infection are called reagin and the IgE class of antibody involved in hypersensitivity is also known as reagin. Reagin, which arises as a result of treponemal infection, reacts with cardiolipin as well as alcohol extractable lipids from other organs from many animal species. Beef cardiolipin is the usual antigen source for nontreponemal (heterospecific) tests used to diagnose syphilis, yaws, pinta, and bejel. Nontreponemal tests such as the Venereal Disease Research Laboratory (VDRL) and rapid plasma reagin (RPR) 18-mm circle card tests use beef cardiolipin, phosphatidylcholine, and cholesterol as antigen. VDRL and RPR tests are inexpensive, easy to perform, and useful in monitoring progress of infection and efficacy of treatment. Nontreponemal tests are screening tests and should be confirmed with treponemal tests so that false-positive reactions may be avoided. Treponemal tests such as the fluorescent treponemal antibody-absorbed and microhemagglutination tests use whole or disintegrated treponemal antigens. The reactivity of serologic tests in each stage of syphilis infection is shown in Table II.

Treponemal infection begins when infectious organisms come into contact with mucous membranes. The narrow width and great motility of treponemes, demonstrated by their ability to migrate into agar and maneuver in viscous environments, facilitates penetration of unbroken skin or mucous membranes. A systemic infection ensues when spirochetes enter the body. Usually about 3 wk after infection (range 1–12), a painless, indurated lesion known as a chancre occurs at the site where spirochetes entered the body. Treponemes are plentiful in chancres, and observation of organisms with tightly coiled spirals, demonstrating corkscrew and flexing motility and ranging in length from

Table II   Sensitivity of Nontreponemal and Treponemal Tests for Syphilis

Stage of syphilis	% Reactive (number tested)				
	VDRL	RPR	TPI	MHA	FTA-ABS
Primary	70.5 (542)	80.6 (283)	57.3 (119)	77.3 (485)	91.5 (695)
Secondary	100　(182)	100　(74)	99.4 (156)	100　(166)	100　(309)
Latent	90.4 (521)	89.7 (195)	96.7 (60)	97.5 (836)	97.1 (681)
Late	68.2 (44)	75.0 (12)	91.7 (12)	94.7 (75)	100　(56)
Congenital	87.5 (7)	100　(10)		100　(17)	100　(25)

FTA-ABS, fluorescent treponemal antibody-absorbed; MHA, microhemagglutination; RPR, rapid plasma reagin; TPI, *Treponema pallidum* immobilization; VDRL, venereal disease research laboratory.

6 to 20 μm corroborates the diagnosis of syphilis. The course of syphilis infection is divided into three stages by most venereologists. The chancre appears during primary syphilis, and its disappearance after a few weeks marks the end of primary syphilis. Secondary syphilis is characterized by widespread lesions on the skin, mucous membranes, and hair follicles. Again, *T. pallidum* can be visualized in exudates from lesions by dark-field microscopy. Syphilis may be transmitted in a nonvenereal fashion through contact with skin lesions or mucous patches. Secondary lesions abate in 4–8 wk and the disease then becomes latent. Epidemiologists divide latent syphilis into an early, infectious period and a late, noninfectious period because the chance of transmission is greater in the first 1–2 yr of latency. The chance of sexual transmission is diminished in late latent syphilis, but *in utero* transmission and congenital syphilis occur readily. Late syphilis is the third or tertiary stage of syphilis and may occur 5–40 yr after the initial infection. Late syphilis results from reactivation of sequestered treponemes. Symptoms correlate with organ systems where cryptic organisms reactivate, i.e., brain and neurosyphilis, eye and blindness, etc. Late lesions are called gummas, painless swellings that progress to ulceration, and can result in dysfunction and disfigurement. Syphilitic lesions elicit macrophage and T-cell responses. Venereal transmission of human immunodeficiency virus (HIV) is facilitated by the presence of syphilitic lesions and the abundance of infiltrative cells (macrophages and lymphocytes), which have receptor sites for HIV.

Syphilis is the most serious of treponemal infections because it affects all organs and causes congenital infections, malformations, stillbirths, and abortions. Yaws, pinta, and bejel are more limited in tissue involvement and rarely, if ever, cause congenital disease. Although immunologically related, the organisms that cause yaws, pinta, and bejel seem incapable of conferring immunity to syphilis. Indeed, antibodies that arise during syphilis infections do not prevent subsequent reinfection: Mothers nursing babies with congenital syphilis can develop primary breast lesions, and patients previously cured of syphilis can be reinfected many times.

Many if not all spirochetes are antigenically related. At least one antigen is apparently common to many treponemes. Unless serum is adsorbed with extracts of Reiter's treponeme (*T. phagedenis* biovar. *reiter*) prior to testing, up to one-third of uninfected individuals have antibodies that react with *T.*

pallidum in treponemal tests for syphilis. Antibodies to syphilis, yaws, pinta, and bejel are indistinguishable by any serological test currently available. Cross-reacting antibodies to the syphilis spirochete in persons with relapsing fever or lyme borreliosis have been documented. Low-level antibodies that react with *Borrelia* and *Treponema* occur in persons with spirochetal peridontal disease. Interpretation of serologic test results requires knowledge of clinical and epidemiologic factors.

The diagnosis of syphilis may be made by observing spirochetes with the morphology and motility of *T. pallidum* by dark-field examination of an exudate from a lesion of primary or secondary syphilis. More often, the diagnosis of syphilis is made serologically because syphilitic lesions are painless and easy to miss or ignore. The sensitivity of classic serologic tests for syphilis is summarized in Table II. Data for congenital syphilis represent persistent reactivity, which indicates infection and not passive placental transfer of maternal antibodies.

Treponema pallidum is extremely sensitive to penicillin. Progression of all stages of syphilis is interrupted and reversed by treatment. Tetracycline, erythromycin, ceftriaxone, and other third-generation cephalosporins are effective alternate treatments. Erythromycin is the least effective and should be used only when all other antibiotics are contraindicated. Penicillin is the treatment of choice because of proven success, because it crosses the blood–brain barrier in neurosyphilis, and because it crosses the placenta sufficiently to cure congenital syphilis *in utero*. Higher doses of penicillin are required for neurosyphilis than for primary or secondary syphilis. The Jarisch–Herxheimer reaction can occur 1–2 hr after administration of penicillin to syphilitics. The reaction is characterized by high fever and other systemic manifestations caused by release of endotoxin due to lysis of spirochetes. The Jarisch–Herxheimer reaction is frightening but is self-limited and abates in 12–24 hr.

Species of *Treponema* other than *T. pallidum* and *T. carateum* are parasitic or pathogenic in humans and animals. *Treponema paraluiscuniculi* causes syphilis in rabbits and has not been cultivated in artificial culture media. *Treponema denticola*, *T. vincentii*, *T. scoliodontum*, *T. refringens*, *T. minutum*, *T. phagedenis*, *T. succinifaciens*, and *T. bryantii* are species that have been grown in artificial culture media using anaerobic conditions. Most species require complex lipids or other growth factors provided by serum.

Treponema denticola, T. vincentii, and *T. scolio-dontum* occur in the human oral cavity. *Treponema vincentii* in combination with fusiform bacteria is associated with trench mouth (Vincent's angina). Dental literature links periodontal disease with the presence of large numbers of spirochetes and anaerobic bacteria on tooth pulp and in deep periodontal pockets. The application of topical antibiotics effective against spirochetes has an ameliorating effect. Spirochetes are only occasionally seen in the absence of periodontal disease. High spirochetal and other anaerobic bacterial counts seem to cause halitosis. Koch's postulates are difficult to satisfy for spirochetes in Vincent's angina, periodontal disease, and halitosis because infections are multiorganismic. Associations between these conditions and spirochetes are certain, but a causal role cannot be definitely established. Diagnosis is usually made through microscopic observation of spirochetes, fusiform and other bacterial forms, pyogenic response (pus formation), and typical clinical lesions. Organisms are rarely isolated and identified because their recovery from oral specimens is expected—even from healthy mouths.

Treponema refringens, T. minutum, and *T. phagedenis* often inhabit the human genital tract. These species are parasitic but have not been implicated in disease. *Treponema succinifaciens* and *T. bryantii* parasitize animal digestive tracts but remain unassociated with disease.

D. Serpula

Treponema (T.) hyodysenteriae and *T. innocens* were placed in a new genus *Serpula* in 1991. These organisms differ from *Treponema* in protein electrophoretic patterns, DNA composition, and, most importantly, DNA homology. *Serpula innocens* has been isolated from the intestinal contents of dogs and swine but is not associated with disease. *Serpula hyodysenteriae* is enteropathogenic, causing swine dysentery, a significant veterinary disease. This organism is usually detected in carrier or dysenteric swine through isolation and identification or microscopic observation in histologic sections of the large intestine. *Serpula hyodysenteriae* is an obligate anaerobe, requires serum as a source of cholesterol or other complex lipids, and is strongly β-hemolytic. Swine dysentery is controlled through antibiotic treatment and detection and quarantine of carriers.

E. Borrelia

Species, vectors, hosts, and distribution of *Borrelia* are shown in Table III. *Borrelia* are pathogenic in humans or animals and are transmitted by insect vectors, usually ticks or lice. Many species of *Borrelia* have not been cultivated, and their classification is based on the vector from which they were observed. Uncultivatable *Borrelia* can be maintained in laboratory tick colonies for many years, especially in ticks that transmit organisms transovarially to their offspring. *Borrelia* that have been cultivated artificially are microaerophilic. *Borrelia* are differentiated from other agents of human spirochetosis by their ability to be stained with Gram or Giemsa stains, whereas *Treponema* and *Leptospira* are not. *Borrelia* are gram-negative organisms. Loose coils of long amplitude characterize *Borrelia,* whereas *Treponema* are more tightly and regularly coiled. *Leptospira* are so tightly coiled that it is difficult to visualize coils in rapidly moving cells. *Borrelia* develop 15–22 periplasmic flagella, whereas *Treponema* have 1–8 and *Leptospira* have 2. In *Borrelia,* periplasmic flagella overlap such that fine-structure analysis of the center of the cell shows 30–44 flagella; periplasmic flagella do not overlap in *Leptospira.* Fine-structure analysis of *Treponema, Serpula,* and *Leptonema* shows cytoplasmic fibrils or intracytoplasmic tubules, which arise near insertion points of the periplasmic flagella and extend in clusters of six to eight along the inner layer of the cytoplasmic membrane. The fact that antigenic variation is common in *Borrelia* is best illustrated by the relapsing fever organisms, which have up to 12 serologic types. Immunity is type-specific, therefore up to 11 relapses can occur if new serotypes arise one at a time. Borrelioses are serious septicemic infections, some of which have long-lasting sequelae. Animals become infected through the bite of an infected vector or by scratching organisms deposited on the skin by the infected vector into the bite site.

Borrelia cause avian borreliosis, relapsing fever, or Lyme disease in humans and animals. *Borrelia burgdorferi* causes Lyme disease, is transmitted by a variety of ticks and possibly fleas and mosquitoes, and appears to have an animal reservoir that includes cats, dogs, rodents, deer, and other wildlife.

The diagnosis of relapsing fever is made by observing *Borrelia* in stained peripheral blood films. Organisms may be so sparse in late relapses that amplification through animal passage may be required for visualization. *Borrelia hermsii, B. park-*

Table III Agents, Vectors, Hosts, and Distribution of Borreliosis

Borrelia Species	Vector	Host	Occurrence
B. anserina	Argas miniatus, perisca, reflexus	Birds	Worldwide
B. brasiliensis	Ornithodoros brasiliensis	Rodents	Brazil
B. burgdorferi	Ixodes dammini, others	Humans, animals	Worldwide
B. caucasica	O. verrucosus	Rodents, humans	Middle East
B. crocidurae	O. erraticus	Rodents, humans	Africa, Asia
B. dugesii	O. dugessi	Rodents	Mexico
B. duttonii	O. moubata	Humans	Africa
B. graingeri	O. graingeri	Rodents, humans	Africa
B. harveyi	Unknown	Monkeys	Africa
B. hermsii	O. hermsii	Rodents, humans	North America
B. hispanica	O. erraticus	Rodents, humans	Mediterranean
B. latyschewii	O. tartakovskyi	Rodents, humans, reptiles	Iran, Asia
B. mazzottii	O. talaje	Rodents, humans, armadillos	Mexico, Central America
B. parkeri	O. parkeri	Rodents, humans	United States
B. persica	O. tholozani	Rodents, humans, bats	Middle East, Asia
B. recurrentis	Pediculus humanus subsp. humanus	Humans	Worldwide
B. theileri	Rhipicephalus decoloratus	Horses, ruminants	Austrailia, Africa
B. tillae	O. zumpti	Rodents	North America, Africa
B. turicatae	O. turicatae	Rodents, humans	North America
B. venezuelensis	O. rudis	Rodents, humans	Central and South America

eri, *B. turicatae*, *B. hispanica*, *B. recurrentis*, and *B. burgdorferi* are the only species that have been grown in artificial culture media. Observation of *Borrelia* is sufficient to confirm the diagnosis of relapsing fever. Knowledge of vector species and the geographical locale where the infection was acquired can suggest which species of *Borrelia* is likely to be involved. Tetracycline is the treatment of choice. Jarisch–Herxheimer reactions have been reported.

The etiology of Lyme disease was shown to be *B. burgdorferi*, a newly described species, following an outbreak of inflammatory arthritis in children in New Lyme, Connecticut. The characteristic skin lesion that occurs at the portal of entry (erythema chronicum migrans) was described as early as 1908 in residents of the Black Forest of Germany. Pathogenicity of Lyme spirochetes parallels that of syphilis in that the infection becomes systemic upon entry, a primary lesion develops at the site of infection, and late manifestations depend on the organ system in which the spirochetes are sequestered. Late manifestations may be cardiac, neurologic, musculoskeletal, and/or ophthalmic in nature. Prompt treatment with penicillin, tetracycline, cefotaxime, or ceftriaxone prevents the occurrence of late manifestations. Symptoms resolve slowly if treatment is initiated after the onset of late manifestations.

Borrelia burgdorferi has been isolated sporadically from the blood and skin of infected patients. A monoclonal fluorescent antibody specific for *B. burgdorferi* is used to confirm the identity of isolates. Isolation is not successful often enough to make culture an important method of diagnosis in Lyme disease. The detection of antibodies to *B. burgdorferi* is a more dependable diagnostic method than culture but is fraught with undersensitivity and cross-reactions. The major antigens of *B. burgdorferi* appear to be lipoprotein, as is the case with most spirochetes. Serologic cross-reactions occur between spirochetes such that serologic results must be evaluated in light of clinical and epidemiologic factors in each case.

III. Leptospiraceae

A. *Leptospira*

Leptospira are tightly coiled spirochetes with one or both ends hooked. Two species are recognized: *Leptospira interrogans* is pathogenic for humans and animals, and *Leptospira biflexa* is saprophytic. *Leptospira interrogans* occurs as 23 or more serovars in animal disease but using serovars *copenhageni, grippotyphosa, canicola, wolfii, pomona, djatzi, autumnalis,* and *potoc* as test antigens allows serodiagnosis of most leptospirosis infections because of shared antigens among *Leptospira* serovars.

Leptospira are aerobic and grow in serum-enriched media (Stuart or Fletcher), bovine serum albumin-enriched media (Ellinghausen), and some protein-free media. *Leptospira* utilize long-chain fatty acids or alcohols as carbon and energy sources. Carbohydrates and amino acids are not utilized. Ammonium salts are required as the nitrogen source. Pyrimidines are synthesized but not incorporated from external sources, allowing *Leptospira* to be resistant to 5'-fluorouracil (5-FU). Most bacteria are inhibited by 5-FU so that 5-FU is an effective selective agent for *Leptospira*. Optimal incubation temperature is 28–30°C. Slight uniform turbidity develops in liquid media, and a disk of turbidity develops near the top of semisolid media as a result of positive chemotaxis to oxygen. *Leptospira* migrate through 0.22-μm pores in cellulose acetate membrane filters. Many leptospira migrate through 2% agar. Migration through membrane filters and agar can be used in conjunction with 5-FU to select *Leptospira* from contaminated environments.

Leptospira interrogans is pathogenic in animals and humans. Humans are an accidental host infected through contact with leptospiruric animals or contaminated soil or water. *Leptospira* survive in alkaline freshwater for at least 3 mo. Transmission may be increased by extended dry weather because organisms are concentrated as ponds recede or because more animals converge on fewer watering places. Leptospires penetrate the skin, mucous membranes of the oropharynx or esophagus, or the eye. Swimming in ponds that serve as watering holes for farm animals has been documented as a source of exposure to leptospirosis, but this practice is far more widespread than the disease. The spectrum of disease varies from a mild febrile illness to a severe hepatic and renal infection. *Leptospira* seem to have an affinity for the convoluted tubules of the kidney and are shed during acute and convalescent carrier stages of the disease.

Diagnosis of leptospirosis acute human infection may be made by isolating organisms from blood during the first week of illness and from urine after the first week. Organisms may be shed in urine for several months. No more than 1–2 drops of blood or urine should be used per 5 ml of culture medium to dissipate inhibitors that arise during infection. Cultures are incubated at 28–30°>C in the dark and are examined each week for 4–6 wk by dark-field microscopy. Chronic leptospirosis is significant in domestic animals, causing chronic renal involvement, loss of milk production, abortion, and stillbirth. Organisms are more difficult to recover from chronic disease—no more than 60% of cultures are positive despite 26 wk of incubation with blind subcultures and addition of fresh media. Leptospires are recognized through visualization of tightly coiled, hooked, highly motile organisms. Microscopic agglutination with select serovar antisera can be used to type isolates. Complete serologic typing is done at leptospira reference centers such as the Centers for Disease Control, Atlanta, Georgia.

Serologic diagnosis may be made by demonstrating increasing leptospiral antibodies during the course of the disease. Serologic tests include microscopic agglutination of live leptospires, macroscopic agglutination of formalin-killed leptospires, or enzyme-linked immunosorbent assays using heat-extracted *L. interreogans* serovar *icterohaemorrhagiae* antigens. Serologic methods are most sensitive in the diagnosis of acute disease. Insignificant leptospiral titers are found in 50–75% of animals with chronic renal infection.

Leptospira are sensitive to many antimicrobial agents *in vitro*. Penicillin, erythromycin, and tetracycline have been used to treat leptospirosis. To be effective, antibiotics must be initiated during the first few days of infection. The outcome of patients in which antibiotics are delayed until after severe hepatic or renal involvement occurs is similar to the natural history of untreated leptospirosis. Dialysis may be lifesaving in severe cases. In chronic animal leptospirosis, return of milk production is not hastened by antimicrobial therapy. Animal vaccines are available, appear to prevent serious infection, and reduce but not eliminate the urinary carriage and shedding.

B. *Leptonema*

Leptonema illini was previously *Leptospira illini* with uncertain taxonomic status because of dissimilarities with *Leptospira*. *Leptonema* was recognized as a new genus in the family Leptospiraceae in 1987. *Leptonema illini* is nonpathogenic. *Leptonema* are morphologically similar to *Leptospira* but differ in DNA composition, DNA homology, flagellar insertion organelles, and presence of cytoplasmic tubules.

Bibliography

Alexander, A. D. (1985). Leptospira. *In* ''Manual of Clinical Microbiology,'' 4th ed. (E. H. Lennette, A. Balows, W. H. Hausler, Jr., and H. J. Shadomy, eds.), pp. 473–478. American Society for Microbiology, Washington, D.C.

Burgdorfer, W. (1985). Borrelia. *In* ''Manual of Clinical Microbiology,'' 4th ed. (E. H. Lennette, A. Balows, W. H. Hausler, Jr., and H. J. Shadomy, eds.), pp. 479–484. American Society for Microbiology, Washington, D.C.

Canale-Parola, E. (1984). The spirochetes. *In* ''Bergey's Manual of Determinative Bacteriology. (N. R. Krieg and J. G. Holt, eds.) pp. 38–70. Williams and Wilkins, Baltimore, Maryland.

Fitzgerald, T. J. (1985). Treponema. *In* ''Manual of Clinical Microbiology,'' 4th ed. (E. H. Lennette, A. Balows, W. H. Hausler, Jr., and H. J. Shadomy, eds.), pp. 485–489. American Society for Microbiology, Washington, D.C.

Huber, T. W. (1984). *Clin. Microbiol. News*. **6,** 47–50.

Smibert, R. M. (1985). Anaerobic spirochetes. *In* ''Manual of Clinical Microbiology,'' 4th ed. (E. H. Lennette, A. Balows, W. H. Hausler, Jr., and H. J. Shadomy, eds.), pp. 490–494. American Society for Microbiology, Washington, D.C.

Statistical Methods for Microbiology

Martin Alva Hamilton
Montana State University

I. Measurement
II. Descriptive Statistics
III. Probability Models
IV. Statistical Inference

Glossary

Analysis of variance Methods of statistical inference used for studying a wide range of linear hypotheses; also, computational procedures for partitioning the total variability of a continuous response variable into relevant components

Contingency table analysis A body of methods for analyzing cross-classified attribute variable data that accomplishes for attribute data what analysis of variance does for continuous variable data

Linear regression analysis Methodology for the case where the mean of the response variable is a function of the predictor variables and some unknown parameters such that the function is linear in the parameters; contrasts with nonlinear regression analysis where the function is nonlinear in the parameters

Multivariate analysis A body of methods for analyzing simultaneous measurements on many variables, including classification methods, discriminant analysis, correlation analysis, multivariate analysis of variance, and factor analysis

Nonparametric methods Statistical methods for estimation and significance testing that are valid when the probability distribution of the random variable is more vaguely specified than required for conventional statistical methods; many are applicable to ordinal data; also known as distribution-free methods

STATISTICS is a mathematical science directed primarily at three issues—data summarization, uncertainty assessment, and experimental design. Statistical techniques for data summarization include statistical graphics, empirical modeling techniques, and statistical measures for describing random phenomena (e.g., measures for describing typical values, dispersion, or association).

Uncertainty assessment includes both deductive reasoning by means of mathematical probability theory and inductive reasoning by means of statistical inference. Statistical inference procedures provide a framework for using observed data to infer something about the mechanisms governing the system under investigation and to attach measures of uncertainty to those inferences. Significance testing and confidence intervals are the most used inferential tools. The conceptual basis for the theory of statistical inference is a major concern of philosophers of science.

The statistical theory of experimental design provides guidelines for data collection and tools for optimization. The guidelines for data collection are devised to eliminate sources of bias and to reduce variability. Tools for optimization show how to determine which alternative designs are relatively free of bias, achieve a desired small level of uncertainty in its conclusions, and require the least resources. Because experimentation strategies differ greatly among the many subfields of microbiology, it is not feasible to discuss statistical aspects of experimental design in this article.

The goals of this article are to present some basic terminology and concepts of statistical reasoning, discuss some statistical techniques that are frequently misapplied, and describe a few promising new statistical methods. The focus is on the perceived needs and interests of microbiologists. Microbiologic investigations can present unconventional statistical problems. For example, if the experimental units are microbes, then an experiment can involve units in the millions, a potential advantage because the "sample size" is so large. Unfortunately, this advantage is seldom realized because the individual microbes are usually not observable, instead data pertain to ensembles. Microbiological assay techniques may produce grouped or censored measurements; that is, the investigator can determine that the response value is within some range of numbers, the exact value being indeterminable. Statistical methods for microbiology necessarily include analyses appropriate for ensembles and for censored or grouped data. Consequently, "off-the-shelf" statistical methodology will not always be suitable for microbiological applications.

All of the methods discussed in this article are based on "frequentist" foundations, in contrast to "Bayesian" foundations. Bayesian theory provides a structure by which one can modify prior beliefs in light of new data. The Bayesian approach answers scientific questions with direct mathematical probability calculations (deductive reasoning), in contrast to frequentist statistical inference methods for inductive reasoning. There is tension in the statistical community concerning the relative merits of frequentist methods versus Bayesian methods, especially with respect to uncertainty assessment in science; however, because the frequentist approach predominates in microbiology, Bayesian methods will not be described in this article.

As research becomes ever more specialized, it will be unlikely that an individual possesses the technical knowledge required to perform all stages of a project; consequently, collaboration is often required. Because few microbiologists choose to become experts in statistics, collaboration or consultation with a statistician may well benefit microbiological investigators.

I. Measurement

Statistical thinking revolves around data. The investigator must determine what factors should be measured and how those measurements should be obtained. Obviously, knowledge of microbiology is required for these critical determinations. This article will not consider microbiologic strategies for collecting data or for selecting the variables to be analyzed. It is directed at the task of matching statistical methodology to the specified data and research goals.

Each characteristic the investigator has chosen to measure is a variable, and variables describing the experimental end point are called response variables. Variables thought to influence or cause changes in the response variables are called predictor variables or factors, and the investigator usually purposely manipulates one or more of the predictor variables. Variables that are statistically associated with the response variables and with the predictor variables but are not of interest to the experimenter are called confounding variables or factors. It is important to use experimental designs and methods of analysis that prevent confounding variables from affecting the interpretation of the relationships between the predictor variables and the response variables.

It is helpful to differentiate between deterministic and random variables. If the experimenter knows exactly the value for a variable before the experimental results are observed, then it is a *deterministic variable*. In an assay of antimicrobic activity of a disinfectant, the experimenter will decide what concentration of the disinfectant to use; the disinfectant concentration variable is deterministic.

If the experimenter does not know before hand what the variable's value will be but can only list the possible values that variable is a random variable (stochastic variable). In order to work with a random variable, one must consider the possible outcomes for that variable. Different statistical techniques have been devised for random variables having the different types of outcomes. Almost any random variable can be classified as, at least, one of the following types: binary, discrete, continuous, ordinal, or attribute.

A binary variable can take on only two values. For example, in an experiment where a smear has been placed on a Petri dish and incubated, the outcome might be recorded as positive or negative, depending on whether bacterial growth did or did not occur.

A discrete variable can take on two or more possible numerical values but not so many as to form a continuum. The most common discrete variable is a count variable where the possible values are nonnegative whole numbers. An example would be the number of colonies growing on a Petri dish.

A continuous variable can take on numerical values that conceptually lie on a continuum. For example, if a test tube sample were incubated, one might measure the intensity of light that could pass through the tube as an indication of microbial density. For any two light intensity values, the outcome could conceivably lie between them. Thus, light intensity is a conceptually continuous variable. The continuum is "conceptual" because instruments have only finite accuracy, and although the true values of the variable lies on a continuum, the recorded values are discrete.

It is convenient to identify two different types of continuous variables with the distinction depending on the microbiologist's method of comparing two values of the variable. If a continuous variable is such that one would compare duplicate observations by taking the difference (subtracting one value from the other), then the variable has a linear scale. Incubation time, for example, is usually considered to be measured on a linear scale. If the comparison would be based on a ratio of the values, then the variable has a ratio scale. (These definitions of scale differ from those used in the behavioral sciences.) If the researcher views 0.5 and 2.0 as being equally far from 1.0, a ratio scale is indicated. The ratio scale allows only positive values. Measures of the sizes of microbial populations, for example, are usually treated as having a ratio scale.

Sometimes quantitative variables in microbiology are impossible to measure accurately; instead, censored observations may be recorded. For example, if the number of colonies counted on a Petri dish is recorded as "greater than 250," the observation has been right-censored at 250. If a biochemical measurement is recorded as "less than detectable," and it is known that the limit of detection is 2 ppm, then the observation has been left-censored at 2 ppm. An observation may be both left- and right-censored, in which case, the observation is said to be interval-censored or grouped. For example, in an experiment to determine the minimal inhibitory concentration (MIC) of an antimicrobic agent, it might be observed that a concentration of 0.1 mg per dl inhibits organisms, but a concentration of 0.01 mg/dl does not. Any MIC between 0.01 and 0.1 is consistent with the observations; no finer determination is possible.

An ordinal variable is one for which the possible values are ordered but not given a quantitative value. When a technician makes a visual judgment that a test tube is clear, somewhat cloudy, or very cloudy, there is no quantitative scale of cloudiness but there is an ordering.

An attribute (or nominal or categorical) variable is a nonquantitative, nonordinal variable that can take on only a finite number of outcomes. For example, the species of an organism is an attibute; there is not necessarily a meaningful ordering of species. A code number might be assigned to each species, but the code numbers should not be statistically analyzed.

Although the basic principles of statistical reasoning are the same for all types of data, the technical details depend on the type of variable, the measurement scale, presence of censoring or grouping, etc. In particular, different statistical techniques are used with quantitative variables, such as continuous or count variables, than for nonquantitative variables, such as ordinal and attribute variables.

II. Descriptive Statistics

Suppose a number of realized values of a continuous response variable have been recorded. As part of the analysis, the investigator might peruse a list of all the values. To communicate the data to others, that list could be published. When the data are voluminous, however, lists become impractical. Even if a set of data is small, there may be some aspects that the investigator wishes to give special emphasis. It is important, therefore, to summarize the data in a way that succinctly conveys its important features, and the method of summarization must fit the specified scientific goals. Fortunately, some standard statistical techniques are suitable for a variety of purposes. A statistical summary could be a visual presentation (e.g., a graph), a tabular presentation, or a list of a few summary measures (e.g., the median and the range).

A. Statistical Graphics

Statistical graphs are used to detect patterns, to help determine the appropriate statistical analysis, and to communicate the results. Advances in computer graphics enable one to produce sophisticated graphs on desktop computers. A hazard of modern computer software is that it entices the analyst to create beautiful, eye-catching graphics, perhaps to the detriment of effective scientific communication. There are two easy rules that help the analyst maintain a focus on scientific communication: First, add no lines or details to a graph unless those additions convey information, and second, eliminate all lines or details that can be erased without sacrificing information. Cross-hatched, three-dimensional bar charts may be visually attractive, but those embellishments may detract from the main points. The third dimension, the cross-hatching, and even the sides of the bars, usually can be eliminated without loss of information.

1. One-Dimensional Plots

Figure 1 displays 250 measurements of the intensity of light that has passed through a bacterially contaminated liquid. Light intensity is inversely related to the amount of contamination. The bottom display in Fig. 1 is a jittered one-dimensional plot in which each point has been randomly jittered in a vertical direction so that points do not overlap. The horizontal axis shows the exact value of each point; the direction vertical has no meaning. A one-dimensional plot shows all the data points and effectively communicates important features of the data, such as the range, clustering, and outliers. The plots are also called *line plots* and *dot plots*.

2. Density Traces

The top display in Fig. 1 is an example of a computer-generated density trace. For any point on the light intensity scale (horizontal axis), the probability density (vertical axis) is proportional to the number of measurements near that value. The analyst specifies "near" by entering a numerical value into the computer program. The small wiggles in both tails of the density trace of Fig. 1 are caused by the smallest and largest light intensities. The size of each wiggle shows how much individual observation contributes to the height of the density trace. The

density trace is a smooth, improved version of a histogram.

A density trace is especially effective at displaying the extent to which the data are asymmetrically distributed and the mode (or modes) of the distribution. The *mode* is that value of the variable where the data are most dense; that is, where the density trace reaches a maximum. If there are multiple modes, the density trace will clearly reveal them. Fig. 1 shows that the light intensity data are symmetrically distributed and that the mode is slightly less than 0.30.

3. Cumulative Percentage Plots

Table I shows the MIC of an antifungal drug for an assay using twofold dilutions of the drug. This assay produces grouped data. For example, if the organism grows in the presence of drug concentrations of

Table I Quantitative *in Vitro* Susceptibility Dilution Test Data

Strain #	MIC (μg/ml)	Strain #	MIC (μg/ml)
CA-1	1.16	CA-29	0.58
CA-3	1.16	CA-30	0.58
CA-4	1.16	CA-31	0.58
CA-5	1.16	CA-32	0.29
CA-6	0.58	MSU 81–11	0.29
CA-7	0.29	MSU 81–12	0.58
CA-9	0.58	MSU 82–48	0.29
CA-10	0.58	MSU 82–74	2.31
CA-13	0.58	MSU 82–121	0.29
CA-14	0.58	MSU 83–27	0.58
CA-16	1.16	MSU 83–39	0.58
CA-17	0.29	MSU 83–42	0.58
CA-18	0.58	MSU 83–44	0.58
CA-19	1.16	STRAIN 11	0.29
CA-21	1.16	STRAIN 9938	0.58
CA-23	1.16	STRAIN 394	0.58
CA-28	0.58	MSU 80–60	0.58

Antifungal antimicrobic: Amphotericin B
Type of susceptibility test: Macro broth-dilution
Fungus: *Candida albicans* (Robin) Berkhout
    34 clinical isolates—obtained from hospitals throughout the United States
Minimum inhibitory concentrations (MIC) measured at 24 h
Medium: Synthetic amino acid medium, fungal (SAAMF)
Temperature of incubation: 35°C
Inoculum concentration: 10^4 blastoconidia/ml (ungerminated)
    Concentrations of drug evaluated: ≤0.14–≥18.47 μg/ml (twofold serial dilutions)
[Data courtesy of M. Rinaldi, Department of Pathology, University of Texas Health Science Center, San Antonio.]

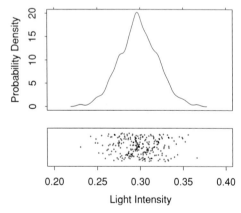

Figure 1 Jittered one-dimensional plot and associated probability density trace for 250 measurements of light intensity as recorded by a fiber optic sensor that monitors changes in frequency modulated infrared light penetrating a bacterially contaminated liquid. The measurements were calculated as log (l_0/l), where l_0 is the intensity of the incident radiation and l is the emergent intensity. (Data courtesy of Z. Lewandowski, Center for Interfacial Microbial Process Engineering, Montana State University, Bozeman.)

1.16 μg/ml and less but is inhibited by concentrations of 2.31 μg/ml and greater, the recorded MIC is 2.31 μg/ml. Any MIC in the interval (1.16, 2.31) is consistent with this observation. The MIC data are interval censored or grouped and that fact must be taken into account during statistical analyses.

An example of a cumulative percentage plot is shown in Fig. 2. Let c denote a drug concentration that was used in the assay. Figure 2 was constructed from Table I by plotting the proportion of strains having an MIC less than or equal to c (vertical axis) against c (horizontal axis). One point is plotted for each concentration used in the assay. The plotted points are connected by straight lines, a method that amounts to spreading the MICs uniformly over the interval into which they were grouped. The cumulative percentage plot facilitates calculation of percentiles. The dotted lines show how to graphically determine the median MIC (0.43 μg/ml, by linear interpolation) and the 90th percentile (0.94 μg/ml). When the data are grouped, I believe the cumulative percentage plot is the best possible way to visualize the data.

A cumulative percentage plot is also effective when the data are not grouped. It is constructed by arranging the data in ascending order, plotting the ith smallest value on the horizontal axis against $100i/(n+1)$ on the vertical axis, i=1,2, . . . , , and connecting the points with straight lines. (Some statisticians prefer to plot $100i/n$ or $100(i-0.5)/n$ on the vertical axis.)

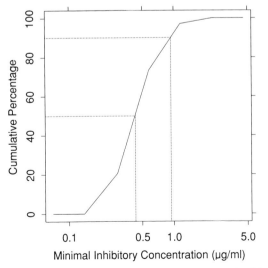

Figure 2 Cumulative percentage plot for the data of Table I using a log scale for the MIC's.

B. Descriptive Measures

It is standard practice for data summaries to present a measure, of location and dispersion.

1. Measuring Location (Indicating the Typical Value)

There are many possible measures of "typical value" for a set of quantitative data. In microbiology, the most common measures are the mean, the geometric mean, and the median. Quantitative typical value calculations are not appropriate for attribute data and are rarely appropriate for ordinal data.

The mean is the arithmetic average. For the 250 light intensity measurements of Fig. 1, the mean is 0.298. Traditional methods of statistical inference are based on the mean. If the distribution of measurements is symmetric, then the mean and the median (defined later) are identical. The mean is the measure to use if the ultimate goal is to calculate a total. For example, suppose one wanted to estimate the total biomass in a large tank of liquid where the available data are biomass determinations in each of several small volume samples taken from the tank. If the sample mean biomass were v units per ml., then one could approximate the total biomass by multiplying v times the number of milliliters in the tank. The median cannot be used in a similar manner to approximate the total.

Deficiencies of the mean are that it is greatly influenced by a few unusual values, and it is incalculable if any of the extreme values are censored. For the data {1,2,3,4,5}, the mean is 3; changing the last observation 5 to 50 increases the mean to 12. For the data {1,2,3,4,>10}, where >10 indicates that the last observation is right-hand censored at 10, the mean must be larger than 4 but how much larger is indeterminable.

To calculate the mean for grouped data, first replace each recorded value by the midpoint of the interval within which the value lies, then average. The sample mean for the data of Table I is 0.53 μg/ml. Note that this is different from the mean 0.71 that is found by averaging the recorded MICs, a technique that improperly ignores the fact that the data are grouped.

The median is the middle value (the 50th percentile). The median is usually the location measure of choice when the distribution of the variable is highly skewed (asymmetric) or the data are censored. Distinct advantages of the median are that it is not influenced by a few censored or outlying observations and it can often be calculated in situations where the mean cannot. The median is 3 for all of these data sets: {1,2,3,4,5}; {1,2,3,4,50}; and {1,2,3,4,>10}.

For grouped data, one can identify an interval within which the median lies. To determine a specific value for the median, I recommend calculations based on the cumulative percentage plot. For the data of Table I and Fig. 2, the median is 0.43 μg/ml. The light intensity data of Fig. 1 have a symmetric distribution and the median is 0.298, the same as the mean, at least to 3 decimal places.

When a continuous variable has a ratio scale, it may be appropriate to log-transform the observations before calculating the mean. If one calculates the mean of the log-transformed data and then takes the antilog of that mean to produce a value on the original measurement scale, the result is the *geometric mean*. Specifically, denote the n points in the data set by $x_1, x_2, \ldots, x_n$ and the geometric mean by gm. Then

$$ gm = antilog \left[\frac{1}{n} \sum_{i=1}^{n} \log(x_i) \right]. $$

The geometric mean will always be smaller than the mean. If the log-transformed data have a nearly symmetric distribution, then the geometric mean and the median will be nearly identical. The geometric mean is popular in microbiology, probably because many variables have the ratio scale and the log-transformed data often have symmetric distribution. As was the case with the mean, the geometric mean is sensitive to anomalous observations, and it cannot be calculated if there are censored observations.

For grouped data, the geometric mean is calculated using the midpoint of the interval on the log scale. For the data of Table I, the geometric mean is 0.43 μg/ml. Note that because the cumulative percentage plot of Fig. 2 is nearly symmetric (on the log scale), the median and geometric mean are the same to two decimal places.

2. Measuring Dispersion (Variability)

Denote the ith value in the data set of n values by x_i, i = 1, 2, . . . , n. Denote the typical value (mean, median, or other measure) by T. Then a simple model for the data is given by the equation $x_i = T + d_i$, where d_i is the "deviation" of the ith datum from the typical value ($d_i = x_i - T$). The larger the deviations, the greater the dispersion of the data.

The variance is the average of the squared deviations from the mean. Letting M denote the mean and Var denote the variance, the formula is

$$ Var = \frac{1}{n} \sum_{i=1}^{n} (d_i)^2 = \frac{1}{n} \sum_{i=1}^{n} (x_i - M).^2 $$

The variance is awkward because it is in terms of squared units.

The standard deviation (SD) is the square root of the variance. The SD has the same units as the data. The standard deviation of the data {1,2,3,4,5} is 1.41; the standard deviation of the data {1,2,3,4,50} is 19.03. For the light intensity data of Fig. 1, the standard deviation is 0.0225. (If the x_is are sample data, many statisticians advocate dividing by n-1, rather than n, in the formula for the variance).

For a continuous variable on a ratio scale, one can find the standard deviation of the log-transformed data. The antilog of that standard deviation, sometimes called the geometric standard deviation, is the geometric mean of the ratios $r_i = x_i/gm$, where gm is the geometric mean of the x_is.

3. Interval Measures of Dispersion

The range interval runs from the minimum value to the maximum value. The range for the light intensity measurements of Fig. 1 is (0.231, 0.366). A deficiency of the range interval is that an outlying value makes the range very wide, indicating more dispersion than evident in the bulk of the data.

The interquartile range interval contains the middle 50% of the observations. It can be found by reading the 25th percentile (the first quartile) and the 75th percentile (the third quartile) from the cumulative percentage plot. The interquartile range interval for the light intensity data is (0.284, 0.313).

In medical microbiology, the interval containing the middle 95% of the observations is often of interest. For a conceptually continuous diagnostic measurement taken on a healthy group of individuals, this interval is called the normal range. The endpoints of the normal range, the 2.5th and 97.5th percentiles, can be read from the cumulative percentage plot.

An interval containing a known percentage of the observations is a 100% content range where P is some specific proportion, $0 \leq P \leq 1$. The normal range is the special case of P = 0.95. The endpoints of the interval, the $100(1-P)/2$ percentile and the $100(1+P)/2$ percentile, can be read from the cumulative percentage plot. Only intervals with content near 100% are greatly influenced by a few unusual values. In most applications, the 100P% content interval, with $0.5 \leq P \leq 0.95$, is a better measure of dispersion than the range.

III. Probability Models

Assessing uncertainty in studies involving random variables requires a probability model for the random variables. Consider a single, continuous random variable denoted by X. Before the data are collected, one should be able to specify a set of possible values for X. The set is called the sample space. A completely specified probability model allows one to calculate the probability of any particular outcome. The form of the probability model could be based on historical data, a mathematical derivation using knowledge of the mechanisms that produce X, or the investigator's experience and expert judgment. The main components of probability modeling are (1) stating fundamental assumptions, (2) choosing a model that is compatible with those assumptions, and (3) testing the model's goodness-of-fit to real data.

A. Fundamental Assumptions

Many diverse fundamental assumptions are plausible. Among them, the assumption of independence is of singular importance because almost all standard textbook methods for statistical inference are based on the assumption that realizations of a random variable are independent. Suppose n observations of X are to be recorded; Denote the potential observations by $X_1, X_2, \ldots, X_n$. The two variables X_i and X_j and said to be independent if the probability model for X_i is the same whether or not one knows the realized value of X_j. The probability model is greatly simplified if the n random variables X_1, . . . ,X_n each has the same probability model and are independent, in which case the observations are said to be independently and identically distributed (i.i.d.). One must carefully study the methods by which the data are gathered in order to decide whether the i.i.d. designation is appropriate.

The decision to assume independence depends primarily on the subject expert knowledge of the measurement process and material. Correlation analysis is the main statistical tool for using data to assess independence. If two variables are correlated, they cannot be independent. (Converse rea-

soning is not correct, however, because two variables may be uncorrelated but dependent.) Although such correlation analyses are beyond the scope of this article, two examples of lack-of-independence will be cited.

Suppose one is conducting immunologic research using mice and X is the count from an autoradiographic analysis of liver tissue. If the data are 100 measurements of X, made up of 10 readings for each of 10 mice, the 100 values probably are not independent. The intramouse correlation coefficient may well be significantly large indicating that the 10 readings on the same mouse liver are dependent.

If the density of microbes in a solution is being observed at consecutive points in time, observations at adjacent time points may not be independent. For such data, it is important to know the minimum time between observations required for the measurements to be essentially independent. The statistical evaluation of independence could be based on a serial correlation analysis.

If independence is incorrectly assumed, then the probability model, the statistical analysis based on that model, and the scientific conclusions based on the analysis may also be incorrect.

B. Probability Models for Continuous Random Variables

When the sample space of X is a continuum, the normal or Gaussian distribution is the most commonly used model. The normal probability model depends on two parameters or constants, denoted by μ and σ, where μ is the mean and σ is the standard deviation of X. Important characteristics of a normal distribution include (1) the distribution is symmetric around μ, (2) the density trace is bell-shaped (similar to Fig. 1), (3) X takes on a value between $\mu-\sigma$ and $\mu+\sigma$ with probability approximately equal to two-thirds, and (4) X takes on a value between $\mu-2\sigma$ and $\mu+2\sigma$ with probability approximately equal to 0.95. The normal probability model is usually appropriate if the continuous variable X arises as the sum or average of a large number of small contributions.

For an X that takes on only positive values and arises as the product of a large number of small contributions, microbiologists often use a normal probability model for log(X), in which case, X is said to have a *lognormal distribution*.

More information about the normal, lognormal, and other distributions for continuous random variables can be found in the books cited in the Bibliography.

C. Probability Models for Discrete Random Variables (Counts)

Counting variables are particularly important in microbiology, and there are two popular models for a counting variable, the Poisson and negative binomial distributions. Let Y be the counting variable and suppose the sample space for Y is the set of integers 0,1,2, . . . (no upper limit).

The Poisson distribution for Y depends on a single constant, denoted by λ. The Poisson can be called the distribution of complete randomness as the following example demonstrates. Consider a large amount of solution containing a single microbe species. Let u denote the volume of a small aliquot of the solution, where u is so small that the aliquot can contain only 0 to 1 microbe; that is, the chance that it could contain 2 or more microbes is essentially 0. Suppose the density of microbes in the solution is λ per u (i.e., the chance that an aliquot of volume u contains exactly one microbe is λ). Assume that the microbes are randomly distributed in the solution so that a sample, taken from anywhere in the solution and having volume u, has the same probability λ of containing exactly one microbe. Further assume that two distinct samples taken from the solution yield microbe counts that are independent. Under these conditions, if Y is the number of microbes in a random sample of volume V (same units as u), then the probability that Y equals the integer y is

$$P(y) = \frac{e^{-\lambda V}(\lambda V)^y}{y!}.$$

The mean and variance of Y are both equal to λV. The Poisson distribution is widely used in microbiology. In limiting dilution assays, for example, the number of viable organisms in a test tube is often modeled using a Poisson distribution.

The negative binomial distribution depends on two constants, denoted by m and c, and the equation for the probability that Y equals integer y is

$$P(y) = \frac{\Gamma(y + c^{-1})}{y!\,\Gamma(c^{-1})} \left(\frac{cm}{1 + cm}\right)^y \left(\frac{1}{1 + cm}\right)^{c^{-1}},$$

where Γ is the gamma function. The mean and variance of Y are m and m(1+cm), respectively. The

limit of the negative binomial probability P(y) as c approaches 0 is the Poisson probability model with m taking the place of λV. Thus, the Poisson model is a special case of the negative binomial model.

D. Goodness-of-Fit

Because statistical methods for assessing uncertainty are based on the assumed probability model, it is worthwhile to check whether or not the data seem to be distributed as the probability model predicts. This check, which often involves both graphical analyses and significance testing, is known as goodness-of-fit assessment.

Consider, for example, the task of determining the goodness-of-fit of the Poisson model. Suppose there are n replicates of a colony counting experiment where the observable random variables are the number of colonies counted on each of n plates. For a Poisson distribution, the mean and variance of the n observations should be equal. It is not unusual, however, for the variance to be considerably larger than the mean, in which case, the data are said to be "overdispersed." To see if the variance is statistically significantly larger than the mean, the usual procedure is to calculate the test statistic

$$D = \frac{\sum_{i=1}^{n}(y_i - \bar{y})^2}{\bar{y}},$$

where

$$\bar{y} = \frac{1}{n}\sum_{i=1}^{n} y_i.$$

Under the null hypothesis that the counts are independent realizations of a Poisson random variable, the statistic D has a chi-squared sampling distribution with n−1 degrees of freedom. One would reject the Poisson model if D were too large relative to that chi-squared distribution. For example, this is a typical statement, "As a check on the appropriateness of the Poisson distribution for the data [colony counts], 80 replicates of the experiment were run using the same bacterial population incubated in the same medium. There was no obvious departure of the data from the Poisson distribution. The ratio of the sample variance to the sample mean was 0.95; the chi-squared test was not significant (1-sided p-value = 0.6)."

In situations where the data are overdispersed and

the Poisson model seems inappropriate, the negative binomial model might fit the data adequately. Of course, one should check the negative binomial model for goodness-of-fit before analyses are based on that model.

IV. Statistical Inference

Suppose there is a measure of special interest to the investigator. Examples of such a measure are the mean number of colony forming units, the density of organisms in a solution, or the median minimum inhibitory concentration of an antimicrobic drug. Suppose further that the measure can be expressed as a function of the parameters of the probability model for the observable data. In this situation, two possible aims are to use the data to guess the true value of the measure for the process under investigation and to determine what values of the measure are compatible with the data. The first aim is called point estimation and the second is significance testing or confidence interval estimation.

A. Point Estimation and Associated Uncertainty

1. Overview

There are many strategies for choosing estimators. For example, one popular method, known as maximum likelihood estimation, chooses the value of the parameter for which the observed data are most likely. Regardless of the estimation method, the estimator is calculated from the observations that are random variables, and therefore, the estimator is also a random variable. Consequently, the estimator has a probability distribution. One hopes the estimator's distribution is narrowly dispersed around the true parameter value. Two properties of that distribution, bias and standard error, are often used to indicate the quality of an estimator. The bias is the difference between the mean of the estimator's distribution and the parameter. If the bias is zero, then the estimator is said to be unbiased, which indicates that it is exactly correct on the average. The standard error is the standard deviation of the estimator's distribution. An estimator having a small standard error is reliable in the sense that it would usually vary little if the experiment were replicated.

2. Special Case: The Mean

An important situation worth special discussion is when the sample mean is used as the estimator of the true mean. Statistical theory has shown that the sample mean is unbiased and its standard error is $\sigma/\sqrt{n}$, where σ is the standard deviation of an individual observation and n is the number of observations. Because the numerical value of σ is usually unknown, it is estimated by the sample standard deviation, denoted by s. The abbreviations SE (for standard error) or SEM (for standard error of the mean) are often used to denote the realized value of $s/\sqrt{n}$, which approximates the true standard error of the sample mean. A small SEM indicates that the sample mean is a reliable estimator of the true mean.

Statistical theory asserts even more about the sample mean. If the distribution of the data has a finite standard deviation σ (not a stringent requirement in practice) and n is not too small, then the distribution of the sample mean is approximately normal. Denote the realized sample mean by $\bar{x}$. For the light intensity data of Fig. 1, $\bar{x} = 0.298$, s = 0.0225, and SEM = 0.00142, information that can be used to assess variability in the data and uncertainty of the estimate.

To describe variability in the data, $\bar{x}$ and s can be used to calculate an approximate 100 P% content interval. The interval with endpoints $\bar{x}-s$ and $\bar{x}+s$ estimates the interval $(\mu-\sigma, \mu+\sigma)$, which is a 100·⅔% content interval if the underlying data follow a normal distribution. The interval with endpoints $\bar{x}-2s$ and $\bar{x}+2s$ estimates the interval $(\mu-2\sigma, \mu+2\sigma)$, which is a 95% content interval (the normal range) if the underlying data follow a Gaussian distribution. For the light intensity data, which appears to be Gaussian distributed, the normal range is (0.253, 0.343).

To describe the uncertainty of the estimate, an approximate confidence interval estimate for the true mean can be calculated as follows. Regardless of the underlying distribution of the data, if n is sufficiently large, the absolute difference between the estimator and the parameter has about a two-thirds chance of being less than SEM and about a 95% chance of being less than 2×SEM. Thus, one can be 95% confident that the interval $(\bar{x}-2·SEM, \bar{x}+2·SEM)$ contains the true mean μ. For the light intensity data, the 95% confidence interval for μ is (0.295, 0.301). (Confidence intervals are also discussed in Section IV,B,2.)

When there are multiple treatments and the treatment means are the parameters of interest, a popular visual presentation is a plot of the means with "error bars" emanating from each mean. All too frequently, an explanation of the error bars is not provided. The error bars could be the range, the interquartile range, $\pm$SD, $\pm$SEM, a prediction interval, a 95% confidence interval for the true mean, or something else. The most common error bars are $\pm$SD or $\pm$SEM, which indicate the variation of individual measurements or the uncertainty of the sample mean, respectively.

3. Bootstrap Approach to Estimating the Standard Error

Although statistical theory tells us how to find an approximate standard error for the sample mean, it is not always clear how to approximate the standard error for other sample statistics. In the absence of a standard error formula, a computer simulation technique called bootstrapping might be used to assess the standard error of the estimator. Suppose the data have been observed. The bootstrap procedure starts with a probability model for the observations. The probability model can be either the exact distribution if it is known or, if it is unknown, the cumulative percentage plot of the observed data, the latter being a surrogate for the true distribution. Next one draws many independent (bootstrap) samples from the distribution (or surrogate distribution), each sample of the same size as the observed data set. (Usually 500 to 2000 samples will provide a sufficiently accurate standard error.) For each bootstrap sample, the associated estimate is calculated. In this fashion a list of estimates is created. The standard deviation of the list approximates the standard error of the estimator. In my experience, this technique works particularly well when the observations are continuous random variables, but not quite so well with discrete variables. The bootstrap is a relative of jackknife and cross-validation techniques.

B. Significance Testing

Statistical significance testing helps the investigator determine whether or not an observed effect could be caused by chance. If the effect is too large to be attributable to chance, then the investigator has some basis for concluding that the effect is real. Consider an experiment where the effect is defined as the difference between the mean number of adhering bacteria in a control group and in a group treated with an agent having potential antiadherence

properties. Even if the agent had no true antiadherence properties, the means in the two groups could be somewhat different because of inherent variability in the experiment. If the difference is substantially larger than can be attributed to inherent variability, then one has confidence that the agent is effective.

Specific testing methods (t-tests, Wilcoxon tests, chi-squared tests, etc.) are beyond the scope of this article. The choice of a correct statistical method depends on the definition of "effect" and on the type of data (binary, ordinal, etc.). New experiments and new ways to measure an effect may well require new statistical methodology. Nevertheless, some important generic features of significance testing do deserve comment.

1. The Generic Significance Test

Every significance test entails three steps: stating the null hypothesis, choosing the test statistic, and calculating the p-value. Almost always, the null hypothesis is an assumption of "no effect." For the antiadherence assay example, the experimenter has in mind the general null hypothesis that the agent has no more antiadherence properties than the control; the specific null hypothesis usually tested is that the true mean number of adhering bacteria for the control group is equal to the true mean number for the group treated with the agent.

The test statistic quantifies the discrepancy between the data actually observed and the data expected if the null hypothesis were true. It is helpful to know the null value of the test statistic, that is, the numerical value for the statistic if the observed data exactly agree with the null hypothesis. For the antiadherence assay example, the test statistic might be the difference between means (observed mean number of adhering bacteria for the control group minus observed mean number of adhering bacteria for the agent-treated group) divided by an estimate of inherent variability (estimated standard error of the difference). The null value for this statistic is zero.

The result of the significance test is summarized by the p-value, which measures, on a scale running from zero to one, the extent to which the data agree with the null hypothesis. A p-value near zero indicates that the data discredit the null hypothesis. The p-value is actually a probability, the probability of observing such an unusual value for the test statistic if the null hypothesis is true. Here "unusual" means departing from the null value as far as does the realized value.

2. Significance Tests and Confidence Intervals

There is a correspondence between significance tests and confidence intervals that leads to a useful interpretation of confidence intervals. Suppose we are interested in a parameter denoted by θ, say, for which we want a $100(1-\alpha)\%$ confidence interval estimate, where α is a specified error rate (usually $\alpha = 0.01$ or 0.05). Suppose we know how to conduct a significance test of the null hypothesis that $\theta = \theta_O$, where θ_O is any specific numerical value of interest. Let us say that "the data discredit the null hypothesis" if and only if the p-value is less than α. Then the $100(1-\alpha)\%$ confidence interval estimate of θ is made up of all numerical values of θ_O for which the data do NOT discredit the null hypothesis $\theta = \theta_O$.

For the light intensity data of Fig. 1, the 95% confidence interval for the mean μ is (0.295, 0.301), which indicates that the observed data do not discredit the null hypothesis $\mu = \mu_O$ for any μ_O between 0.295 and 0.301, or equivalently, that any hypothesized value for the true mean less than 0.295 or greater than 0.301 is discredited by the observed data. In this case $\alpha = 0.05$ and a p-value less than 0.05 is required to discredit the null hypothesis.

3. Complications in Interpreting Significance Tests

The p-value is calculated assuming that the null hypothesis is correct. It is often informative to do some calculations assuming that the null hypothesis is incorrect. Such calculations require specification of an alternative hypothesis, a task that requires some pondering. Usually, the relevant alternative is associated with the smallest effect of practical importance.

Even if the observed data yield a large p-value, one cannot automatically conclude that the data confirm the null hypothesis. It may turn out that a relatively large p-value is quite likely even if the alternative hypothesis is true; that is, the data do not possess sufficient power to discriminate between the null and alternative hypotheses. On the other hand, if a small p-value is quite likely when the alternative hypothesis is true, then the fact that the realized p-value is large allows us to conclude that the data corroborate the null hypothesis for all practical purposes.

Statistical significance does not imply practical significance. By publishing the estimates, as well as the p-values, investigators can let the readers decide if statistically significant results are of practical importance.

Acknowledgment

This project was partially supported by the Center for Interfacial Microbial Process Engineering at Montana State University, a National Science Foundation Engineering Research Center. Thanks to Robert Boik and Kirk Johnson for their helpful comments.

Bibliography

Cleveland, W. S. (1985). "The Elements of Graphing Data." Wadsworth, Monterey, California.

D'Agostino, R. B., and Stephens, M. A. (1986). "Goodness-of-fit Techniques." Marcel Dekker, New York.

Efron, B., and Tibshirani, R. (1991). *Science* **253**, 390–395.

Freedman, D., Pisani, R., Purves, R., and Adhikari, A. (1991). "Statistics," 2nd ed. Norton, New York.

Haaland, P. D. (1989). "Experimental Design in Biotechnology." Marcel Dekker, New York.

Ilstrup, D. M. (1990). *Clin. Microbiol. Rev.* **3**, 219–226.

Kotz, S., Johnson, N. L., and Read, C. B. (1982–1988). "Encyclopedia of Statistical Sciences," Vol. 1–9. Wiley, New York.

Morrison, D. F. (1990). "Multivariate Statistical Methods," 3rd ed. McGraw-Hill, New York.

Press, S. J. (1989). "Bayesian Statistics: Principles, Models, and Applications." Wiley, New York.

Sterilization

Seymour S. Block
University of Florida

Glossary

Gravity displacement autoclave In an effort to prevent the presence of air from giving a deceptive reading in an autoclave, this employs the principle that cold air is twice as heavy as steam and is displaced downward as steam is introduced into the chamber and is removed through an outlet at the bottom

HEPA filters High-Efficiency Particulate Air filters that are used in laboratory rooms, hospital isolation rooms, and other settings where requirements are very high for air free of microorganisms and other suspended particles

Megarad Unit of energy absorbed per gram of material irradiated by a cobalt-60 source or an electron beam generator

Prevacuum autoclave Uses a high-vacuum pump in conjunction with a condenser system to remove air and increase the efficiency of steam penetration

STERILIZATION is part of the daily routine of microbiological laboratories, surgical theaters, and food production plants. It is any process or procedure designed to entirely eliminate microorganisms from a material or medium. It should not be confused with disinfection, sanitization, pasteurization, antisepsis, or other processes that are intended to kill or inactivate microorganisms but may not necessarily entirely destroy or eliminate them. Sterilization may be accomplished by the use of heat, chemicals, radiation, or filtration.

I. Introduction

Sterilization is a vital part of today's civilization that we take for granted. Before it existed, food could not be kept without spoiling except in unappetizing form by drying or salting. Surgical operations resulted in infection and death in a very large percentage of cases. In war, more men died from infection and disease than directly from battle injuries. Today we buy cartons of fresh milk and fruit juices that keep for months without refrigeration and bottles of draught beer that have not been heated to diminish the flavor. Flash heating, chemical sterilants, and filtration sterilization bring about these modern-day marvels. In medical and dental practice, the instruments used in surgery as well as the gloves, gowns, drapes, dressings, and solutions are rendered sterile by steam, dry heat, or gaseous chemicals. Pharmaceutical companies sterilize many products using radiation sterilization. Microbiological laboratories require that cultures, containers, media, and equipment be sterilized so only the desired organisms that are inoculated will grow and all others will be eliminated. Hospital waste is made noninfectious by processes such as steam sterilization and incineration.

Sterilization, as we practice it, had its beginning little more than 100 years ago. We must credit the pioneer scientists of the last half of the nineteen-century—men like Louis Pasteur, Joseph Lister, Robert Koch, Charles Chamberland, and John Tyndall—for many of the great discoveries that made it happen. Certain empirically based practices, however, had their beginnings much earlier. In biblical and medieval times, fire was used to destroy clothes and corpses of diseased persons. In 1718, Joblot showed that a hay infusion could be sterilized by boiling it for 15 min and then sealing the container. In 1810, Nicolas Appert applied this principle to the practice process of canning food.

II. Kinetics of Microbial Death

Sterilization in the traditional sense is defined as a process that destroys or removes all microorganisms. It is conceived as an absolute. A product or a substance is either sterile or not sterile; it cannot be partially sterile. Even one live bacterium in a 10,000-gal tank would make it unsterile. But in the real world, we seldom deal with absolutes, and our problem here is in our inability to determine if and when all of the organisms have been destroyed, or are rendered incapable of reproduction. There are many reasons for this problem. For one, we cannot produce the growth conditions of nutrition, temperature, pH, oxidation-reduction potential, osmotic concentration, etc., necessary for all of the different species that may be infecting the product. For another, individuals of a population of the same species grown under the same conditions vary widely in their ability to survive treatments designed to kill them so that conditions that terminate some or most microorganisms may not kill them all. A few cans of food among the millions that undergo rigorous sterilization procedures are found to contain live microorganisms. Because of this inability to ascertain with certainty when all organisms have been destroyed, we now regard sterilization in terms of the mathematical probability of organisms surviving the treatment. For a product to be deemed sterile it would have to have a very low probability of survival of its original living inhabitants.

A study of the kinetics of microbial death provides us with information with which to make quantitative measurements of the sterilization process. The following are terms and concepts of importance in this study.

A. Bioburden

Bioburden is the initial population of living microorganisms in the product or system being considered.

B. Microbial Death

Microbial death is the inability of microbial cells to metabolize and to reproduce when given favorable conditions for reproduction. The organisms may still retain some of the aspects of life, but when they can

no longer multiply and increase in numbers they cannot spread disease or cause food spoilage or other damage. This definition is useful in practice because microbial reproduction is the method generally employed for detecting and counting microorganisms. The mechanism of death is believed to result from the interaction of the sterilizing agent with critical proteins, nucleic acids, enzymes, or other vital agents within the cell.

C. Biocide

Biocide is a chemical or physical agent intended to produce the death of microorganisms.

D. Death Rate

Death rate is the rate at which a biocidal agent reduces the number of cells in a microbial population that are capable of reproduction. This is determined by sampling the initial population and populations during and following the treatment and making counts following dilution and incubation of the survivors on growth media. The death rate of microorganisms resembles a unimolecular or first-order bimolecular chemical reaction; i.e., the organisms die in direct proportion to their concentration at any given time. If a sterilizing agent kills 90% of the organisms per minute, the number killed in each succeeding minute will become less as fewer microbes survive the process. Assume that 1 million cells were going to be sterilized using this agent. After the first minute, 900,000 (90%) would be killed, leaving 100,000. After the second minute, 90,000 would be killed, leaving 10,000. After the third minute, 1000 would survive; 100 after the fourth minute; 10 after the fifth minute; 1 after the sixth minute; and after the seventh minute there would be only 1 chance in 10 that there would be a survivor. A plot of time versus survivors appears in Fig. 1, which shows the rapid decrease in numbers at first and then the progressively slowing decrease in the following minutes. When time (or dosage in the case of radiation sterilization) is plotted against the logarithm of the number of survivors, as in Fig. 2, a straight line is obtained.

E. Death Rate Constant

Death rate constant, or velocity constant, is the slope of the logarithmic survivor plot. The velocity

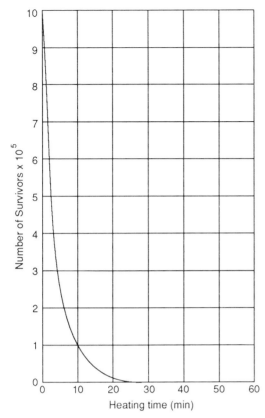

Figure 1 Survivor curve with the number of survivors and heating time plotted on an arithmetic scale. [From Block, S. S. (ed.) (1991). "Disinfection, Sterilization and Preservation," 4th ed. Lea & Febiger, Malvern, Pennsylvania.]

constant, K, is calculated by the equation:

$$K = \frac{1}{t}\log_{10}\frac{N_o}{N_1},$$

where t = time in minutes, N_o = initial number of organisms, and N_1 = number of survivors at time t. The greater the value of K, the steeper the slope and the greater the rate of kill.

F. *D* Value

D value stands for decimal reduction time and is the time required in minutes at a specified temperature to produce a 90% reduction in the number of organisms. It is the reciprocal of the velocity constant when the rate of kill is exponential. It is found to be useful in practical work because it is expressed in units of time instead of rate. In radiation sterilization, the *D* value is expressed in terms of dose rather than of time. Since 90% reduction in number of viable organisms is 1 $\log_{10}$ reduction, the *D* value can be

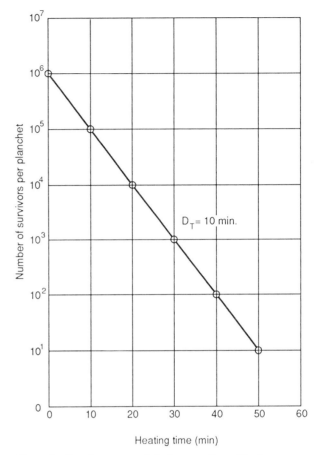

Heating time (min)

Figure 2 Survivor curve plotted on semilogarithmic paper with the number of survivors plotted on the logarithmic scale. [From Block, S. S. (ed.) (1991). "Disinfection, Sterilization and Preservation," 4th ed. Lea & Febiger, Malvern, Pennsylvania.]

termed the log reduction time. Mathematically it is

$$D_{\text{value}} = \frac{t}{\log N_o - \log N_u},$$

where t = time to destroy 90% of the initial population, N_o = initial population, and N_u = surviving population.

The *D* value may be derived graphically, as in Fig. 2, where the time for each log reduction is 10 min. *D* values are used to describe the effectiveness of a sterilization process, by giving the number of *D* values to achieve a desired reduction in microbial population. For most situations, a notation of 6 *D*, indicating a 6 log reduction of the initial population, is considered a satisfactory sterilization treatment, although conditions exist where less or greater protection is called for.

The *D* value plotted in Fig. 2 is constant over time,

and this is usually the case with death rate curves plotted logarithmically. However, there are death rate curves that are nonlogarithmic, but usually for only a portion of the curve. Clumping of bacterial cells can produce a flat shoulder in the initial part of the curve. A mixture of organisms may give a terminal lag in the curve where the sensitive organisms are killed first and the more resistant ones are left. A curious looking curve is one in which there is an initial hump in the curve due to an increase in population. This occurs where spores are heat-activated. This results in higher counts than with the initial unheated population and continues as long as the rate of activation exceeds the rate of killing.

G. Z Value

Z value is the temperature difference producing a 10-fold change in D value. The D value is determined at a specified temperature, but the death rate and D value vary with temperature. It is useful to know how the D value decreases with increasing temperature. The Z value may be plotted on semi-log paper, as with the D value, with the D value in minutes on the log scale plotted against the temperature on the linear scale. The equation for the Z value is

$$Z = \frac{t_1 - t_2}{\log D_2 - \log D_1},$$

where t_1 = higher temperature, t_2 = lower temperature, D_2 = D value at the lower temperature, and D_1 = D value at the higher temperature. Z values for different bacteria treated at temperatures from 70° to 130°C with steam average about 10°C with a range of 6–14°C and an average of around 25°C with a range of 10–70°C for dry heat. [*See* Temperature Control.]

H. Q_{10} Value

Q_{10} value is a temperature coefficient, namely the ratio of microbial death rate at one temperature to the rate at another 10°C higher. The Q_{10} value is used as an expression of microbial death kinetics and has found considerable use in the food industry for canning problems. It is expressed as

$$\log Q_{10} = \frac{10}{t_1 - t_2} \log \frac{k_2}{k_1},$$

where t_1 = higher temperature and t_2 = lower temperature.

The Q_{10} value is related to the Z value as follows:

$$Z \text{ (in °C)} = \frac{10}{\log Q_{10}} \text{ and } Z \text{ (in °F)} = \frac{18}{\log Q_{10}}$$

I. F Value

F value is the expression for a heat treatment at any temperature that is equal to a heat treatment at a given temperature and Z value for a certain period of time. In practice, the F_o value is used. It is defined as a lethality factor equating the equivalent time in minutes at various temperatures required to produce a given sterilization result in saturated steam at 121°C, where the Z value is 10°C. It is commonly used in the pharmaceutical and food industries as a measure of sterilization effectiveness. It permits minimizing heat exposure for heat-sensitive products while retaining the reliability of their sterility. This is accomplished by integrating temperatures <121°C, for example, during the heat-up and cool-down phases, and equating these integrated values to time at 121°C. Vitamins and other heat-sensitive products can thus be sterilized satisfactorily at lower temperatures.

J. Thermal Death Time and Thermal–Chemical Death Time

Thermal and thermal–chemical death times are expressions referring to the times required to kill a specified microbial population, distributed uniformly among replicate samples, upon exposure to a thermal or thermal–chemical sterilizing agent under specified conditions. Typical values with highly resistant spores are 15 min at 121°C for steam sterilization and 120 min with about 500 mg/liter ethylene oxide gas concentration at 54.4°C and 60% relative humidity.

K. Inactivation Factor

The inactivation factor is the degree of reduction or percentage kill of a microbial population by a particular treatment. It expresses the severity of the treatment process. It is calculated by dividing the initial viable count by the final extrapolated count after treatment, using highly resistant test organisms such as spores. With a process having an experimentally determined inactivation factor of 10^6, every organism has a probability of 10^{-6}, or one-in-a-million chance, of survival. If the initial population (bioburden) were 10^9, 1000 (10^3) survivors would be anticipated when the inactivation factor is 10^6. If, however, the bioburden were only 10^5 microbes, there would be a probability of approximately 0.90, or a 90% chance, that all organisms would be destroyed. Thus a treatment safe for one product would not be safe for another, depending on its level of contamination. This is illustrated in Table I, which gives the probability of achieving sterilization when increasing populations are treated by processes of increasing severity, as measured by their inactivation factors. Table I shows that as the inactivation factor becomes greater and greater, the probability of achieving sterility, the certainty of having no survivors, becomes infinitely closer to 100%, but this is never actually reached.

L. Sterility Assurance

Sterility assurance is the calculated probability that a microorganism will survive sterilization and leave some of the treated articles unsterile. It is measured as the "degree of sterility," which is the inactivation factor divided by the average number of organisms per article or bioburden, and is the inverse of the final count. If units of a product have an average degree of contamination of 10^6 microbes per sample and the inactivation factor for the treatment is 10^8, the degree of sterility would be $10^8/10^6 = 10^2$, indicating a risk of 1 unsterile article for every 100 processed. This, no doubt, would be unacceptable in most cases. A risk of 1 unsterile article in 1 million is considered acceptable for medical equipment. With an inactivation factor of 10^8, a risk of 1 unsterile article in 1 million would require an initial contamination with an average number of organisms no greater than 10^2 per sample. Thus it is important that the bioburden be minimized in order to meet rigid

requirements for sterility assurance. If an emulsion wax for automobiles is sterilized only to maintain a longer shelf-life for the product, it may not be worth the extra cost of treatment for strict sterility assurance. However, quite a different situation occurs in the case of nonacid canned foods, which are likely to be contaminated with spores of the toxin-producing *Clostridium botulinum*. Here a high degree of sterility assurance is essential. Assume that the food is steam-sterilized at 121°C and the process has a log reduction factor (D value) of 0.21 min. If each can in the pack contains only one spore, a minimum treatment of 12D ($12 \times 0.21 = 2.52$ min) is required to provide sterility assurance. In actual practice, an additional 4 min is given, allowing for greater contamination with *C. botulinum* or with spores of spoilage organisms having greater heat resistance. In regular microbiological laboratory or hospital practice where the acceptable probability of a surviving microorganism might be 10^{-6}, assume a contamination level of 10^5 where the articles are to be steam-sterilized. For sterility assurance spores of the steam heat-resistant organism *Bacillus stearothermophilus* are employed with steam sterilization at 121°C. The D value for spores of this organism is 1.8 min. In this example, the minimum holding time at this temperature would be 5D (9 min) plus 6D (10.8 min) or 19.8 min to overcome the contamination and achieve the degree of sterility desired. The actual sterilization cycle will be longer, because to this time of 19.8 min must be added the time necessary for the articles to heat up to 121°C, which will vary with the size and contents of the load and the drying and cooling time. In cases where the product will not stand the rigorous or long duration treatment of the sterilization process, it is necessary to modify the manufacturing procedure or give prior treatment de-

Table I Relationship between Inactivation Factor, Percent Kill, and Probability of Achieving Sterilization of Different Initial Populations of Bacteria

Inactivation factor	% Kill	Probability of no survivors in an initial population of:				
		100	1000	10,000	100,000	1,000,000
10^1	90	0.00	0.00	0.00	0.00	0.00
10^2	99	0.37	0.00	0.00	0.00	0.00
10^3	99.9	0.90	0.37	0.00	0.00	0.00
10^4	99.99	0.99	0.90	0.37	0.00	0.00
10^5	99.999	0.999	0.99	0.90	0.37	0.00
10^6	99.9999	0.9999	0.999	0.99	0.90	0.37
10^7	99.99999	0.99999	0.9999	0.999	0.99	0.90

Probability of 0.90 = 90% chance of no survivors. Probability of 0.9999 = 99.99% chance of no survivors.
[From Rubbo, S. D., and Gardner, J. F. (1965). "A Review of Sterilization and Disinfection." Year Book Medical Publishers, Chicago.]

signed to reduce the bioburden so that a milder or shorter treatment is possible. With cleanliness, purified water, filtered air, laminar air flow, washing the product, use of masks and gloves by the handlers, use of ultraviolet light, or antiseptics, the bioburden of products might be reduced prior to sterilization. Tyndallization is another possibility.

M. Tyndallization

Tyndallization is a process that consists of heating materials to be sterilized in steam at 100°C for 20–45 min/day for 3 successive days. The bacteria are killed at this temperature while the spores that survive are heat-activated and tend to germinate in the moist environment and room temperature that follows the heatings. The vegetative forms that result from germination are then killed when the medium is reheated to 100°C the following days.

III. Sterilization with Heat

Over the years, heat has proved to be the most popular method of sterilization. It is the most economical, safe, and reliable method. Heat is believed to kill microorganisms by denaturation and coagulation of their vital protein systems. Oxidation and other chemical reactions are also greatly accelerated as the temperature is increased, roughly doubling for every rise of 10°C. There are principally two methods of thermal sterilization: moist heat (saturated steam) and dry heat (hot air) sterilization. Moist heat has the advantage of being more rapid and requiring lower temperatures. It is used for sterilization of many materials except those that are harmed by heat, particularly heat in combination with moisture. Dry heat is employed for materials such as metal instruments that are corroded by moist heat as well as powders, ointments, and dense materials that are not readily penetrated by steam. Because dry heat is effective only at considerably higher temperatures and longer times than moist heat, dry heat sterilization is restricted to those items that will withstand these higher temperatures.

Experiments on the effect of heat and moisture on egg albumin shows why dry heat requires higher temperatures than moist heat. Albumin containing 50% water is coagulated at 56°C, with 25% water at 80°C, and with 0% water at 170°C. It is evident that water assists heat in coagulating protein and from experiments with microorganisms the temperature

for sterilization is also found to be inversely related to the moisture present. Moist heat sterilization is much faster than dry heat sterilization. With dry heat the time for sterilization is 120 min at 160°C or 45 min at 260°C, whereas with moist heat sterilization is accomplished at 121°C in 15 min, at 134°C in 3 min, and at 150°C in 1 min. The reason for this difference between dry heat and wet heat is the amount of heat transferred to the material. Condensed steam supplies the latent heat of vaporization as well as the sensible heat, which amounts to 524 calories per gram at 121°C with more than four times as much heat as latent heat than as sensible heat, whereas hot air supplies only the sensible heat of 1 calorie per gram for each degree that it is cooled.

A. Moist Heat Sterilization

Moist heat sterilization is carried out in a pressure vessel in a manner similar to cooking with a home pressure cooker. Water boils at 100°C at sea level but a greater temperature is required to kill resistant bacterial spores in a reasonable length of time. A temperature of 121°C with a holding time of 15 min, excluding the time necessary for reaching that temperature and then cooling down to ambient, is an accepted standard. The gauge steam pressure of 15 pounds per square inch at this temperature aids in the penetration of the heat into the material being sterilized. Further aiding in penetration is the almost 2000-fold contraction in volume of steam when it condenses to liquid water upon contacting the colder materials being sterilized. This creates a strong negative pressure that pulls more steam into the materials. It is important that the steam be saturated with water for maximum heat transfer. If the steam is supersaturated, i.e., if the temperature of the steam exceeds the dew point, namely the temperature at which it condenses to water, it will be less efficient because it will transfer less latent heat than would be possible if it were saturated with water at that temperature. It is also important that the steam be free of air and other gases, because other gases, while exerting pressure, will not supply the latent heat of vaporization of water. Air must be removed from the sterilizer and from the materials to be sterilized. This is the first part of the sterilization cycle with a pressure steam sterilizer, called an autoclave. Other parts of the cycle are heating the contents to be sterilized (the load) to the desired temperature and steam pressure, holding it there for the required time, exhaustion of steam from the sterilizer chamber, cooling the load and pulling a vacuum to assist in drying the sterilized materials, and finally bringing the chamber to atmospheric pressure opening the door and removing the sterile contents.

If a pressure gauge is used to indicate autoclave chamber temperature, the presence of air will give a deceptive reading of the temperature since the air exerts pressure in addition to the steam pressure. Two types of autoclaves attack this problem of air removal. The downward (gravity) displacement autoclave (Fig. 3) employs the principle that cold air is about twice as heavy as steam and is displaced downward as steam is introduced into the chamber and is removed through an outlet at the bottom. Removing air from the load is much more difficult than removing it from the chamber, especially when the load is made up of porous materials such as tightly wrapped textile packs and containers in which pockets of air may be trapped. With the downward displacement autoclave, it is important not to overload the chamber and to place the contents so that there is a free flow of steam around the individual items to better remove residual air. Hot air at this temperature (121°C) would require 60 hr to ensure killing resistant spores rather than 15 min with steam.

The prevacuum autoclave was developed to answer this problem of air removal. This autoclave uses a high-vacuum pump in conjunction with a condenser system to remove air and increase the efficiency of steam penetration. It removes about 98% of the air within 4 min, after which steam penetration is almost instantaneous. This greatly reduces the oxidation of rubber and textile articles and permits a higher temperature of 132°C (27 pounds gauge pressure) and a reduced holding time of 4 min at this temperature. With this autoclave, the overall heating–sterilization–cooling–drying cycle in hospitals has been reduced from 45 min to as little as 20 min (Fig. 4). These times are only approximate and the sterilization cycles may be longer with increased heat-up and pulsing time as well as drying time due to the size and make-up of the load. In this type of autoclave, steam if first injected into the chamber followed by a vacuum and a series of 3 or 4 pulses of steam pressure and vacuum over a period of about 9 min to remove all the air trapped in surgical packs and heat the chamber and its contents. This is followed by the 4 min period at 132°C after which there is as little as 7 min of vacuum drying, the vacuum being then released to atmospheric pressure with filtered air. For rapid turnaround of operating room instruments a 12 min express cycle is available with 4 min at 132° but a shorter heat-up pulsing period (Fig. 5). With solutions or other materials that can-

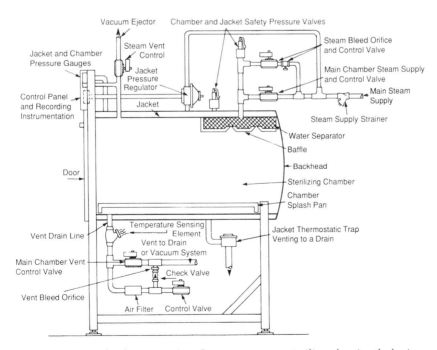

Figure 3 Longitudinal cross-section of a steam pressure sterilizer showing the basic features. (Courtesy of Joslyn Valve Company.)

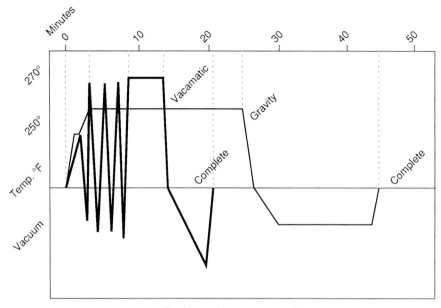

Pre-Vac cycle versus gravity

Figure 4 Comparison of illustrative cycles with prevacuum and gravity displacement autoclaves. (Courtesy of American Sterilizer Company.)

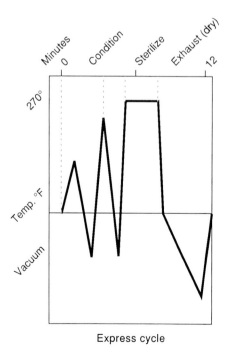

Express cycle

Figure 5 Express (flash) cycle for rapid sterilization of instruments in hospital operating rooms. (Courtesy of American Sterilizer Company.)

not be subject to a high vacuum a special cycle is used. With modern autoclaves, the cycles are controlled automatically with a microprocessor and instrument recording and a printout of the details of each operation is available.

For sterilized liquids in any autoclave, the chamber pressure must be reduced slowly to atmospheric pressure and the liquid allowed to cool below the boiling point at atmospheric pressure before opening the door to prevent the solution from boiling over. If the quantity of solution is large, cooling may take considerable time. To shorten the cooling process, some autoclaves are equipped with a water spray. Compressed air is also introduced to make up for the steam pressure lost due to condensation upon cooling. The modern sterilizer is equipped with automatic controls and a recording temperature gauge to record chamber temperature during the run. There is a steam jacket around the chamber to prevent cooling and condensation. Thermocouples should be placed inside densely packed materials to be certain that they are penetrated by steam and reach sterilizing temperatures.

B. Dry Heat Sterilization

The dry heat sterilizer is even more economical, safe, and reliable than the steam-pressure sterilizer, but it is much slower and therefore less popular. It is economical because the equipment is simpler and less expensive, safe because there is no steam under pressure, and reliable because the high temperature and long duration can be counted on to heat all parts of the load to a temperature that no organism can long withstand. Some materials that are dense or well insulated may not reach sterilization temperatures in the short times of steam sterilization and are better treated in a dry heat sterilizer. Among the items often treated in a dry heat sterilizer are glass syringes and needles, lab glassware, surgical knives and scissors, drills, dental tools, cotton pledgets, oils, waxes, and powders. The hot air oven is the most common form of dry heat sterilizer. It usually employs electric heating and a blower fan to circulate air and distribute the heat evenly to all parts of the chamber. The loading should allow free air circulation so that the contents reach and maintain necessary sterilization temperatures. As in moist heat sterilization, articles must be covered or wrapped to prevent recontamination following sterilization. The wrapping as well as the articles themselves must be able to withstand the temperatures employed without decomposition. A stainless steel box or aluminum foil or mylar wrapping may be used. At a temperature of 150–160°C the holding time is usually 2 hr, and at 180°C 30 min. Dry heating at 160°C for 2 hr is considered equivalent to moist heating at 121°C for 10–15 min. The heat-up time will vary depending on the material, its size, and conductivity. The whole cycle may vary from 1 hr to several hours. An advantage of dry heat sterilization is that pyrogens are destroyed; also the articles are left dry and ready to use. Disadvantages are that it cannot be used for aqueous solutions or materials such as rubber and textiles that are deteriorated in air at high temperatures and especially that it is slow compared to moist heat sterilization. An attempt to circumvent two of these problems has been made with the high-vacuum infrared oven, which operates at 280°C and completes a sterilization cycle in 15 min. Conversely, some dry heat ovens operated as low as 140°C for heat-sensitive materials take many hours for achieving sterilization. Articles should be cleaned prior to dry heat sterilization because organic matter can interfere with the action of dry heat, increasing markedly the necessary exposure time.

Incineration is another form of dry heat sterilization that has enjoyed wide use for treating waste containing infectious microorganisms from hospitals and medical laboratories. This treatment combusts all organic materials at 600–1000°C, converting them to gases. However, there are reports of microorganisms in the effluent gases and solid residue from incinerators. This is due to improper operation or design of the incinerator because no living organism can survive such temperatures even for a second. [*See* INFECTIOUS WASTE MANAGEMENT.]

IV. Sterilization with Chemicals

Relatively few chemicals are capable of performing sterilization and have the additional properties of stability, safety, lack of color, etc., that make them marketable. Some of these may be dispensed as gases, which gives them the ability to penetrate deep into materials; others are used as liquids in applications where the volatility of gases would not be suitable.

A. Gaseous Sterilants

Certain medical and industrial instruments of intricate construction containing plastics, organic cements, and coatings that will not tolerate heat sterilization may be treated with gases capable of killing microorganisms.

1. Ethylene Oxide

Of these gases, ethylene oxide has proved to be the most successful. The articles sterilized with ethylene oxide include endoscopes, optical instruments, heart–lung machines, electrical and electronic equipment, cardiac catheters and pacemakers, oxygen tents, dental handpieces, and microsurgical accessories. Advantages of ethylene oxide sterilization are that materials are treated at a low temperature in an air-dry state; thus they are not damaged by the treatment. There is good penetration of porous materials, and complex equipment made up of many materials can be effectively sterilized. Disadvantages are that the gas is highly toxic and leaks can be dangerous. Furthermore, the process is slow and costly compared to steam sterilization. Humidity control is required for effective sterilization and the materials must be degassed after sterilization. Ethylene oxide is flammable and explosive, but as used for sterilization, mixed with freons or carbon dioxide, that danger is averted. It is an alkylating agent that alkylates the amino, sulfhydryl, carboxyl, and hydroxyl groups of vital metabolites of the microorganisms. It inactivates all microorganisms, spores being more resistant than vegetative forms. Spores of *Bacillus subtilis* var. *niger* are the most resistant spores. Factors that affect ethylene oxide sterilization are gas temperature, relative humidity, concentration, and the moisture condition of the microorganisms.

Ethylene oxide is a sterilant at room temperature but the long time factor makes sterilization at that temperature impractical. However, with the rate of sporicidal action doubling for each 10°C rise, the time for sterilization at 55°C would be one-eighth of that at 25°C. The temperature range of 45–60°C is nondestructive to most materials and is therefore used for ethylene oxide sterilization. The relative humidity is adjusted to 98%. A relative humidity of 30% or greater is necessary for effective antimicrobial activity but, microorganisms must be hydrated to be inactivated by ethylene oxide within a practical time. Spore inactivation increases as the concentration of ethylene oxide gas increases from 50 to 500 mg/liter, but there is no appreciable increase as the gas concentration is increased further. In practice, however, a concentration of 800–1200 mg/liter is used to ensure penetration and allow for absorption of the gas by the packaging material.

In a sterilization cycle with ethylene oxide, the following steps are taken. The material to be sterilized is placed in the autoclave and a vacuum is drawn. Steam is injected and then 4 steam-vacuum pulses remove air while the condensed steam assists in the penetration of the sterilant into hard-to-penetrate materials. The chamber is then filled with ethylene oxide gas at 10 lbs per square in. and 55°C for a period of 1 hr and 45 min. The gas is then exhausted, filtered air is released into the chamber, followed by two pulses of air and vacuum, and 20 min of an air wash is given with filtered sterile air run in and out of the chamber. After approximately a 3.5 hr cycle, the chamber is opened and the contents are transferred to an aeration cabinet at 55°C for 12–72 hr as needed, to remove any residual ethylene oxide that is adsorbed or dissolved in the sterile materials and packaging. The toxic ethylene oxide can be disposed of safely by catalytic combustion in an apparatus, which converts it to carbon dioxide and water vapor.

2. Formaldehyde

Formaldehyde has been used for many years as a fumigant. More recently it has been used in conjunction with low pressure steam at 70–80°C as a steri-

lant for many of the same type of instruments as sterilized with ethylene oxide, providing they can tolerate the somewhat higher temperature. The temperature cannot be lowered below 70°C because formaldehyde will polymerize to solid paraformaldehyde. Like ethylene oxide, formaldehyde is an alkylating agent, is toxic, and has a broad spectrum of biocidal action. It has a strong odor, which helps in its detection.

The sterilizer used for this low temperature steam and formaldehyde sterilization is similar to the prevacuum steam sterilizer but is operated at negative pressure with steam at 70–80°C. A heated jacket helps to keep the temperature constant. Formaldehyde is generated by heating formalin, a 37% aqueous solution of formaldehyde. A concentration of 5 g/liter or more is maintained in the chamber. Sterilization time runs from 1 to 2 hr depending on the load. The cycle is similar to that with ethylene oxide with evacuation of air and introduction of steam and formaldehyde; holding at 70–80°C for the sterilization time; removing the vapor; and drying and aerating the load.

3. Other Gases

Beta-propiolactone is an alkylating agent that is highly antimicrobial. It has been reported to be 4000 times more active than ethylene oxide and 25 times more effective than formaldehyde. A vapor concentration of 2–4 mg/liter at 24°C and 70% relative humidity provides sterilization in 2 hr. However, it does not penetrate solids readily, and therefore it is useful mainly in sterilizing surfaces in large spaces, such as rooms. It is toxic and a carcinogen.

Hydrogen peroxide has found new application in the form of its vapor. Today aseptic packaging systems for food items such as nonrefrigerated milk and fruit juices are using hydrogen peroxide as a sterilant for the food-contact surfaces. A 30% solution of hydrogen peroxide is vaporized at 125°C and the surface to be treated is exposed to the vapor for 3–4 sec at 90°C. Because of its low toxicity, a sterilizer based on hydrogen peroxide vapor has been developed and is of considerable interest to pharmaceutical, food, and biotechnology companies.

Peracetic acid, the peroxide of acetic acid, is a powerful sterilant and its vapor has been used for sterilization in applications like those of hydrogen peroxide. Chlorine dioxide vapor is a strong oxidant and microbiocide, and its use for sterilization of medical instruments has recently been investigated.

B. Liquid Sterilants

It is obvious that in many situations where sterilization is required, a liquid solution is more convenient to use than a highly volatile gas. For example, in medical and dental operations, heat-sensitive instruments may be soaked in a liquid sterilant rather than sterilized in an ethylene oxide sterilizer. The capital expense is much less and there is less hazard and less trouble. There are many chemical disinfectants but few chemicals that act as sterilizing agents. A disinfectant destroys microorganisms but not necessarily all of them. Bacterial spores and some parasite cysts are very resistant to chemical attack and may not be inactivated by disinfectants. They are killed, however, by sterilants that are therefore called sporicides. Actually, the same chemical may be both a disinfectant and a sterilizing agent. This will depend on its concentration of use, the time of contact, the temperature, and the presence of interfering substances such as blood. For a scalpel used in invasive surgery, sterility is imperative; in cleaning a hospital toilet, disinfection is sufficient—sterilization would be overkill. In medical application, critical devices that enter sterile tissue or the vascular system, or apparatus through which blood will flow, must be sterilized. In addition, equipment exposed to pathogens or used in germ-free rooms should be sterilized. Aseptic packaging for foods also requires sterilization. For medical instruments, a soak of 6–10 hr is usually given but this may be shortened if the temperature is raised.

Those chemicals recognized as sterilants are glutaraldehyde, hydrogen peroxide, peracetic acid, chlorine dioxide, and formaldehyde. All of these are disinfectants when used under other conditions, as mentioned previously. Chemicals that are effective disinfectants but are not included in the category of sterilants are alcohols, chlorine compounds, phenolic compounds, iodine compounds, and quaternary ammonium compounds.

1. Glutaraldehyde

A 2% solution of glutaraldehyde is used in hospitals and medical laboratories for sterilization of instruments such as cystoscopes and anesthetic equipment. The stock solution is supplied on the acid side to prevent polymerization but when ready for use an activator is added to bring the pH to 8–8.5. A soak of 10 hr is recommended, although research has shown that spores are inactivated in 3 hr. It should be noted that, as with heat sterilization and all other methods,

the recommended treatment is an overtreatment that provides a considerable margin of safety to allow for slow penetration, etc. Alkaline glutaraldehyde may be used for 2–4 wk, after which it must be discarded because the concentration is considerably reduced, even if the solution is not used. Glutaraldehyde is much more stable on the acid side, but although acid and neutral glutaraldehyde products have been developed they have not been as popular as the alkaline solution. To prolong the useful life of the solution, instruments should be clean and dry when immersed. Irritating fumes are produced when glutaraldehyde is heated. It does not have the problem of metal corrosion of the oxidant-type sterilizers.

2. Hydrogen Peroxide

Hydrogen peroxide in a 3% solution is well known as a skin antiseptic, but in a 6% concentration it is effective as a disinfectant and at 20% or greater it has found use as a sterilizing agent. Temperature increase, ultraviolet light, and copper ions stimulate the production of active hydroxyl radicals, which are believed to be responsible for the microbiocidal action. Hydrogen peroxide is normally quite stable but is readily decomposed by the enzyme catalase. In high concentration, it is a strong oxidant and must be used with caution because it is irritating to the skin and eyes. It does not harm the environment because it breaks down to oxygen and water. It has found application in sterilizing milk cartons for aseptic filling, for the sterilization of soft contact lenses and acrylic resin implants, and as a spray for mechanical ventilators. When used as a sterilizing bath, objects are soaked for 6 hr and the bath may be reused for 6 wk.

3. Peracetic Acid

Peracetic acid, the peroxide of acetic acid, is a very powerful oxidant and sterilant, more powerful antimicrobially than hydrogen peroxide. It is stabilized with hydrogen peroxide, acetic acid, and sulfuric acid with the concentration of hydrogen peroxide and acetic acid being possibly greater than the peracetic acid. Experiment has shown it to be a more potent sporicide than glutaraldehyde, hydrogen peroxide, formaldehyde, ethylene oxide, and several other chemicals, killing spores at concentrations from as little as 0.05%. It is used as a disinfectant in the food-processing and beverage industries. As a sterilant it is used for treating animal isolators and entry locks for germ-free animal experimentation. It is also used as a terminal disinfectant and sterilant for stainless steel and glass tanks, piping, tank trucks, and railroad tankers in the handling and transportation of fluid foods and beverages. In the preparation of pharmaceuticals, it permits the cold sterilization of emulsions, hydrogels, ointments, and powders. Like strong hydrogen peroxide, peracetic acid should be used with caution. Because it breaks down to give mainly acetic acid and oxygen, it has no residual toxicity or environmental hazard.

4. Chlorine Dioxide

Chlorine dioxide differs from chlorine, hypochlorites, and organic chlorine germicides in that it does not combine with ammonia or amines and is not reduced in activity in alkaline solutions. It is a stronger oxidant than chlorine and a stronger sporicide, acting more rapidly. It is not a stable compound and is generated where it is used; however, a stable complex has been developed. For sterilizing baths, objects should be soaked for 6 hr and the solution discarded daily. It has been used for gnotobiotic work. Chlorine dioxide has found its major use to date as a disinfectant for treatment of potable water in municipal plants and process water in paper mills. Because it does not produce trichloromethanes, it has a more favorable environmental aspect than other chlorine compounds. Like other strong oxidants, it is corrosive to metal and some plastics.

5. Formaldehyde

As a sterilizing agent, formaldehyde is used in a concentration of 6–8%. It is not considered as potent a sporicide as glutaraldehyde and owing to its many undesirable properties like odor and irritation, polymerization, and possible carcinogenicity, it has been displaced by glutaraldehyde as a liquid sterilant and is used primarily as a vapor sterilant. It is still used, however, to sterilize certain hemodialysis equipment.

V. Radiation Sterilization

Radiation sterilization is an optional treatment for heat-sensitive materials. This includes ultraviolet light and ionizing radiations.

A. Ultraviolet Light

Ultraviolet light is chemically active and causes excitation of atoms within the microbial cell, particularly the nucleic acids, producing lethal mutations, thus making the organism unable to reproduce. Bright sunlight has some microbiocidal activity due to ultraviolet radiation. The range of the ultraviolet spectrum that is microbiocidal is 2400–2800 angstroms. The sterilizing dose is measured as the product of intensity and time of exposure and is a function of the distance from the object, being inversely proportional to the square of the distance as in all electromagnetic radiations. There is a great difference in the susceptibility of microorganisms to ultraviolet radiation, with *Asperigillus niger* spores being 10 times as resistant as *Bacillus subtilis* spores, 50 times as resistant as *Staphylococcus aureus* and *Escherichia coli,* and 150 times as resistant as influenza virus.

Because most materials strongly absorb ultraviolet light, it lacks penetrating power and its applications are limited to surface treatments, with the exceptions of pure air and water, which permit some penetration. But any turbidity in the water or particles in the air greatly reduce the distance penetrated. Ultraviolet lamps in the air ducts in schools have been shown to prevent outbreaks of contagious diseases and ultraviolet has been used to purify drinking water and produce sterile water for surgery. Use of ultraviolet light in food storage rooms allows meat enzymes to tenderize the meat while the ultraviolet rays and ozone generated by them from the oxygen in the air prevent the growth of surface bacteria. With air and water, banks of lamps are employed to supply sufficient dosage to the fluid in the time it takes to flow past them since the energy of ultraviolet radiation is relatively low.

Much higher energy, 100 to millions of times greater, is generated by ionizing radiations. These include gamma-rays, high energy X-rays, and high-energy electrons.

B. Ionizing Radiations

Ionizing radiations, unlike ultraviolet rays, penetrate deep into the atom knocking out electrons, causing ionization. They may directly target the DNA of the cells or produce active ions and free radicals that react indirectly with DNA and the vital constituents of the microorganism preventing their functioning. Gamma radiation, used more often than x-rays or high-energy electrons for purposes of sterilization, are generated by radioactive isotopes, cobalt-60 being the usual source. Gamma-rays have the advantage of being completely reliable because they are generated continuously and there can be no mechanical breakdown. However, there are some disadvantages: the gamma-ray sterilizer cannot be shut off; there is a potential safety hazard; the cobalt-60 source itself is relatively expensive and has a half-life of only 5.3 years; and, more important, the sterilizer requires a major installation costing millions of dollars. Like x-rays, gamma-rays from a cobalt-60 source readily penetrate materials (2 ft of water) so that products may be in packages inside boxes and still be reached by the sterilizing rays. On the other hand, high-energy electrons, called beta-rays, can penetrate only 1 in. of water; therefore, they are useful only for shallow materials or surface treatment. Gamma radiation requires many hours of exposure for sterilization, but the high-energy electrons take only seconds. In the case of the electron accelerator that produces high-energy electrons, it may cost as little as one-tenth as much as the gamma-ray sterilizer, but it is a sophisticated piece of equipment that can generate unwanted x-rays. The energy of the gamma source is fixed by the emission of cobalt-60. The energy generated by the electron beam can be increased but reaches a limit because the treated materials could become radioactive. A sterilizing dose from a cobalt-60 source or from an electron beam generator is measured in megarads, a unit of energy absorbed per gram of material irradiated. Vegetative bacteria are sensitive to ionizing radiations taking a dose of 0.02–0.2 megarads to kill them whereas some bacterial spores, *Bacillus pumilus* in particular, are very resistant, taking a dose of 0.2–0.4 megarads to inactivate them. Certain factors influence the sensitivity of microorganisms. Organic matter such as dried serum and grease films protect the microbes, and anoxic conditions increase resistance two to five times. Radiation sterilization is usually carried out at room temperature but increases in temperature decrease resistance. Freezing the substrate produces a significant increase in resistance.

Because it is so penetrating, gamma radiation is used for treating many of the type of heat-sensitive products that are treated by gaseous sterilization. All sorts of medical materials and equipment, pharmaceuticals, biologicals, and laboratory equipment are on this list. In addition, industrial materials such as wool and hair contaminated with anthrax spores and hides are treated. Some plastics are unstable to the dosages of gamma-rays employed, but polyethylene, polystyrene, phenolics, and epoxies present no problems. Some chemicals and solutions are unstable and cannot be sterilized in this way. Glass turns color but is not otherwise affected.

For years there has been great interest in gamma radiation for sterilization and pasteurization of food and a great deal of research has been done, especially by the military. The fear of producing toxic radiolytic products in the food has held back the exploitation of this method except for a few materials such as spices and mushrooms.

In treating the products in a gamma-ray irradiator, sealed packages of items such as surgical scalpels, scissors, and dializers are packed in cardboard cartons and placed on a conveyor belt and circulated around the source so that they are irradiated from both sides for a sufficient time to receive a dosage of about 2.5 megarads, equal to 8D values for *B. pumilus* spores. With electron accelerators, flat sealed packages of materials like sutures and syringes on a conveyor belt pass under the thin window from which the beam is emitted.

VI. Sterilization by Filtration

Filtration is a useful method for sterilizing liquids and gases that excludes microorganisms rather than destroying them. It does not require heat, chemicals, or radiation; therefore, it does not damage sensitive substances nor pose a hazard to the operator. It removes unwanted particles in addition to microorganisms. The capital cost is low and many filters are disposable so no maintenance is required.

A. Depth Filters

There are essentially two types of filters, depth filters and membrane filters. The depth filter for sterilization was introduced by Pasteur and Chamberlin in France in 1884. It was a hollow unglazed ceramic candle made by sintering sand and clay. The Seitz filter, a pad developed a few years later in Germany, was made of asbestos fibers and was more convenient to use. This was followed by larger, fibrous filters using glass wool as well as asbestos, cotton, and wool, which were supported in rigid frames and could pass large volumes of liquids or gases. These filters are only partly dependent on the size of the filter's pores. They are aided by the irregular twisting and turning of the fibers and open channels through the filter by which the particles are slowed

1μ

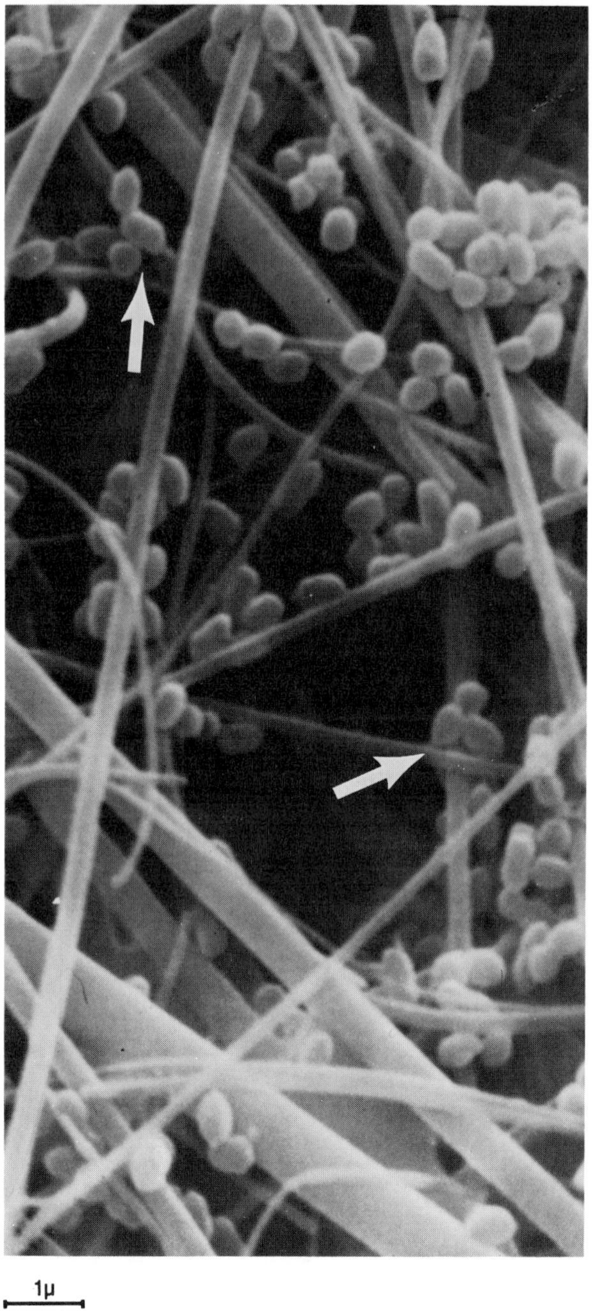

Figure 6 Scanning electron micrograph of *Pseudomonas diminuta* ATCC 19146 on a glass fiber-depth filter. Arrows indicate examples of bacterial cells adsorbed to fibers. [Courtesy of Millipore Corporation.]

and entangled. In addition there are electrostatic forces resulting in the adsorption of particles to the fibers. This is illustrated in Fig. 6, which shows bacteria entrapped by the fibers of a depth filter.

B. Membrane Filters

The membrane filter was introduced in 1922 but did not come into general use until decades later. It can be made of polymers like nitrocellulose, nylon, dacron, and teflon or, in some cases, of porous stainless steel. It differs from the depth filter, which entraps the particles. The membrane filter screens them out, collecting them on the surface of the filter as seen in Fig. 7. These filters depend largely on the size of the pores to determine their screening effectiveness. However, electrostatic forces are also at work for it has been demonstrated that a membrane filter having an average pore size of 0.8 μm will retain particulate matter as small as 0.05 μm. For removing bacteria, a pore size of 0.2 μm is commonly used. For retention of viruses and mycoplasmas, pore sizes of 0.01–0.1 μm are needed. Cocci and bacilli, run in size from about 0.3 to 1 μm in diameter; most viruses are 0.02–0.1 μm, with some as large as 0.25 μm. Most airborne microorganisms, however, are carried on dust particles larger than they are.

Membrane filters are easily clogged by particles larger than the pore diameter and must be cleaned or replaced. They are more reliable than depth filters because if the correct pore size is used bacteria, spores, or viruses can be sure to be excluded. Because the membrane is so thin, less of the filtrate is retained and lost than with depth filters. Because there is less adsorption, there is less change in the composition of the filtrate. Where the problem of clogging of the membrane occurs, it is customary to use a coarser prefilter to screen out the larger particles. The prefilter may be a membrane filter of 0.8–1.2 μm but a depth filter is an option because it is not so dependent on pore size and is not readily clogged, due to its depth.

Depth filters have little resistance to flow because the pore volume is only 20–30%. This makes the depth filter valuable in the purification and sterilization of air and compressed gases. Membrane filters are used for the commercial production of a number of pharmaceutical solutions, among which are parenterals, ophthalmic solutions, antibiotics, and heat-sensitive injectables. Serum for use in bacterial

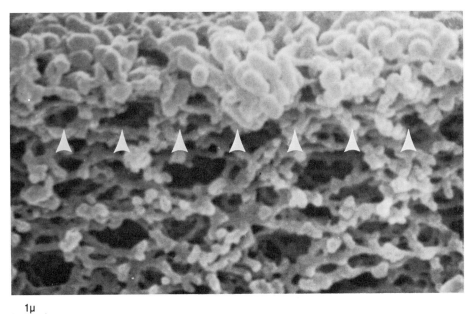

1μ

Figure 7 Cross-sectional electron micrograph of a 0.22-μm pore size membrane filter challenged with *Pseudomonas diminuta* ATCC 19146. Arrows indicate interface of membrane surface and internal polymer structure. A layer of bacterial cells is shown trapped at the surface. [Courtesy of Millipore Corporation.]

and viral culture media must be sterilized by filtration as well as some sugars that are unstable when heated. Even intravenous solutions that are sterilized with heat receive preliminary filtration to remove live or dead bacteria and other particular matter that could produce pyrogens when heated. In the clarification of wines, membrane filters of 0.45–1.2 μm are used to remove yeast cells. The filters themselves are steam-sterilized at 121°C for 30 min. Membrane filtration is valuable also in testing pharmaceutical and medical products for sterility. After passing the product through the filter, the filter is washed to remove preservatives or microbial inhibitors and then cultured on growth media for the presence of bacteria, fungi, or yeasts. Water and other materials are also tested for microbial content by this technique.

C. Filtration of Air

The large-scale production of sterilized air became necessary with the advent of penicillin and other antibiotics that are produced by aeration of large volumes of solution. Even a very small amount of contamination of the growth medium with fast-growing microbes could prove to be very costly. A

high efficiency of removal, 99.99–100%, is required because large volumes of air are employed in the fermentation. Filtration provides the only economical way of supplying this quantity of sterile air. For hospital operating theaters and isolation rooms for immune-suppressed patients, ultraclean air is also a necessity. Assembly rooms where microelectronic equipment such as computer chips are made represent still another area demanding air free of dirt and all particulate matter.

Fibrous depth filters, granular depth filters, paper sheet filters, and membrane filters are all employed to meet these stringent demands. Finely spun glass fibers or resin-treated wool are used to make depth filters for industrial fermentation purposes. The glass fibers are of uniform diameter of 0.06–6 μm. The fibrous materials are packed into large containers held under pressure from both ends. Packing should be uniform to prevent channeling, especially along the sides of the container. Depth filters are usually sterilized by steam for a period of 2 hr, but this may cause channeling and requires drying off any water that is produced because a wet filter permits organisms to penetrate. Therefore dry heat sterilization at 160–180°C is preferred if possible.

Low-velocity filters, which run at 0.2–0.5 ft/sec

are favored over high-velocity filters because they are not much affected by changes in air velocity as with high-velocity filters. They are economical to run because the pressure drop across them is low. Fine fibers are more efficient than coarse ones, the optimum size for glass wool being 6 μm or less. They must not be packed too densely or there will be resistance to air flow. A packing density of 17–25 lb/cubic ft is employed. The thickness of an air filter is determined by the reliability and sterility assurance desired. If a certain thickness removes 90% of the microbes in 1 in., 90% of those remaining will be removed in the second inch, and 90% in the third. This logarithmic relationship assumes that there is uniform fiber thickness, packing density, and constant air velocity. Glass wool filters operating at a low velocity of <0.5 ft/sec will achieve 99.99% removal at a depth of 3 in.

High-efficiency particulate air (HEPA) filters are employed where requirements are very high for air free of microorganisms or other suspended particles. This includes laboratory rooms where work on contagious pathogens is conducted, isolation rooms in hospitals, industrial premises where sterile pharmaceuticals are produced, gnotobiotic modules for germ-free animal research, and ultraclean rooms for assembly of special electronic components. HEPA filters have efficiencies of 99.97–99.997% retention of particles 0.3 μm or larger. Actually, bacteriophages as small as 0.05 μm have been shown to be removed. HEPA filters are used to help meet the class 100 clean room standard of not more than 3.5 airborne particles per liter of 0.5 μm or greater size. Laminar air flow systems and ultraviolet lamps are frequently used in conjunction with banks of HEPA filters. HEPA filters are made of large paper sheets constructed from glass or cellulose fibers up to 5 μm in diameter. These sheets are pleated around supporting structures, sealed at all edges to prevent leakage, and placed in a container or frame. With this construction they have minimal resistance to air flow. They are usually used with a prefilter to improve efficiency and extend the useful life to 3–4 yr.

Granular carbon of 1–2 μm is sometimes used in filters to provide sterile air for industrial fermentations. These are large, low-velocity filters with a pressure drop of 1–5 lb/in.^{2}. Membrane filters can also be used for air purification, but their high pressure drop restricts their use to small-scale applications such as supplying sterile air to tissue culture systems.

VII. Testing Sterilizing Agents

It is necessary to know whether or not the sterilizing agent is effectively doing the job intended for it. Each method of sterilization is different and each requires its own test procedures.

For heat sterilization, the first requirement is knowledge that the temperature is recording correctly. Furthermore, the operator must know that the temperature is reaching all parts of the load and is maintained for the desired length of time. Recording thermometers (and barometers for steam sterilizers) are employed for the chamber and thermocouples may be buried inside the load. Paper strips treated with chemicals that change color at the required temperature may be used. If, in a steam autoclave, a container is tightly closed and receives no steam, it may reach the correct temperature, but this will not ensure that sterilization will occur. To give this assurance requires the use of biological testing in the form of heat-resistant spores. With moist heat, spores of *B. stearothermophilus* are used, and with dry heat sterilizers, spores of *B. subtilis* var. *niger* are selected. The spores are dried on paper treated with nutrient medium and chemicals. After the sterilization treatment, they are incubated for germination and growth and a color change indicates whether they have or have not been activated. This method may take several days of incubation whereas the chemical and physical methods are immediate, but the biological tests are more dependable.

For chemical sterilization, there are color indicator tapes for ethylene oxide and formaldehyde, which show whether or not these gases have penetrated in sufficient quantity at the prescribed temperature to provide sterilization. But here, as with other methods, the biological tests are preferable. With ethylene oxide, strips treated with *B. subtilis* var. *niger* are employed, whereas with formaldehyde *B. stearothermophilus* is used. In the case of liquid sterilants, spores of *B. subtilis* var. *niger* and *Clostridium sporogenes* are both tested. The spores are dried onto carriers that may be porcelain or stainless steel penicylinders or suture loops. They are exposed to the solution of the chemical sterilant at the desired concentration, given a specified temperature and time of immersion, after which they are incubated in a rich medium for germination and growth. These tests are replicated many times and if any of the replicates show growth the candidate sterilant fails the test.

In radiation sterilization, the radiation dose can be measured by color indicator discs, by chemicals that undergo oxidation-reduction followed by chemical analysis, or by pellets that darken in proportion to the dosage received. In the biological testing, it is found that *B. pumilus* is more resistant to radiation than the species employed in other methods and is therefore used for this test.

With filtration sterilization, membrane filters may be readily tested for passage of microorganisms of different sizes. Spores of *B. subtilis* var. *niger,* cells of *Pseudomonas diminuta,* bacteriophages, and other viruses give a range of sizes. The bubble point test can also be used. It correlates the pore diameter with the air pressure required to cause the first bubble to break through a filter. Depth filters may be tested by the passage of selected organisms or by the penetration of aerosols made up of chemical dusts of known particle size. For example dry particles of sodium chloride can be detected and quantitatively determined with a hydrogen flame photometer.

Bibliography

Avis, K. E., and Akers, M. J. (1986). Sterilization. *In* "The Theory and Practice of Industrial Pharmacy," 3rd ed. (L. Lachman, H. A. Lieberman, and J. L. Kanig, eds.). Lea & Febiger, Malvern, Pennsylvania.

Block, S. S. (ed.) (1991). "Disinfection, Sterilization, and Preservation," 4th ed. Lea & Febiger, Malvern, Pennsylvania.

Borick, P. M. (ed.) (1973). "Chemical Sterilization." Dowden, Hutchinson & Ross, Stroudsburg, Pennsylvania.

Gardner, J. F., and Peel, M. M. (1986.) "Introduction to Sterilization and Disinfection." Churchill Livingstone, Melbourne, Australia.

Marino, F. J., and Benjamin, F. (1986). Industrial sterilization. *In* "Pharmaceutical Dosage Forms: Parenteral Medications," Vol. 2, 2nd ed. (K. E. Avis, L. Lachan, and H. A. Lieberman, eds.) Marcel Dekker, New York.

Phillips, G. B., and Miller, W. S. (eds.) (1973). "Industrial Sterilization." Duke University Press, Durham, North Carolina.

Sulfide-Containing Environments

Rutger de Wit
Centro de Investigación y Desarrollo, Barcelona

I. Sulfide in Metabolism
II. Overview of Sulfide-Containing Ecosystems
III. Taxonomy and Ecological Physiology of Sulfide-Consuming Microorganisms
IV. Communities and Interactions in Sulfide-Containing Environments

Glossary

Anoxygenic photosynthesis Light-driven metabolic reduction of small carbon compounds (often CO_2) using electrons derived from a compound other than water, and, consequently, oxygen is not produced; typical electron donors include H_2S, other reduced sulfur compounds, H_2, and small organic substrates; this type of photosynthesis has been found only among bacteria

Benthos Community consisting of organisms growing on the sediment

Electron acceptor Compound that is reduced in a metabolic redox reaction

Electron donor Compound from which electrons are derived in a metabolic redox reaction, thus resulting in the oxidation of the electron donor

Meromixis Condition in which stratification of the water mass in a lake is maintained during the whole year, often due to a solute concentration gradient

Microbial mat Sediment ecosystem with very high populations densities of microorganisms; often, the top millimeters comprise a clearly laminated structure: (1) the top layer is dominated by oxygen-producing phototrophs, especially cyanobacteria, (2) the anoxic bottom layer is rich in sulfide as a result of the degradation of organic matter by fermenting and sulfate-reducing bacteria, and (3) phototrophic and/or chemotrophic sulfur bacteria sandwich in between the top and bottom layers, forming a fine lamina at the oxygen–sulfide interface

Syntrophy Simultaneous growth of two or more populations that depends on interspecies crossfeeding (product of species A serves as substrate for species B)

Trophic mode (-trophy) Designation of the different nutritional types (Fig. 1)

SULFIDE-CONTAINING ENVIRONMENTS are found in aquatic and sediment ecosystems and are inhabited by predominantly prokaryotic communities. In some places, geochemically formed sulfide emerges into the biosphere, but more commonly sulfide is biologically produced in anoxic environments by sulfate-reducing bacteria. Sulfide is a toxic compound, but it is used as a substrate for growth in certain bacterial groups. These bacteria comprise the following:

1. Phototrophic sulfur bacteria that oxidize sulfide in a light-driven reaction and use the electrons to reduce CO_2 for subsequent biosynthesis.
2. Chemotrophic sulfur bacteria that oxidize sulfide in a chemical reaction with oxygen or nitrate as electron acceptors and thereby obtain metabolically useful energy.

Hence, phototrophic sulfur bacteria bloom in sulfide-containing environments if sufficient light is present, whereas chemotrophic sulfur bacteria proliferate where sulfide and oxygen coexist.

I. Sulfide in Metabolism

A. Sulfur in Living Organisms and Assimilatory Sulfate Reduction

Sulfur is an essential element for all living organisms. The content of sulfur in biomass averages about 1%, but some variation is species-specific or caused by environmental conditions. Most of the

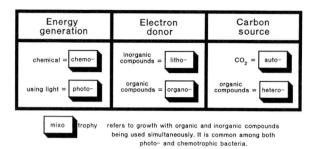

being used simultaneously. It is common among both
photo- and chemotrophic bacteria.

Figure 1 General nutritional types found among bacteria.
The bacteria encompass a much broader spectrum of
nutritional types than that found among animals and green
plants. Consequently, the concepts of heterotrophy and
autotrophy originally established for plants and animals are
insufficient to characterize the diverse types found among
microorganisms. Presently, combined summary designations
are commonly used in microbiology and refer to the
mechanism used for the generation of metabolic energy, the
electron donors, and the carbon source. Hence, a purple
sulfur bacterium growing in the light on sulfide and CO_2 is a
photolithoautotroph, and a *Thiobacillus*, which generates
energy from sulfide oxidation and uses CO_2 as the only
carbon source, is a chemolithoautotroph. Bacteria that
respire or ferment organic compounds, such as *Escherichia
coli*, are correctly designated as chemoorganoheterotrophs.

It is not always necessary to mention both the electron
donor and the carbon source. Almost all organisms that use
organic compounds as electron donors also use organic
compounds as the carbon source, and, consequently,
these organisms are referred to as chemoorgano- or
photoorganotrophs.

In many organisms, lithotrophy is coupled to autotrophy;
these organisms are sometimes designated as chemolitho- or
photolithotrophs. However, among the phototrophic and
chemotrophic sulfur bacteria, this rule of thumb frequently
does not hold, because an extremely wide variety of
metabolic types has been found. Therefore, the more
elaborate designation in this figure is especially valuable
when dealing with these organisms.

When dealing with large groups of organisms, shorter
designations can be used if such groups contain different
metabolic types. The major distinctive feature among the
sulfide-oxidizing bacteria is the energy-generating
mechanism. Accordingly, the main groups are designated as
phototrophic sulfur bacteria and chemotrophic sulfur
bacteria.

sulfur in organisms occurs in proteins, namely as a
component of cysteine and methionine, and as a
component in biologically important cofactors such
as coenzyme A, thiamine, biotin, lipoic acid, and the
ferredoxins. In these substances, sulfur occurs in its
reduced state with valence -2, or -1 as, for example, in disulfide bonds. However, in oxic environments, the thermodynamically stable and most
abundant sulfur source occurs in an oxidized state as
sulfate (valence of sulfur $= +6$). All green plants,
fungi, and most bacteria can reduce sulfate to the

level of sulfide for biosynthetic purposes; this
process is referred to as assimilatory sulfate reduction. Animals and many heterotrophic bacteria obtain organic reduced sulfur compounds from their
food. Not unexpectedly, some bacteria highly specialized to sulfide-containing environments cannot
perform assimilatory sulfate reduction but, rather,
take sulfide directly from their environment.

B. Sulfide Formation by Sulfate-Reducing Bacteria

In anoxic environments of low redox potential, hydrogen sulfide is the thermodynamically stable sulfur compound. Hence, a specialized group of bacteria known as the sulfate-reducing bacteria can
obtain metabolically useful energy from the anaerobic oxidation of small organic molecules or H_2
coupled to the reduction of sulfate to sulfide. This
process, referred to as dissimilatory sulfate reduction, generates many times more sulfide than required for biosynthesis. Consequently, when actively growing, sulfate-reducing bacteria discharge
huge amounts of sulfide into their anoxic environments. However, sulfate-reducing bacteria obviously require sulfate. In many freshwater ecosystems, this ion is very scarce (<0.3 mM), thus not
allowing for the proliferation of a significant population of sulfate-reducing bacteria.

The degradation of organic material in anoxic environments is a rather complex process in which
different and mutually interacting bacteria interfere.
Biopolymers are first converted by fermenting bacteria into small organic molecules and hydrogen.
Subsequently, these products are substrates for
sulfate-reducing bacteria and methanogenic bacteria. Sulfate-reducing bacteria outcompete methanogenic bacteria if sufficient sulfate is available.
Consequently, in sulfate-rich anoxic environments,
all reducing equivalents from degradable biomass
finally flow into dissimilatory sulfate reduction and
appear as sulfide in the environment. Some sulfide
also appears as a consequence of a desulfuration of
reduced sulfur from organic compounds. However,
it can be calculated that this quantity is small with
respect to the quantities formed by dissimilatory
sulfide reduction (e.g., 1 kg of biomass would yield
10 g of H_2S by desulfuration and 570 g of H_2S by
sulfate reduction, provided that sulfate is not
growth-limiting). Thus, desulfuration is a major
source of sulfide only in methanogenic freshwater
environments. Due to abiotic and biotic processes,

inorganic sulfur compounds of intermediary oxidation states (e.g., elemental sulfur, polysulfides, thiosulfate) appear in the environment. Some sulfide can be formed from sulfur by the dissimilatory sulfur-reducing bacteria. These bacteria are metabolically similar to the sulfate-reducing bacteria but can only use sulfur instead of sulfate as the electron acceptor.

C. Sulfide Consumption by Phototrophic and Chemotrophic Sulfur Bacteria

The sulfide present in the environment is probably consumed by many organisms as a source of sulfur for biosynthetic purposes; however, the use of sulfide as an electron donor by lithotrophic sulfur bacteria is usually quantitatively predominant. The oxidized sulfur compounds formed by these bacteria are not incorporated in organic molecules but, instead, accumulate as sulfur in intra- or extracellular globules or are discharged into the environment. The major redox reactions performed by these organisms are summarized in Table I and Fig. 2. The synthesis of reduced cell material from CO_2 with H_2S as the electron donor is an exergonic reaction, as demonstrated for glucose in Table I. Consequently, lithoautotrophic sulfur bacteria need an energy-generating system to pursue growth on sulfide and CO_2. The nature of this energy-generating system is the distinctive feature among the different groups of sulfur bacteria. In phototrophic bacteria, energy is generated from light, which is harvested by photosynthetic pigments consisting of bacteriochlorophylls and carotenoids. These pigments provide these bacteria with colors that, at high population densities, are visible to the naked eye. Colorless chemotrophic sulfur bacteria obtain energy from a chemical reaction, which, in fact, represents a respiratory type of metabolism; the electrons from sulfide are transposed to oxygen or nitrate (see Table I). Phototrophic sulfur bacteria grown in the light use all electrons from sulfide for CO_2 fixation, but, in contrast, chemolithotrophic sulfur bacteria fuel most of the reducing equivalents into respiration, thus obtaining less biomass per amount of sulfide consumed (see Fig. 2).

Most sulfur bacteria can use small organic compounds as a carbon source for biosynthesis. However, many of these substrates, such as acetate, are still too oxidized and need metabolic reduction prior to incorporation in biomass. Sulfide is used as an electron donor in this energy-requiring reaction.

Such growth is referred to as lithoheterotrophy and is common among both phototrophic and chemotrophic sulfur bacteria.

1. Phototrophic Sulfur Bacteria

The famous Dutch microbiologist Van Niel stressed the similarities between oxygenic plant-type photosynthesis and the sulfide-dependent bacterial type of photosynthesis. His generally accepted proposal that CO_2 is reduced with water or hydrogen sulfide as the respective electron donors led to a breakthrough in research on photosynthesis. Light of the appropriate wavelength is absorbed by the photosynthetic pigments and processed in the photosystems, whereas CO_2 fixation occurs in a so-called dark reaction, which requires adenosine triphosphate (ATP) as metabolic energy and the reduced cofactors reduced nicotinamide adenine dinucleotide (NADH) or NADPH as electron donors. Plants, eukaryotic algae, cyanobacteria, and prochlorophytes perform oxygenic photosynthesis in which oxygen is produced from water mediated by two photosystems coupled in a series that comprises the well-known Z-scheme. In contrast, phototrophic sulfur bacteria only contain one photosystem. The light-energy harvested by carotenoids or bacteriochlorophyll is subsequently transposed to a minority of bacteriochlorophyll molecules localized in the reaction centers. An electron from the reaction center bacteriochlorophyll molecule is thus excited to a higher energy state and enters in an electron transport chain (see Fig. 2). Phototrophic sulfur bacteria belong to two distinct groups, namely the purple sulfur bacteria and the green sulfur bacteria (see Table III).

Purple sulfur bacteria contain bacteriochlorophyll *a*, or bacteriochlorophyll *b* in a few species, and carotenoids. In purple sulfur bacteria, the generation of metabolic energy is mediated by photosynthesis through cyclic electron transport. Nicotinamide adenine dinucleotide (NAD^+) is reduced through an energy-consuming reversed or "uphill" electron flow. Thus, light is only indirectly involved in NAD^+ reduction. Apparently, reversed electron flow is required because the redox potential of the first photosynthetically reduced compound is not sufficiently negative to reduce NAD^+.

Green sulfur bacteria contain bacteriochlorophyll *c, d,* or *e* and carotenoids as light-harvesting pigments and a small amount of bacteriochlorophyll *a* localized in the reaction center. In contrast to purple sulfur bacteria, photosynthesis in green sulfur bacteria mediates both energy generation through cyclic

Table I   Standard Free Energy of Metabolic Reactions at pH = 7 ($\Delta g°'$)

Bacterial group	Reaction	kJ/reaction	kJ/mol O_2
Aerobic chemoorganotrophic bacteria	Glucose + 6 O_2 → 6 CO_2 + 6 H_2O	−2865	−477
Chemolithotrophic sulfur bacteria	H_2S + $\frac{1}{2}$ O_2 → $S°$ + H_2O	−204	−407
	$S°$ + $1\frac{1}{2}$ O_2 + H_2O → SO_4^{2-} + 2 H^+	−583	−388
	H_2S + 2 O_2 → SO_4^{2-} + 2 H^+	−786	−393
	H_2S + 1.6 NO_3^- → SO_4^{2-} + 0.8 N_2 + 0.8 H_2O + 0.4 H^+	−736	
All lithoautotrophic sulfur bacteria	3 H_2S + 6 CO_2 + 6 H_2O → glucose + 3 SO_4^{2-} + 6 H^+	+506	

electron transport and direct NAD^+ reduction through linear electron transport. The electrons consumed by linear electron transport are replenished from sulfide oxidation.

2. Chemotrophic Sulfur Bacteria

Chemotrophic sulfur bacteria do not contain photosynthetic pigments and are often referred to as

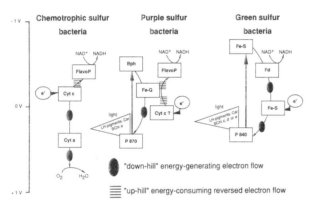

Figure 2   Electron transport in chemotrophic and phototrophic sulfur bacteria. Electrons are derived from reduced sulfur compounds (see Table I). Note that (1) in chemotrophic sulfur bacteria, most of the electrons are used for respiration (''down-hill) flow), while a part is used to reduce nicotinamide adenine dinucleotide (NAD^+), (2) in purple sulfur bacteria, energy is generated by photosynthetic cyclic electron transport, while NAD^+ is reduced through reversed linear electron flow, and (3) in green sulfur bacteria, photosynthesis directly mediates both cyclic electron transport and NAD^+ reduction. BChl, bacteriochlorophyll; Bph, bacteriophaeophytine; Car, carotenoids; Cyt, cytochrome; Fd, ferredoxin; Fe-Q, iron–ubiquinone complex; Fe-S, iron–sulfur protein; FlavoP, flavoprotein; LH-pigments, light-harvesting pigments; NADH, reduced nicotinamide dinucleotide; P 840, reaction center molecule of BChl a with absorption maximum at 840 nm; P 870, reaction center molecule of BChl a with absorption maximum at 870 nm. ●, ''down-hill'' energy-generating electron flow; ≡, up-hill energy-consuming electron flow.

colorless sulfur bacteria. Energy is generated in a redox reaction with oxygen as the electron acceptor, but nitrate is an alternative electron acceptor in some species. The electrons from sulfide are accepted by cytochrome c and subsequently run downhill through the cytochrome section of the respiratory chain coupled to energy formation. Because NAD^+ cannot be directly reduced by reduced cytochrome c, reversed or uphill energy-consuming electron flow occurs in a mode similar to that occurring in the purple sulfur bacteria. As in plants and purple sulfur bacteria, CO_2 reduction subsequently occurs through the Calvin cycle.

D. Sulfide Toxicity

Sulfide is a toxic compound for all living organisms; surprisingly, this is also true for sulfide-producing and sulfide-consuming organisms. However, the sensitivity to sulfide is extremely variable among different species: While most aerobic organisms are seriously inhibited by 0.1 mM sulfide, some green sulfur bacteria grow well up to 10 mM sulfide. Several characteristics probably contribute to the sensitivity to sulfide.

For oxygenic photosynthesis, it has been clearly demonstrated that sulfide dramatically blocks the activity of photosystem II (the water-splitting photosystem). An exception to this rule is found in some cyanobacteria from sulfide-containing environments that possess a slightly modified photosystem II.

Another characteristic of sulfide is its strong tendency to precipitate metal ions forming highly insoluble salts, especially with Fe^{2+}, Zn^{2+}, and Mn^{2+}. As a result, these cations, which are required in many metabolic reactions, are highly inaccessible.

The deleterious effects of sulfide are pH-dependent. Hydrogen sulfide is a weak acid and oc-

Table II Main Natural Ecosystems with Sulfide-Containing Environments

Ecosystem	Geographical occurrence	Most conspicuous sulfide-oxidizing populations[a]
Inland Environments		
1 Stratified lakes water-rich in SO_4^{2-}	Karstic regions, glacier valleys	Planktonic plate of purple and green SB at depth
2 Inland salt lakes	Endorheic[b] basins in semiarid climates	Benthic purple SB and *Beggiatoa* (microbial mats)
3 Eutrophic lakes and sediments	Ubiquitously	Chemotrophic or purple SB dependent on conditions
4 Local accumulations of organic matter	Ubiquitously	Dependent on conditions
5 Volcanic lakes	Volcanic regions	Acid-tolerant thiobacilli
6 Sulfide-containing springs	Volcanic and karstic regions	Purple SB and filamentous chemotrophic SB (thermoacidophilic archaebacteria)
Coastal Environments		
7 Tidal sediments; sediments of coastal lagoons	Ubiquitously on nonrocky coasts	On fine sand: laminated ecosystems with purple SB; on clay: chemotrophic SB
8 Stratified lagoons	Both fresh- and seawater inflow	Planktonic plate of green SB at depth
9 Evaporative hypersaline environments	Seawater inflow or seepage in (semi)arid regions	Laminated sediment ecosystem with purple or chemotrophic SB
Marine Environments		
10 Stratified watermasses	Seas and fjords surrounded by land (e.g., Black Sea)	Planktonic plate of chemotrophic/green SB
11 Hydrothermal vents	Ocean floor tectonicus spreading center	Chemotrophic SB in (1) the vent, (2) symbiotic association with inverbrates, and (3) the benthos

[a] SB, sulfur bacteria.
[b] Endorheic basins do not pertain to river catchment areas.

curs at neutral pH both in undissociated form and as HS^-. The undissociated acid diffuses passively through the cellular membrane, and, consequently, sulfide is more toxic at lower pH values.

The growth response of sulfide-oxidizing bacteria as a function of the sulfide concentration is an optimum curve (Fig. 3). At low concentrations, growth is limited by sulfide, and, consequently, the growth rate increases hyperbolically with increasing substrate concentration until a maximum value is reached. In contrast, at high concentrations, the toxic effects of sulfide determine the growth rate, and, consequently, the growth rate decreases with increasing substrate concentration. Above a given sulfide concentration, growth is not possible.

II. Overview of Sulfide-Containing Ecosystems

At some places, sulfide chemically formed in hot, deep layers of the earth (magma) emerges into the biosphere through volcanic activity. In such places, unique microbial ecosystems are found. However, most of the sulfide in the biosphere is formed by sulfate-reducing bacteria. Sulfate-reducing bacteria become active in habitats that have the following three prerequisites:

1. Sufficient input of organic matter
2. Restricted input of oxygen
3. Sulfate present in sufficient amounts

The coincidence of both conditions 1 and 2 allows the establishment of required anoxia.

Generally, sulfide-containing environments are not closed ecosystems but, rather, are in close contact with oxidized environments. Sulfide tends to migrate into the oxidized environments, while oxidized compounds such as oxygen and nitrate migrate in the opposite direction. As a result, most ecosystems with sulfide-containing environments contain gradients and can be divided into the anoxic sulfide-containing habitats, the oxygen–sulfide coexistence zone, and a completely oxidized habitat.

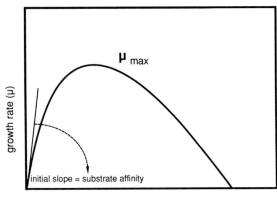

Figure 3 Relation between the specific growth rate of a sulfide-oxidizing lithotrophic bacterium and the sulfide concentration. The substrate affinity (initial slope of the curve) is decisive for competition for sulfide at a very low concentration: Species with a low affinity are outcompeted by species with higher affinities. Note the strong toxic effect of sulfide at high concentration.

Table II gives an overview of the different natural ecosystems where sulfide-containing habitats are found. The table also mentions the most conspicuous sulfide-oxidizing populations typical for the different ecosystems; however, see Sections III and IV for a detailed discussion of the specific ecological factors determining the occurrence or absence of different populations and the community structure. This section concentrates on a description of the geological and the geographical aspects defining the ecosystems on a global scale.

Anoxic conditions occur in wet sediments and in aquatic environments. In wet sediments, oxygen transport is limited to slow molecular diffusion, and, most often, oxygen consumption exceeds oxygen input. In contrast, in aquatic environments, atmospheric and photosynthetically produced oxygen is rapidly transported downward due to mixing. However, mixing diminishes when the water body becomes stratified due to a density gradient. These density gradients are a reflection of a temperature gradient or of a solute concentration gradient. Temperature gradients occur only in certain periods of the year, especially in summer. Hence, thermally stratified lakes are completely mixed once or twice a year. In contrast, solute concentration gradients are normally persistent throughout the year, and, consequently, the lake is stratified year round. This phenomenon is referred to as meromixis. The hypolimnion in thermally stratified systems and the

monimolimnion in meromictic systems are the deep stagnant water layers that do not mix with the overlying water. These layers normally are or become anoxic with the exception of very oligotrophic ecosystems.

In many freshwater systems, sulfate reduction is restricted due to the very low sulfate concentrations (often <0.3 mM). Even if this small amount were completely reduced to sulfide, it is still unlikely that free sulfide appears because most often it will be completely bound to iron. In contrast, the sulfate concentration in marine environments is nonlimiting for sulfate-reducing bacteria (e.g., the sulfate content of normal seawater of 33% salinity is 28 mM). Hence, sulfide-containing environments are extremely common in coastal environments, but their occurrence in inland systems is determined by geological factors or local conditions.

From the aforementioned characteristics, it becomes clear that stratified inland lakes with sufficient sulfate are likely to contain sulfide in the hypolimnion. Deep lakes in glacier valleys contain sufficient sulfate if gypsum is present in parent material or in sediments originated from eroded rocks. Lakes with sulfide and phototrophic bacteria are found in gypsum–carbonate karstic systems. Both thermally stratified and meromictic lakes are especially common and have been extensively studied in the eastern karstic regions of Spain.

Endorheic basins are areas that do not pertain to a river catchment area and are common in semiarid regions. Rainwater filters through the ground and accumulates in the lowest parts of these basins, often forming permanent or epimeral lakes that are subjected to strong evaporation. As a result, salts, nutrients, and frequently sulfate accumulate in these lakes. Inland salt lakes are extremely variable with respect to ionic composition as a reflection of the local geological conditions. Sulfide is most often present in the sediments of these lakes. Some of the deeper lakes are meromictic and harbor a sulfide-containing monimolimnion. Examples of meromictic inland salt lakes with planktonic phototrohic sulfur bacteria are Big Soda Lake (United States) and Lake Chiprana (Spain).

In some freshwater habitats, originally low in sulfate, the actual presence of sulfide is a direct consequence of eutrophication and/or contamination. In some cases, human, industrial, and agricultural activities result in direct discharge of sulfide in natural ecosystems. Eutrophication contributes both to increased sulfate levels and to increased biomass, which upon death represents organic material available for degradation. The emissions of sulfur oxides into the atmosphere due to the common use of fossil fuels by humans resulted in continuously increasing sulfate concentrations in freshwater systems during the last century. The effects of eutrophication and sulfate deposition are clearly reflected in the development of Lake Vechten (The Netherlands). During the last 40 years, total sulfur in this lake progressively increased and presently moderate sulfide concentrations occur in the hypolimnion, allowing planktonic phototrophic sulfur bacteria to proliferate.

The local accumulation of organic material also creates sulfide-containing habitats. The sulfide can be formed mainly by desulfuration of organic compounds. An extreme example of such an environment is a humanmade anaerobic sewage treatment plant. The sulfide formed in these systems is a nuisance.

Sulfide of geochemical origin is found in volcano lakes and thermal springs and geysers. Acid volcano lakes are found in Japan, while sulfide springs occur in Yellowstone National Park (United States), Iceland, New Zealand, and the Soviet Union. Some of the springs are extremely hot and acidic.

Sulfide-containing environments are especially common in coastal environments. The sediments occurring in the tidal fringe and in coastal lagoons almost always contain sulfide. Meromictic coastal lagoons are formed when fresh continental water stratifies on salty seawater. Such environments are known from the Mediterranean and tropical regions and generally are inhabited by green sulfur bacteria.

Evaporative hypersaline environments are found along the seacoast in semiarid and arid climates. Often they represent ephemeral shallow lagoons with sulfide mainly localized in the sediment. Solar Lake (Egypt) has been extensively studied. This lake is stratified during the winter period when its hypolimnion contains 1.5 mM sulfide. Both benthic and planktonic phototrophic sulfur bacteria proliferate in this lake. Many of the coastal hypersaline ecosystems have been converted into salterns. These systems are maintained and manipulated by humans to obtain salt from seawater, but they still closely resemble the original ecosystems. Sulfide is prominent in the sediments.

Stratified seawater masses sometimes occur in fjords and small seas with a strong influence from the continent. Under these circumstances, seawater diluted by rivers stratifies on undiluted seawater. The

best-known example is the Black Sea, which is continuously stratified and inhabited by chemotrophic and green sulfur bacteria.

Extremely spectacular underwater sulfide-containing ecosystems were discovered in the mid-1970s. At the Galapagos Rift ocean tectonic spreading center, hydrothermal vents were found at depths of approximately 2500 m. Hot fluid containing geochemically formed sulfide is mixed with oxygen containing seawater and emitted at a high flow rate into the ocean. The coexistence of sulfide and oxygen provides an ideal habitat for chemotrophic sulfur bacteria, and in fact populations have been found in three types of ecological niches: within the vent system itself, in symbiotic associations with invertebrate animals, and on various surfaces within the plume of the hydrothermal fluid.

The chemotrophic sulfur bacteria are the primary producers in this ecosystem, thus, constituting the base of the food chain. Hence, this unique deep-sea ecosystem is independent of phototrophic organisms and, instead, thrives on geochemical energy.

III. Taxonomy and Ecological Physiology of Sulfide-Consuming Microorganisms

A. Phototrophic Sulfur Bacteria

1. Taxonomy

The phototrophic sulfur bacteria are divided in two main groups: the purple sulfur bacteria and the green sulfur bacteria. This distinction is based on pigment composition and organization of the photosynthetic apparatus (see Section I.C.1 and Fig. 2). The families and genera with their major characteristics are listed in Table III.

The purple sulfur bacteria comprise two families, namely Chromatiaceae and Ectothiorhodospiraceae. Pigment composition and photosynthetic apparatus are similar to those of the purple nonsulfur bacteria. All purple sulfur bacteria can oxidize sulfide completely to sulfate; however, upon exposure to sulfide, the sulfide is initially oxidized to zero-valence sulfur, which appears as highly refractile globules. Subsequently, when sulfide has dropped to very low levels, the sulfur is oxidized to sulfate. The sulfur globules appear intracellularly in

Table III　Different Groups, Families, and Genera of Phototrophic Sulfur Bacteria and Their Major Characteristics

Genus	Morphology	Motility[a]	Gas vacuoles
Purple Sulfur Bacteria			
Family Chromatiaceae (pigments: bacteriochlorophyll a or b^b and carotenoids; division by binary fision; gram-negative; CO_2 fixation by the Calvin cycle; sulfur globules inside the cells[c])			
Chromatium	Rods to ovoid cells	+	−
Lamprobacter	Rods	+	+
Thiodictyon	Rods	−	+
Thiospirillum	Spirilloid cells	+	−
Thiocystis	Spherical cells	+	−
Thiocapsa	Spherical cells	−	−
Lamprocystis	Spherical cells	+	+
Amoebobacter	Spherical cells	−	+
Thiopedia	Spherical cells in platelets	−	+
Family Ectothiorhodospiraceae (pigments: bacteriochlorophyll a or b^b and carotenoids; division by binary fission; gram-negative; CO_2 fixation by the Calvin cycle; sulfur globules outside the cells[c])			
Ectothiorhodospira	Spirilloid to curved cells	+	+ or −
Green Sulfur Bacteria			
Family Chlorobiaceae (pigments: bacteriochlorophyll c, d, or e plus a small amount of bacteriochlorophyll a and carotenoids; chlorosomes[d]; division by binary or binary and ternary fission; gram-negative; CO_2 fixation by reverse citric acid cycle; sulfur globules outside the cells[c])			
Chlorobium	Straight or curved rods	−	−
Pelodictyon[e]	Straight, curved or ovoid cells	−	+
Prosthecochloris	Starlike with extrusions, prosthecae	−	−
Ancalochloris	Starlike with extrusions, prosthecae	−	+
Chloroherpeton	Long unicellular flexible rods	Gliding	+

[a] +, motility by flagella unless otherwise indicated; −, motility absent.

[b] Only a few species.

[c] Zero-valence sulfur is formed during the oxidation of sulfide and occurs as characteristic refractile globules visible with the light microscope.

[d] Chlorosomes are intracellular nonunit membrane enclosed vescicles in green phototrophic bacteria that contain bacteriochlorophyll c, d, or e and underlie the cytoplasmic membrane where bacteriochlorophyll a and the reaction center pigments are located. In contrast, in purple sulfur bacteria, the photosynthetic pigments are located in intracytoplasmatic membrane systems that are continuous with the cytoplasmic membrane.

[e] *Pelodictyon* cells often form tridimensial nets.

the Chromatiaceae and extracellularly in the Ecto-thiorhodospiraceae. The Chromatiaceae comprise 9 genera with 24 described species; the Ectothiorhodospiraceae only comprises the genus *Ectothiorhodospira* with 6 species.

The green sulfur bacteria only comprise the family Chlorobiaceae with 5 genera and 24 described species. Sulfur globules are always outside the cell. Pigment composition and organization of the photosystem are very similar to those in the green filamentous bacteria (Chloroflexaceae).

The recent introduction of molecular biological techniques, especially 16-S-ribosomal RNA sequencing, has contributed to new insights on the phylogeny of phototrophic sulfur bacteria. The purple sulfur bacteria belong to a phylogenetically homogeneous group, albeit, strikingly, nonphototrophic bacteria such as *Escherichia coli* also belong to this group. This group, known as the gamma-subdivision of the purple bacteria, combines with the purple nonsulfur bacteria, the sulfate-reducing bacteria, the thiobacilli, *Beggiatoa*, and the typical gram-negative chemoorganotrophs to form this large phylogenetic unit within the eubacterial kingdom. Surprisingly, the phylogenetically homogeneous group of the green sulfur bacteria is only distantly related to both the purple bacteria and to the green filamentous bacteria.

2. Species in Extreme Environments

As becomes clear from Table II, sulfide is found in hypersaline, acidic, and thermal environments. Phototrophic sulfur bacteria appear to be absent in strongly acidic environments but can be conspicuously present in thermal and hypersaline environments. *Chromatium tepidum* from a geyser in

Yellowstone National Park (United States) grows optimally at 50°C and can grow up to 57°C. *Chlorobium*-like cells have been observed in New Zealand thermal microbial mats, and a new thermophilic species, *Chlorobium tepidum* (temperature optimum 48°C) was described. *Chromatium salexigens* has been isolated from Mediterranean salterns and grows in media up to 15% salinity. The six species of the genus *Ectothiorhodospira* fall into two distinct groups. These species are marine or slightly halophilic species (up to 7% salinity), while *Ectothiorhodospira halophila*, *Ectothiorhodospira halochloris*, and *Ectothiorhodospira abdelmalekii* only grow in extremely salty and alkaline solutions (salinity >12%; pH about 9). The alkaliphile *Ectothiorhodospira* sp. form dense planktonic blooms in inland salty soda lakes.

3. Interaction with Light

The way light is converted into metabolic energy is described in Section I.C.1 and is depicted in Fig. 2. Phototrophic sulfur bacteria grow on visible and near-infrared light; however, not all wavelengths are equally used, depending on the pigment composition. *In vivo* absorption spectra of different phototrophic bacteria are shown in Fig. 4. *In vivo* spectra are very different from those obtained when the pigments are extracted with organic solvents. These differences occur because the physiological active pigments have absorption characteristics corresponding to their localization and function within the photosynthetic apparatus. Several *in vivo* peaks correspond to bacteriochlorophylls: one at approximately 400 nm and one or two (bacteriochlorophyll *a*) in the infrared region. Carotenoids have *in vivo* absorption maxima between 450 and 600 nm. Numerous different carotenoids have been found among the different species, but a given species only contains one or some specific carotenoids. For example, some *Chlorobium* species contain isorenieratene as the major pigment and are in fact brown.

Figure 4 nicely demonstrates the different groups of phototrophic organisms specialized for use of different and complementary wavelengths regions. This elegant example of niche-differentiation allows the coexistence or cooccurrence of representatives of the various groups (discussed in Section IV). However, some additional features are very important. First, light is not only absorbed by organisms but also by the substrate. While green light penetrates best in aquatic systems, red and infrared light

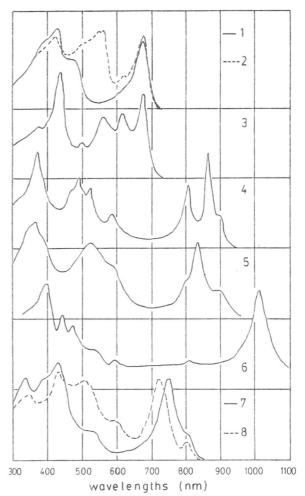

Figure 4 *In vivo* absorption spectra of different phototrophic microorganisms. 1, green algae; 2, red algae; 3, cyanobacteria; 4, purple sulfur bacteria containing bacteriochlorophyll *a* and the carotenoid spirolloxanthine; 5, purple sulfur bacteria containing bacteriochlorophyll *a* and the carotenoid okenone; 6, purple sulfur bacteria containing bacteriochlorophyll *b*; 7, green sulfur bacteria containing bacteriochlorophyll *c* and the carotenoid chlorobactene; 8, green sulfur bacteria containing bacteriochlorophyll *e* and the carotenoid isorenieretene (actually brown). [From Caumette, P. (1989). *In* "Microbial Mats. Physiological Ecology of Benthic Microbial Communities" (Y. Cohen and E. Rosenberg, eds.), pp. 283–304. American Society for Microbiology, Washington, D.C.]

does so in sandy sediments. Consequently, planktonic phototrophic bacteria harvest the light mainly with their carotenoids, while benthic phototrophic bacteria predominantly use their chlorophylls for this purpose. Second, it should be noted that phototrophic bacteria in the absence of other energy sources need a minimum amount of light to maintain cell integrity and ensure survival. These energy costs are about 10 times lower for green sulfur bacteria than for purple sulfur bacteria. Consequently, green sulfur bacteria can grow at lower light intensities than purple sulfur bacteria.

4. Additional Metabolic Options

Hydrogen and, most often, reduced sulfur compounds such as thiosulfate, polysulfides, and elemental sulfur are excellent electron donors for phototrophic growth of purple and green sulfur bacteria. In addition, a variety of carbon compounds is photoassimilated by purple sulfur bacteria. These include short-chain fatty acids and alcohols, but even sugars are photoassimilated by some species. While, the large *Chromatium* species are nutritionally more restricted and need sulfide, the smaller representatives are extremely versatile. Photoorganotrophic growth of *Chromatium vinosum* and other *Chromatium* species capable of assimilatory sulfate reduction is sulfide-independent. Photoassimilation of organic compounds, especially of acetate, also occurs in green sulfur bacteria. However, in contrast to the small *Chromatium* species, sulfide is always required by *Chlorobium* as an electron donor. The use of sulfide as an electron donor in combination with an organic compound as the carbon source is referred to as photolithoheterotrophy (see Fig. 1 and Section I.C).

Originally, purple sulfur bacteria were considered to be obligate anaerobes and obligate phototrophs. However, in the early 1970s, it was found that *Thiocapsa roseopersicina* and *Amoebobacter roseus* can grow in the presence of oxygen. Under these circumstances, the synthesis of photosynthetic pigments is repressed, but energy can be generated by oxygen respiration with sulfide as an electron donor, similar to the chemotrophic sulfur bacteria. Later it was demonstrated that this capacity of chemolithoautotrophic aerobic growth is widespread among the small members of *Chromatium* as well as in several other genera of the Chromatiaceae. In contrast, *Ectothiorhodospira* species can only grow in the presence of oxygen with organic substrates, while members of the Chlorobiaceae are completely incapable of growing in the presence of oxygen.

5. Growth Kinetics and Interspecies Competition

The small *Chromatium, Ectothiorhodospira,* and *Chlorobium* all exhibit similar maximal growth rates of about 0.1 hr^{-1}, which corresponds to a doubling time of 7 hr. These species are opportunists that are easily enriched from natural samples when employing nutrient-rich growth conditions. In contrast, large-celled *Chromatium* such as *Chromatium weissii* and several representatives of other genera grow more slowly. The competitive advantage of the large-celled *Chromatium* with respect to its smaller counterparts has been elegantly demonstrated in a series of experiments using the chemostat. *Chromatium weissii* not only has a lower μ_{max} but also a lower affinity for sulfide than the small *Chromatium vinosum* (see Fig. 3 for kinetic parameters). Thus, under constant conditions, *C. vinosum* grows faster than *C. weissii* at any sulfide concentration. However, the natural environmental conditions are fluctuating: Sulfide concentrations rise at night in the absence of photosynthesis and drop in the morning. It was found that *C. weissii* has a twofold greater uptake capacity for sulfide than *C. vinosum*. Sulfide taken up at high rate is only oxidized to intracellularly stored sulfur and is further oxidized when sulfide concentrations are very low. Hence, *C. weissii* is better equipped to take advantage of the temporal occurrence of high sulfide concentrations.

Members of Chlorobiaceae and Ectothiorhodospiraceae, which store sulfur extracellularly, have a 5- to 10-fold higher affinity for sulfide than Chromatiaceae (see Fig. 3). Thus, Chromatiaceae are expected to lose the competition when sulfide is growth-limiting. To test this hypothesis, it was essential to use chemostat culture techniques. In such a culture, the losing species is drastically washed out, and, consequently, only the competitive species can maintain stable population densities. Most surprisingly, when cultivating a mixture of *Chlorobium* and *Chromatium* under sulfide limitation, the outcome is not a monoculture of *Chlorobium* but, rather, a stable coexistence of both organisms (see Fig. 5). This result indicates that, beside sulfide, another substrate is influencing the growth rates. The extracellular sulfur of *Chlorobium* appeared to be tightly bound to the cell, thus being unavailable for *Chromatium*. However, this extracellular sulfur continuously reacts with sulfide forming soluble polysulfides. In principle, both species can grow on polysulfide, but, whereas this capacity is constitutive in *Chromatium,* it is not expressed by *Chlorobium* in the presence of sulfide (see Fig. 6). Thus, the polysulfide being formed in the sulfide-limited culture is exclusively used by *Chromatium,* which allows this species to coexist with *Chlorobium.*

It is still difficult to appreciate the ecological relevance of both the common capability to use organic substrates among the phototrophic sulfur bacteria

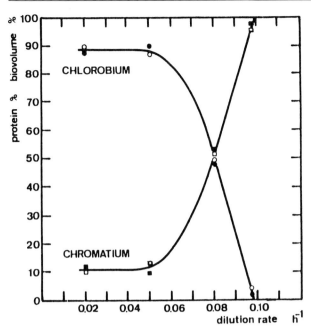

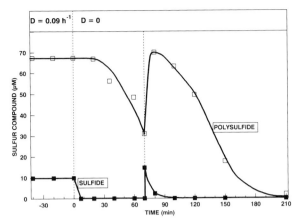

Figure 5 The steady-state outcome of competition experiments between the purple sulfur bacterium *Chromatium vinosum* and the green sulfur bacterium *Chlorobium limicola* in sulfide-limited continuous cultures at various dilution rates. During the steady state, both organisms grow at a specific rate equal to the dilution rate. In previous experiments, it was found that *C. limicola* has a sixfold higher affinity for sulfide than *C. vinosum*. Hence, it was expected that *C. limicola* outcompetes *C. vinosum* completely at dilution rates up to 0.09 hr^{-1}. To explain the unexpected results one must presume that substrate other than sulfide was also influencing the growth rate. [From Van Gemerden, H., and Beeftink, H. H. (1981). *Arch. Microbiol.* **129**, 32–34.]

Figure 6 Experiments performed in a chemostat culture of the green sulfur bacterium *Chlorobium limicola*. Before the manipulation at t = 0, the culture fed with sulfide-containing medium was in steady state (dilution rate = 0.09 hr^{-1}). In the culture vessel, both sulfide and polysulfide were found. At t = 0, the feed of fresh medium was shut off. While sulfide rapidly disappeared, the consumption of polysulfide only started after a lag period. At t = 80 min, a shot of sulfide was given to the culture, resulting in the formation of polysulfide. From this experiment, it was concluded that (1) polysulfide is formed from sulfide and extracellular sulfur, and (2) the consumption of polysulfide is not constitutive in *Chlorobium* but must be derepressed in the absence of sulfide. [Redrawn and kindly supplied by the authors. From Visscher, P. T., and Van Gemerden, H. (1988). *In* "Green Photosynthetic Bacteria" (J. M. Olson, J. G. Ormerod, J. Amesz, E. Stackebrandt, and H. G. Trüper, eds.). Plenum Press, New York.]

and the chemolithoautotrophic capabilities of some members of the Chromatiaceae. Using these additional metabolic options only, phototrophic sulfur bacteria always attain much lower growth responses than the specialized bacteria. Phototrophic sulfur bacteria are perhaps especially competitive under mixotrophic conditions (compare Section III.B.5); however, it is also possible that these additional metabolic options only serve to ensure survival during adverse periods.

B. Chemotrophic Sulfur Bacteria

1. Taxonomy

The families and genera of the chemotrophic sulfur bacteria and their main characteristics are listed in Table IV. Representatives of the chemotrophic sulfur bacteria are often taxonomically not related and,

in fact, have merely been grouped together as being nonphototrophic sulfide oxidizers. These bacteria are also known as colorless sulfur bacteria (see also Section I.C.2). The phylogenetical heterogenity has been confirmed by 16-S ribosomal RNA sequencing techniques. Chemotrophic sulfide oxidizers are found in both the archaebacterial and eubacterial kingdoms. Archaebacterial sulfide oxidizers are extremely thermoacidophilic. The best-known eubacterial families are the Thiobacilliaceae and the Beggiatoceae. However, the taxonomic position of many species is still uncertain: They have not been assigned to families and many species have not been cultivated yet.

2. Species in Extreme Environments

The archaebacterial sulfide oxidizers of the family Sulfolobaceae are extremely thermoacidophilic and grow at pH 1–6 and at temperatures up to 85–96°C, depending on the species. These bacteria have been isolated from hot sulfide-containing acidic springs. In addition, *Thermothrix* is a eubacterial species

Table IV Different Groups, Families, and Genera of Chemotrophic Sulfur Bacteria and Their Major Characteristics

Genus	Morphology	Motility[a]	Sulfur[2] inclusions
Eubacterial Chemotrophic Sulfur Bacteria			
Family Thiobacilliaceae (all species can obtain useful energy from the oxidation of reduced sulfur compounds and can fix CO$_2$ by the Calvin cycle; gram-negative)			
Thiobacillus	Rods	+ or −	−
Thiomicrospira	Spirilloid cells	+	−
Thiosphaera	Spherical cells sometimes in chains	−	−
Appendaged bacteria (related to *Hyphomicrobium*)			
Thiodendron	Vibrio-shaped cells with stalks	+ or −	−
Uncertain affilation (thermophilic organisms capable of chemolithoautotrophic growth; filamentous cells are produced under oxygen-limiting conditions; gram-negative)			
Thermothrix	rods and filamentous cells	+[c]	−
Family *Beggiatoaceae* (multicellular filaments; gram-negative)			
Beggiatoa	Single filaments	Gliding	+[d]
Thioploca	Filaments occur in bundles enclosed by a slimy sheath	Gliding	+
Thiothrix	Attached rigid filaments within a sheath which produce gonidia	Gliding[e]	+[d]
Uncertain affilation (organisms not capable to generate energy from the oxidation of reduced sulfur compounds but benefit from the oxidation of sulfide by detoxification of hydrogen peroxide)			
Macromonas	Straight or curved rods	+	+[f]
Uncertain affilation (no axenic cultures are available)			
Achromatium	Very large spherical to ovoid cells	Rotating[g]	+[f]
Thiovulum	Large round to ovoid cells	+	+
Thiospira	Spirilloid	+	+
Thiobacterium	Rods embedded in gelatinous masses	−	+
Archaebacterial Chemotrophic Sulfur Bacteria			
Family Sulfolobaceae (extremely thermoacidophilic; peptidoglycan absent from the cell wall; contain isoprenyl ether membrane lipids, calditol, and caldariella quinone)			
Sulfolobus[h]	Coccoid, highly irregular	−	−
Acidianus[h]	Coccoid, highly irregular	−	−

[a] +, motility by flagella unless otherwise indicated; −, motility absent.

[b] Zero-valence sulfur is formed during the oxidation of sulfide and occurs as characteristic refractile globules visible with the light microscope.

[c] Only rods are motile.

[d] By light microscopy, the sulfur globules appear to be internal, but, in fact, they occur outside the cytoplasm within invaginations of the cytoplasmatic membrane.

[e] Only the gonidia are motile by gliding.

[f] *Macromonas* and *Achromatium oxaliferum* also contain inclusions of CaCO$_3$-crystals.

[g] Slight rotating movement by means of slime production.

[h] *Acidianus* and *Sulfolobus* are different, because only the former organism can also reduce sulfur to sulfide.

adapted to growth at high temperatures up to 80°C albeit at neutral pH.

Many thiobacilli can grow at low pH values, which, for example, explains their predominance in acidic volcano lakes. Four species even grow best between pH 2 and 4. It should be born in mind that the acidity often is a consequence of their own metabolic activity, because the formation of sulfate from sulfide results in proton release (see Table I). In contrast, in phototrophic sulfur bacteria, this pH effect is counterbalanced by the fixation of CO$_2$ (weak acid), which tends to raise the pH.

The sulfide oxidizers observed in and around hydrothermal vents on the ocean floor (see Table II) are adapted to withstand pressures >200 bar (=2 · 10^7 Pa).

3. Ecological Niche in Oxygen–Sulfide Gradients

Chemotrophic sulfur bacteria are the oxygen–sulfide gradient organisms "par excellence." However, *Thiobacillus denitrificans*, *Thiobacillus delicatus*, *Thiomicrospira denitrificans*, *Thiosphaera pantotropha*, and a strain of *Beggiatoa* can use nitrate as an alternative electron donor under anaerobic conditions. In natural ecosystems, the niches for the denitrifying bacteria are always close to the oxic–anoxic interface, because nitrate is formed in the oxic zone. All chemotrophic sulfur bacteria can compete successfully with the chemical reaction of oxygen and sulfide. Motility and chemotactic or chemophobic responses facilitate the bacteria to locate themselves within the gradient.

4. Metabolic Options

Beside sulfide, other reduced sulfur compounds such as thiosulfate, sulfur, and polythionates are appropriate electron donors for many chemotrophic sulfur bacteria. *Thiobacillus ferrooxidans* can grow with Fe^{2+} as the electron donor at low pH and converts it into Fe^{3+}.

Many chemotrophic sulfur bacteria are not obligately dependent on inorganic compounds as the electron donor and on CO_2 as the only carbon source. Several species can use organic compounds. In fact, the chemotrophic sulfur bacteria comprise a wide spectrum of metabolical adaptations, which can be classified into four nutritional groups (see also Table I):

1. Obligate chemolithoautotrophic species cannot use organic compounds.
2. Facultative chemolithoautotrophic species are very versatile: both CO_2 and organic compounds can be used as a carbon source; both sulfide and organic compounds can be used as electron donors.
3. Chemolithoheterotrophic species are heterotrophic (CO_2 not used) but can obtain energy from sulfide oxidation.
4. Chemoorganotrophic species exclusively grow on organic compounds and no energy can be obtained from sulfide oxidation. These species probably benefit from the oxidation of sulfide by detoxification of hydrogen peroxide.

Some of the facultative chemolithoautotrophs and chemolithoheterotrophs perform best under mixotrophic conditions.

A century ago, the observation of sulfur globules in conspicuous *Beggiatoa* filaments inspired the great Russian microbiologist Sergei Winogradsky to formulate the concepts of chemolithoautotrophic growth. Later, his concept was confirmed for easily cultivalable thiobacilli, but it remained highly controversial for the *Beggiatoa*. Only recently has it been demonstrated that some marine strains of *Beggiatoa* are really capable of chemolithoautotrophic growth.

5. Growth Kinetics and Interspecies Competition

Chemostat culture experiments were essential to elucidate the ecological relevance of the different metabolic options occurring among the chemotrophic sulfur bacteria (see Section III.B.4). It was observed, that specialist chemolithoautotrophic and chemoorganotrophic bacteria perform best when sulfide or organic substrates, respectively are the only substrates. However, versatile mixotrophic species can reach high population densities when both sulfide and organic substrates are growth-limiting. Hence, it was concluded that the ratio of sulfide with respect to organic substrates available as nutrients determines which metabolic option is selected in the ecosystem.

C. Strategies of Cyanobacteria to Cope with Sulfide

In the mid-1970s, a spectacular discovery revealed that the cyanobacterium *Oscillatoria limnetica* was capable of anoxygenic photosynthesis. In the presence of sulfide, photosystem II is completely blocked, and, consequently, oxygen production stops immediately. However, after a lag period, sulfide is used as the electron donor in a type of photosynthesis driven only by photosystem I. Thus, this anoxygenic type of photosynthesis is very similar to that of phototrophic sulfur bacteria.

Later studies revealed that several other cyanobacteria can perform sulfide-dependent anoxygenic photosynthesis (see Table V). However, it was also found that some cyanobacteria have a photosystem II that is tolerant to sulfide (see also Section I.D). Most cyanobacteria do not tolerate sulfide. The tolerant species employ one of the following strategies to maintain growth in sulfide-containing environments:

1. Photosystem II completely blocked, but anoxygenic photosynthesis inducible.

Table V Phototrophic Bacteria, except Phototrophic Sulfur Bacteria, Capable of Anoxygenic Photosynthesis with Sulfide as the Electron Donor[a]

Group and genus or species	Main trophic type
Cyanobacteria	
At least 17 filamentous and coccoid species (only nonheterocystous species)	Photolithoautotrophy through oxygenic photosynthesis
Purple nonsulfur bacteria	
Rhodobacter div. sp., *Rhodopseudomonas palustris, sulfoviridis Rhodomicrobium vanniellii*	Photoorganotrophy (anaerobic)
Green filamentous bacteria	
Chloroflexus	Photoorganotrophy[b] (anaerobic) or chemoorganotrophy (aerobic)
Oscillochloris	Photoorganotrophy (anaerobic)

[a] In these organisms, sulfide oxidation is of minor importance.

[b] Some strongly green-pigmented strains that are obligate photolithoautotrophs dependent on sulfide have been isolated.

2. Photosystem II tolerant to sulfide, but anoxygenic photosynthesis does not occur.
3. Photosystem II slightly tolerant to sulfide and anoxygenic photosynthesis inducible. Thus, both types might operate simultaneously.

The ubiquitous mat-forming *Microcoleus chthonoplastes* employs the third strategy. For this species, it is crucial to maintain a low level of oxygen production, because it needs this compound for the synthesis of certain fatty acids.

D. Sulfide Oxidation by Photoorganotrophic Bacteria

Table V lists species of purple nonsulfur and green filamentous bacteria, which can photooxidize sulfide. This metabolic option is normally of minor importance. However, in the vicinity of sulfide-containing springs, some deeply green-colored *Chloroflexus* strains that are obligate photolithoautotrophs have been isolated.

IV. Communities and Interactions in Sulfide-Containing Environments

Sulfide-containing environments are found in many different ecosystems (see Table II and Section II). In these ecosystems, the anoxic sulfide-containing habitats are in close contact with oxidized environments (see Section II). Most often, the oxidized habitat occurs above the anoxic habitat, and, consequently, the resulting oxygen–sulfide gradients run vertical and parallel to the light gradient. The different phototrophic and chemotrophic sulfide-oxidizing populations stratify along these gradients in accordance with their ecological niches. These phenomena occur both in aquatic and in sediment ecosystems, albeit, strikingly, the vertical scale is some orders of magnitude smaller in the sediment ecosystems. The idealized schemes of the gradients and the localization of the different populations are shown in Fig. 7 for both the aquatic ecosystem and the sediment ecosystem. The localization of the different population is, thus, a reflection of the gradients, but, at the same time, the metabolic activities of the populations actually determine the shape of the oxygen–sulfide profile. The effect of the populations is clearly demonstrated when comparing day and night profiles. The catabolic processes such as oxygen respiration and sulfate reduction are light-independent. In contrast, photosynthetically mediated oxygen production and sulfide consumption are light-dependent. As a result, the oxygen–sulfide interface shifts downward during the day and upward during the night.

In principal, representatives of the different groups of phototrophic organisms need not compete for light because their absorption spectra are complementary (see Fig. 4). However, abiotic characteristics also modify the light available at depth in an ecosystem. As stated before (Section III.C.3), planktonic phototrophic bacteria mainly rely on the light harvested by the carotenoids, while benthic species rely on the red to infrared light harvested by chlorophylls. For Lakes in Wisconsin (United States), it has been demonstrated how the actual color of the lakewater influences the phototrophic community. Surprisingly, the absorption spectra of the brown-colored *Chlorobium* species is in the carotenoid region very similar to the absorption spectrum of *Chromatium*. In Spanish karstic lakes, it has been demonstrated that brown *Chlorobium* cannot

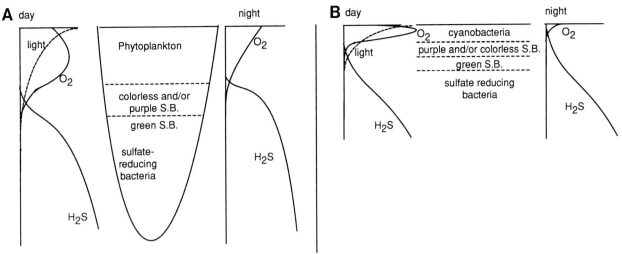

Figure 7 (A) Stratified lake. (B) Sediment ecosystem (microbial mat). Idealized schemes of the population distributions and the physicochemical profiles found in stratified ecosystems with anoxic bottom layers containing sulfide. Striking similarities are found among aquatic and sediment ecosystems, despite the large differences of the vertical scale (planktonic plates of phototrophic bacteria in lakes are 1 to several decimeters thick and may be found at great depth; microbial mats only comprise a few millimeters of the top layer of the sediment). Not all mentioned groups are always present. In fact, the occurrence of green sulfur bacteria in microbial mats is uncommon. Note the diurnal up and down shifts of the oxygen–sulfide interface. S.B., sulfur bacteria; colorless S.B., chemotrophic sulfur bacteria.

occur below a plate of *Chromatium;* however, brown *Chlorobium* is especially adapted to grow at extremely low light, which does not support growth of *Chromatium*.

Figure 8 shows characteristic profiles of oxygen, sulfide, and light together with the rates of oxygen production and the localization of a population of the phototrophic sulfur bacterium *Chromatium* in a microbial mat in Baja California (Mexico). Oxygen is mainly produced by cyanobacteria in this system. The cooccurrence of sulfide and oxygen places the oxygen-tolerant purple sulfur bacteria in competition with the chemotrophic sulfur bacteria. In quite similar mats, as described in Fig. 8, it was observed that the chemotrophic *Beggiatoa* dominated if the oxygen–sulfide interface received <0.1% of 825 nm incidence light.

As already mentioned, the oxygen–sulfide gradients shift up and down during a diurnal cycle, because photosynthetic oxidation processes are light-dependent while catabolic reduction processes are light-independent (see Fig. 7). Diurnal migration patterns have been described for benthic *Chromatium* and *Beggiatoa* and several planktonic sulfur bacteria. Surprisingly, the nonmotile purple sulfur bacterium *Thiocapsa roseopersicina* is especially common in the fluctuating oxic–anoxic habitats of the microbial mat. Exposure to oxygen seems to be

disadvantageous due to repression by O_2 of pigment synthesis. The organism can grow chemolithoautotrophically, but its capacity is insufficient to compete with specialized chemolithotrophic sulfur bacteria under such conditions. The clue was found when *T. roseopersicina* was subjected in laboratory chemostat cultures to a simulation of naturally occurring conditions (see Fig. 9). During the oxic day periods, pigment synthesis is indeed repressed; however, during the anoxic night periods, pigment synthesis occurs at high rates. In fact, this bacterium can actually grow as a phototroph due to pigment synthesis occurring at night.

Apart from negative or neutral interactions among different species, some positive or mutually beneficial interactions have been found. Some green sulfur bacteria live in symbiosis with sulfur- or sulfate-reducing bacteria forming consortia, which are apparently stable. The arrangement of the different cells within the complex is often remarkably constant; combinations have been assigned a Latin name and a description if they were a single species (e.g., *Chlorochromatium, Pelochromatium, Chloroplana*). The symbiosis is based on syntrophy: The chemoorganotrophic partner produces sulfide while this compound is recycled to sulfur by the phototrophic partner.

Sulfide-containing environments are inhabited by

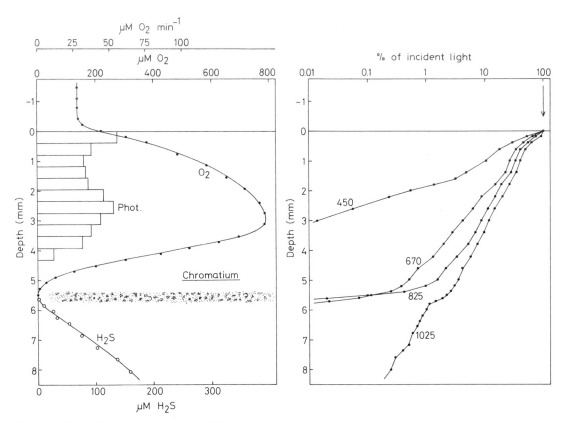

Figure 8 Vertical profiles of oxygen, sulfide, oxygenic photosynthesis (Phot.), and light of different wavelengths in an illuminated microbial mat in Baja California (Mexico). A layer of the purple sulfur bacterium *Chromatium vinosum* was found as indicated. [From Jørgensen, B. B., and Des Marais, D. D. (1986). *FEMS Microb. Ecol.* **38,** 179–186.]

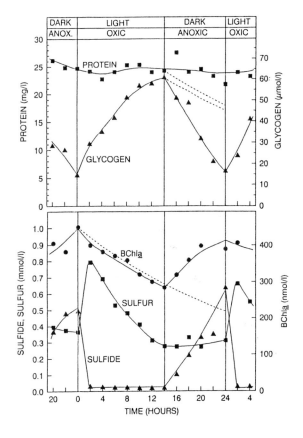

Figure 9 Experimental simulation of the fluctuating conditions experienced by the purple sulfur bacterium *Thiocapsa roseopersicina* in its natural habitat of the microbial mat. A sulfide-limited culture (dilution rate = 0.03 hr^{-1}) was exposed to light plus oxic and dark plus anoxic alternations. Previous to the data sampling the culture was grown at this regimen for a week. Bacteriochlorophyll *a* was washed out during the day, thus indicating that no synthesis occurred. In contrast, bacteriochlorophyll *a* was synthesized at a high rate during the night with glycogen as the energy and carbon source. During the night, the sulfide concentrations rose because this compound was not consumed and continuously fed to the culture. [From de Wit, R., and Van Gemerden, H. (1990). *Arch. Microbiol.* **154,** 459–464.]

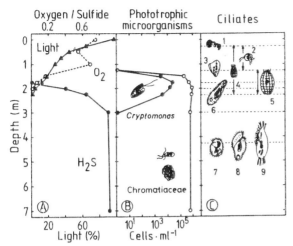

Figure 10 Vertical distribution of physicochemical parameters and microbial community in Lake Cisó (Spain), 15 June 1984. (sulfide in mM; dissolved oxygen concentration must be multiplied by 10 to get mg/l). *Cryptomonas* is a sulfide-tolerant eukaryotic alga. Ciliates: 1, suctorians; 2, *Vorticella*; 3, *Stentor*; 4, *Coleps* spp.; 5, *Coleps hirtus*; 6, *Paramecium* spp.; 7, *Plagiopyla ovata*; 8, *Metopus es*; 9, hypotrich. [From Dyer, B. D., Gaju, N., Pedrós-Alió, C., Esteve, I., and Guerrero, R. (1986). *BioSystems* **19**, 127–135.]

predominantly prokaryotic communities. However, some specialized eukaryotes might be present. Figure 10 resumes the biotic communities observed in Lake Cisó (Spain). *Cryptomonas phaseolus* is a cryptomonad alga ubiquitously found in the oxygen–sulfide overlap zone. Sulfide-tolerant eukaryotes are especially well presented among the protozoic ciliates. The anaerobic sulfide-loving species feed on the dense bacterial populations and are presumably able to generate energy by fermentation.

Acknowledgment

This work was supported by a grant from the European Community (STEP-program).

Bibliography

Caumette, P. (1989). Ecology and general physiology of anoxygenic phototrophic bacteria in benthic environments. *In* "Microbial Mats. Physiological Ecology of Benthic Microbial Communities" (Y. Cohen and E. Rosenberg, eds.), pp. 283–304. American Society for Microbiology, Washington, D.C.

Gorlenko, V. M., Dubinina, G. A., and Kuznetsov, S. I. (1983). The ecology of aquatic micro-organisms (Chapter 5—Microorganisms reducing and oxidizing sulphur compounds). *In* "Die Binnengewässer," Band XXVIII, p. 252. Schweitzerbart'sche Verlagsbuchhandlung, Stuttgart.

Jørgensen, B. B. (1988). Ecology of the sulphur cycle: Oxidative pathways in sediments. *In* "The Nitrogen and Sulphur Cycles" (J. A. Cole and S. J. Ferguson, eds.), pp. 31–63. Cambridge University Press, New York.

Kuenen, J. G. (1989). Comparative ecophysiology of the nonphototrophic sulfide-oxidizing bacteria. *In* "Microbial Mats. Physiological Ecology of Benthic Microbial Communities" (Y. Cohen and E. Rosenberg, eds.), pp. 349–365. American Society for Microbiology, Washington, D.C.

Madigan, M. T. (1988). Microbiology, physiology, and ecology of phototrophic bacteria. *In* "Biology of Anaerobic Microorganisms" (A. J. B. Zehnder, ed.), pp. 39–111. John Wiley, New York.

Van Gemerden, H. (1983). *Ann. Microbiol. (Paris)* **134b**, 73–92.

Van Gemerden, H., and de Wit, R. (1986). Strategies of phototrophic bacteria in sulphide-containing environments. *In* "Microbes in Extreme Environments" (R. A. Herbert and G. A. Codd, eds.), pp. 111–127. Academic Press, London.

Van Gemerden, H., and de Wit, R. (1989). Phototrophic and chemotrophic growth of the purple sulfur bacterium *Thiocapsa roseopersina*. *In* "Microbial Mats. Physiological Ecology of Benthic Microbial Communities" (Y. Cohen and E. Rosenberg, eds.), pp. 313–319. American Society for Microbiology, Washington, D.C.

Sulfur Cycle

Warren A. Dick

The Ohio State University

Glossary

Assimilatory reduction Chemical reduction of an element for the purpose of introducing the element into cellular material

Atmosphere Gaseous envelope surrounding the earth's surface

Biosphere Portion of the earth's mass contained in living organisms (plant, animal, and microbial)

Dissimilatory reduction Chemical reduction of an element in an energy-yielding reaction; element does not become incorporated into cellular material

Hydrosphere Aqueous portion of the earth's surface

Immobilization Assimilation of inorganic nutrients or elements into an organic form during cell growth; it is the reverse of mineralization

Lithosphere Solid crust of the earth's surface

Mineralization Conversion of an organic complex of an element to its inorganic state (e.g., the conversion of organic sulfur to sulfate); it is the reverse of immobilization

Pedosphere Layer of soil lying on the earth's surface

Soil microbial biomass Living portion of the soil organic matter excluding plant roots and soil animals

Weathering Breakdown of the earth's rock to form soil

SULFUR CYCLING is a natural environmental process that prevents the accumulation of all of the earth's sulfur into one specific compound within a single ecosystem. The transformation reactions involved represent a continuous flow of sulfur-containing compounds among the various components of the earth (Fig. 1). Sulfur is an essential element for the growth of plants and animals. It can exist as a gas, liquid, or solid, in organic or inorganic forms, and in soluble or insoluble forms. It occurs in valence states ranging from -2 to $+6$. It is the 8th most abundant element in the solar atmosphere and the 14th most abundant element in the earth's crust. The reactions of the sulfur cycle alter the chemical, physical, and biological states of sulfur and its compounds so that sulfur cycling can occur. Besides cycling among the major components of the earth's environment, there is also an internal cycling within each of the major components. Many reactions of the sulfur cycle are mediated by microorganisms.

I. Components of the Global Sulfur Cycle

Sulfur is found in several large environmental components associated with the earth (Table I). The lithospheric component is the largest and contains roughly 95% of this element. The second largest component is the hydrosphere and the earth's oceans contain approximately 5% of the total sulfur. The other components in which sulfur is found together comprise <0.001% of the remaining sulfur.

The annual fluxes between each of these components (Fig. 2), given in Tg/yr of sulfur (1 teragram, or Tg, is equivalent to 10^9 kg), indicates a rapid movement of sulfur from one component to another. The combined effect of this movement is called the sulfur cycle. [*See* SULFIDE-CONTAINING ENVIRONMENTS.]

A. Lithosphere

The lithosphere is the earth's mantle rock and it is assumed to be the ultimate source of all crustal sul-

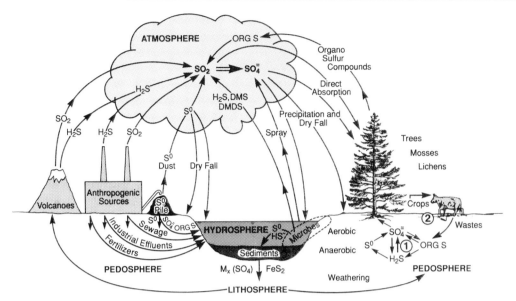

Figure 1   Simplified version of the overall sulfur cycle in nature. [From Krouse, H. R., and McCready, R. G. L. (1979). Biogeochemical cycling of sulfur. *In* "Biogeochemical Cycling of Mineral-Forming Elements" (P. A. Trudinger and D. J. Swaine, eds.), p. 402. Elsevier Scientific Publishing Company, New York.]

fur. Export of sulfur from the lithosphere occurs as volcanic activity, weathering, and groundwater leaching. Although solution chemistry plays an important role in weathering solid surfaces, evidence indicates that microorganisms are directly involved by catalyzing oxidation/reduction reactions of sulfur minerals.

The activities of humans are responsible for most of the recent sulfur flux from the lithosphere. Fuel

Table I   Amount of Sulfur Contained in Various Components of the Earth

Component	Amount of sulfur (kg)
Atmosphere	4.8×10^9
Lithosphere	24.2×10^{18}
Hydrosphere	
Sea	1.3×10^{18}
Freshwater	3.0×10^{12}
Marine organisms	2.4×10^{11}
Pedosphere	
Soil	2.6×10^{14}
Soil organic matter	0.1×10^{14}
Biosphere	8.0×10^{12}

[From Stevenson, F. J. (1986). "Cycles of Soil," p. 287, John Wiley and Sons, New York; and Trudinger, P. A. (1986). "Sulfur in Agriculture" (M. A. Tabatabai, ed.). p. 2. American Society of Agronomy, Madison.]

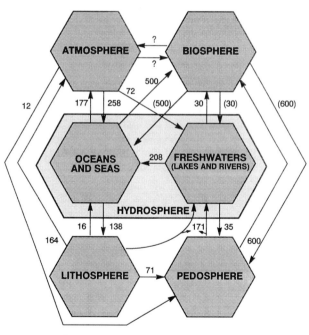

Figure 2   Fluxes of sulfur among the various components of the earth. Numbers are in Tg per year of sulfur and the flux from the lithosphere to the atmosphere includes the contribution from the pedosphere. [From Trudinger, P. A. (1986). Chemistry of the sulfur cycle. *In* "Sulfur in Agriculture" (M. A. Tabatabai, ed.), p. 3. American Society of Agronomy, Madison, Wisconsin.]

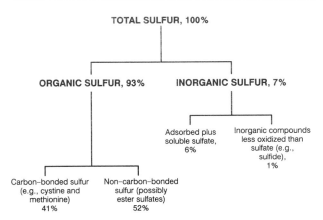

Figure 3 Average distribution of the forms of sulfur in soil. [Reprinted from Freney, J. R. (1967). Sulfur-containing organics. In "Soil Biochemistry" (A. D. McLaren and G. H. Peterson, eds.), p. 235. Courtesy of Marcel Dekker, New York.]

combustion and smelting, effluent from the chemical industry, and fertilizer production account for export of 113, 28, and 28 Tg sulfur per year. Volcanic activity also exports a substantial amount of sulfur from the lithosphere, expecially in respect to land areas bordering on or occurring as islands in the Pacific Ocean.

B. Pedosphere

Covering the earth's crust is a thin layer or skin called soil. Soils are products of rock weathering, and the initial sulfur in soils is derived from weathering and from atmospheric sources. Additional sulfur is also added to soil as fertilizer. The sulfur in the pedosphere is extensively modified, however, by chemical and microbiological processes.

The mean concentration of sulfur in the surface meter of soils is 0.05%. Typically, organic sulfur compounds are predominant in surface soils (Fig. 3). Deeper in the soil, inorganic forms of sulfur become more prevalent.

In well-drained, aerated soils inorganic sulfur occurs as soluble sulfate, calcium sulfate, and sulfate absorbed to clays, iron and aluminum oxides. In poorly aerated, waterlogged soils, tidal swamps, bogs, and rice paddies, a significant quantity of sulfur accumulates as sulfide. Upon drainage and introduction of oxygen to these soils, oxidation of the reduced sulfides to sulfates rapidly occurs.

Sulfates are water-soluble and readily migrate or leach through soil. Leaching of sulfate is the primary means by which sulfur is exported from the pedo-

sphere. Sulfate that reaches the groundwater may be considered part of the lithospheric sulfur component.

Organic sulfur in soils is poorly characterized and appears to be chemically diverse. Two major forms of sulfur exist in the soil organic matter or humus (Fig. 3). Carbon-bonded sulfur compounds include the amino acids cysteine, cystine, and methionine. Non-carbon-bonded sulfur compounds are presumed to be primarily ester sulfates such as phenolic sulfates and sulfated polysaccharides. An ester-sulfate has its sulfur atom bonded to oxygen.

The decomposition of organic sulfur compounds in soils often leads to the formation of trace amounts of organic sulfur gases. The fate of the sulfur gases produced in soil include escape to the atmosphere, solubilization in waters, and a return to the oxidized form via chemical and biological reactions. [*See* ORGANIC MATTER, DECOMPOSITION.]

C. Biosphere

The sulfur in the biosphere is in rapid flux. It is imported from the hydrosphere, lithosphere, and pedosphere and it becomes fixed into protein, vitamins, and other compounds necessary for life. A small amount of this sulfur becomes fixed in the soil organic matter, but the majority is cycled back and returned to its place of origin.

In some ecosystems, the pedosphere and biosphere work in concert to create a cycle by which sulfur is effectively conserved. An example of an ecosystem in which a very tight cycling of sulfur occurs is the tropical rainforest, where the majority of plant nutrients, such as sulfur, are contained in the plants of the forest. If the forest is cut and the plants allowed to decay, the sulfur contained in the plants is converted in the pedosphere via microbiological activity to sulfates which are leached from the soil and lost. However, if sufficient number of new plants are immediately allowed to regrow in the cleared area, the sulfur in the original plant material that is released in the soil and converted to sulfate is incorporated into new plant biomass and conserved.

D. Atmosphere

A large number of sulfur compounds have been detected in the atmosphere including sulfur dioxide (SO_2), hydrogen sulfide (H_2S), volatile organic sulfur compounds, and sulfate aerosols. The source of these sulfur compounds are volcanic gases (8%),

Table II Global Atmospheric Sulfur Concentrations, Pool Sizes, and Residence Times in Remote Areas of the World

Sulfur compounds	Concentration[a]	Pool size (Tg sulfur)	Residence time (days)
Sulfate	0.1–0.56	0.26	7
Sulfur dioxide	45–340	0.25	4
Carbonyl sulfide	100–560	2.2	160
Carbon disulfide	70–370	0.6	45
Dimethyl sulfide	58	0.021	0.75
Hydrogen sulfide	0–320	0.041	1.5

[a] Sulfate concentration in micrograms per cubic meter while all others are in parts/10^{12} (v).

[From Trudinger, P. A. (1986). "Sulfur in Agriculture" (M. A. Tabatabai, ed.), p. 12. American Society of Agronomy, Madison, Wisconsin.]

aeolian dust (6%), biogenic gases (12%), and activities of humans (33%).

Concentrations of gaseous sulfur compounds in the atmosphere vary and often can be orders of magnitude greater in locations where high emissions occur. The concentrations of sulfur compounds in the atmosphere (Table II) are for remote areas of the world.

The residence time of most sulfur gases in the atmosphere is very short (Table II) and depends on their chemical stability. The residence time is an important parameter because not only does it affect the concentration of sulfur gases in the atmosphere but also the amount of environmental impact they may have. The oxidation of sulfur gases in the atmosphere is complex and is affected by sunlight intensity, temperature, humidity, and the nature of the particulate.

Sulfur dioxide is primarily produced by human activities, whereas microbial activity is primarily responsible for the production of atmospheric carbonyl sulfide (COS), carbon disulfide (CS_2), and dimethyl sulfide [$(CH_3)_2S$] and these gases may well comprise the major component of biogenic sulfur in the atmosphere. Carbonyl sulfide is also formed by reaction of carbon disulfide with hydroxy radicals and this form of sulfur is estimated to account for 65% of the total atmospheric sulfur component.

The end-products of oxidation of hydrogen sulfide and volatile organic sulfur gases are sulfur dioxide and sulfate. However, sulfate is primarily introduced to the atmosphere by sea spray and in aeloian dust.

Removal of sulfur from the atmosphere occurs by both wet and dry precipitation mechanisms. The chemical form of the sulfur deposited depends on the origin and location of the sulfur. In industrial regions, much of the sulfate is deposited in rain as sulfuric acid. This phenomenon called "acid rain" is thought to contribute to the acidification of lakes and soils in sensitive regions of the world. In more remote regions, sulfate is deposited primarily as salts of calcium, ammonium, sodium, and magnesium.

Of the total amount of sulfur exported from the atmosphere, 82% occurs as wet deposition (of which the majority falls on the oceans), 10% occurs as dry deposition, and 8% is absorbed out of the atmosphere as sulfur dioxide.

E. Hydrosphere

The oceans and seas are the major repositories of hydrospheric sulfur, which exists primarily in the form of the soluble sulfate ion. Most of the sulfate is derived from rivers and rainfall. Minor amounts of reduced sulfur compounds exist in the oceans with concentrations greatest in stagnant coastal areas. The water overlying the sediments may also contain substantial amounts of reduced sulfur compounds that can diffuse upward.

Episodic emissions of hydrogen sulfide into marine waters occur from ocean vents and these waters are enriched in phytoplankton and other marine life, the sulfide serving as a source of electrons (energy) for biological growth.

Rivers and lakes of the continents represent a lesser amount of hydrospheric sulfur although concentrations can be high in localized areas. Weathering of pyrite rocks and other sulfide minerals are a source of this sulfur but increasingly anthropogenic sources have become a major contributor of sulfur in lakes and rivers, especially in urban areas.

Table III Microorganisms Involved in Oxidation of Reduced Sulfur Compounds

Genus	Comments
Autotrophs	
Sulfolobus	Facultative autotroph Thermophilic acidophilic
Thiobacillus	Genus includes strict autotrophs, facultative autotrophs, mixotrophs, and acidophiles Some species also oxidize Fe^{3+} *T. denitrificans* grows anaerobically with nitrate
Beggiatoa	Aerobic to microaerophilic Facultative autotroph
Heterotrophs	
Arthrobacter, Bacillus, Micrococcus, Pseudomonas, Mycobacterium	Sulfur-oxidizing bacteria
Saccharomyces, Debarymyces	Sulfur-oxidizing fungi
Photosynthetic bacteria	
Chromatium, Thiocapsa, Thiocystis, Thiosarcina, Thiospirillum	Elemental sulfur is deposited intracellularly during sulfide oxidation
Amoebobacter, Thiopedia, Lamprocystis, Thiodictyon	Contain gas vacuoles with sulfur within the peripheral part of the cell
Clathrochloris, Chlorobium, Ectothiorhodospira, Pelodictyon, Prosthechloris	Elemental sulfur is deposited extracellularly during sulfide oxidation

[From Trudinger, P. A. (1986). Chemistry of the sulfur cycle. *In* "Sulfur in Agriculture" (M. A. Tabatabai, ed.). p. 6. American Society of Agronomy, Madison, Wisconsin.]

Sulfur is removed from the hydrosphere by sea spray and by emission of volatile sulfur gases via the activity of microorganisms.

II. Transformation Processes

Physical, chemical, and biological processes are all involved in the cycling of sulfur. Physical processes include the formation of aeolian dust, wet and dry deposition from the atmosphere, and the crystallization and dissolution of minerals. Chemical processes include changes in valency of sulfur minerals and precipitation with other metals. Since sulfur in the form of amino acids are part of all living matter, all organisms play a role in the biological sulfur cycle. The major types of reactions involved in sulfur cycling in natural environments are oxidation, reduction, mineralization, immobilization, and volatilization. These reactions will be discussed separately.

A. Sulfur Oxidation

Sulfide will form minerals with a large number of metallic elements, many of which are stable in the environment. When exposed to water and air, these sulfide (or polysulfide) minerals are transformed by oxidation reactions manifested by the generation of acidity, the conversion of the sulfur moiety to sulfate, and the solubilization of the metallic portion of the mineral.

The involvement of microorganisms in the oxidation of sulfide minerals has been unequivocally demonstrated. As oxidation proceeds, a succession of microbial populations is established, each population preparing the way for the development of the next step in the overall oxidation process. Because acid is one product of the oxidation reaction, the extent to which oxidation of the sulfide mineral proceeds depends largely on the sequential development of microbial populations capable of living at the constantly changing acidity levels.

Microorganisms capable of oxidizing one or more chemical forms of sulfur (Table III) illustrate the wide diversity of species involved. The most common microorganisms associated with sulfur oxidation are *Thiobacillus* spp., and some iron-oxidizing microorganisms such as *Thiobacillus ferroxidans* and *Metallogenium* spp. can also oxidize sulfur.

Sulfur-oxidizing bacteria of the genus *Thiobacillus* are non-spore-forming, gram-negative rods about 0.3 μm in diameter and 1–3 μm long. Most species are motile by polar flagella. The majority of *Thiobacillus* are obligate aerobes; i.e., they have an absolute requirement for oxygen to live and complete their life-cycle. Typical oxidation reactions catalyzed by *Thiobacillus* are (sulfur valence states are given in parentheses)

$$
\begin{array}{ccccccccc}
\text{SH}^- & \rightarrow & \text{S}_o & \rightarrow & \text{S}_2\text{O}_3{}^{2-} & \rightarrow & \text{S}_4\text{O}_6{}^{2-} & \rightarrow & \text{SO}_4{}^{2-} \\
\text{sulfide} & & \text{elemental} & & \text{thiosulfate} & & \text{tetrathionate} & & \text{sulfate} \\
& & \text{sulfur} & & & & & & \\
(-2) & & (0) & & (-2, +6) & & (+1.67 \text{ to } +3) & & (+6).
\end{array}
$$

In laboratory systems, sulfur-oxidizing bacteria often produce thiosulfate and polythionates from sulfide and elemental sulfur. These reactions are limited in the natural environment except in places containing high concentrations of both reduced sulfur compounds and oxidizing bacteria.

Most agricultural soils contain only 100–200 thiobacilli per gram of soil, but numbers can rapidly increase to millions per gram upon the addition of reduced sulfur compounds. Environmental factors that influence sulfur oxidation include soil type, soil pH, sulfur particle size, temperature, soil moisture content, and the type of reduced sulfur compound.

The principal ways microorganisms facilitate the oxidation of sulfide minerals are summarized as follows:

1. Excretion of surface-active materials, such as phospholipids by *Thiobacillus* spp., affects the degree of aggregation in slurries and increases the rates of mass transfer of reactants, such as oxygen, at phase boundaries.

2. Changing the pH of the microenvironment imposes a different stoichiometry upon the overall reaction. For example, elemental sulfur may be formed during the nonbiological oxidation of sulfide minerals, whereas sulfuric acid is formed by sulfur-oxidizing microbes. The resulting acidity then further enhances oxidation of the sulfide mineral.

3. Sulfide minerals are generally prone to electrochemical processes that can rapidly lead to the coating of the mineral surface by sulfur. Sulfur oxidizers remove this layer providing a fresh surface for mineral dissolution to proceed.

4. Direct enzymatic mediated attack by microorganisms on either or both of the sulfide and metallic ions in mineral sulfide lattices.

5. The so-called "indirect" mechanisms whereby oxidation involves attack by a reactant

that is cyclically regenerated by the sulfur-oxidizing microorganisms. An example is the formation of ferric ions by *T. ferroxidans* and the iron, in a subsequent chemical reaction, oxidizes the sulfide in the mineral.

Several species of *Thiobaccili* and *Thiomicrospira* can utilize nitrate in place of oxygen as the terminal electron acceptor when grown under anaerobic conditions. In addition, *T. ferroxidans* can utilize reduced iron and *Thiothioparus* can use thiocyanide as electron donors besides reduced sulfur compounds. Recently, research has indicated that several species thought to be obligate autotrophs can also use some organic carbon compounds as either electron acceptor or donor.

Sulfolobus acidocaldarious is a thermophilic, acidophilic, lobed sphere capable of oxidizing elemental sulfur, hydrogen sulfide, and reduced iron. *Sulfolobus* can be found at concentrations of 10^8 cells/ml in acid hot springs all over the world.

Beggiatoa is a filamentous, gliding bacterium usually containing intracellular sulfur globules. It was the study of these bacteria that led to the development of the important concept that energy generation can be linked to inorganic oxidations. An interesting habitat where *Beggiatoa* can be found is within the vicinity of roots of plants such as rice and cattail growing in flooded soils. *Beggiatoa* oxidizes hydrogen sulfide, thereby reducing its toxic effect on the plants.

Two principal groups of photosynthetic sulfur oxidizers are the green and purple sulfur bacteria. The green sulfur bacteria (Chlorobiniaceae) are generally found in areas of high hydrogen sulfide concentrations (4–8 mM) while the purple sulfur bacteria (Chromatiaceae) are usually in areas of intermediate hydrogen sulfide concentrations (0.8–4 mM). The so-called purple nonsulfur bacteria (Rhodospirillaceae) are found at low hydrogen sulfide concentration environments (0.4–2 mM). These bacteria are obligate anaerobes and can occupy only restricted niches in the natural environment. In certain lakes, for example, where sulfate is reduced in the bottom waters and sediments, photosynthetic bacteria may be sharply stratified at the hydrogen sulfide–oxygen boundary at a depth determined by light penetration into the water.

Photosynthetic sulfur bacteria are thought to have played a role, in concert with sulfate reducers, in the formation of elemental sulfur deposits. The sulfate reducers produce hydrogen sulfide, which can then

Table IV Dissimilatory Sulfate-Reducing Bacteria

Genera	*Desulfobacter, Desulfobulbus, Desulfococcus, Desulfonema, Desulfosarcina, Desulfotomaculum, Desulfovibrio*
General characteristics	Strict anaerobes Grow at mildly acidic to mildly alkaline pH Generally mesophilic, but some species thermophilic
Substrates	Most sulfate reducers will also reduce sulfite and thiosulfate Some species reduce elemental sulfur Organic matter utilization varies with genus and species As a group, capable of completely oxidizing fatty acids from C_1-C_{18}, lactate, pyruvate, low molecular weight alcohols, and some aromatic compounds
Habitats	Anaerobic sediments of freshwater, brackish water, and marine environments, thermal regions, marine and animal intestines

[From Trudinger, P. A. (1986). Chemistry of the sulfur cycle. In "Sulfur in Agriculture" (M. A. Tabatabai, ed.), p. 4. American Society of Agronomy, Madison, Wisconsin.]

be photosynthetically oxidized to elemental sulfur. Purple sulfur bacteria and sulfate reducers often grow together tightly coupled so that a small amount of sulfur may be cycled many times from one bacteria to another. One could imagine a complete stable ecosystem based on the photosynthetic sulfur bacteria producing oxidized sulfur and other bacteria reducing the oxidized sulfur compounds. Sulfur compounds could conceivably replace oxygen as the primary movers of the carbon cycle and this sort of cycle has been termed a sulfuretum.

A large number of heterotrophic organisms have been found to oxidize sulfur including members of the genera *Arthrobacter, Bacillus, Micrococcus, Pseudomonas, Mycobacterium*, some actinomycetes, and the fungi *Saccharomyces* and *Debaryomyces*. Apparently no energy is made available to the microorganisms as a result of the sulfur oxidation reaction. Their role in natural environments is difficult to assess because many other organisms are present that are also capable of oxidizing reduced sulfur compounds. In the case of heterotrophic activity, the end-product of oxidation often is not sulfate but some other sulfur compound, most commonly thiosulfate. [*See* HETEROTROPHIC MICROORGANISMS.]

Sulfur in organic compounds, such as the sulfur amino acids, can also be oxidized. The products of these reactions are sulfate, hydrogen sulfide, and volatile sulfur compounds.

B. Sulfur Reduction

Sulfur usually accounts for 0.4–0.8% of the microbial cell's dry weight. Almost all of this sulfur is in the form of proteins and polypeptides as the amino acids cysteine, cystine, and methionine. The valence state of this sulfur is -2. Small amounts of other sulfur-containing compounds such as biotin, thiamin, and coenzyme M as well as sulfate esters of polysaccharides of structural materials also occur in microorganisms.

Most microorganisms are capable of using sulfate as a sole sulfur source for growth, and in so doing they reduce it to hydrogen sulfide intracellularly and then replace the hydroxyl groups on serine and homoserine with sulfhydryl groups. These reactions are termed assimilatory because virtually all of the sulfur taken up and reduced by the microorganisms is incorporated into amino acids. In contrast, dissimilatory reduction refers to reactions whereby sulfate or some other oxidized sulfur compound is used as a terminal electron acceptor and the sulfur is not incorporated into cellular material.

The initial step in assimilatory sulfate reduction is the activation of sulfate by adenosine triphosphate (ATP). This reaction is believed to be common to all sulfate reducers. Subsequent reduction steps to produce sulfide vary according to whether the reduction is assimilatory or dissimilatory and the type of microorganism involved.

During dissimilatory sulfate reduction, the sulfate serves as a substitute for oxygen as a terminal electron acceptor. ATP, the currency of living cells, is synthesized during the process to provide energy for microbial growth. The anaerobes *Desulfovibrio* and *Desulfotomaculum* spp. are believed to be the main contributors of dissimilatory sulfate reduction, although other species are also involved (Table IV).

Microorganisms are completely capable of reducing sulfate under temperatures and pressures tolerated by living organisms while there are no known chemical methods that can accomplish the same thing. Sulfate reduction has been observed in a wide variety of habitats and requires anaerobiosis, organic matter, a rather narrow pH range of 5.5 to slightly alkaline, and the presence of sulfate although other partially oxidized sulfur compounds can also be microbiologically reduced. Optimum temperatures for dissimilatory sulfate reduction are 10–40°C.

C. Mineralization/Immobilization of Sulfur

Sulfur is constantly being converted from mineral to organic forms and vice versa during plant and microbial growth. When the soil is well aerated, organic sulfur is mineralized (via oxidative reactions) to form sulfate. Concurrently, sulfate is assimilated (taken up) by plants and microorganisms to form biomass in a process called immobilization.

Increases in organic sulfur content in the soil (i.e., immobilization) will only occur when conditions are suitable for the accumulation of organic matter or humus. Favorable conditions include frequent additions of carbon-rich plant residues to the soil and an environment in which microbial growth can rapidly occur (i.e., nuetral pH, warm temperatures, and adequate soil moisture levels).

The net rates of immobilization or mineralization depends on the differences in the relative rates of the two opposing processes:

$$\text{organic sulfur} \underset{\text{immobilization}}{\overset{\text{mineralization}}{\rightleftharpoons}} \text{inorganic sulfur (SO}_4^{2-})$$

Some observations that have been made concerning mineralization and immobilization of sulfur in soil include:

1. The amount of sulfur mineralized does not appear to be directly related to soil type, soil pH, or mineralizable nitrogen.

2. Sulfur mineralization follows a number of patterns, i.e., an initial immobilization followed by net mineralization in later stages, a steady rate of mineralization with time, an initial rapid mineralization rate followed by a slower mineralization rate, and a rate of mineralization that decreases with time.

3. Mineralization of sulfur in the presence of plants is greater than in the absence of plants, probably due to the greater proliferation of microorganisms in the plant root rhizosphere.

4. Mineralization is affected by those factors that affect microbial growth such as temperature, moisture, pH, and availability of food supply.

5. More sulfur is mineralized when the soils are dried and then rewetted than when they are incubated without first undergoing a drying process.

The amount of sulfur mineralized also depends on the carbon : sulfur ratio of the material added to the soil. When the carbon : sulfur ratio is <200, there is a net release of mineral sulfur as sulfate to the soil. When the carbon : sulfur ratio is >400 there is a net removal of mineral sulfur as sulfate from the soil. In between these two extremes there is a balance where there is neither a net gain nor loss of mineral sulfur from the soil. Carbon : sulfur ratios of 200 and 400 correspond to sulfur contents in plants of about 0.25 and 0.50%, respectively.

The sulfur fraction in soil in natural environments that is most rapidly cycled is contained in the microbial biomass. The microbial biomass is defined as the living portion of the soil organic matter, excluding plant roots and soil animals. The carbon : sulfur ratio for bacteria is approximately 85 and for fungi it ranges from 180 to 230. The sulfur content of most microorganisms is 0.1–1.0% on a dry weight basis.

The quantities of sulfur in soil that reside in the microbial biomass have been estimated by fumigating the soil to lyse all of the microbial cells so that the sulfur contained in the cells is released. The released sulfur is then extracted from the soil and the amount measured is used to calculate the microbial biomass sulfur concentration. The equation used for biomass sulfur calculations is

$$B = F/k_s,$$

where B is the amount of microbial biomass sulfur, F is the amount of sulfate ion released from microorganisms by the fumigant, and K_s is the percentage of the biomass sulfur released by the fumigant and extracted from soil.

From about 1 to 3% of the organic sulfur in soils can be accounted for by the microbial biomass. This sulfur fraction is extremely labile with cycling times on the order of 1–10 yr, whereas the sulfur cycling times in other components of the environment often occurs on a time scale of thousands of years. In natural ecosystems, the majority of sulfur that is

Table V Volatile Sulfides Released from Organic Matter by Microorganisms

Metabolic process	Genera involved
Cysteine, cystine, and homocysteine ⟶ H_2S	Proteus Escherichia Bacillus Propionibacterium Pseudomonas
S-methylcysteine ⟶ CH_3SH	Neurospora Pseudomomas Saccharomyces Scopulariopsis
Methionine ⟶ CH_3SH	Clostridium Pseudomonas Proteus Sarcina Streptomyces Micromonospora Aspergillus Fusarium Candida Escherichia Achromobacter Propionibacterium Schizophylum Scopulariopsis
Methionine ⟶ $(CH_2)_3S$	Pseudomonas Schizophylum
S-methylmethionine ⟶ $(CH_2)_3S$	Scopulariopsis

[Adapted from Krouse, H. R., and McCready, R. G. L. (1979). Biogeochemical cycling of sulfur. In "Biogeochemical Cycling of Mineral-Forming Elements" (P. A. Trudinger and D. J. Swaine, eds.). p. 416. Elsevier Scientific Publishing Company, New York.]

utilized for plant growth is thought to have originated from the microbial biomass.

D. Volatilization of Sulfur Compounds

The transformation of organic sulfur compounds often leads to the formation of trace amounts of sulfur gasses (Table V), many of which are extremely odorous. These gases include mercaptans, alkyl sulfides, and hydrogen sulfide. Hydrogen sulfide is a common constituent of poorly drained soils, of swamp gases, and of putrifying organic remains. Volatile organic compounds have been implicated in the stimulation of repression of certain plant pathogenic fungi. They may also inhibit the conversion of ammonium to nitrate, an important reaction of the nitrogen cycle.

A variety of aerobic, facultative, and strictly anaerobic heterotrophs are capable of forming volatile sulfur gases. Generally, the amino acid cysteine cystine) is the substrate for hydrogen sulfide formation and methionine for the production of alkyl sulfides.

Much interest has been shown in recent years concerning the flux of sulfur gasses between the soil and atmosphere. These gasses are undesirable in the atmosphere because they have the potential to adversely affect climate through their aerosol-forming properties. Whereas pollution activities generally give rise to sulfur dioxide as a volatile source of sulfur, microbiological activity primarily releases its volatile sulfur in the form of hydrogen sulfide.

III. Environmental Consequences and Technological Applications

Oxidation of sulfide minerals is detrimental to the environment when large amounts are exposed to the air and water during mining. Large quantities of sulfuric acid are formed, which are transported off-site in what is called acid mine drainage. The acid inhibits plant growth and aquatic life and increases concentrations in water of many heavy metals.

In some regions of the world, draining areas such as marine sediments (e.g., the polders of The Netherlands) also results in the oxidation of sulfur and the formation of sulfuric acid that must be neutralized before the soils can be farmed. These acid sulfate soils are commonly called "cat clays." The reverse reaction in several tropical areas is also of agricultural significance. Flooding an acid sulfate soil causes reduction of the sulfate and a reversal of the acidification process. The pH of the soil rises and rice can then be grown on the flooded soil that was previously too acidic for plant growth.

Reduced sulfur compounds are commonly added to alkaline soils so that sulfuric acid is produced and the soil pH is decreased to a level acceptable for plant growth. Blueberries, for example, require the soil to be acid and elemental sulfur can be added to lower the soil pH, thus improving blueberry production. Lowering the pH to 5.0 is also important in controlling potato scab (*Streptomyces scabies*) and sweet potato rot (*Streptomyces ipomaeae*). Elemental sulfur is also added to alkaline soils to remove sodium ions and increase water movement through the soil. Other sulfur compounds are also used as fertilizer materials to improve crop production in areas of the world where sulfur is limited.

Corrosion of metals is another process in which microorganisms of the sulfur cycle are involved. Contributions to the corrosion process by microorganisms include

1. the formation of mineral acids, especially sulfuric acid, which dissolves mineral;
2. the formation of organic acids;
3. changing the electrode potential of the metal surface and creating microgalvanic cells; and
4. producing hydrogen sulfide, which is itself corrosive.

The key reaction in anaerobic corrosion is catalyzed by sulfate-reducing bacteria, which causes electrons to be stripped from the metal and added to sulfate to produce hydrogen sulfide (Fig. 4). The conditions in soils at which optimum anaerobic corrosion occur are a pH >5.5, a low concentration of oxygen, and a plentiful supply of sulfate and organic matter. [*See* CORROSION, MICROBIAL.]

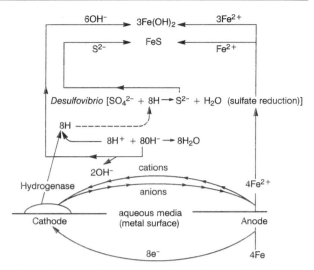

Figure 4 Anaerobic corrosion of metallic iron. [From Zajic, J. E. (1969). "Microbial Biogeochemistry," p. 227. Academic Press, New York.]

Sulfur dioxide when dissolved in water produces a solution that is acidic. Pollution by humans has greatly increased the sulfur dioxide levels in the atmosphere, which dissolves in water and is deposited on the earth as rainfall. This acidic rainfall has been implicated in the deterioration of the growth of some forests in North America and Europe. The acid rain also has a detrimental effect on structures and other artifacts such as ancient art works, causing etching and destroying their original beauty.

Many of the volatile sulfur compounds are extremely odorous and may be produced under natural conditions where organic matter is present at high concentrations and aeration is restricted. Farmyard feedlots and improperly managed composting facilities are examples of places where offensive sulfur gases are generated.

Bibliography

Fischer, U. (1988). Sulfur in biotechnology. *In* "Biotechnology, Vol. 6B, Special Microbial Processes" (H. J. Rehm, ed.), pp. 463–496. VCH Verlagsgesellschaft, Weinheim, German Federal Republic.

Grant, W., and Long, P. E. (1985). Environmental microbiology. *In* "The Natural Environment and the Biogeochemical Cycles" (The Handbook of Environmental Chemistry Ser., Vol. 1, Pt. D). (O. Hutzinger, ed.). pp. 125–237, Springer-Verlag, Berlin.

Howarth, R. W. (1984). *Biogeochemistry* **1,** 5–27.

Jorgenson, B. B. (1983). The microbial sulfur cycle. *In* "Microbial Geochemistry" (W. E. Krumbein, ed.), pp. 91–124. Blackwell Scientific Publications, Oxford.

Kreig, N. R., and Holt, J. G. (eds.) (1989). "Bergey's Manual of Systematic Bacteriology." Williams and Wilkins, Baltimore, Maryland.

Nakas, P. (1986). The role of microflora in terrestrial sulfur cycling. *In* "Microfloral and Faunal Interactions in Natural and Agroecosystems" (M. J. Mitchell and J. P. Nakas, eds.), pp. 285–316. Martinus Nijhoff Publishers, Dordrecht, The Netherlands.

Paul, E. A., and Clark, F. E. (1989). Sulfur transformations in soil. *In* "Soil Microbiology and Biochemistry," pp. 233–250. Academic Press, New York.

Stevenson, F. J. (1984). The sulfur cycle. *In* "cycles of Soil: Carbon, Nitrogen, Phosphorus, Sulfur, Micronutrients," pp. 285–320. John Wiley and Sons, New York.

Tabatabai, M. A. (1986). "Sulfur in Agriculture." American Society of Agronomy, Madison, Wisconsin. 688 pp.

Sulfur Metabolism

Larry L. Barton
The University of New Mexico

Glossary

Assimilatory sulfate reduction Synthesis of sulfur-containing organic compounds from sulfate
Disproportionation reaction Compound serves as both the electron donor and electron acceptor
Dissimilatory sulfate reduction Use of sulfate as the final electron acceptor with the generation of hydrogen sulfide

SULFUR is an important element for microorganisms because it is required for a variety of metabolic activities. Essential sulfur-containing compounds for microbial metabolism include cysteine, methionine, biotin, lipoic acid, thiamine pyrophosphate, coenzyme A, and S-adenosylmethionine. While most microorganisms can synthesize sulfur-containing amino acids and related organic compounds from sulfate or sulfide, it is the special physiological groups of aerobic and anaerobic bacteria that are involved in the respiratory-coupled transformation of inorganic sulfur compounds. Considerable diversity exists in transport activities of the various sulfur compounds and in the types of metabolic products formed by sulfur-transforming microorganisms.

I. Metabolism of Cysteine and Methionine

A. Assimilatory Sulfate Reduction

In the aerobic environment, numerous sulf-oxy anions may be produced; however, most of these anions are chemically reactive and are not an appropriate source of sulfur for microbial systems. Sulfate, the most stable of the sulf-oxy anions and the most oxidized form of sulfur with a valence of $+6$, is used by bacteria, algae, fungi, and plants in an assimilatory sulfate reduction system. Protozoans, like higher animals, do not appear to possess an assimilatory reduction process for sulfate but utilize cysteine or methionine as a source of sulfur.

The first step in sulfate utilization is the activation by adenosine triphosphate (ATP) to produce adenylyl sulfate, also known as adenosine phosphosulfate (APS). The structure of APS is given in Fig. 1A and the pathway for sulfate utilization is summarized in Fig. 2. The equilibrium of the reaction catalyzed by

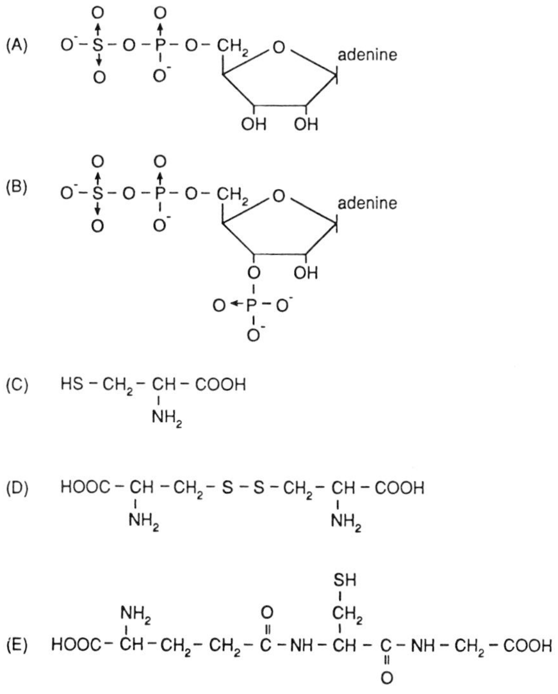

Figure 1 Structures of activated molecules and sulfur-containing amino acids. (A) Adenosine-5-phosphosulfate, (B) 3-phosphoadenosine-5-phosphosulfate, (C) cysteine, (D) cystine, (E) reduced glutathione.

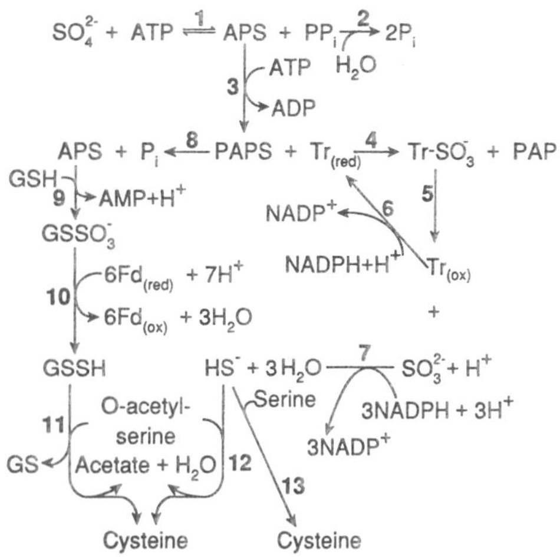

$$SO_4^{2-} + ATP \xrightarrow{1} APS + PP_i \xrightarrow{2} 2P_i$$

[Figure 2 pathway diagram, left column]

Figure 2 Metabolic pathways for the reduction of sulfate. Numbers near the arrows correspond to the enzymes for that reaction and are as follows: 1, sulfate adenylyltransferase or ATP sulfurylase; 2, inorganic pyrophosphatase; 3, adenylyl sulfate kinase or APS kinase; 4, phosphoadenylyl sulfate reductase or PAPS : thiol sulfotransferase; 5, chemical reaction; 6, NADPH : thioredoxin oxido-reductase; 7, sulfite reductase; 8, PAPS reductase; 9, APS : glutathione sulfotransferase; 10, ferredoxin : glutathione thiosulfonate reductase; 11, O-acetylserine sulfurylase for GSSH; 12, O-acetylserine sulfhydrase for HS⁻; and 13, cysteine synthase or serine sulfhydrase. ADP, AMP, and ATP, adenosine di-, mono-, and triphosphate; APS, adenosine-5-phosphosulfate; Fd, ferredoxin; GS, glutathione; GSSH, glutathione with sulfide added through a S-S linkage; GSSO₃⁻, glutathione with sulfite added through a S-S linkage; P_i, inorganic phosphatase; PAPS, phosphoadenylyl sulfate; PP_i, inorganic pyrophosphate; Tr, thioredoxin.

ATP sulfurylase (Fig. 2, reaction 1) lies in the direction of sulfate formation from APS with a reaction constant of $K_{eq} \sim 10^{-8}$. For APS formation to proceed, the equilibrium of the reaction must be shifted as a result of inorganic pyrophosphatase (Fig. 2, reaction 2) or APS kinase (Fig. 2, reaction 3) activity. Phosphoadenylyl sulfate (PAPS; Fig. 1B) results from APS kinase activity (Fig. 2, reaction 3) and has been demonstrated in cell extracts of numerous bacteria, fungi, and algae. With the production of PAPS, either of two avenues can be employed for the synthesis of cysteine (Fig. 1C).

In one system, termed the PAPS pathway, a 12-kDa dithiol protein called thioredoxin (Tr) may be involved (Fig. 2, reaction 4). After the formation of Tr-SO₃, the unstable intermediate sulfite (SO_3^{-2}) is produced and is subsequently reduced by NADH to HS⁻. A strain of *Escherichia coli* that lacks Tr may use glutaredoxin (Gr) in this PAPS pathway. The enzyme O-acetylserine sulfhydrase (Fig. 2, reaction

12) catalyzes the formation of cysteine from HS⁻ and O-acetylserine.

A second system is the APS reductive pathway, which requires glutathione (GSH) in the transfer of the sulfur group from APS to O-acetylserine with the ultimate formation of cysteine (Fig. 2, reactions 9–11). Reduced ferredoxin is required for the reduction of GS-SO₃⁻, and neither Tr nor NADPH are capable of participating in this APS pathway. The final step in the synthesis of cysteine from GSSH or HS⁻ requires the presence of O-acetylserine and involves O-acetylserine sulfhydrase, which would display specificity for GSSH (Fig. 2, reaction 11).

B. Control of Cysteine Synthesis

The synthesis of cysteine requires the coordination of two major processes: the generation of sulfide from sulfate (Fig. 2) and supply of the carbon skeleton for cysteine from the pathway of serine biosynthesis (Fig. 3). A system for serine synthesis involving phosphorylated intermediates of 2-

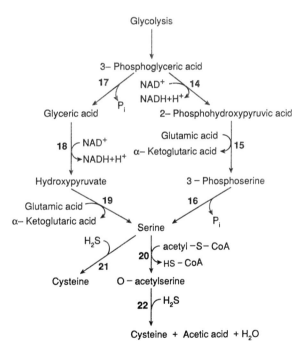

Figure 3 Metabolic pathways for biosynthesis of cysteine. Numbers correspond to the enzymes and are as follows: 14, phosphoglyceric acid dehydrogenase; 15, phosphoserine aminotransferase; 16, phosphoserine phosphatase; 17, phosphoglyceric acid phosphatase; 18, glyceric acid dehydrogenase; 19, serine aminotransferase; 20, serine O-acetyltransferase; 21, cysteine synthase (see Fig. 2, reaction 13); and 22, O-acetylserine sulfhydrase (see Fig. 2, reaction 12).

phosphohydroxypyruvic acid and 3-phosphoserine is employed by *Haemophilus influenza, Desulfovibrio desulfuricans, E. coli,* and *Salmonella typhimurium* while the pathway with intermediates of glyceric acid and hydroxypyruvic acid is preferred by other organisms. Transcriptional regulation of genes coding for enzymes producing serine from 3-phosphoglyceric acid by repressor/inducer systems would not appear likely because serine is a precursor for numerous metabolic compounds. However, serine is known to be an allosteric effector for phosphoglycerate dehydrogenase (Fig. 3, reaction 14) and is proposed to regulate the activity of phosphoglycerate phosphatase (Fig. 3, reaction 17). High concentrations of serine would decrease the activity of both phosphoglycerate dehydrogenase and phosphoglycerate phosphatase while low levels of serine would not inhibit the activity of these enzymes. O-acetylserine is produced from serine by serine O-acetyltransferase (Fig. 3, reaction 20), and this enzyme is regulated by cysteine, which serves as a negative allosteric effector to reduce enzyme activity when cysteine is present.

Cysteine production may result from the condensation of H_2S and serine (Fig. 3, reaction 21). Although this reaction has been proposed for yeast, additional research is needed to confirm this activity.

The availability of cysteine in *E. coli, Salmonella,* and *Chlorella* has a marked impact on the expression of genes coding for enzymes of the assimilatory sulfate-reduction pathway (Fig. 4). Cysteine regulates synthesis of ATP sulfurylase, APS kinase, PAPS : thiol sulfotransferase, sulfite reductase, and sulfate permease (Fig. 4, reactions 1, 3, 4, 7, and 23, respectively) by serving as a repressor at the transcription level. The absence of cysteine accumulation in the cell is known to permit transcription of genes coding for ATP sulfurylase, APS kinase, sulfite reductase, and sulfate permease.

Because SO_3^{-2} and HS^- are toxic anions, the reactions leading to their production must be highly regulated. In those bacteria that do not have a sulfite reductase, sulfite accumulation may contribute to cell death. Under conditions where pantothenic acid is deficient, *Saccharomyces sake, Saccharomyces cerevisiae,* and *Saccharomyces carlsbergensis* will produce moderately high levels of HS^- from sulfate in the synthesis of cysteine. Presumedly, this is due to the requirement of pantothenic acid for the production of O-acetylhomoserine, which is a precursor in methionine synthesis. The enzymes of sulfate metabolism are suggested to be localized in the mitochondria of *Euglena* and in the chloroplasts of *Chlorella.* Compartmentalization of the sulfate metabolizing enzymes in organelles of eukaryotic microorganisms would present an added dimension of cellular control.

In a few instances, other inorganic compounds may serve as the sulfur source for microorganisms. *Salmonella typhimurium* and *Aspergillus nidulans* will synthesize S-sulfocysteine from thiosulfate and L-serine by the enzyme S-sulfocysteine synthase. With the reduction of S-sulfocysteine by Gr or Tr,

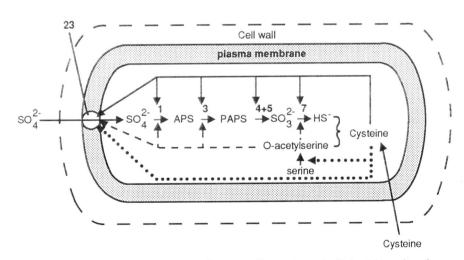

Figure 4 Regulation of assimilatory-reduction sulfate pathway in *Escherichia coli* and related bacteria. Solid line indicates repression by cysteine, dotted line reflects feedback inhibition by cysteine, and dashed line shows the induction of sulfate reduction enzymes by O-acetylserine. Numbers (1, 3–5, and 7) refer to reactions listed in Fig. 2. The presence of a sulfate permease (23) would participate in transport across the plasma membrane.

cysteine plus sulfite would be generated, thereby completing the chain of events for assimilation of thiosulfate into organic compounds. Other sulfur sources that may be used by bacteria include polythionates (Fig. 5A), hydrogen sulfide, and elemental sulfur (Fig. 5B). Many anaerobic bacteria appear to be able to synthesize cysteine from sulfide by the reactions listed in Fig. 6. Methanogens are highly versatile in that they can use the following as sole sulfur sources: methanethiol, ethanethiol, methyl sulfide, ethyl sulfide, carbonyl sulfide, carbon disulfide, dimethyl sulfoxide, thiosulfate sulfite, and elemental sulfur. Very few strains of methanogenic bacteria are capable of assimilatory sulfate reduction. [*See* METHANOGENSIS.]

C. Synthesis of Methionine

Methionine synthesis in microorganisms is associated with the aspartic acid family of amino acids. In *E. coli,* the conversion of aspartic acid to β-aspartylphosphate is catalyzed by aspartic kinases I, II, and III. Methionine and lysine repress transcriptional activity leading to the synthesis of aspartic kinases II and III, respectively, whereas repression of aspartic kinase I is multivalent requiring the presence of both threonine and isoleucine. Aspartic kinase in *Bacillus polymyxa, Bacillus licheniformis, Rhodobacter sphaeroides,* and *Rhodopseudomonas capsulatus* has only one form. In *Bacillis polymyxa* and *Rhodopseudomonas capsulatus* aspartic kinase is subject to cooperative feedback inhibition, whereas in *Rhodobacter sphaeroides* and *B. licheniformis* it is inhibited by aspartate 4-semialdehyde. Methionine, threonine, and isoleucine serve to repress the isoenzymes of homoserine dehydrogenase. Some evidence suggests that aspartic kinase and homoserine dehydrogenase are physically associated in the bacterial cell. Methionine represses transcription leading to the formation of

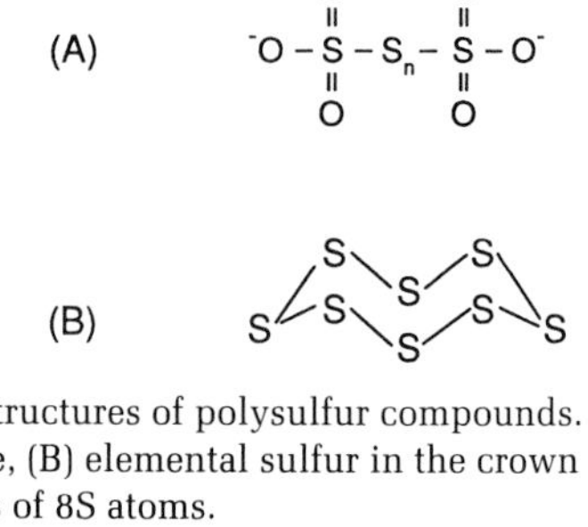

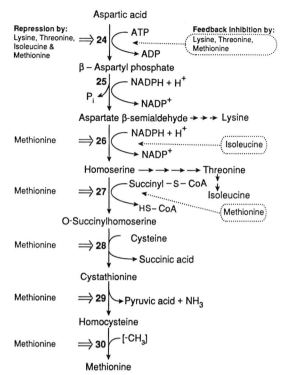

Figure 6 Metabolic pathway for synthesis of methionine from aspartic acid. Enzymes involved include the following: 24, β-aspartokinase; 25, aspartate semialdehyde dehydrogenase; 26, homoserine O-succinyltransferase; 27, cystathionine γ-synthase; 28, cystathionine β-lyase; 29, cystathionase II; and 30, homocysteine methyltransferase. Points of enzyme repression through inhibition at the transcription level are noted as ⇒. Amino acids regulating enzyme activity of the pathway through feedback inhibition are shown by ·····>.

homoserine O-succinyltransferase, cystathionine γ-synthase, cystathionine β-lyase and homocysteine methyltransferase (Fig. 6, reactions 27–30). S-adenosylmethionine controls homoserine O-succinyltransferase (Fig. 6, reaction 26) by feedback inhibition. Cysteine plus O-succinylhomoserine are enzymatically transformed to produce cystathionine and succinic acid (Fig. 6, reaction 28). Homocysteine results from cystathionine and with the acquisition of a methyl group, usually from methyl-5,6,7,8-tetrahydrofolic acid (coenzyme F), methionine is produced. In bacteria, methionine is synthesized from cysteine while in animal metabolism the converse is true in that cysteine is produced from methionine.

D. Catabolism of Sulfur Compounds

It may be necessary for bacteria of fungi to use organic sulfur compounds as a carbon and/or energy

(A) $^-O-\overset{\overset{O}{\|}}{\underset{\underset{O}{\|}}{S}}-S_n-\overset{\overset{O}{\|}}{\underset{\underset{O}{\|}}{S}}-O^-$

(B)

Figure 5 Structures of polysulfur compounds. (A) Polythionate, (B) elemental sulfur in the crown arrangement that consists of 8S atoms.

source by catabolic processes. The sulfur amino acid cysteine is commonly degraded by cysteine desulfurase to produce pyruvic acid (Table I, reaction 1). Cystine (Fig. 1D) may be converted to cysteine by cystine reductase in the presence of NADH (Table I, reaction 2). Generally, cystine is not an essential amino acid for microorganisms, although anaerobic sulfur-reducing bacteria may use cystine as a final acceptor of electrons for growth with the production of cysteine. Methionine catabolism is a multistep sequence (Table I, reactions 3), which not only results in cysteine production but also provides for the generation of α-hydroxybutyric acid, a readily metabolizable carbon unit.

II. The Role of Sulfur in Respiration

A. Dissimilatory Sulfate Reduction

Sulfate is used as an electron acceptor to support growth of a unique physiological group of organisms known as the sulfate-respiring or dissimilatory sulfate-reducing bacteria. In the pathway for the dissimilatory reduction of sulfate, the activation of sulfate to APS (Fig. 7, reaction 31) is the same as in the assimilatory sulfate-reduction process (Fig. 2, reaction 1). Major points distinguishing the dissimilatory pathway from the assimilatory pathway would include the production of HSO_3^- in the absence of PAPS and the accumulation of great quantities of hydrogen sulfide. Bisulfite, and not sulfite, has been determined to be the active species for the enzyme that transfers $6e^-$ from reduced ferredoxin or flavodoxin to the sulfur in sulfite with the formation of sulfide (Fig. 7, reaction 34).

Several bisulfite reductases have been characterized from the sulfate-reducing bacteria. These en-

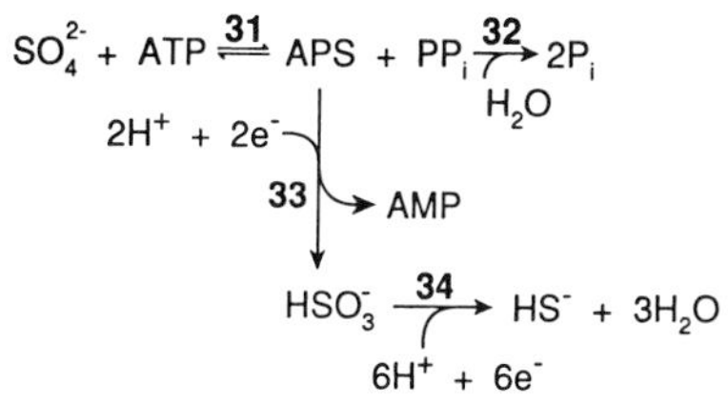

Figure 7 Dissimilatory-sulfate reduction by bacteria. Enzymes involved are the following: 31, ATP sulfurylase (see Fig. 2, reaction 1); 32, inorganic pyrophosphatase (see Fig. 2, reaction 2); 32, APS reductase; and 34, bisulfite reductase. For abbreviations, see Fig. 2 legend.

zymes all share the following characteristics: located in the cytoplasm of the cell, have a high spin state of siroheme iron, an $\alpha_2\beta_2$ structure and four (4 Fe-4 S) clusters per molecule. The bisulfite reductases are distinguished from each other by several physiochemical characteristics including molecular weight and wavelength of absorption. Characteristics of the various sulfite reductases and sulfur metabolizing enzymes are given in Table II.

Additional reactions characteristic of the sulfate-respiring bacteria would include the reduction of thiosulfate by hydrogen or formic acid to sulfide (Table III, reaction D). This reduction of thiosulfate is exergonic as is evident from the level of energy released in the reaction. Of the various sulf-oxy anions subject to reduction by the sulfate-reducing bacteria, sulfite provides the highest level of free energy and is sufficient to account for bacterial growth on hydrogen or formic acid as an energy source with sulfite as the electron acceptor. It should be noted that 1 mole of sulfate requires the equivalent of 2 moles of ATP for the activation of sulfate to APS because the inorganic pyrophosphate is hydrolyzed to phosphate (Fig. 7, reaction 32).

The energetics of the sulfate-reducing bacteria have been extensively examined. These anaerobic bacteria are capable of coupling ATP synthesis to the inward movement of protons across the plasma membrane. The mechanism of this "anaerobic oxidative phosphorylation" would appear to follow the hypothesis of Peter Mitchell. Considerable variation in the type of electron and proton carriers exists among the sulfate-respiring bacteria, especially with respect to the reducing power to drive the expulsion of protons outward across the plasma membrane. The number of environmentally derived electron donors for these bacteria exceeds 75 different types of molecules. It appears that with the presence of periplasmic, membrane, and cytoplasmic hydrogenases a "hydrogen cycle" may occur under certain condi-

Table I Microbial Catabolism of Sulfur-Containing Organic Compounds

(1) Cysteine + $H_2O \longrightarrow$ pyruvic acid + H_2S + NH_3
(2) Cystine + NADPH + $H^+ \longrightarrow$ 2 cysteine + $NADP^+$
(3) Methionine + ATP $\longrightarrow$ S-adenosylmethionine + PP_i + P_i

S-adenosylmethionine $\xrightarrow{-CH3}$ S-adenosylhomocysteine

S-adenosylhomocysteine $\longrightarrow$ homocysteine + adenosine

Homocysteine + serine $\longrightarrow$ cystathionine

Cystathionine + $H_2O \longrightarrow$ cysteine + α-ketobutyric acid + NH_3

Table II　Enzymes Metabolizing Inorganic Sulfur Compounds that Have Been Well Characterized

Enzyme	Characteristics	Location[a]	Organisms[b]
ATP sulfurylase	150 kDa 50- and 53-kDa subunits $\alpha_2\beta$ structure K_M (APS) = 0.17 mM K_M (PP$_i$) = 0.13 mM	C	*Archaeoglobus fulgidus* (*Thiobacillus ferrooxidans*)
APS reductase	220 kDa 20- and 72-kDa subunits 12 nonheme Fe/molecule 1 FAD/molecule	C	*Desulfovibrio vulgaris* (*Desulfovibrio gigas*)
Bisulfite reductase	225 kDa, α,β structure 545,393 nm absorption bands 2 siroheme/molecule	C	*Desulfovibrio baculatus* (*Desulfotomaculum nigrificans*, *Thermodiscus commune*, *Desulfovibrio gigas*)
Rhodanese (thiosulfate : sulfur transferase)	45 kDa K_M = 3.3 mM for thiosulfate	C	*Rhodopseudomonas palustris*, (*Thiobacillus denitrificans*, *T. ferrooxidans*, *Chromatium* sp., *Chlorobium vibriforme*, *Desulfotomaculum nigrificans*, *Acinetobacter caloaceticus*, *Neisseria gonorrhea*)
Sulfite : cytochrome *c* oxidoreductase	44 kDa, interacts with cytochrome c_{551}	P	*Thiobacillus versutus*
Sulfite reductase	27.2 kDa 590,405 nm absorption bands 1 (4 Fe − 4 S) cluster/molecule 1 siroheme/molecule	C	*Desulfovibrio vulgaris* (*Desulfuromonas acetoxidans*)
Sulfur reductase	200 kDa with 85 kDa subunits, contains Fe-S	M	*Wolinella succinogenes* (*Chromatium vinosum*, *Desulfuromonas acetoxidans*)
Sulfur reductase	13 kDa, contains tetraheme cytochrome c_3	P	*Desulfovibrio gigas* (*Desulfovibrio baculatus*, *Desulfovibrio salexigens*, *Desulfovibrio africanus*)
Thiosulfate : cytochrome *c* oxidoreductase	134 kDa with 45 kDa subunits	P	*Thiobacillus tepidaris* (*Pseudomonas* sp.)
Thiosulfate reductase	133 kDa K_M of 0.26 mM for thiosulfate	C	*Proteus mirabilis* (*Desulfotomaculum nigrificans*, *Desulfovibrio vulgaris*, *Desulfovibrio gigas*, *Rhodopseudomonas pastris*, *Neisseria gonorrhea*, *Chlorella fusca*)
Thiosulfate system	Enzyme A: 16 kDa plus enzyme B: 63 kDa	P P	*Thiobacillus versutus*

　[a] C, cytoplasmic; P, periplasmic; M, plasma membrane.
　[b] Characteristics of enzymes listed are from bacteria not placed in parentheses. Enzymes have been purified from those organisms in parentheses and their characteristics may differ slightly from the characteristics listed here.
　For abbreviations, see Fig. 2 legend.

tions to charge the membrane and serve as a cyclic form of proton metabolism.

Relatively unique reactions involving disproportionation of sulfite or thiosulfate are found in a few sulfate-reducing bacteria. Sufficient energy is released in these reactions (Table III, D.2.) to provide growth on sulfite or thiosulfate. In the absence of other electron donors or acceptors, *Desulfovibrio sulfodismutans* will reduce the sulfane group, outer sulfur atom in thiosulfate, to sulfide while oxidizing the central sulfur atom of thiosulfate to sulfate. One molecule of sulfite will be reduced to sulfide while

Table III Energetics of Sulfur Reactions and Bacteria that Utilize the Energy for Growth[a]

A. Oxidation of elemental sulfur

1. Aerobic chemolithotrophs

   $2S^o + 3O_2 + 2H_2O \longrightarrow 2SO_4^{2-} + 4H^+$ (-182.4 kJ/reaction)

   Acidianus brierleyi,* *Acidianus infernas*,* *Desulfurolobus ambivalens*,* *Sulfolobus acidocablarius*,* *Sulfolobus solfataricus*,* *Thiobacillus thiooxidans*, marine *Beggiatoa* spp.

B. Oxidation of sulfide and thiosulfate

1. Aerobic chemolithotrophs

   $2H_2S + O_2 + H^+ \longrightarrow 2S^o + 2H_2O$ (-124 kJ/reaction)

   Beggiatoa alba

   $S_2O_3^{2-} + O_2 + 2H^+ \longrightarrow S^o + SO_4^{2-} + 2H^+$ (-13.4 kJ/reaction)

   Thiobacillus thioparus

2. Aerobic chemolithotrophs

   $H_2S + fumarate \longrightarrow S^0 + succinate$ (-53 kJ/reaction)

   Wolinella succinogenes

3. Anoxygenic phototrophs (e^- generating reactions for photosynthesis)

   $S^{2-} \longrightarrow S^o + 2e^-$ (-4.9 kJ/reaction)

   or

   $S_2O_3^{2-} + 5H_2O \longrightarrow 2SO_4^{2-} + 8e^- + 10H^+$ (-10.9 kJ/reaction)

C. Reduction of elemental sulfur

1. $H_2 + S^o \longrightarrow HS^- + H^+$ (-27.9 kJ/reaction)

   and/or

   $Formate + S^o \longrightarrow H_2S + CO_2$ (-37 kJ/reaction)

   Acidianus brierleyi,* *Acidianus infernus*,* *Beggiatoa alba, Campylobacter* sp. (DSM 806), *Desulfovibrio africanus, Desulfovibrio baculatus, Desulfovibrio gigas, Desulfovibrio salexigens, Desulfuromonas acetexigens, Desulfuromonas acetoxidans, Desulfuromonas succinoxidans, Desulfurolobus ambivalens*,* *Spirillum* 5195, *Sulfolobus* sp.,* *Pyrobaculum islandicum*,* *Pyrodictium brockii*,* *Pyrodictium occulum*,* *Thermodiscus maritimus*,* *Thermoproteus neutrophilus*,* *Thermoproteus tenax*,* *Wolinella succinogenes*

2. $Acetic\ acid + 2H_2O + 4S^o \longrightarrow 2CO_2 + 4H_2S$ (-39 kJ/reaction)

   Desulfuromonas acetoxidans, Desulfuromonas acetexigens, Desulfuromonas succinoxidans, Desulfurella acetivorans

3. Organic substrates $+ S^o \longrightarrow H_2S + end$-products

   $Pyruvate + 3H_2O + 5S^o \longrightarrow 5H_2S + 3CO_2$ (-14.3 kJ/reaction)

   Caldococcus litoris,* *Desulfurococcus amylolyticus*,* *Desulfurococcus mobilis*,* *Desulfurococcus mucosus*,* *Pyrobaculum islandicum*,* *Pyrobaculum organotrophum*,* *Pyrococcus furiosus*,* *Pyrococcus woesei*,* *Staphylothermus marinus*,* *Thermococcus celer*,* *Thermodiscus maritimus*,* *Thermofilum pendens*,* *Thermoproteus tenax**

D. Reduction of sulfate and related compounds

1. Dissimilatory reduction systems

   $4H_2 + SO_4^{2-} + 2H^+ \longrightarrow H_2S + 4H_2O$ (-168.8 kJ/reaction)

   $3H_2 + HSO_3^- \longrightarrow HS^- + 3H_2O$ (-171.7 kJ/reaction)

   $4H_2 + SSO_3^{2-} \longrightarrow 2HS^- + 3H_2O$ (-174.1 kJ/reaction)

   Desulfotomaculum geothermicum, Desulfotomaculum guttoideum, Desulfotomaculum kuznetsovii, Desulfotomaculum nigrificans, Desulfotomaculum orientis, Desulfotomaculum ruminis, Desulfobacter curvatus, Desulfobacter hydrogenophilus, Desulfobacterium autotrophicum, Desulfobacterium catecholicum, Desulfobacterium macestii, Desulfobacterium vacuolatum, Desulfobulbus elongatus, Desulfobulbus propionicus, Desulfurococcus niacini, Desulfurococcus multivorans, Desulfomonas pigra, Desulfonema limicola, Desulfurolobus variabilis, Desulfovibrio africanus, Desulfovibrio baculatus, Desulfovibrio carbinolicus, Desulfovibrio desulfuricans, Desulfovibrio gigas, Desulfovibrio giganteus, Desulfovibrio salexigens, Desulfovibrio simplex, Desulfovibrio sulfodismutans, Desulfovibrio vulgaris, Thermodiscus commune,* *Thermodiscus mobile**

   Organic compounds used as e^- donors include lactate and pyruvate.

   $2\ Lactate + SO_4^{2-} \longrightarrow 2acetate + H_2S + 2HCO_3^-$ (-168.8 kJ/reaction)

   $4Pyruvate + SO_4^{2-} \longrightarrow 4acetate + H_2S + 4HCO_3^- + H^+$ (-356.1 kJ/reaction)

   Desulfotomaculum acetoxidans, Desulfotomaculum antarticum, Desulfotomaculum geothermicum, Desulfotomaculum sapomandens, Desulfobacter latus, Desulfobacter postgatei, Desulfobacterium anilini, Desulfobacterium indolicum, Desulfobacterium phenolicum, ***Desulfomicrobium apsheronum****, Desulfovibrio baarsii, Desulfovibrio sapovorans, Archaeoglobus fulgidus*,* plus organisms listed in section D.1. of this table.

2. Disproportionation reactions

   $S_2O_3^{2-} + H_2O \longrightarrow SO_4^{2-} + HS^- + H^+$ (-21.9 kJ/mol $S_2O_3^{2-}$)

   and

   $4\ SO_3^{2-} + H^+ \longrightarrow 3\ SO_4^{2-} + HS^-$ (-58.9 kJ/mol SO_3^{2-}) *Desulfovibrio sulfodesmutans*

---

[a] Organisms identified with * are archaebacteria while those without * are eubacteria.

three molecules of sulfite will be oxidized to sulfate.

The major isotopes in naturally occurring sulfur compounds are S-32 and S-34. The sulfate-respiring bacteria display an isotope effect in that enzymes prefer the lighter sulfur isotope. Geologically reduced sulfur has a S-32 : S-34 ratio of 22.2 whereas biologically reduced sulfur has a S-32 : S-34 ratio of 23.1. Examination of isotopes in sulfur deposits reveals that sulfate respiration occurred in early evolutionary development. For example, the sulfur domes of Louisiana and Texas are of bacterial origin, and sulfur enriched in S-32 has been found in rocks that are $2–3.5 \times 10^9$ years old.

B. Metabolism Involving Sulfur

Elemental sulfur may serve as either an electron donor or an electron acceptor by specific groups of bacteria. For the most part, anaerobic heterotrophic and chemolithotrophic bacteria would reduce elemental sulfur to sulfide. Autotrophic, aerobic bacteria would oxidize sulfur to sulfate while phototrophic bacteria may use sulfide as an electron donor for the photosynthetic process with production of sulfate and elemental sulfur as an intermediate. There are, of course, several exceptions to these generalizations.

Examples of chemolithotrophic and heterotrophic bacteria that utilize elemental sulfur are given in Table III (sections A–C). Anaerobic Eubacteria or Archaebacteria requiring elemental sulfur may synthesize cysteine from the sulfide produced or use sulfur as a final electron acceptor with the accumulation of hydrogen sulfide in the environment. Respiratory-coupled reduction of elemental sulfur in *Wolinella succinogenes*, *Spirillum* 5175, *Desulfuromonas acetoxidans* is mediated by a membrane-bound enzyme (Fig. 8). Sulfur reductase of *W. succinogenes* appears reversible with sulfide oxidation coupled to fumarate reduction where elemental sulfur and succinate are produced as end-products. The reduction of sulfur by *Desulfovibrio gigas*, *D. baculatus* DSM 1743 and Norway 4, *D. salexigens* British Guiana, and *D. africanus* Benghazi is attributed to a tetraheme cytochrome c_3 and by *Chlorobium thiosulfatophilum* is attributed to a soluble flavocytochrome c_{553}. In either case of sulfur reduction, electrons may be generated from cellular metabolism. The specific electron transport promotes a charge differential across the plasma membrane by a chemiosmotic mechanisms and accounts for growth-promoting activities.

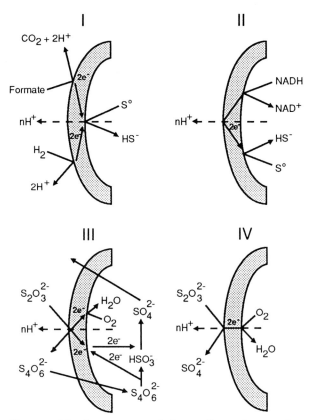

Figure 8 Membrane-associated electron flow in bacteria that metabolize inorganic sulfur compounds. (I) H_2 or formate/S^o respiration in *Wolinella succinogenes*, (II) S^O as electron acceptor in *Desulfuromonas acetoxidans*, (III) thiosulfate oxidation in *Thiobacillus tepidarus*, and (IV) thiosulfate oxidation in *Thiobacillus versutus*.

The oxidation of elemental sulfur to sulfate is observed with *Thiobacillus thiooxidans*, *Thiothrix nivea*, *Beggiatoa alba*, and marine strains of *Beggiatoa*, all of which are autotrophic bacteria. *Chromatium vinosum*, a phototrophic bacterium, may initially produce sulfur from sulfide, but as electron donors become depleted the transformation of sulfur to sulfate may occur.

Sulfide at 0.4–8 mM serves as electron donor in all Chromatiaceae and *Ectorhizospira* sp. Elemental sulfur accumulates inside the cells of the Chromatiaceae but outside the cells of the members of the Ectothiorhodospiraceae. *Chromatium* may use elemental sulfur as an electron donor in photosynthesis when sulfide is absent, and when in the dark elemental sulfur can serve as an electron acceptor with the production of sulfide. Most of the phototrophic bacteria readily use thiosulfate as an electron donor. Sulfate is not readily assimilated into organic compounds by the purple and green sulfur bacteria, but these organisms prefer to use elemental sulfur, sulfide, sulfite, or thiosulfate as sulfur sources. However, the purple nonsulfur bacteria readily assimilate sulfate.

The production of sulfate from sulfite by Chromatiaceae and Ectothiorhodospiraceae members is by a sulfite-acceptor oxidoreductase that bypasses APS as an intermediate. One member of the purple photosynthetic bacteria, *Rhodospirillum sulfidophilus,* is capable of oxidizing sulfide to sulfate without elemental sulfur as an intermediate.

III. Oxidation of Sulfur Compounds

A. Inorganic Compounds

Elemental sulfur, sulfide, thiosulfate, and tetrathionate are environmental sources of inorganic sulfur compounds oxidized by microorganisms. Thiosulfate is a metabolic intermediate in the production of tetrathionate from trithionate by *Thiobacillus tepidarius* (Fig. 9, reactions 35 and 36). Numerous heterotrophic bacteria and yeast readily convert thiosulfate to tetrathionate; however, growth coupled to the transformation of tetrathionate to sulfur appears restricted to the thiobacilli.

Sulfite serves as an intermediate in the production of sulfate from thiosulfate (Fig. 9, reaction 43). The liberation of the sulfane group results in sulfide (Fig.

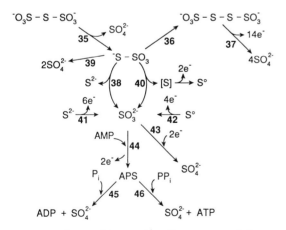

Figure 9 Oxidation of inorganic sulfate compounds by microorganisms. Enzymes required are as indicated: 35, trithionate hydrolase; 36, thiosulfate : cytochrome c oxidoreductase; 37, tetrathionate-oxidizing enzyme system; 38, thiosulfate reductase; 39, enzyme A + enzyme B; 40, rhodanese + sulfite oxidase; 41, siroheme sulfite reductase; 42, sulfur oxygenase; 43, sulfite : cytochrome c oxidoreductase; 44, APS reductase (see Fig. 2, reaction 2; Fig. 7, reaction 33); 45, ADP sulfurylase; and 46, ATP sulfurylase (see Fig. 2, reaction 1; Fig. 7, reaction 31).

9, reaction 38) or sulfur (Fig. 9, reaction 40) production depending on the enzymes employed. In the periplasm of *Thiobacillus versutus,* two sulfite molecules are produced from one thiosulfate molecule by enzymes that are associated with cytochromes c_{551} and $c_{552.5}$. The oxidation of sulfite to sulfate is found in members of the Chlorobiaceae and Chromatiaceae. In some bacteria, the presence of a siroheme sulfite reductase (Fig. 9, reaction 41) would reflect sulfite as a possible intermediate in sulfide oxidation to sulfate. The oxidation of elemental sulfur to sulfate would involve either the "APS system" (Fig. 9, reactions 42 and 44–46) or direct reduction of sulfite (Fig. 9, reaction 43).

Sulfate production from elemental sulfur occurs in numerous bacteria (Table II, section A). For the most part, whether or not the immediate substrate for sulfur oxidation is S_8 in the crown formation or polythionates, which may be mixed in with the bacterially produced hydrophobic sulfur, is unclear.

B. Organic Compounds

Dimethyl sulfoxide (Table IV) is used by specific strains of *Thiobacillus* and *Hyphomicrobium* according to the reaction scheme given in Fig. 10, reaction II. *Hyphomicrobium* consumes the formaldehyde that is generated through the serine pathway or oxidizes it to CO_2, which would be fixed by *Thiobacillus* via the Calvin cycle.

Desulfurization of fossil fuels by microorganisms is attributed to specific oxidation of sulfur compounds. Over 175 organic sulfur compounds have been identified in crude oils from Wasson, Texas. The sulfur level approaches 1.85% (w/v) of these oils and would include thiols, sulfides, disulfides, and thiophenes. Dibenzothiophene, which accounts for 70% of organic compounds in some Texas oils and 40% of organic sulfur in Middle East oils, is metabolized by numerous bacteria with the liberation of sulfur as sulfate in aerobic systems and as hydrogen sulfide in anaerobic conditions. Fossil fuels contain 20-thiophenecarboxylic acid (2-thenoic

Table IV Oxidation of Organic-Containing Sulfur Compounds

(1) Dimethylsulfoxide + NADPH + H$^+$ ⟶ dimethylsulfide + H$_2$O + NADP$^+$
(2) Dimethylsulfide + NADH + H$^+$ ⟶ methanethiol + formaldehyde
 Methanethiol + H$_2$O ⟶ formaldehyde + H$_2$S
(3) Thiophene-2-carboxylic acid + ATP + 3 H$_2$O ⟶ α-ketoglutaric acid + ADP + H$_2$S + P$_i$

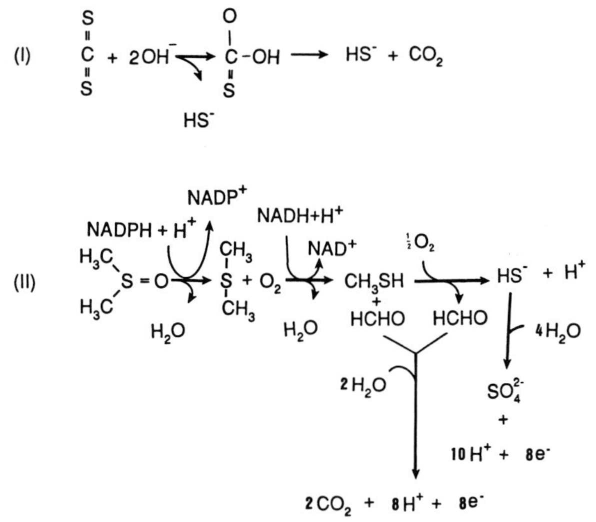

(I)

$$\underset{\substack{\parallel \\ S}}{\overset{S}{C}} + 2OH^- \longrightarrow \underset{\substack{\mid \\ S \\ \mid \\ HS^-}}{\overset{O}{C}}-OH \longrightarrow HS^- + CO_2$$

(II)

Figure 10 Enzymatic degradation of organic sulfur compounds. (I) Carbon disulfide is transformed to hydrogen sulfide and carbon dioxide; (II) dimethylsulfoxide is reduced to dimethylsulfide, which is reduced to methanethiol and formaldehyde by NADH requiring reactions. Sulfide could presumedly be oxidized to sulfate.

acid), which can be degraded by several different types of bacteria (Table IV, reaction 3). [*See* SULFUR MICROORGANISMS IN THE FOSSIL FUEL INDUSTRY.]

IV. Transport Systems

A. Binding Proteins

In environments where the solute is dilute, gram-negative bacteria will produce binding proteins specific for the needed nutrient. *Salmonella typhimurium* produces a 32-kDa binding protein that has a K_D of 3×10^{-5} M for sulfate. In the "cystine general system" of *E. coli*, a 27-kDa binding protein is involved with a binding affinity (K_D) of 1×10^{-7} M for L-cystine. The cystine-binding protein of *E. coli* has a broad specificity for diamino-dicarboxylic acids, and diaminopimelic acid will inhibit cystine binding with a K_i of 17×10^{-6} M. In both cases of cystine transport, the binding protein resides in the periplasm of the bacterial cell. Based on the work with the sulfate-binding protein, it would appear that there are about 10^4 molecules of binding protein per cell. The binding proteins concentrate the solute at the periphery of the plasma membrane and transmembrane movement would appear to require energy, perhaps in the form of ATP. The K_M values for sulfate and L-cystine transport systems are

4×10^{-6} and 1×10^{-8}, respectively, which reflect higher affinity than the binding of the solute to the binding protein.

B. Membrane-Specific Systems

Numerous sulfur compounds enter microbial cells by systems that require energy and are highly specific for the solute. Sulfate transport has been examined in several different microorganisms and a summary of these transport systems is given in Table V. Sulfate may be used by either assimilatory or dissimilatory reduction processes. In sulfate-reducing bacteria, the transport of sulfate is driven by proton symport, which follows the chemiosmotic principles of transport. Sulfide moves across membranes by diffusion and not by an active transport process. A

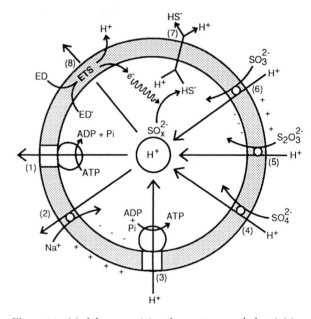

Figure 11 Model summarizing the proton-coupled activities in sulfate-reducing bacteria. (1) ATPase activity that contributes to the charge across the membrane by pumping H^+ outward. (2) Na^+/H^+ anteport system. (3) Proton-driven ATP synthetase reaction. Proton-driven symport with sulfate (4), thiosulfate (5), or sulfide (6) accounting for transmembrane uptake of sulfur-oxy anions. (7) Sulfide exits across the membrane as H_2S. (8) The electron transport system (ETS) obtains electrons from external or internal oxidations of electron donors ED or ED', respectively. Electrons are diverted to the dissimilatory sulfate reduction system and protons are pumped outward by the ETS, attributed in part to the quinone cycle. [From Fauque, G., LeGall, J., and Barton, L. L. (1991). Sulfate-reducing and sulfur-reducing bacteria. *In* "Variations in Autotrophic Life" (J. M. Shively and L. L. Barton, eds.), pp. 271–338. Academic Press, London.]

Table V Characteristics of Sulfate Transport in Microorganisms

Organism	K_M (10^6)	Comments
I. Bacteria		
Anacystis nidulans	75	Inhibited by selenite and thiosulfate
Desulfovibrio desulfuricans	?	Driven by Δ pH across plasma membrane
Desulfobacter postgatei	200	Uptake ceases at 5–20 μM
Desulfovibrio salexigens	76.7	V_{max}, 2.6 μmoles/mg dry weight/hr
Desulfovibrio sapovorans	7.3	V_{max}, 0.8 μmoles/mg dry weight/hr
Desulfovibrio vulgaris (Hildenborough)	32.2	V_{max}, 1.1 μmoles/mg dry weight/hr
Desulfovibrio vulgaris (Marburg)	4.8	V_{max}, 1.5 μmoles/mg dry weight/hr
Escherichia coli	50	Inhibited by selenate
Paracoccus denitrificans	?	Driven by Δ pH across the plasma membrane
Salmonella typhimurium	3.6	Inhibited by selenite, chromate, sulfite, and thiosulfate
Thiobacillus ferrooxidans	380	Cells at pH 2.2 bind 1 μmole sulfate/mg dry weight
II. Fungi		
Aspergillus nidulans	75	Transports Group VI anions
Neurospora crassa		
Conidia	200	Termed Permease I, encoded by *cys-13* $^{+}$ locus
Mycelium	8	Termed Permease II, encoded by *cys-14* locus
Penicillium notatum	6.3	Kinetics indicates Ca^{2+} and H^+ added to carrier before sulfate; selenate, molybdate, and thiosulfate are transported on this system
III. Algae		
Chlorella pyrenoidosa	3.9	Inhibited by anions of Group VI with K_i for chromate of $4.1 \times ^{-7}$ M.
Monochrysis lutheri	3.2	Inhibited by chromate or selenate but not tungstate, molybdate, nitrate, or phosphate

model reflecting the transport-coupled activities in sulfate-reducing bacteria is presented in Fig. 11. While the chemiosmotic transport of sulfur compounds has not been demonstrated conclusively in *E. coli* or *S. typhimurium*, proton-driven transport systems are widely distributed in bacteria.

In the transport of methionine by *S. typhimurium* and *E. coli*, two distinct permeases have been demonstrated. One is a general permease that recognizes histidine, phenylalanine, tyrosine, tryptophan, and methionine with an apparent K_M of 10^{-4} M and the other is highly specific for methionine with a K_M of 10^{-7} M. The driving system for this active transport has not been established.

Active transport systems for sulfur compounds function in eukaryotic microorganisms; however, the mechanism of transmembrane movement is less clear than with bacteria. Due to the solute specificity and high affinity, these transport processes are referred to as carrier systems. With respect to sulfate, carrier systems have been demonstrated in *Neurospora*, *Penicillium*, and *Aspergillus* species. A kinetic evaluation of sulfate transport in *Penicillum notatum* revealed a specificity for H^+ as well as a divalent cation such as Ca^{2+}; these are bound by the carrier before sulfate. In *Penicillium* and *Asper-*

gillus, thiosulfate, selenate, and molybdate appear to be transported on the sulfate transport carrier system. On the other hand, sulfite and tetrathionate appear to require a distinct transport system. In the filamentous fungi, methionine and thiosulfate repress the active transport of sulfate while L-djenkolic acid and L-cysteic acid derepress the sulfate transport system. Carrier systems of high specificity would explain the transport of numerous organic sulfur compounds (Table VI).

V. Diversity of Sulfur Metabolism

A. Complex Sulfur Compounds of Metabolic Importance

Sulfur-containing organic compounds may serve as vitamins or provide cells with the component for important activities. Lipoic acid (Fig. 12A) thioctic acid, is a coenzyme involved in acyl transfer reactions with the acyl group attached to the sulfur atom closest to the carboxyl of lipoic acid. Release of lipoic acid from proteins by a "lipoic acid-splitting enzyme" was first demonstrated in the protozoan *Tetrahymena pyriformis*. Lipoic acid is a microbial

Table VI Specific Solute Transport Systems in Microorganisms

Compound	Organism	K_M
Biotin	Baker's yeast	1×10^{-6}
Choline O-sulfate	Neurospora crassa	?
Cysteic acid	Neurospora crassa	7×10^{-6}
Cysteine	Penicillium chrysogenum	1.4×10^{-4}
Cystine	Escherichia coli	
General system		3×10^{-7}
Specific system		2×10^{-8}
Cystine	Penicillium chrysogenum	2×10^{-5}
Glutathione	Penicillium chrysogenum	1.7×10^{-5}
Methionine	Baker's yeast	1.2×10^{-5}
Methionine	Neurospora crassa	2.3×10^{-6}
Methionine	Penicillium chrysogenum	1×10^{-5}
S-adenosylmethionine	Baker's yeast	3.3×10^{-6}
Thiosulfate	Aspergillus nidulans	75×10^{-6}

growth factor that may be involved in oxidative caboxylation of α-keto acids.

Thiamine (Fig. 12B), as thiamine pyrophosphate (TPP), plays an important role in conversion of substrates to "active aldehydes," which result in release of free aldehydes or the transfer to an appropriate receptor. TPP is important in carboxylase

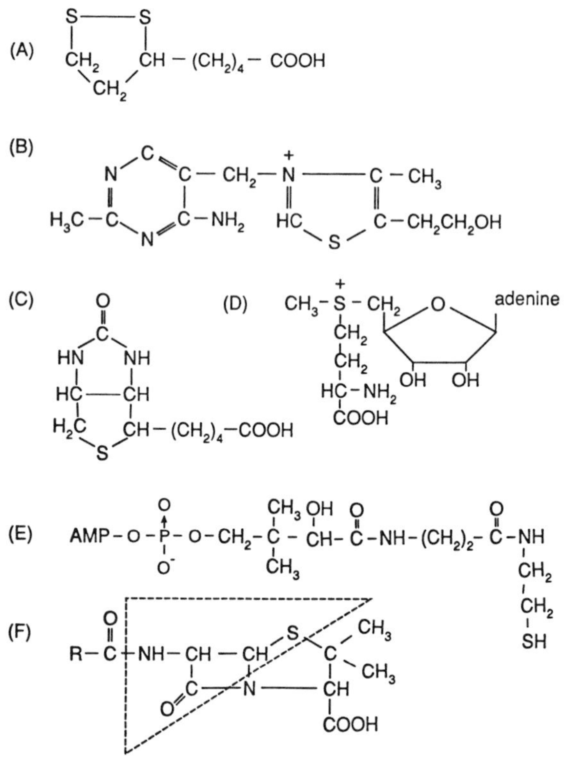

Figure 12 Structures of important sulfur-containing compounds. (A) Lipoic acid, (B) thiamine, (C) biotin, (D) S-adenosylmethionine, (E) coenzyme A, and (F) penicillin core structure.

reactions, enzymatic conversion of pyruvate to acetyl-CoA, decarboxylation of α-keto acids, and transketolase reactions.

Biotin (Fig. 12C) (vitamin H) is synthesized from cysteine, pimelic acid, and carbamyl phosphate. Biotin was initially found to be involved in aspartic acid metabolism, decarboxylation of succinic acid, and synthesis of oleic acid. Biotin is important in numerous carboxylation reactions with ATP hydrolysis providing the energy for these reactions.

S-adenosylmethionine (Fig. 12D) is essential as a methyl-group donor in numerous biosynthetic reactions. The charged sulfonium compound is unstable and reacts with acceptor molecules to produce phosphatidylcholine, creatine, ε-N-methyl lysine, 5-methyl cytosine and N-methyl adenine. Methylation and demethylation of "methylatable chemotactic proteins," by S-adenosylmethionine, is associated with chemotactic response in bacteria.

Coenzyme A (Fig. 12E) is composed of adenosine 3'-phosphate 5'-diphosphate, pantothenic acid, and β-mercaptoethylamine. This molecule is important in activating acyl groups in transfer of acyl groups to metabolites. Its presence is essential for all cell systems.

Glutathione, the tripeptide γ-glutamylcysteinylglycine (Fig. 1E), has been reported present in several different types of bacteria. The thiol group of reduced GSH helps to maintain cysteine in the reduced state by nonenzymatic reactions. GSH is important in chemical reduction of peroxides and reduction of toxic metals or metalloids.

In the family of penicillins, the essential structure for inhibition cross-bridge formation in peptidoglycan biosynthesis is the 6-aminopenicillic acid moiety (Fig. 12F). The synthesis of this core segment of penicillin is derived from cysteine and valine. Unlike the other compounds listed in this section where the presence of sulfur is essential for molecular activity, sulfur in 6-aminopenicillic acid does not enter into subsequent chemical reactions but serves only to provide appropriate molecular structure for inhibition of bacterial cell wall synthesis.

Many strains of bacteria and yeast produce thiols that are important in nucleic acid metabolism. Examples of these compounds would include *S*-methylcytokinins, 6-(4-hydroxy-3-methyl-2-butenyl-amino)-2-methylthio-9-β-D-ribofuranosylpurine and 6-(3-methyl-2-butenylamino)-2-methylthio-9-β-D-ribofuranosylpurine or the thionucleotide 5(6)-carboxymethoxy-methyl-2-thiouracil. While the physiological activities of these compounds are known, the mechanisms of synthesis and breakdown are unexplored.

B. Sulfatases

Several bacteria and fungi produce an enzyme that hydrolyzes sulfate from 2-hydroxy-5-nitrocatechol sulfate, *p*-nitrophenyl sulfate, or a variety of alkalyl-, aryl-, or carbohydrate esters. Microbial production of sulfatases is repressed in media where adequate levels of inorganic sulfate exist and is derepressed when sulfur is limiting. The arylsulfatases are either bound to the cell wall of gram-positive bacteria and fungi or are in the periplasm of gram-negative bacteria. The control of arylsulfatase appears complex, with tyramine suggested to derepress sulfatase production in *Enterobacter aerogenes* and *Klebsiella aerogenes*. The enhancement of sulfatase activity by addition of tyramine to a culture of *Aspergillus oryzae* may be due to production of a phenylsulfotransferase in addition to induction of the sulfatase.

C. Lipids

A broad array of sulfur-containing lipids is found in microorganisms, but none is found in the membranes of viruses. Present in many different types of photosynthetic microorganisms is sulfoquinovosyldiacylglycerol (Fig. 13A).

Lipids containing both chlorine and sulfur (Fig. 13B) have been reported in over 30 different algae

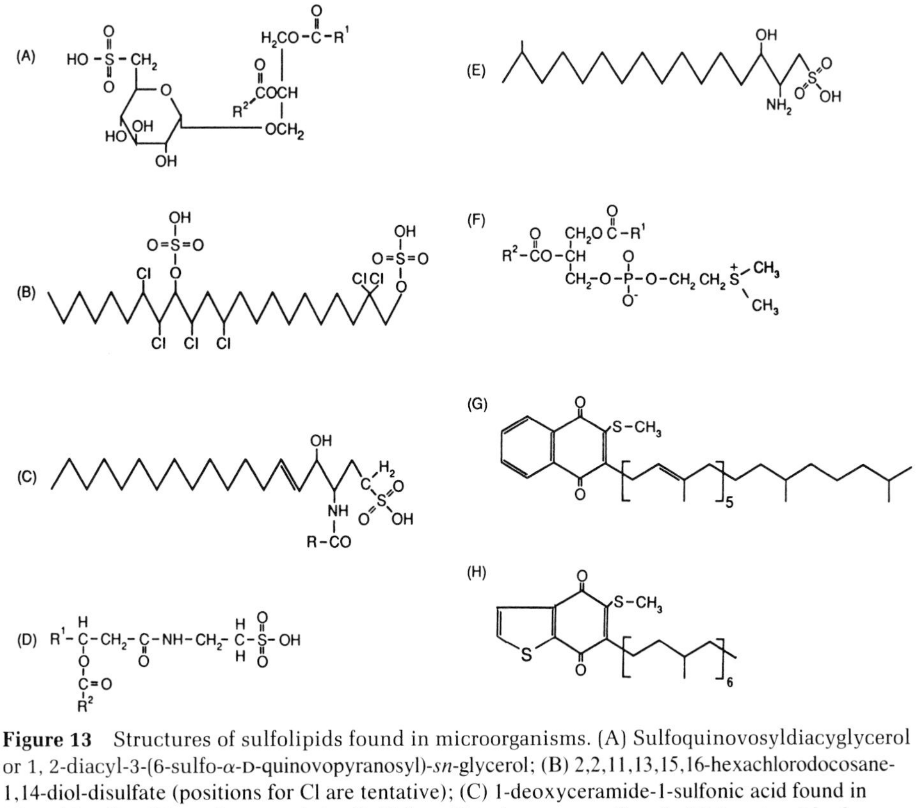

Figure 13 Structures of sulfolipids found in microorganisms. (A) Sulfoquinovosyldiacyglycerol or 1, 2-diacyl-3-(6-sulfo-α-D-quinovopyranosyl)-*sn*-glycerol; (B) 2,2,11,13,15,16-hexachlorodocosane-1,14-diol-disulfate (positions for Cl are tentative); (C) 1-deoxyceramide-1-sulfonic acid found in *Nitzschia alba*; (D) taurine-containing lipid from *Tetrahymena pyriformis*; (E) 2-amino-3-hydroxy-15-methyl-hexadecane-1-sulfonic acid or capnine from the gliding bacteria of the *Cytophaga–Flexibacter* group; (F) phosphatidylsulfocholine from *Nitzschia alba*; (G) methionaquinone or 2-methylthio-3-VI,VII-tetrahydro-heptaprenyl-1,4-napthoquinone; and (H) caldariellaquinone.

belonging to the Chlorophyceae, Chrysophyceae, Phaeophyceae, Rhodophyceae, and Euglenophyceae and also in numerous cyanobacteria. Other examples of sulfur-containing lipids would include capnine, sulfur-containing quinones, taurine-containing lipid, phosphatidyl sulfocholine, and ceramide sulfonate (Fig. 13C–F). Unusual sulfur-containing napthoquinones, such as methionaquinone (Fig. 13G), have been isolated from thermophilic hydrogen bacteria. A heterocyclic quinone, caldariellaquinone (Fig. 13H), is a major lipid in *Sulfolobus solfataricus*, the sulfur-oxidizing thermophilic archaebacterium. The unique structures of these lipids and the great distribution throughout the microbial world suggest fundamental roles for these compounds. To date, little is known about the biosynthetic pathways of these special lipids but it is assumed that the oxidized sulfur compounds originate from PAPS.

Considerable interest was generated with the demonstration of 2,3,6,6′-tetra-acyltrehalose 2′-sulfate and 2,3,6-triacyltrehalose 2′-sulfate as sulfolipids in *Mycobacterium tuberculosis*. The speculation that these trehalose-containing sulfur lipids may protect the bacteria from lethal phagocytic action by prevention of lysosome fusion with the phagosome has been proposed; however, the functioning of sulfur-lipid in promoting intracellular survival awaits further research.

D. Production of Sulfur Compounds

Numerous thiols and sulfides are produced by microorganisms from decomposition of organic sulfur compounds. Methanethiol (methylmercaptan), dimethylsulfide, and dimethyldisulfide are produced by a vast array of fungi and algae as intermediates in the conversion of *S*-methylcysteine to cysteine and methionine or from the metabolism of allylcysteine, allylcysteine sulfoxide, and sulfonium compounds. These volatile compounds are end-products of bacterial decomposition of sulfur-rich plants and are considered to inhibit the growth of phytopathogenic fungi including *Botrytis allii*, *Colletrichum circinans*, and *Aphanomyces euteiches*. Spoilage of fish, poultry, and beef with the production of volatile sulfur end-products has been attributed to numerous bacterial species while the fermentation of beer by bottom yeast are also known to release undesirable volatile sulfur compounds. In *Lentinus edodes*, the shiitake mushroom, a characteristic aroma is attributed to lenthionine (1,2,3,5,6-pentathiepane), which is produced from lentinic acid. Not all of the volatile sulfur compounds are undesirable because methane thiol, a product of bacterial metabolism in dairy products, is an important component in the aroma of cheddar cheese.

Several sulfur compounds are produced in eukaryotic microorganisms by enzyme processes that have received little attention. *Daedelina juniperina*, a wood-rotting fungus, produces 5-(prop)-1-ynyl)-thienyl-2-aldehyde; *Saccharomyces cerevisiae*, a yeast, produces β-methyllanthionine and *Coprinus atramentarius*, a mushroom, produces tetraethylthiuram disulfide. *Tilletia caries*, the stinking smut of wheat, produces djenkoic acid, which accumulates in the fungal spores. 6-Aminopenicillinic acid (Fig. 12F), the active component of penicillin, produced by *Penicillium chrysogenum*, and phalloidin, a mushroom toxin produced by *Amanita phalloides*, are examples of bioactive sulfur-containing peptides. Dimethylpropiothetin, a sulfonium compound, is produced by numerous green and red algae and may function to promote cellular osmotic balance. Taurine, cysteic acid, and D-cysteinolic acid are produced by fungi as well as by red, brown, and green algae, whereas choline sulfate is found in red algae and asexual spores of *Aspergillus niger*. Polysaccharides containing galactose-6-sulfate occur in *k*-carrageenin and are produced by *Chondrus crispus* (Irish moss), whereas fucoidin, a product of brown algae, contains fucose-4-sulfate.

E. Iron–Sulfur Centers

Many enzymes contain iron and sulfur in a specified arrangement, referred to as a Fe-S cluster, in which the iron is bonded through four cysteinyl-sulfur ligands. Sulfur as sulfide is bound to the iron atoms and due to this inorganic interaction is referred to as ''acid-labile'' sulfur. These iron–sulfur prosthetic groups may be formed as either a Fe_2-S_2 cluster or Fe_4-S_4 cluster in enzymes or electron transport proteins. The models proposed for the Fe_2-S_2 and Fe_4-S_4 clusters (Fig. 14) have been based on the X-ray diffraction studies by numerous investigators.

Ferredoxins are classical examples of iron–sulfur proteins that participate in numerous oxidation-reduction reactions that typically have a low redox potential of −180 to −455 mV. Plantlike ferredoxins, with a Fe_2-S_2 cluster, are found in the bacteria *Agrobacterium tumefaciens*, *Azotobacter vinelandii*, *E. coli*, *Halobacterium halobium*, and *Pseudomonas putida* and in the alga *Nostoc* strain MAC. Most common among the bacteria is the Fe_4-S_4 ferredoxin, which may occur as one, two, three,

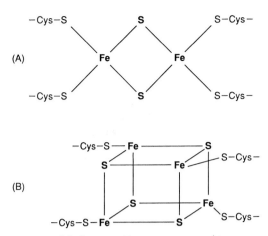

Figure 14 Model of iron–sulfur centers in proteins.

or four clusters per ferredoxin molecule. Electron paramagnetic resonance spectroscopy has proven useful in evaluation of the iron–sulfur centers in ferredoxin-mediated reactions. A partial list of bacteria with the Fe_4-S_4 ferredoxin includes *Azotobacter vinelandii*, *B. polymyxa*, *Bacillus stearothermophilus*, *Chlorobium limicola*, *Chromatium vinosum*, *Clostridium acidi-urici*, *Clostridum pasteurianum*, *Clostridum thermoaceticum*, *D. desulfuricans*, *D. gigas*, *Mycobacterium flavum*, *Rhodospirillum rubrum*, and *Peptococcus aerogenes*.

Iron and sulfur are important elements in many enzymes involved in oxidation-reduction reactions. Examples of enzymes with acid-labile sulfur include nitrogenase, nitrate reductase, formate dehydrogenase, xanthine dehydrogenase, sulfite reductase, succinate dehydrogenase, bisulfite reductase, glutamate synthase, dihydroorotate dehydrogenase, APS reductase, trimethylamine dehydrogenase, *w*-hydroxylase, pyruvate-ferredoxin oxidoreductase, and hydrogenase. The hydrogenase genes from *Desulfovibrio* have been cloned in *E. coli;* the resulting protein is active, indicating the presence of the necessary iron–sulfur centers in the enzyme. Biosynthesis and insertion of iron–sulfur centers into the apoprotein are under study in several laboratories.

VI. Pertubations on Sulfur Systems

The nature of molecular interactions between inhibitors and sulfur compounds may be attributed to chemical analogs of sulfate or to interaction with thiol groups.

A. Molybdate

In certain conditions, molybdate may enter cells on a sulfate transport system or act competitively as an inhibitor for sulfate entry. At pH levels of 5.0 or lower, molybdate would exist as $HMoO_4^{3-}$ and presumedly would not compete with sulfate in solute transport systems. With either assimilatory or dissimilatory sulfate reduction, molybdate interferes with the formation of APS. Molybdate plus ATP in the presence of the enzyme ATP sulfurylase (Fig. 2, reaction 1; Fig. 7, reaction 31) produces an unstable molecule of adenosine phospho-molybdate, which quickly degrades to adenosine monophosphate plus molybdate. Thus, the cells are deprived of reduced sulfur compounds.

B. Chromate

The toxicity of chromate observed in bacteria that require sulfate either for assimilatory or dissimilatory reduction may be attributed to competition with the sulfate transport system. If the concentration of chromium is equal or greater than sulfate levels, chromium could serve as an inhibitor and now allow sulfate into the cell. On the other hand, low concentrations or chromate may enter the cells by the sulfate transport system and be toxic to the cells by activities distinct from sulfate metabolism.

C. Selenium

Selenate or selenite may enter cells by the transport systems used by sulfate or sulfite. At appropriate ratios of selenate and sulfate, selenate could be metabolized by ATP sulfurylase with the putative formation of adenosine phospho-selenate. This would interfere with metabolic systems requiring APS or PAPS leading to the formation of reduced sulfur. There is the possibility that seleno-cysteine or seleno-methionine may be synthesized in high levels of selenide and the resulting seleno-amino acids would be inappropriate substitutes for cysteine and methionine in proteins. However, it must be noted that specific microbial enzymes do require seleno-cysteine, which is inserted into the protein by a specific codon. A delicate balance exists between selenium requirement and toxicity. Research is needed to understand the mechanisms for regulation of selenium metabolism and the specificity between sulfur and selenium in biosynthetic systems.

Bibliography

Amesz, J. (1991). Green photosynthetic bacteria and heliobacteria. *In* "Variations in Autotrophic Life" (J. M. Shively and L. L. Barton, eds.), pp. 99–120. Academic Press, London.

Dahl, C., Koch, H.-G., Kueken, O., and Trüper, H. G. (1990). *FEMS Microbiol. Lett.* **67,** 27–32.

Dawes, E. A., and Large, P. J. (1982). Class II reactions: Synthesis of small molecules. *In* "Biochemistry of Bacterial Growth" (J. Mandelstam, K. McQuillen, and I. Dawes, eds.), pp. 159–184. Blackwell Scientific Publications, Boston.

Drews, G., and Imhoff, J. F. (1991). Phototrophic purple bacteria. *In* "Variations in Autotrophic Life" (J. M. Shively and L. L. Barton, eds.), pp. 51–98. Academic Press, London.

Fauque, G., LeGall, J., and Barton, L. L. (1991). Sulfate-reducing and sulfur-reducing bacteria. *In* "Variations in Autotrophic Life" (J. M. Shively and L. L. Barton, eds.), pp. 271–338. Academic Press, London.

Foght, J. M., Fedorak, P. M., Gray, M. R., and Westlake, D. W. S. (1990). Microbial desulfurization of petroleum. *In* "Microbial Mineral Recovery" (H. L. Ehrlich and C. L. Brierley, eds.), pp. 379–408. McGraw-Hill, Publishing, New York.

Kadota, H., and Ishida, Y. (1972). *Annu. Rev. Microbiol.* **26,** 127–138.

Kelly, D. P. (1988). Oxidation of sulfur compounds. *In* "The Nitrogen and Sulfur Cycles" (J. A. Cole and S. J. Ferguson, eds.), pp. 65–98. Forty-second Symposium of the Society for General Microbiology, Cambridge University Press, Cambridge.

Kotyk, A., and Janacek, K. (1975). "Cell Membrane Transport," pp. 191–212. Plenum Press, New York.

Kuenen, J. G. (1989). Colorless sulfur bacteria. *In* "Bergey's Manual of Systematic Bacteriology," Vol. 3 (J. T. Staley, M. P. Bryant, N. Pfennig, and J. G. Holt, eds.), pp. 1834–1837. Williams & Wilkins, Baltimore.

Le Faou, A., Rajagopal, B. S., Daniels, L., and Fauque, G. (1990). *FEMS Microbiol. Lett.* **75,** 31–62.

Peck, H. D., Jr., and Lissolo, T. (1988). Assimilatory and dissimilatory sulfate reduction: Enzymology and bioenergetics. *In* "The Nitrogen and Sulfur Cycles" (J. A. Cole and S. J. Ferguson, eds.), pp. 99–132. Forty-second Symposium of the Society for General Microbiology, Cambridge University Press, Cambridge.

Poole, R. K. (1988). Chemolithotrophy. *In* "Bacterial Energy Transduction" (C. Anthony, ed.), pp. 183–230. Academic Press, London.

Ratledge, C., and Wilkinson, S. G. (1988). Fatty acids, related and derived lipids. *In* "Microbial Lipids" (C. Ratledge and S. G. Wilkinson, eds.), pp. 23–53. Academic Press, New York.

Richmond, D. V. (1973). Sulfur compounds. *In* "Phytochemistry," Vol. 3 (L. P. Miller, ed.), pp. 41–73. Van Nostrand Reinhold, New York.

Smith, D. W., and Strohl, W. R. (1991). Sulfur-oxidizing bacteria. *In* "Variations in Autotrophic Life" (J. M. Shively and L. L. Barton, eds.), pp. 121–146. Academic Press, London.

Widdel, F., and Pfennig, N. (1984). Dissimilatory sulfate- or sulfur-reducing bacteria. *In* "Bergey's Manual of Systematic Bacteriology," Vol. 1 (N. R. Krieg and J. G. Holt, eds.), pp. 663–679. Williams & Wilkins, Baltimore.

Wood, B. J. B. (1988). Lipids of algae and protozoa. *In* "Microbial Lipids" (C. Ratledge and S. G. Wilkinson, eds.), pp. 807–867. Academic Press, New York.

Yoch, D. C., and Carithers, R. P. (1979). *Microbiol. Rev.* **43,** 384–421.

Sulfur Microorganisms in the Fossil Fuel Industry

Andrea Maka and Douglas J. Cork
Institute of Gas Technology, Illinois Institute of Technology

Glossary

Char Carbonaceous product formed from the gasification of coal

Coal Complex material formed from plant substances

Fossil fuels Coal, coal char, oil, oil shale, and natural gas

Hydrogen sulfide Toxic gas released when fossil fuels are combusted

Microbial desulfurization Removal of sulfur by bacterial action

Oil shale Kerogen material from which synthetic fuels are produced

Sulfur Nonmetallic element present in fossil fuels occurring in either the inorganic or organic form

Sulfur dioxide Toxic gas released when fossil fuels are combused

SULFUR is a major process stream constituent produced in various energy and chemical industries. When a fossil fuel is burned, it releases sulfur compounds such as hydrogen sulfide and sulfur dioxide into the environment. These compounds must be eliminated or reduced from the environment. Sulfur microorganisms are being developed and utilized to remove the sulfur from these fossil fuels prior to combustion, in the abatement of hydrogen sulfide and sulfur dioxide generated by burning the fossil fuel, and in the cleanup of sulfate wastes generated from chemical and physical processes that desulfurize fossil fuels.

I. Introduction

In today's world, the concern to protect the environment is immense. Governments and industries throughout the world are trying to establish research, development, and implementation initiatives to prevent environmental pollution. A major environmental pollution problem is the disposal and treatment of sulfur. Sulfur is a major process stream constituent produced in the food, fertilizer, mining, chemical, and fossil fuel industries. In the fossil fuel industry, when a fossil fuel is burned it releases sulfur compounds such as hydrogen sulfide and sulfur dioxide, which are toxic gases that have been associated with respiratory disorders. In addition, when the sulfur dioxide enters the upper atmosphere, it can form dilute sulfurous acid, which can fall to the earth as acid rain. Acid rain is thought by some to damage vegetation and wildlife, especially in the northeastern United States, Canada, and Central Europe.

Concern about acid rain has led to the passage by the U.S. Congress of the Clean Air Act Standard program. The program has resulted in the initiation of research to develop technologies that will result in the elimination or reduction of these toxic sulfur compounds. This article discusses the research being conducted with microorganisms to eliminate or reduce the presence of these sulfur compounds in the environment. Sulfur microorganisms are being developed and utilized to remove the sulfur from the fossil fuels coal, coal char, and oil shale, in the cleanup of hydrogen sulfide and sulfur dioxide generated by burning fossil fuels, and in the cleanup of sulfate wastes generated from chemical and physical processes that desulfurize fossil fuels. [*See* SULFIDE-CONTAINING ENVIRONMENTS.]

II. Microbial Desulfurization of Coal, Coal Char, and Oil Shale

A. Microbial Desulfurization of Coal

Coal is the most abundant fossil fuel in the United States and comprises about 75% of the total resources of fossil fuels. In the United States today, it is responsible for almost 60% of the electricity generated. The coal reserves at present consumption rates are expected to last well into the twenty-second century.

1. Coal Structure

The sulfur in coal is in both inorganic and organic forms. A major compound of the inorganic material is FeS_2, which is present either as pyrite or marcasite, depending on its structure. The inorganic material is quite separate from the chemical structure of the coal.

The exact forms of the organic sulfur compounds present in coal are not known, but major groups have been classified and are thiol, sulfide, disulfide, and thiophene groups. The organic sulfur is an integral part of the coal molecule and, thus, is difficult to remove.

2. Sulfur Removal from Coal

To meet the Clean Air Act Standard for sulfur emissions, about 90% of the sulfur from coal must be removed. Sulfur can be removed from coal either prior to combustion, during combustion, or after combustion. Physical and chemical technologies exist that are applied prior to combustion that can remove 30–50% of the pyritic sulfur or 10–30% of the total sulfur. Also, some microorganisms (*Thiobacillus ferrooxidans, Thiobacillus thiooxidans,* and *Sulfolobus acidocaldarius*) can remove the pyrite from coal. These microorganisms oxidize the iron pyrite at low pH (2.0–3.5), as a source of metabolic energy, producing ferric iron and sulfate. The overall microbial oxidation of pyrite is

$$4\ FeS_2 + 15\ O_2 + 2\ H_2O \longrightarrow 2\ Fe_2(SO_4)_3 + 2\ H_2SO_4\ (1)$$

The microbial removal of sulfur from coal is capable of removing 90% or more of the inorganic sulfur within a few days. Yet a commercial process based on a microbial approach has not been realized because the physical and chemical technologies are more rapid and less expensive.

Although these physical and chemical technolo-gies remove the pyritic sulfur, they cannot remove the organic sulfur. Microorganisms are being developed that will remove the organic sulfur from the coal prior to combustion.

Because the coal molecule is such a complex structure, sulfur microorganisms are being developed using model organic sulfur compounds. These are compounds that are structurally similar to the organic sulfur present in the coal. A more detailed understanding of the ability of sulfur microorganisms to utilize and metabolize organically bound sulfur can be obtained from studies that employ model compounds. Enrichment culture techniques are typically used to isolate for microorganisms capable of organic sulfur removal from model compounds. The chemical dibenzothiphene (DBT) is generally regarded as a good model compound representative of organic sulfur found in coal. It is used as the substrate of choice in most enrichment culture experiments. Microorganisms (*Pseudomonas* spp., *Rhizobium* spp., *Acinetobacter* spp., and *Beijerinckia* spp.) have been isolated that are capable of utilizing DBT as their sole source of carbon, energy, and sulfur. Also, microorganisms (*Pseudomonas* spp.) have been isolated that can use DBT as a cometabolite, i.e., degrading DBT while growing on an alternative substrate. These organisms have the capacity to degrade DBT by hydroxylating the benzoid component of the compound to a dihydroxy-dihydro-derivative, cleaving the ring between the diol substituent and then producing in sequence 4-[2-(3-hydroxy)-thionaphthenyl]-2-oxo-3-butenoic acid and 3-hydroxy-2-formylbenzo-thiophene (Fig. 1, pathway B).

An alternative pathway to DBT metabolism has recently been postulated, the so-called sulfoxide–sulfone–sulfonate–sulfate, or 4S, pathway (Fig. 1, pathway A). In this thermodynamically favorable set of reactions, the sulfur is released from the organic frame as a sulfate ion, no carbon–carbon bonds are disrupted, and the carbon frame is only minimally oxidized. Microorganisms, which include the Institute of Gas Technology's *Rhodococcus rhodochrous* and ARTECH's *Pseudomonas* designated CB1 have been isolated with the ability to perform this series of oxidations. As shown in Fig. 1, it is pathway A of the sulfur microorganisms that is desired. In this pathway, the sulfur from the coal is solubilized while leaving the carbon moiety of the coal intact. Consequently, oxidization of the coal carbon is minimal, and coal retains its calorific value. [*See* SULFUR METABOLISM.]

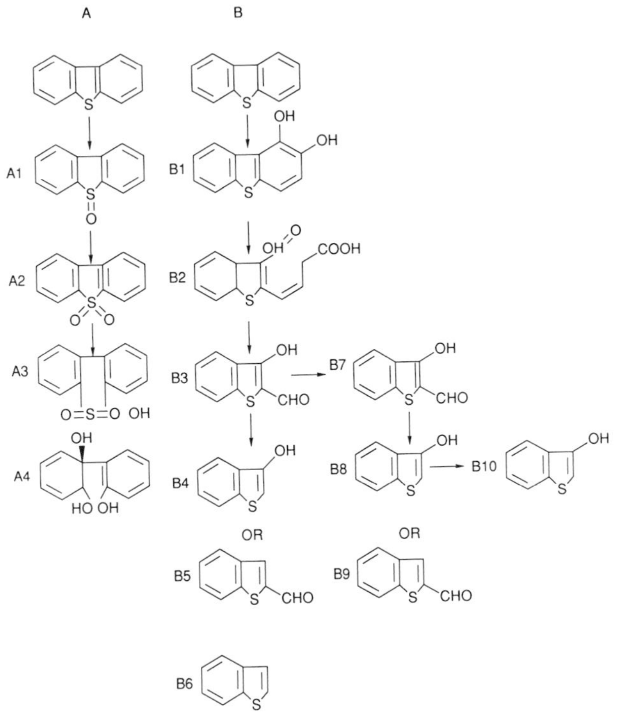

Figure 1 Proposed pathways of dibenzothiophene degradation.

The isolation of microorganisms that remove the organic sulfur from coal has been reported. Microorganisms reported capable of organic sulfur removal include *Pseudomonas* spp., *Thiobacillus* spp., and *Sulfolobus* spp. The amount of organic sulfur removal varies depending on the organism utilized and the type of coal. It has recently been reported that 91% of the organic sulfur from a bituminous coal (Illinois #6) was removed by a mixed culture (the Institute of Gas Technology's IGTS7).

B. Microbial Desulfurization of Coal Char

Because of the promising results obtained in the microbial desulfurization of coal, sulfur microorganisms are being utilized or developed to remove the inorganic and organic sulfur from other fossil fuels such as coal char and oil shale. The microorganism *Thiobacillus ferrooxidans* can remove the inorganic pyrrhotitic sulfur from char but not the organic sulfur. It has been reported that the Institute of Gas Technology's mixed culture, IGTS7, which removes organic sulfur from coal, can remove the organic sulfur from char.

C. Microbial Desulfurization of Oil Shale

Oil shale, which is a significant energy resource for the production of synthetic fuels in the United States, also has a high sulfur content, which is present in both inorganic and organic forms. The inorganic sulfur is present as pyrite. The organic sulfur occurs in C-S-C bonding, in which the carbon atoms may be either unsaturated or saturated. This three-atom grouping may be part of either acyclic or cyclic systems. Some of the sulfur groups in oil shale are thiols, disulfides, and thiophenes. The microorganism *T. ferrooxidans* can remove the pyrite from the shale but not the organic sulfur. Also, microorganisms (unidentified mixed cultures) have been developed that can transform the sulfur in the oil shale to cellular organic sulfur compounds. These microorganisms were developed using the enrichment culture technique in which the oil shale served as the only avaliable source of sulfur.

The study of microbial processes for the removal of organic sulfur from coal, coal char, and oil shale is in its infancy, with coal being the most extensively studied. Each researcher is studying a different microorganism or strain and utilizing different sources of the fossil fuels. Consequently, there have been essentially no independent confirmations of research results in this field. The main problem in this research is the goal: 90% or more reduction in the total sulfur content of the fossil fuel. There is no microbial treatment reported that can achieve this goal. Perhaps the best way to achieve this goal is through genetic engineering. In this approach, bacterial cultures proven to possess desulfurization abilities relevant to the fossil fuels will be used to isolate and amplify the genes or enzymes responsible for desulfurization. Once all the genes involved in the sulfur-specific metabolic pathway have been identified and cloned, the level of activity of that pathway can be increased by increasing the number of copies of the genes, increasing the amount of expression for each gene and/or altering the gene to produce a more active/efficient product. All of these strategies can be used until higher levels of desulfurization activity are achieved. In addition to achieving high levels of organic desulfurization activity, it may also be desirable to transfer this metabolic pathway to other species of microorganisms because of conditions present in a treatment process. The ultimate goal of these genetic studies would be to yield a microbial culture that is stable and of practical utilization in an economically viable desulfurization process. The overall economics and the ease with which a microbial desulfurization process can be integrated into the industry's current practices will be the deciding factors in the development of a successful process.

III. Microbial Removal of Hydrogen Sulfide, Sulfur Dioxide, and Sulfate Wastes

In addition to investigating sulfur microorganisms for their effectiveness in removing the sulfur compounds from fossil fuels prior to combustion, other sulfur microorganisms are being investigated for their effectiveness in cleaning up the sulfur compounds, hydrogen sulfide, and sulfur dioxide generated when the fossil fuels are burned. Also, sulfur microorganisms are being investigated for their effectiveness in cleaning up the sulfate wastes generated by the physical and chemical technologies that are in use for hydrogen sulfide and sulfur dioxide removal. Research is being conducted to improve the gas cleanup technologies to make them more environmentally acceptable and economically attractive. A number of these technologies do not reduce the hydrogen sulfide and sulfur dioxide concentrations to the levels acceptable by the Clean Air Act Standard program. Microbial processes can be applied as a polishing step to these existing technologies for tailgas cleanup.

A. Microbial Cleanup of Hydrogen Sulfide

Several microorganisms including *Thiobacillus thioparus*, *Thiobacillus denitrificans*, and *Chlorobium limicola forma thiosulfatophilum* can metabolize hydrogen sulfide and are proposed for application in microbial processes that remove hydrogen sulfide.

Thibacillus thioparus is a nonphotosynthetic, aerobic, chemolithotrophic organism capable of converting hydrogen sulfide to sulfate. Elemental sulfur is also produced but only under sulfide inhibition. The elemental sulfur is excreted to the medium by the bacteria and therefore could be separated from the bacterial cells and medium.

Thiobacillus denitrificans is also a nonphotosynthetic, chemolithotrophic organism capable of utilizing hyrogen sulfide. The organism can oxidize

the hydrogen sulfide under sulfide-limiting conditions to sulfate in batch and continuous cultures. Hydrogen sulfide is an inhibitory substrate; therefore, sulfide-limiting conditions must be maintained. When sulfide inhibitory levels are present, elemental sulfur is produced. Conversion of hydrogen sulfide to sulfur occurs under aerobic and anaerobic conditions. Under aerobic conditions, there is a lower biomass yield per unit oxidation of hydrogen sulfide than under anaerobic conditions because of inhibition of growth by oxygen. However, under aerobic conditions, the maximum loading of the biomass is two to three times higher than that observed for anaerobic conditions. Under anaerobic conditions, nitrate may be used as a terminal electron acceptor with reduction to elemental nitrogen.

In contrast to the nonphotosynthetic, chemolithotrophic organisms *T. thioparus* and *T. denitrificans*, there is the photosynthetic, anaerobic, autotrophic organism. *C. l. f. thiosulfatophilum*, which is also capable of utilizing hydrogen sulfide. *Chlorobium* converts the hydrogen sulfide to elemental sulfur, which is excreted to the medium by the bacteria. Therefore, the sulfur can be separated from the cells and medium. The conversion of the hydrogen sulfide to elemental sulfur is economical, attractive from the standpoint of an industrial process because the elemental sulfur can be marketed for use in manufacturing a myriad of products. For every mole of hydrogen sulfide converted, one half mole of carbon dioxide is fixed as cellular biomass and intracellular starch. Light is necessary for this reaction as shown in Equation 2:

$$2n \ H_2S + n \ CO_2 \xrightarrow{light} n \ (CH_2O) + 2n \ S^0 + n \ H_2O \quad (2)$$

Although the light energy requirement of this organism makes a hydrogen sulfide removal process uneconomical, a detailed mathematical model exists for optimal elemental sulfur formation.

Based on *T. ferrooxidans* ability to oxidize iron sulfide (see Equation 1), a process, BIO-SR, for hydrogen sulfide has been commercialized in Japan. This process is being used by chemical plants and refineries. The BIO-SR entails two stages: biological and chemical. In the first stage, which is chemical, hydrogen sulfide is oxidized to elemental sulfur by a solution of ferric sulfate. At the same time, the ferric sulfate is reduced to ferrous sulfate. The elemental sulfur is removed from the solution by a separator.

In the second stage, which is biological, the ferrous sulfate is oxidized back by the microorganism

T. ferrooxidans in the presence of air to ferric sulfate. The oxidized solution is cycled back to the first stage for reuse. This bacterial reaction requires low pH (i.e., 2.2), which may be corrosive in contrast to the other bacterial reactions with *T. thioparus*, *T. denitrificans*, and *Chlorobium*, which requires neutral pH.

B. Microbial Cleanup of Sulfur Dioxide and Sulfate Wastes

Microbial systems are also being proposed to remove sulfur dioxide. Several promising chemical and physical technologies, using adsorbents, which are regenerated, exist or are under development for sulfur dioxide removal. To regenerate these adsorbents, the sulfur dioxide must be disposed or recovered. Microbial systems are proposed for application in microbial processes that dispose of sulfur dioxide. The strict anaerobic sulfate-reducing bacteria (SRB) are proposed in a process to reduce the sulfur dioxide to hydrogen sulfide, which then can be chemically or microbiologically converted to elemental sulfur. The SRB utilize sulfate as terminal electron acceptor with reduction to sulfide.

In one proposed process, the SRB, *Desulfovibrio desulfuricans*, and a mixed anaerobic heterotrophic culture convert the sulfur dioxide to hydrogen sulfide. The hydrogen sulfide is then converted by a chemical process to elemental sulfur. In theory, pure cultures of SRB could be used to reduce sulfur dioxide to hydrogen sulfide. However, these cultures would require strict control of oxygen in the culture media and the addition of redox-poising agents. This would adversely affect the economics of a large-scale operation due to the difficulty in maintaining strict anaerobic conditions and cost of reagents. In addition, pure cultures of SRB would require very specific carbon sources, which could represent significant costs in comparison to conventional sources of carbon for fermentation. The use of mixed cultures of SRB and heterotrophic anaerobes allows for the use of a carbohydrate as the ultimate source of carbon for the SRB. In addition, the need for redox-poising agents is eliminated. These benefits could make microbial reduction of sulfur dioxide economically attractive.

In another proposed process, the acetate utilizing SRB, *Desulfobacter*, converts the sulfur dioxide to hydrogen sulfide. The hydrogen sulfide is then converted to elemental sulfur by *Chlorobium*. This two-stage microbial process is also being proposed for the disposal of sulfate wastes such as gypsum (Ca-SO$_4$: H$_2$O), which can arise from the coal desulfurization industry. The important stoichiometric reactions for sulfate disposal and conversion to sulfur are summarized in Equations 3 and 4:

$$CH_3COOH + SO_4^{2-} \xrightarrow{\text{Desulfobacter}} H_2S + 2\ HCO_3^{2-} \quad (3)$$

$$n\ H_2S + 0.5n\ CO_2 \xrightarrow{\text{Chlorobium}} 0.25n\ (CH_2O) + n\ S^0 + 0.5n\ H_2O \quad (4)$$

The study of microbial processes for the removal of hydrogen sulfide, sulfur dioxide, and sulfate wastes, similar to microbial processes for the removal of organic sulfur from fossil fuels, is in its infancy. Most of the microbial processes are in the developmental stages. Only the BIO-SR is commercialized. Further fundamental research with these microorganisms is required for optimal kinetics of hydrogen sulfide formation. In addition, the investigation of the kinetics, microbial physiology, metabolism, and genetics of these sulfur microorganisms should result in economically viable alternatives to conventional chemical desulfurization processes.

Bibliography

Cork, D. J., and Kenevan, J. R. (1990). *Dev. Ind. Microbiol.* **31**, 127–131.

Couch, G. R. (1987). *IEA Coal Res.* (*London*) **ICTIS/TR38**, 1–52.

Fulkerson, W., Judkins, R. R., and Sanghvi, M. K. (1990). *Sci. Am.* 129–135.

Kilbane, J. J. (1988). "Biodesulfurization of Coal." Institute of Gas Technology Symposium on Gas, Oil, and Coal Biotechnology, New Orleans.

Kilbane, J. J. (1988). "Sulfur-Specific Microbial Metabolism of Organic Compounds.' Bioprocessing of Coals Workshop, Tyson Corner, Virginia.

Kilbane, J. J., and Bielaga, B. A. (1989). Molecular biological enhancement of coal biodesulfurization. *In* "5th Annual Coal Preparation, Utilization, and Environmental Control Contractors Conference." Pittsburg, Pennsylvania.

Krawiec, S. (1990). *Dev. Ind. Microbiol.* **5**, 337–354.

Maka, A. (1990). Microbial removal of hydrogen sulfide from gas streams. *In* "Proceedings: 1990 First International Symposium on the Biological Processing of Coal, P19–P24." Palo Alto, California.

Maka, A., and Cork, D. J. (1990). *J. Ind. Microbiol.* **5**, 337–354.

Maka, A., and Cork, D. J. (1990). *Dev. Ind. Microbiol.* **31**, 99–102.

Maka, A., and Srivastava, V. J. (1990). "Micriobial Desulfurization of Char." Institute of Gas Technology Symposium on Gas, Oil, Coal, and Environmental Biotechnology, New Orleans.

Maka, A., Akin, C., Punwani, D. V., Lau, F. S., and Srivastava,

V. J. (1989). Micriobial desulfurization of eastern oil shale: Bioreactor studies. *In* "1989 Eastern Oil Shale symposium, Proceedings," pp. 471–479.

Sublette, K. (1986). *Biotech. Bioeng.* **17,** 543–564.

Sublette, K. (1990). Microbial reduction of sulfur dioxide as a means of by-product recovery from regenerable processes for flue gas desulfurization. *In* "Proceedings: 1990 First International Symposium on the Biological Processing of Coal." Palo Alto, California. **5,** 51–64.

Sublette, K. L., and Sylvester, N. D. (1987). *Biotech. Bioeng.* **29,** 249–257, 753–758.

Torrens, I. M. (1990). *Environment* **32,** 11–32.

Tursman, J. F., and Cork, D. J. (1989). *Biol. Waste Treat.* 273–285.

Susceptibility

Daniel Amsterdam
State University of New York at Buffalo

Glossary

Anti-infective Compound that possesses antimicrobial activity against either bacteria, fungi, viruses, or protozoa

Antimicrobial agent Compound that interferes with the growth and metabolism of microorganisms; antimicrobial agents can be of a natural or synthetic origin and are termed antibacterial if there are activities directed against bacteria, antifungal against fungi, antiviral against viruses, etc; antimicrobial agents used to treat infections are called chemotherapeutic agents

Antiseptic Chemical substance that interferes with sepsis (i.e., prevents the growth or action of microorganisms either by destroying the microorganisms or inhibiting their growth and metabolism); antiseptic agents are associated with use on the body

Chemotherapy Treatment of disease with a chemical substance called a chemotherapeutic agent

Disinfectant Substance, usually chemical in nature, that kills replicating or growing forms of disease producing micro-organisms; term is applied to substances used on or for inanimate objects; disinfection is the process of destroying infectious agents

Sanitizer Agent that is capable of reducing a microbial population in an environment to safe levels as judged by Public Health standards. A sanitizer is usually a chemical agent that is capable of killing more than 95% of the growing bacteria and yet produce no harm to individuals. Sanitizers are used and applied in all aspects of the dairy and food industry and on utensils in restaurants. The process of disinfection could accomplish sanitization; however, the chemical compounds used in disinfection are frequently toxic for individuals.

Susceptibility Susceptibility has two definitions and the definition that will be used here relates to the state of bacteria or other microorganisms being prone to the action of antimicrobial agents. In the interactions between the human host and microorganism, susceptibility would be the state of being open to disease, specifically the capability of being infected, often caused by a lack of appropriate immunologic mechanisms immunity.

SUSCEPTIBILITY IS the term applied in this chapter that refers to the state of microorganisms being affected or acted on by the inhibitory activity of antimicrobial agents. (Another meaning of susceptibility refers to the state of being open to disease, specifically the capability of being infected as a result of deficient or absent immune mechanisms.) Microorganisms are susceptible to physical and chemical agents that inhibit their growth and metabolism. Filtration devices acting as a barrier can limit microbial access or remove them from a particular environment. A large number of chemical compounds have the ability to inhibit the growth of microorganisms and, in some instances, kill them. These chemicals are divided into a number of categories according to their chemical origin and mode of action and are defined as disinfectants, antiseptics, and antimicrobial agents. Antiseptics and disinfectants are used in the food and beverage industry and the hospital environment for controlling microbial populations under varied circumstances.

Antimicrobial agents have targeted activity against different types of microorganisms and are used based on the evaluation of susceptibility testing in the laboratory. No one agent is best for any and all purposes and all kinds of microorganisms. Because of the variety of conditions under which these compounds are used, it is not surprising that there are differences in their modes of action and as a result differences in the susceptibility and resistance among microbial species. Research and experience have demonstrated that certain kinds of chemicals are more appropriate and effective in particular situations.

We do not live in a sterile environment. Microorganisms abound in the air we breathe, the seas that surround us, and in the earth below us, down to a depth of several feet. It has been estimated that the top 6 inches of fertile soil contains more than 2 tons of fungi and bacteria per acre. Viruses seed the air we breathe, bacteria and spores grow on body surfaces, and other organisms exist in hair follicles and skin cavities. In the deeper tissues of the body, even within cells, microbial forms may be dormant. We continually ingest microorganisms in the food we eat and the beverages that we drink. Yet, disease caused by the invasion of microorganisms is not a common occurrence. Although some struggles occur with considerable frequency, the body's defenses readily repulse or come to terms with most microbial invaders. However, on occasion, we need to use chemical or physical agents to control the growth or reduce the number and activity of the microbial flora. The major reasons for controlling microorganisms are to prevent transmission of disease and infection, to prevent contamination or growth by undesirable microorganisms, and to impede deterioration and spoilage of food stuffs and other materials by microorganisms.

Frequently, the literature uses the term "sensitivity" interchangeably with susceptibility. However, this is inappropriate because sensitivity is more properly applied to the phenomenon of allergenicity or intolerance that the body exhibits in response to certain irritable compounds. Thus, susceptibility describes the interaction of an antimicrobial agent and its target bacterium, virus, fungus, etc. If the microbe succumbs to the activity of the antimicrobial agent, it is considered susceptible. If it does not succumb, it would be considered resistant.

I. Historical Highlights in the Development of Antimicrobial and Chemotherapeutic Agents

Human's need for using antimicrobial agents probably stems as early as the quest for the New World, when explorers were seeking spices that would serve to promote foods by inhibiting growth of microorganisms. Perhaps the earliest application of chemotherapeutic agents can be traced back to South American Indians when they ate bark from the cinchona tree to obtain the source of quinine, which is an inhibitor of the malarial parasite. The modern era of antimicrobial agents is credited to Alexander Fleming with the discovery of penicillin at St. Mary's Hospital in London in 1928. Although antimicrobial agents may have been known for hundreds of years, it was not until the commercial success and development of Fleming's discovery of penicillin that the world acknowledged the utility and benefits of antimicrobial agents in warding off disease-producing microorganisms. Fleming observed that not only was a contaminated mold growing in a culture dish that had previously been carelessly open to the air, but other microorganisms in that dish were undergoing autolysis (inhibition). Fleming concluded that the mold later identified as a strain of *Penicillium notatum* had produced a diffusible substance capable of killing staphylococci and other bacteria (see Fig. 1).

Syphilis is the first known disease for which a chemotherapeutic substance was used. Before that, mercury had been used as early as 1495 to treat syphilis. But it was not until 1910 when the arsenical compound known as Salvarsan was synthesized by Paul Ehrlich that a specific drug capable of curing a disease without danger to the patient was developed. As a result of the effectiveness of Erlich's discovery, other investigators, specifically Domagk, showed in 1935 the therapeutic activity of a group of compounds known as sulfonamides. The sulfonamide that Domagk discovered in 1935 in Germany and its derivatives are still used today and are effective in the treatment of infection caused by meningococci and other microorganisms.

Effective chemotherapeutic agents must possess several attributes. The most important is that of selective toxicity (i.e., it must be a substance that has low toxicity for the host cells and is highly toxic for the offending organism). Next, the agent needs to

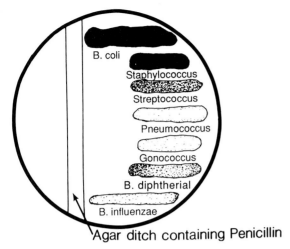

B. coli

Staphylococcus

Streptococcus

Pneumococcus

Gonococcus

B. diphtherial

B. influenzae

Agar ditch containing Penicillin

Figure 1 Schematic representation of Fleming's "ditch plate," in which a solution of penicillin was placed in the ditch and several bacterial species were streaked perpendicular to the ditch. Absence of growth of the five species excepting "*B. coli*" (now *Escherichia coli*) and "*B. influenzae*" (now *Haemophilus influenzae*) near the diffused penicillin indicates the susceptibility of the bacteria to the activity of penicillin. [Reproduced, with permission, from Amsterdam, D. (1988). Principles of antibiotic testing in the laboratory. *In* "Laboratory Diagnosis of Infectious Diseases Principles and Practice, Vol. 1, Bacteria, Mycotic and Parasitic Diseases" (A. Balows, W. J. Hausler, Jr., M. Ohashi, and A. Turano, eds.), pp. 23. Springer-Verlag, New York.]

come in contact with the parasitic microbe by penetrating the cells and tissue of the host in concentrations that are efficacious and, as noted above, not toxic. Last, it needs to demonstrate the appropriate pharmacodynamics that leave the host unaltered in its natural defense mechanisms as in phagocytosis and production of antibodies.

Antibiotics or antimicrobial agents are considered a unique kind of chemotherapeutic agent usually obtained from living organisms. The word "antibiotic" has come to refer to a metabolic product of one organism that in a minute amount is detrimental or inhibitory to the growth of other organisms. It has been known for many years that antagonisms can exist between microorganisms growing in a common environment. The term "antibiosis" was defined in the late 17th century by Vuillemin as a condition in which one creature destroys the life of another to sustain its own. That destruction can be considered as the susceptibility of one group for another. Waksman in 1945 is credited with the current use of the term "antibiotic" as applying to chemical substances of microbial origin, which in small amounts exert an antimicrobial activity.

II. The Meaning of Susceptibility

The terms "susceptibility" and "resistance" are two perspectives that describe the overall response of microorganisms to physical and chemical agents. From the microbiologic perspective, "susceptible" and "resistant" may be used to characterize the capability (or lack of it—incapability) to multiply in the presence of a given concentration of chemical substance or dosage of physical agent under a defined set of conditions. Within a single population of microbial cells, individual cells may be unresponsive-resistant to a certain concentration of antimicrobial agent, although other individual cells might be susceptible to that concentration. Generally, when the major proportion of individual cells within that population-culture are inhibited by a specific concentration or dosage, that strain may be said to be susceptible. However, when most cells in a population are capable of multiplying-growing under a defined set of conditions, it is described as resistant. The conditions under which microorganisms express or develop resistance is described in another article in this encyclopedia. [*See* ANTIBIOTIC RESISTANCE.]

When microorganisms are susceptible to the inhibitory effect of an agent so that they are not killed but do not readily multiply, the agent's activity is classified as bacteriostatic (in the case of bacteria); if the microorganisms are killed and therefore cannot multiply, the agent's activity is referred to as cidal. Depending on the targeted microorganisms, the agent would be referred to as bactericidal (bacteria), fungicidal (fungi), viracidal (viruses), etc.

In the clinical setting, when an antimicrobial agent is used to treat an infected individual, the term "susceptible" takes on broader meaning and is used to predict the microbial outcome (death versus survival) of the microorganisms within the infected individual. As used in microbiology, death is defined as the irreversible loss of the ability to reproduce. Microorganisms that are capable of multiplying (growing) are viable; dead microorganisms do not multiply. Thus, if the microorganism is susceptible to the activity of the antimicrobial agent, it will be unable to multiply within the host and die; if the microorganism is resistant, it will survive and place the host's outcome in jeopardy.

Predictions of susceptibility or its converse, resistance, require laboratory techniques that determine if the selected offending microorganism is capable of reproduction-growth within a defined set of laboratory conditions. If the organism is subjected/exposed to graded concentrations of antimicrobial agents, its susceptibility can be quantified at a specific end point. However, if that end point result is to have clinical significance, the concentration of antimicrobial agents to which the microorganism is susceptible under standardized laboratory conditions must be related to and reflective of the concentration that can be tolerably achieved in the human host.

III. Susceptibility of Microorganisms to Physical and Chemical Agents

A. Physical Agents

Microorganisms are susceptible to heat and radiation, or they can be physically removed by filtration (less than 0.2 μm). Organisms are susceptible to high temperatures (160°C), which inactivate proteins and nucleic acids by breaking their hydrogen bonds. The result is destabilization of the molecules—unfolding of proteins and the separation of double-stranded nucleic acids.

The emission of energy from the electromagnetic spectrum in the form of radiation kills bacterial cells by inducing changes in the nitrogenous bases of their nucleic acids. The amount of energy in an electromagnetic wave is a product of its frequency and physical constant—the Planck's constant—and is therefore proportional to its frequency. Because wavelength is inversely proportional to frequency, radiation with shorter wavelengths possesses greater energy.

Microorganisms can be removed from liquids (and gaseous environments) by filtration. Filters are made from natural sources such as cellulose or diatomaceous earth or are manufactured from chemical polymers. Membrane filters of polycarbonate and cellulose acetate can be made with specific pore

Table I Antiseptics and Disinfectants

Compound class	Mode of action	Application
Alcohol		
Ethanol, isopropanol, benzyl alcohol (50–70% aqueous solution)	Denature proteins and solubilize lipids	Skin antiseptic
Aldehyde		
Formaldehyde (8% solution); glutaraldehyde (2% solution)	Alkylates—reacts with $-NH_2$, $-SH$, and $-COOH$ groups	Disinfectant
Halogen		
Iodine (tincture 2% in 70% ethanol)	Denatures proteins[a]	Skin antiseptic
Chlorine in neutral or acidic solutions	Strong oxidizing agent; reacts with water to form hypochlorous acid (HClO)	Purify drinking water, disinfectant in food industry
Heavy Metal		
Copper sulfate ($CuSO_4$)	Prevents growth of algae and fungi	Algacide and fungicide
Silver nitrate ($AgNO_3$)	Precipitates proteins	Antiseptic for neonates that prevents eye infection
Mercuric chloride ($HgCl_2$)	Inactivate proteins; reacts with SH groups	Disinfectant
Gas		
Sulfur dioxide	Reacts with several cellular components	Food additive in fruit juices
Ethylene oxide	Alkylating agent	Sterilization of heat-sensitive objects (e.g., plastic, lenses, rubber)
Cationic detergent		
Quartenary ammonium compounds	Disrupts cell membrane	Skin antiseptic and disinfectant
Phenol		
Phenol, carbolic acid, lysol, hexylresorcinol, hexachlorophene	Denature proteins and disrupts cell membrane	Disinfectant[b]

[a] Iodophores do not stain.

[b] At low concentration hexachlorophene is used in soap.

sizes. In general, most bacteria can be retained (bacteria adsorb to the filter material) with a pore size of 0.45 μm. However, bacteria vary in size and shape (spheres to rods) and 0.22-μm filters are used to prepare sterile filtrates.

B. Chemical Agents

Microorganisms are susceptible to a variety of chemical agents. Examples of several compounds, their chemical class, mode of action, and application are noted in Table I.

IV. Susceptibility to Antimicrobial Agents and Their Classification

Antibiotics can be classified in several ways. They can be categorized as to their activity, bacteriostatic or bactericidal, they may be grouped on the basis of their chemical structure, or they may be classified on the basis of their mode of action (i.e., the manner in which they manifest their damage on microbial cells). Bacteriostatic and bactericidal activity was previously defined. It is instructive to use the classification by mode of action and chemical structure to organize the relationships among antimicrobial agents. Table II lists several representative antimicrobial agents and codifies their cell target, chemical structure, and anti-infective nature.

V. Evaluations of Antimicrobial Susceptibility

Microorganisms have varying degrees of susceptibility to antimicrobial agents and disinfectants. In some cases the susceptibility of a particular species to a class of antimicrobial agent is predictable. However, because strains of particular species may vary and because the susceptibility of an organism to a given antibiotic may change, especially during the course of treatment, it is necessary for the clinician to know the identity of the offending organism and the spectrum of activity of several different antimicrobial agents that may be expected to be used against that organism to yield the most satisfactory results in treatment. To obtain this information the clinical microbiology laboratory will need to make an accurate microbial diagnosis as to the identification of the organism and to determine the susceptibility to a variety of antimicrobial agents. During the course of therapy, the laboratory may be required to prepare estimates of any variation in the susceptibility pattern of the pathogen to the drug and perhaps to evaluate (assay) the concentration of the administered antimicrobial agent in a specific body compartment (e.g., serum).

A. Disinfectants and Antiseptics

Evaluating the efficacy of these compounds is relatively straightforward. However, no single microbiological test method is suitable for the evaluation

Table II Cellular Targets of Susceptibility for Antimicrobial Agents

Cell target	Agent class	Example	Anti-infective type
Bacterial cell wall	β-lactam	Penicillin, ampicillin, methicillin, cephalothin	Antibacterial
Cell membrane	Glycopeptide	Vancomycin	Antibacterial
Prokaryotes	Polymyxin	Polymyxin B	Antibacterial
Eukaryotes	Polyene	Amphotericin, nystatin	Antifungal
Protein synthesis	Aminoglycoside	Amikacin, gentamicin, streptomycin, tobramycin	Antibacterial
	Macrolide	Erythromycin, clarithromycin	Antibacterial
	Lincomycin	Clindamycin, lincomycin	Antibacterial
Nucleic acid synthesis	Quinolone	Ciprofloxacin, temafloxacin	Antibacterial
		Flucytosine	Antifungal
RNA synthesis	Rifamycin	Rifampin	Antibacterial
DNA synthesis unknown	Purine nucleoside	Acyclovir	Antiviral
	Tricyclic amine	Amantidine	Antiviral
DNA synthesis	Purine nucleoside	Ribavirin	Antiviral
Growth factor analog	Sulfonamide	Sulfanilimide	Antibacterial

of all germicidal chemicals for all applications. In part, the reasons for this are the ease with which these chemical compounds are neutralized by organic substances and the reality that bacteria and other microorganisms can be enclosed in particles and penetration of the chemical may be impeded.

An official [approved by the Food and Drug Administration (FDA) and the Association of Agricultural Chemists] test method is available that determines the phenol coefficient, a ratio used to compare the efficacy of these chemical compounds. The procedure uses a test organism (a specific strain of either *Salmonella typhi* or *Staphylococcus aureus*) under a controlled set of conditions that determine the temperature at which the test is performed, the manner for preparing subcultures, the composition of the test medium (broth) that supports the growth of the test bacteria, and size of the test tubes. Other details are specified in the official procedure. Briefly, the disinfectant/antiseptic being tested is diluted in a series of tubes, and the test organism is added. Similar dilutions are made with phenol. All tubes are incubated (20°C) and at varying time intervals (5–15 min) subcultures of the contents of each tube are transferred into tubes of sterile medium (no organism or disinfectant). The inoculated subculture tubes are incubated and subsequently examined for growth. The greatest dilution of disinfectant killing the test organism in 10 min (but not 5 min) is divided by the greatest dilution of phenol showing the same result. The number obtained by this division is the phenol coefficient of the substance tested.

B. Antimicrobial Agents

1. Disk Diffusion Agar Method

The paper disk agar diffusion method is the most frequently used technique for determining the susceptibility of antimicrobial agents to chemotherapeutic compounds. In part it is an adaptation of Fleming's "ditch plate" method, except in this approach, the antimicrobial agent is impregnated onto paper disks. Paper disks, 6 mm in diameter, containing known quantities of chemotherapeutic agent are placed on the surface of a medium that has been inoculated with the pathogen. After overnight incubation at 35°–37°C, the plates are observed for radial zones of inhibition surrounding the paper disk. The zone of inhibition that appears as a clear area around the disk indicates that the organism was inhibited by the compound that diffused through the agar from the disk.

The procedure that is currently used for the single disk method of susceptibility testing and recommended by the FDA was developed in 1966 by Bauer-Kirby and others. This is a highly standardized technique, and the antimicrobial agent concentration in the disk is specified, as well as the nature of the medium, the size of the inoculum, conditions of incubation, and other details. When the size of the zone of inhibition is measured, three categories ("susceptible," "resistant," or "intermediate") can be determined. These categories correspond to bacteriologic outcome and can be suggestive of the clinical outcome of the patient. When the test is performed according to procedural guidelines, it is possible to determine whether the microorganism is susceptible or not (resistant) to the antimicrobial agent with a high degree of accuracy.

2. Broth Dilution Technique

The susceptibility of microorganisms to antimicrobial agents can also be determined by a dilution technique in broth. In this procedure the least amount of drug in a graded concentration series that is required to inhibit the growth of the organism *in vitro* is determined by visually observing the tube without turbidity. This concentration is referred to as the minimal inhibitory concentration (MIC) (see Fig. 2). The MIC is related to the level of antimicro-

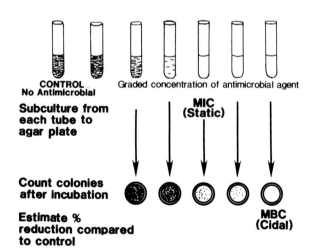

Figure 2 Procedural outline of the broth dilution technique for determining the MIC and MBC. [Reproduced, with permission, from Amsterdam, D. (1988). Principles of antibiotic testing in the laboratory. *In* "Laboratory Diagnosis of Infectious Diseases Principles and Practice, Vol. 1, Bacteria, Mycotic and Parasitic Diseases" (A. Balows, W. J. Hausler, Jr., M. Ohashi, and A. Turano, eds.), pp. 29. Springer-Verlag, New York.]

bial agent that can be achieved in the patient's serum. If a ratio of at least one quarter is achieved (i.e., if the MIC is four times the maximally achieved serum concentration), there is a good prediction of clinical success. The MIC represents a static end point. The minimum bactericidal concentration (MBC) can be determined by subculture of the MIC and tubes containing higher concentrations of antimicrobial agent onto sterile agar medium (see Fig. 2).

There is an inverse linear relationship for a bacterium/antimicrobial agent pair between the MIC and the zone diameter of inhibition as determined by the agar diffusion procedure.

3. Alternative Approaches

Susceptibility of microorganisms to specific antimicrobial agents can be evaluated by demonstrating their resistance. Bacteria can express enzymes that inactivate antimicrobial agents (e.g., penicillinase, a β-lactamase, inactivates penicillin) and therefore would not be susceptible to the inhibitory effect of the drug. Rapid tests (30 sec) are available to detect the presence of this enzyme. Alternatively, advances in molecular microbiology hold forth the promise for direct tests that will target specific and unique nucleotide sequences of the microorganisms as genotypic indicators of resistance to specific chemotherapeutic agents.

The methods discussed above are mainly applied to determine the susceptibility of rapidly growing, aerobic bacteria. Methods for determining the susceptibility of yeasts (fungi) and mycobacteria are currently being developed. Although research techniques are available to evaluate the continued development of new antimicrobial agents directed against fungi, protozoan parasites, and viruses, these techniques are not in routine clinical use.

Bibliography

Amsterdam, D. (1988). *Diagn. Microbiol. Infect. Dis.* **9,** 167–178.

Amsterdam, D. (1988). Principles of antibiotic testing in the laboratory. *In* "Laboratory Diagnosis of Infectious Diseases Principles and Practice, Vol. 1, Bacteria, Mycotic and Parasitic Diseases" (A. Balows, W. J. Hausler, Jr., M. Ohashi, and A. Turano, eds.), pp. 22–38. Springer-Verlag, New York.

Lorian, L. (ed.) (1991). "Antibiotics in Laboratory Medicine," 3rd ed. Williams and Wilkins, Baltimore.

Sande, M. A., Kapusnik-Uner, J. E., and Mandell, G. L. (1990). Chemotherapy of microbial diseases. *In* "The Pharmacological Basic of Therapeutics," 8th ed. (A. G. Gilman, T. W. Rall, A. S. Nies, and P. Taylor, eds.), pp. 1018–1201. Pergamon Press, New York.

Thornsberry, C. (sect. ed.) (1991). Antimicrobial agents and susceptibility tests. *In* "Manual of Clinical Microbiology," 5th ed. (A. Balows, W. J. Hausler, Jr., K. L. Herrmann, H. D. Isenberg, and H. J. Shadomy, eds.), pp. 1059–1201. American Society for Microbiology, Washington, D.C.

Webster, L. T., Jr. (1990). Chemotherapy of parasitic infections. *In* "The Pharmacological Basic of Therapeutics," 8th ed. (A. G. Gilman, T. W. Rall, A. S. Nies, and P. Taylor, eds.), pp. 954–1017. Pergamon Press, New York.

Symbiotic Microorganisms in Insects

A. E. Douglas
University of Oxford

I. Endosymbioses and Ectosymbioses
II. The Mycetocyte Symbioses
III. The Gut Symbionts

Glossary

External transmission Release of microbial symbionts from adult females, usually onto the surface of the egg shell, such that the microorganisms are available for ingestion by (and infection of) hatching larvae

Mycetocyte Specialized insect cell harboring symbiotic microorganisms

Mycetome Group of mycetocytes bounded by an epithelial layer

Nitrogen recycling Metabolism of insect nitrogenous waste products (e.g., uric acid, ammonia) to nutrients (e.g., essential amino acids) that are transferred to, and utilized by, the insect

Symbiont Nonpathogenic microorganism associated with an insect; the growth and reproduction of many insects are impaired in the absence of their symbionts, suggesting that the insects benefit from the association

Transovarial transmission Transfer of microbial symbionts to the cytoplasm of oocytes in the ovaries of the female insect

THE SYMBIONTS OF INSECTS include members of all the microbial kingdoms: many eubacteria (e.g., Enterobacteriaceae), archaebacteria (methanogens), protista (e.g., Parabasalia and Metamonada in the guts of some termites), and fungi (various yeasts). They are believed to occur in representatives of most insect orders and, overall, in approximately 10% of all insect species, including many pests of agricultural and medical importance. Despite the taxonomic diversity of both microbial and insect partners, most symbioses have three characteristics in common. First, the microorganisms possess metabolic capabilities absent in the insect, and insects "use" these capabilities to live on nutritionally poor or unbalanced diets, such as wood, plant sap, or vertebrate blood. Second, both the insect and microbial partners require the association; insects freed of their symbionts grow slowly and produce few or no offspring, whereas many symbiotic microorganisms are unknown in habitats other than the insect and some are apparently unculturable. Finally, many symbionts are transmitted between insects, usually from mother to offspring; both transovarial and external transmission are widespread.

I. Endosymbioses and Ectosymbioses

The symbioses described here are those in which the microbial metabolism utilized by the insect occurs within the body of the insect. These associations are surveyed in Table I. They are often described as endosymbioses, as distinct from ectosymbioses, in which the microorganism is located in the brood chamber or nest of the insect. Examples of insects with ectosymbionts are fungus-growing termites (Macrotermitinae) and leaf-cutting ants (Attini).

Most endosymbionts are located either in the lumen of the insect alimentary tract (the gut symbionts) or in the cytoplasm of insect cells called mycetocytes (the mycetocyte symbionts). Microorganisms are rarely found in other locations, presumably because they are destroyed by the insect's defense system. For example, bacteria released from the mycetocytes into the hemocoel (body cavity) of aphids are lysed rapidly. There are two exceptions to this generality. First, a few species from all major

Table I Survey of Symbioses in Insects

Insect taxon	Location	Taxa	Metabolic capability	Mode of transmission	Incidence of symbiosis
Blattaria (cockroaches)	Hindgut	Various bacteria	—	—	Universal
	MC in fat body	Bacteria *Blattabacterium*	Nitrogen recycling	Transovarial	Universal
Orthoptera (grasshoppers, etc.)	Hindgut	Various bacteria	—	—	Universal
Heteroptera					
Cimicidae (bed bugs)	2 MT in hemocoel	Coccoid bacteria	B vitamin synthesis	Transovarial	Nearly universal
Coreidae (leaf-footed bugs)	Midgut	Various	?	External smearing	Irregular
Lygaeidae (seed-sucking bugs)	Midgut	Various	?	External smearing	Irregular
Pentatomidae (stink bugs)	Midgut	Various	?	External smearing	Widespread
Pyrrhocoridae (fire bugs)	Midgut	Various	?	External smearing	Widespread
Triatominae (kissing bugs)	Midgut	Various, includes *Nocardia rhodnii*	B vitamin synthesis	External smearing	Universal
Homoptera (planthoppers, whitefly, aphids, etc.)	MT in hemocoel	Various bacteria and yeasts	Amino acid synthesis, sterol synthesis	Transovarial	Nearly universal
Isoptera (termites)	Hindgut	Various bacteria	Nitrogen recycling, Nitrogen fixation	—	Universal
		Flagellate protists	Cellulose degradation	Proctodeal trophyllaxis	Lower termites
Anoplura (sucking lice)	MC, variable location	Various bacteria	B vitamin synthesis	Transovarial	Nearly universal
Mallophaga (biting lice)	MC in hemocoel	Various bacteria	?	Transovarial	Irregular
Diptera					
Glossinidae (tsetse flies) Hippoboscidae	MT in midgut epithelium	Bacteria	B vitamin synthesis	Via milk glands to larva	Universal

Taxon	Location	Microorganism	Function	Transmission	Distribution
Nycteribiidae (bat flies)	MT in hemocoel	Various bacteria	—	Transovarial	Universal
Tephritidae (fruit flies, gall flies)	Midgut of larva, cephalic organ of adult in some species	Various bacteria	—	External smearing?	Widespread
Tipulidae (crane flies)	Evaginations of hindgut	Various bacteria	—		Widespread
Coleoptera Anobiidae	MC in midgut caeca	Yeasts	Essential amino acid and B vitamin synthesis	—	Universal
Bostrychidae	2 MT in hemocoel	Variable	—	External smearing	Universal
Cerambycidae	MC, midgut caeca	Yeasts	—	—	Widespread
Chrysomelidae	MC, midgut caeca	Coccoid bacteria	—	—	Irregular
Curculionidae (weevils)	MC, variable location	Various bacteria	—	Transovarial	Widespread
Lucanidae (stag beetles)	Evaginations of midgut or hindgut	Various bacteria	—	—	Universal!?
Scarabaeidae (scarab beetles)	Dilation of hindgut unknown	Various bacteria yeasts	—	—	Widespread
Scolytidae (bark beetles)	Gut	Bacteria	—	—	Universal?
Silvanidae (flat grain beetles)	4 MT in hemocoel	Filamentous bacteria	?	Transovarial	Oryzaephilus only
Hymenoptera Formicidae (ants)	MT in midgut epithelium	Bacteria	?	Transovarial	In all Camponoti, irregular Formicinae

MC, isolated mycetocytes; MT, mycetome.

groups of insects possess "guest" bacteria, present in many or all cells of the body, and, second, a small minority of insects have intercellular symbionts [e.g., the yeasts between the fat body cells of some delphacid planthoppers (Section II.B.1)].

The alimentary tract of most insects bears a substantive and taxonomically diverse microbiota. Evaginations of the midgut, called caeca, and the hindgut are particularly densely colonized. Microorganisms in the caeca are protected from elimination from the gut with the bulk flow of digesta. The hindgut is a particularly favorable habitat because it lacks digestive enzymes and is of neutral pH. Many microorganisms persist for longer periods than the transit time of digesta through the hindgut by adhering to the hindgut wall, but the microbiota is reduced or, in some insects, eliminated when the chitinous lining of the gut wall is shed at each insect molt.

The intracellular symbionts are invariably acquired by endocytosis. Most are retained within the endocytic membrane, with each microbial cell enclosed within an individual membrane (Fig. 1A). This insect membrane is believed to play an important role in the regulated flux of nutrients between the insect cell and microorganisms, but it is not essential to the maintenance of all associations. The bacteria in mycetocytes of tsetse flies, ants, and the weevil *Sitophilus* have "escaped" from the membrane and are in direct contact with the insect cell contents.

Most intracellular microbial symbionts are located in the cytoplasm of cells, whose sole function appears to be to house them. These cells are called mycetocytes. In many insects (see Table I), the mycetocytes are aggregated together and bounded by an epithelial layer, to form a structure known as the mycetome.

II. The Mycetocyte Symbioses

A. Blattaria (Cockroaches)

The mycetocyte symbionts in cockroaches are bacteria. In all cockroach species examined, they are rod-shaped, are $1 \times 1.5–9$ μm, and have a thin (5–10 nm) cell wall, which is gram-variable. Hydrolysates of the bacteria contain *N*-acetylglucosamine and muramic acid, both components of peptidoglycan walls of bacteria. The bacteria are nonmotile and aflagellate. They divide by transverse fission, and both developing cross-walls and mesosomes have been observed.

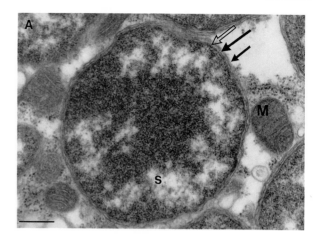

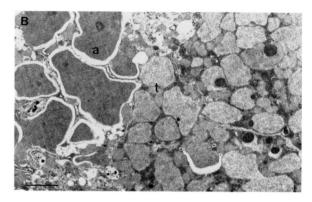

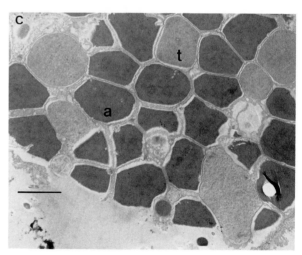

Figure 1 The mycetocyte symbionts of Homoptera (A) A primary symbiont (S) in a mycetocyte of the pea aphid *Acyrthosiphon pisum*, with its cell membrane (⇑) and cell wall (↑) bounded by an insect membrane (↑). The mycetocyte cytoplasm includes mitochondria (M). Negative 91/1692. Bar = 0.5 μm. (B) The vegetative phase of symbionts in the leafhopper *Euscelis incisus*: the *a*-symbiont (a) and *t*-symbiont (t). Negative 86/6368. Bar = 2 μm. (C) The transmission phase of *a*-symbionts (a) and *t*-symbionts (t) in the symbiont ball of a newly deposited egg of the leafhopper *E. incisus*. Negative 87/9645. Bar = 1 μm.

There are many claims in the early literature of cockroach mycetocyte symbionts isolated into axenic culture, and the isolates have been assigned to several known genera including *Bacillus*, *Corynebacterium*, and *Rhizobium*. These isolates are almost certainly not the true symbionts, because they differ markedly in morphology from the bacteria in the mycetocytes.

On the basis of their morphology in the insect, the mycetocyte symbionts have been assigned to a single genus *Blattabacterium*, which includes no other bacteria and is of uncertain systematic position. The only species described is the type species *Blattabacterium cuenoti* (Holland and Favre, 1931) from *Blatta orientalis*. DNA hybridization studies have confirmed that the bacteria in cockroaches of the families Blattidae, Blattellidae, Cryptocercidae, and Blaberidae are closely related. The base composition of the bacterial DNA varies between 21.8 and 28.3 mol% G+C.

The mycetocytes of cockroaches are uninucleate cells of diameter 20–40 μm. The bacteria are very abundant in the cytoplasm; mitochondria, endoplasmic reticulum, and Golgi bodies are sparse. The mycetocyte nuclei are polyploid (e.g., up to 16-ploid in embryos and 256- or 512-ploid in adults of *Periplaneta americana*). The mycetocytes divide during the latter part of each larval stadium; therefore, mycetocytes are most numerous and of minimal size just prior to each molt. The bacteria proliferate throughout larval development, and their division is not synchronized with mycetocyte division. The mycetocytes in most species are in the visceral fat body.

Transovarial transmission of the mycetocyte symbionts has been studied intensively in *Blattella germanica*. The symbionts are absent from the primordial gonads of the early female embryo, but mycetocytes migrate from the fat body to the ovarioles during late embryogenesis. Initially, all the mycetocytes associated with the ovarioles lie in connective tissue external to the tunica propria (the noncellular sheath bounding each ovariole). During larval development, some mycetocytes penetrate through the tunica propria to lie adjacent to the developing oocytes, and, in the final larval instars, the bacteria are exocytosed. They remain external to the oocytes for an extended period, "enmeshed" in the oocytes' microvilli and proliferating rapidly to completely surround the oocytes. The incorporation of the bacteria into the oocyte cytoplasm has not been observed, but they are presumed to be phagocytosed by all regions of the oocyte membrane

during late vitellogenesis, just before the "egg shell" (vitelline membrane and chorion) is laid down.

In the fertilized egg, the bacteria are restricted to the periphery of the deuteroplasm (the yolk region). Their location shifts repeatedly during early embryonic development and associations with insect nuclei have been noted. When the endodermal portion of the digestive tract is formed, the bacteria are included in the gut. From there, they "migrate" toward the presumptive mycetocytes in the epineural abdominal mesoderm. They are phagocytosed by mycetocytes 7–9 days after egg deposition of the ootheca. The mechanisms underlying the complex movements of the bacteria have not been investigated. Most information on the bacteria in cockroaches has come from investigations of the metabolic capabilities of cockroaches treated to eliminate the bacteria. Various methods are available (e.g., omission of Mn and Zn, inclusion of antibiotics, high concentrations of urea, or linoleic acid in the diet), but none are totally effective. The numbers of bacteria in treated insects are drastically reduced, but they gradually "reappear" to repopulate the fat body over several insect generations. The treated insects are small, develop slowly, and produce few offspring. They have exceptionally high levels of uric acid, up to 20 times control insects, and the antibiotic treatment abolishes the ability of cockroaches to incorporate ^{35}S-sulfate into the sulfur amino acids and ^{14}C-glucose into various essential amino acids. These data, together with enzymatic evidence that the bacteria are uricolytic (see earlier), indicate that the bacteria may recycle nitrogen [i.e., the bacteria degrade the insect's chief nitrogenous waste product (uric acid) to ammonia, which is then assimilated into essential amino acids]. The transport of amino acids from the bacteria to insect has not been demonstrated.

B. Homoptera

Mycetocyte symbioses are close to universal in Homoptera. They are present in most species that feed on phloem or xylem sap of plants but absent from several of the groups that feed on intact plant cells, namely the typhlocybine leafhoppers, phylloxerid aphids, and apiomorphine scale insects. The mycetocytes are invariably in the hemocoel but, in other respects, especially the morphology of the microorganisms, organization of the mycetocytes, and details of transovarial transmission, the symbioses vary among the homopteran groups. In particular, there is no homology between the associations in the two superfamilies, the Auchenorrhyncha and Sternorrhyncha, which are considered separately.

1. Auchenorrhyncha (Planthoppers, Leafhoppers, etc.)

H. J. Muller has argued that the ancestral Auchenorrhyncha possessed a single bacterial symbiont, and the various auchenorrhynchan groups have subsequently "acquired" a diversity of additional microorganisms, such that extant species have both the ancestral symbiont and one to five morphologically distinct bacteria or yeasts. The taxonomic position of all the microorganisms is obscure, and the various morphological forms are designated by a single letter. The "ancestral" symbiont is called the *a*-symbiont. It is frequently accompanied by the *t*-symbiont in leafhoppers (Cicadellidae), *w*-symbiont in cicadas (Cicadidae), *b*-and *c*-symbionts in treehoppers (Membracidae), *x*-symbionts in froghoppers (Cercropidae) and *x*-and *H*-symbionts in planthoppers (Fulgoridae); additional microorganisms with distinct morphologies have been described in each insect family. All the symbionts are bacteria except the *H*-symbionts, which are yeasts. The various morphological forms of symbionts occupy structurally distinct mycetocytes, which, in most groups, are aggregated as a single pair of mycetomes associated with the lateral body wall. Exceptionally, the mycetocytes containing each type of symbiont in planthoppers form separate, single or paired mycetomes of characteristic shape and position in the hemocoel.

Detailed structural studies of the symbionts have been conducted only on the *a*- and *t*-symbionts of leafhoppers, especially the dwarf leafhopper *Euscelis incisus* (=*E. plebejus*). In the mycetocytes, these symbionts adopt an unusual morphology, known as the vegetative phase (Fig. 1b). Each *a*-symbiont is an oval-to-elongate cylinder up to 60 μm in length, rolled into loops or spirals within the perisymbiont membrane. The *t*-symbionts are irregularly lobed structures, occasionally resembling rosettes, of dimension 8–15 μm. In the adult female insect, some symbionts adopt the infective morphology, which is rounded, is <10 μm in dimension, and has few structural features. Only the infective stage is believed to be capable of division (but this has not been demonstrated). The infective symbionts are transferred to the posterior pole of each oocyte, and they are evident in the newly deposited egg as a "symbiont ball" lying between the membrane of the egg and egg shell. Within 1–3 days, the symbiont ball becomes

incorporated into the egg, attached to the posterior pole of the developing embryo, and, once the symbionts are distributed to the definitive mycetocytes of the embryo, they adopt the vegetative morphology. Symbiont transmission in other Auchenorrhyncha is believed to be similar to that in leafhoppers.

This traditional interpretation of the mycetocyte symbiosis in the Auchenorrhyncha has been challenged in several respects, as follows.

a. Cladistic analysis of the distribution of symbionts among the Auchenorrhyncha suggests that associations with microorganisms designated by a single letter may have evolved independently several times. Thus, the symbionts represented by one letter may include a taxonomically diverse array of bacteria. DNA studies are needed to resolve this issue.

b. The *a*-symbionts have been interpreted as the degenerate form of the bacterial symbionts. This hypothesis, proposed in 1972, has consistently been ignored over the last 15 yr, but it cannot be disproved by the experimental data currently available.

c. Certain phloem-feeding planthoppers (e.g., *Nilaparvata lugens, Laodelphax striatellus*) lack mycetocytes and symbiotic bacteria, but they have yeasts lying between cells of the fat body. *Laodelphax striatellus* probably derives sterols essential for normal molting from its complement of yeasts. These few species of Asian planthoppers may not be unique. Virtually all research on auchenorrhynchan symbioses has been conducted on temperate European species, and nothing is known about most species from other continents, especially the tropics.

We are ignorant of the metabolic capabilities of the bacterial symbionts in the Auchenorrhyncha, although these microorganisms are undoubtedly required by the insect. For example, the *a*- and *t*-symbionts in *Euscelis incisus* can be eliminated by ligation of the newly deposited egg so that the symbiont ball is separated from the developing embryo. These embryos develop normally to hatching and the symbiont-free larvae feed, but they fail to grow or develop and die within 1–2 wk of hatching. Claims in the literature that the bacteria are essential for osmoregulation and respiratory metabolism in leafhoppers are entirely without experimental foundation, and the proposal that they are required for differentiation of the abdomen arises from deleterious effects of high antibiotic concentrations on the insect.

In the last decade, planthoppers have been used as a source of bacteria capable of antibiotic synthesis. For example, *Bacillus* sp. and an *Enterobacter* sp. from *Nilaparvata lugens* produce polymyxins and a novel compound, andrimid, respectively, and an unidentified bacterium (possibly a pseudomonad) from *Sogatella fucifera* produces pyoluteorin and diacetylphloroglucinol.

2. Sternorrhyncha, Especially Aphids

Modern research on the mycetocyte symbionts in sternorrhynchan Homoptera has concerned aphids of the family Aphididae. Understanding of the microorganisms in other groups has not progressed beyond the light microscope descriptions of entomologists (especially P. Buchner and colleagues) between the 1920s and 1950s: the whitefly (Aleurodidae) with various "cocci and filaments"; the psyllids (Psyllidae) with "elongate tubes" and "delicate rods"; the scale insects and mealy bugs (Coccidae) whose symbionts include a staggering diversity of morphological forms; and aphids of the family Adelgidae with symbionts "shaped like caraway seeds." The potential for reexamination by microbiologists is immense.

Among the aphids, all members of the families, Aphididae and Pemphigidae, possess a bacterial symbiont, frequently referred to as the primary symbiont, in an unpaired mycetome in the abdominal hemocoel. In many species, the primary symbiont is accompanied by one to three morphologically distinct bacteria in separate cells. The primary symbionts are gram-negative bacteria and coccoid bacteria approximately 2.5 μm in diameter, and have a thin cell wall (10 nm thick) containing peptidoglycan. From their 16S ribosomal RNA sequence, they have been assigned to a new genus *Buchnera*, the γ-Proteobacteria, but they are taxonomically distinct from any other known bacteria. As yet, the primary symbionts are unculturable, but they can be maintained for limited periods in simple media, where they respire aerobically and utilize carboxylic acids, such as succinate, and amino acids, such as aspartate, as carbon sources. The genome size of the bacteria is reportedly 1.4×10^{10} daltons, a value five times greater than *Escherichia coli* and larger than any other prokaryote (!); perhaps, the DNA in the primary symbionts is endoreduplicated. Up to 30% of the protein content of these bacteria is represented by a single protein of molecular weight 880 kda, known as symbionin. The amino acid composition of symbionin closely resembles GroEL, a heat-shock protein of *E. coli*.

The primary symbionts of aphids have been implicated in nitrogen recycling. They can utilize ammonia, the principal excretory product of aphids, and the primary symbionts of the green peach aphid *Myzus persicae* synthesize the essential amino acid methionine, which is transported to the insect tissues.

C. Bloodsucking Insects: The Cimicidae, Anoplura, and Diptera Pupiparia

Virtually all insects that feed on vertebrate blood through their life cycle have mycetocyte symbionts: These are the bedbugs (Heteroptera of the family Cimicidae), sucking lice (Anoplura) and Diptera Pupiparia, comprising the Glossinidae (tsetse flies), Hippoboscidae (keds), and Nycteribiidae (batflies). (The Streblidae, a second group of batflies, lack mycetocytes and have bacteria irregularly distributed in the fat body and reproductive system.) An additional group of hematophagous insects, the triatomid bugs (Section III.3), has a substantial gut microbiota, and the only cimicid lacking mycetocyte symbionts, *Primicimex cavernius*, has bacteria associated with the midgut epithelium. Insects that utilize blood solely as adults (e.g., fleas, mosquitoes) lack microbial symbionts.

The symbionts in the various haematophagous insects are believed to provide B vitamins, which are deficient in vertebrate blood, but this has not been demonstrated for any species. A major difficulty in research on these associations is that the microorganisms are apparently unculturable, and their systematic positions are obscure. All are bacteria. The morphology of the symbionts in Anoplura varies widely among insect species, but those in Glossinidae and Cimicidae are more uniform: gram-negative rods up to 9 μm long, and cocci of 1–1.5 μm, respectively. The bacteria in *Cimex* are sensitive to high temperature (bedbugs can be freed of their symbionts by incubation at 36°C), and the bacteria in *Glossina* can be maintained for short periods in medium suitable for mycoplasma cultivation. The *Glossina* symbionts preferentially utilize carboxylic acids, such as succinate and pyruvate, as carbon sources, and they respire aerobically.

The mycetocyte symbionts in the Cimicidae and Anoplura are transmitted transovarially. In ci-

micids, they are transferred from the paired mycetomes in the abdominal hemocoel to the anterior
pole of the eggs via trophic cords that run from
the germarium to each oocyte. The location of mycetomes and mode of transmission varies widely
among anopluran species. In the best-studied species, *Pediculus humanus,* the single mycetome
(sometimes known as the stomach disc) (Fig. 2) is
ventral to the midgut. During the final molt of the
female, bacteria leave the mycetome via a single
"hole" and pass along the ventral surface of the
digestive tract to the ovaries (Fig. 2). They are taken
up by cells at the base of the ovarioles and subsequently transferred to the posterior pole of
each oocyte. The Glossinidae and Hippoboscidae
are unusual in that larvae are retained within
the female reproductive tract until the final larval instar. They feed on secretions of the maternal "milk gland." Whether the mycetocyte symbionts are acquired by the larvae with these
secretions or transovarially by the oocyte is uncertain.

D. Coleoptera (Beetles)

Phytophagous and xylophagous species in 10 families of Coleoptera possess mycetocyte symbionts
(Table I). The microorganisms are bacteria in all
groups except the Anobiidae and Cerambycidae,
which harbor yeasts. None of the bacteria have been
isolated into culture or identified. Various coccoid,
filamentous, and irregularly shaped forms have been
reported, but recent microbial studies are lacking.
The yeast symbionts are culturable on standard
mycological media. The symbiont of the anobiid
Stegobium paniceum has been assigned to the genus
Torulopsis.

The various mycetocyte symbionts may have
evolved from members of the gut microbiota in
beetles and become separated from the gut to varying extents in the different groups. In the anobiids,
cerambycids and chrysomelids, the mycetocytes
contribute to the epithelium lining the midgut caeca.
Their symbionts are transferred to the surface of the
egg shell, during or after egg deposition, and are
ingested by the larval beetle as it hatches. This
process of "external transmission" is the usual
mode of transmission of gut symbionts (Section III).
The mycetomes of other groups (e.g., the silvanid
Oryzaephilus, many curculionids) lie free in the
hemocoel, and the mycetomes of bostrychids are
attached to the gut by long epithelial filaments. Their
symbionts are transmitted transovarially, usually in

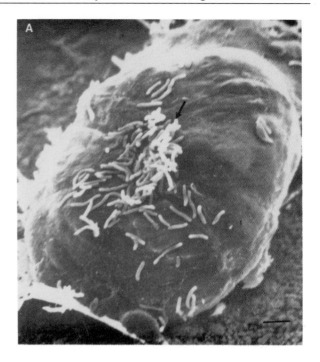

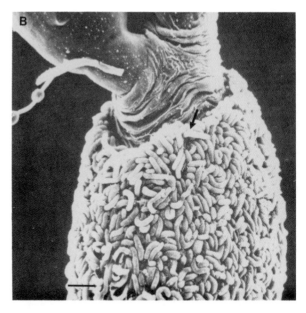

Figure 2 Transovarial transmission in the human body
louse *Pediculus humanus.* (A) The rod-shaped bacteria
(↓) emerging from a single opening in the mycetome.
Bar = 10 μm. (B) The symbionts (↓) accumulating on the
surface of the louse oviduct, after migration from the
mycetome to the reproductive tract. They are subsequently
endocytosed by the cells of the oviduct. Bar = 10 μm.
[Micrographs from Eberle, M. W., and McLean D. L. (1983).
Can. J. Microbiol. **29,** 755–762.]

the adult female. Exceptionally, in the weevil *Sitophilus*, symbionts in the early embryo become associated with the primordial germ cells, whereas a second lineage of symbionts are allocated to the mycetome. The symbionts persist in the germinal rudiments of females, and, once they reach adulthood, they are transferred to the developing oocytes.

Nothing is known about the metabolic capabilities of symbionts in Coleoptera. They are not cellulolytic and none are known to fix nitrogen. Studies with insects deprived of their symbionts suggest that they may variously provide sterols, B vitamins, and essential amino acids.

E. Formicidae (Ants)

The mycetocyte symbionts in ants are large, gramnegative rods, located in the midgut epithelium or in the hemocoel. As with most other mycetocyte symbionts (see earlier), they have not been cultured and their taxonomic position is unknown. They occur in all carpenter ants (Camponoti), several *Formica* species including *Formica fusca*, and *Plagiolepis*. The mycetocyte symbionts may have been widespread, perhaps universal in Mesozoic ants and lost secondarily by many groups.

III. The Gut Symbionts

A. Introduction

The gut microbiota in mammals contribute to the nutrition of herbivores by nitrogen recycling and cellulose degradation. For many years, it has been assumed that microorganisms contribute to phytophagous and omnivorous insects in a similar fashion. This analogy is fallacious. Nitrogen recycling by the gut microbiota is significant in termites, and a minority of termites also have cellulolytic microorganisms (Section III.B) but, for most insects, the gut microbiota is of little nutritional significance. The gut microbiota have also been implicated in the detoxification of plant secondary compounds and insect pheromone production, but these interactions have yet to be demonstrated unambiguously in any association (see Sections III.C.2 and III.E.2).

B. Isoptera (Termites)

The greatest density of microoganisms in the alimentary tract of termites is in the anoxic proximal portion of the hindgut, known as the paunch. In all termites, this region harbors bacteria, at 10^9–10^{10} cells ml^{-1} gut contents. All the bacteria are facultative or obligate anaerobes and they include spirochetes, methanogens, and species of *Enterobacter, Bacteroides, Bacillus, Streptococcus,* and *Staphylococcus*. Between 10 and 20% of the bacteria are culturable. The bacteria are the sole symbionts of the family Termitidae (the higher termites), which account for 75% of all termite species. The paunch of the lower termites, comprising the families Mastotermitidae, Kalotermitidae, Hodotermitidae, Rhinotermitidae, and Serritermitidae, additionally have obligately anaerobic, flagellate protists of the orders Hypermastigida and Trichomonadida (phylum Parabasalia) and Oxymonadida (phylum Metamonada) at densities up to 10^7 cells ml^{-1} (Fig. 3A).

The protist and bacterial symbionts are considered separately.

1. Protists

Approximately 400 species of protists have been reported in the guts of lower termites and a few, including *Trichomitopsis termopsidus* and *Trichonympha sphaerica* (both from *Zootermopsis*) have been brought into axenic culture. These protists are cellulolytic, and they have fastidious culture requirements, including strict anaerobiosis, yeast extract, and heat-killed bacteria from the rumen of cattle (which they phagocytose). They cannot be supported by carbohydrates other than cellulose or by bacteria such as pseudomonads, *Bacillus serratus,* or *E. coli*.

The protists are required by lower termites. They can be eliminated by incubating the termites at elevated oxygen tensions and the protist-free termites (often described as defaunated) cannot survive on high cellulose diets, such as filter paper. The protists degrade cellulose, with carbon dioxide and shortchain fatty acids (SCFAs) such as acetate as the principal waste products of fermentation. The SCFAs are absorbed across the hindgut wall and metabolized as a source of energy by the aerobic tissues of the termite. Cellulase active against crystalline cellulose (i.e., C_1-cellulase) has been demonstrated in *Trichomitopsis* and in mixed populations of protists from *Coptotermes lacteus*. Some of the smaller protists, however, are probably not cellulolytic.

At each molt of the termite, the oxygen tension in the paunch increases dramatically and all the protist symbionts are killed. The termite subsequently acquires a fresh inoculum of symbionts by feeding on a

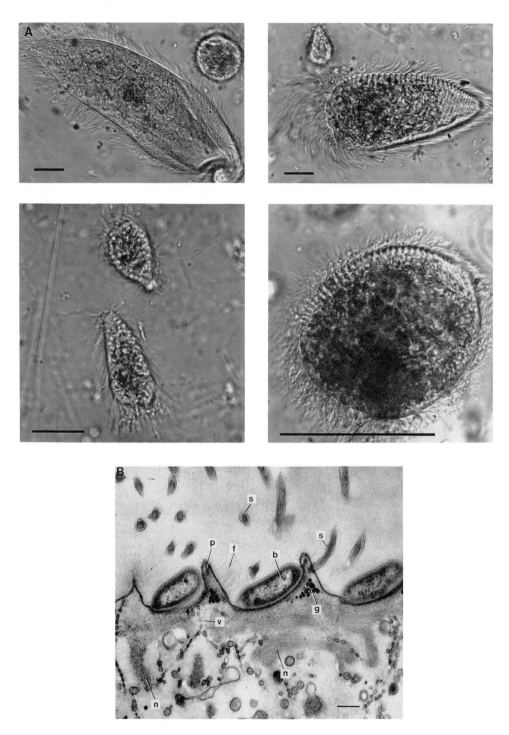

Figure 3 (A) Protists from the paunch of a worker of the caste lower termite *Coptotermes lacteus* Froggatt. Bar = 25 μm. [Photographs by R. T. Czolij and M. Slaytor.] (B) Spirochetes(s) associated with the surface of *Mixotricha darwiniensis*, a hypermastigote from the termite *Mastotermes darwiniensis*. Bar = 0.5 μm. [Micrograph from Cleveland, L. R., and Grimstone, A. V. (1964). *Proc. R. Soc. London B* **159,** 668–686.]

drop of hindgut contents expelled via the anus of another member of the colony, a behavior known as proctodeal trophyllaxis. It has been suggested that the requirement for conspecifics as a source of symbionts was a major selection pressure for the evolution of sociality in termites. However, the protists were lost in the ancestor of the higher termites (for unknown reasons). The latter termites do not exhibit proctodeal trophyllaxis and depend exclusively on regurgitation of masticated ingesta for the distribution of food within the colony.

2. Bacteria

The principal carbon sources utilized by the culturable bacterial symbionts of termites are sugars and organic acids. A few isolates degrade soluble derivatives of cellulose (e.g., carboxymethylcellulose), but none are known to utilize crystalline cellulose. Consequently, cellulolysis in the higher termites (which lack protists) is mediated exclusively by intrinsic enzymes secreted with the saliva or into the midgut. The appreciable concentrations of SCFAs in the paunch of higher termites is generated by bacterial fermentation of sugars and other low molecular weight compounds and by acetogenesis (i.e., fixation of carbon dioxide into acetate). Fermentative metabolism by the various bacteria in both lower and higher termites may be intimately linked. For example, the streptococci preferentially utilize sugars, especially glucose, which they degrade to lactate, and several *Bacteroides* strains are lactate fermentors, with the SCFAs acetate and propionate as major end-products. *Streptococcus lacteus* and *Bacteroides* sp. grow well in co-culture, with glucose as sole carbon source and SCFAs as principal products. Cross-feeding of lactate may also occur between these species in the termite gut.

The bacterial microbiota have been implicated in the nitrogen nutrition of termites, by recycling of the insect nitrogenous waste product uric acid, and by nitrogen fixation. Approximately 10% of bacteria isolated from *Reticulitermes flavipes*, especially *Bacteroides termitides*, *Streptococcus* spp., and *Citrobacter*, are uricolytic. Uricolysis by all isolates is anaerobic, with carbon dioxide, acetate, and ammonia as chief end-products, although uric acid degradation by the isolate of *Streptococcus* was incomplete in the absence of a reductant such as formate. The significance of bacterial uricolysis to the insect was indicated by recovery of radioisotopes and heavy isotopes from insect tissues of termites fed on ^{14}C- and ^{15}N-uric acid, respectively. The nitrogen-fixing bacteria isolated from termite guts include *Citrobacter freundii* and *Enterobacter agglomerans*, but the nitrogen fixation rates vary widely among termite species and with the nitrogen content of the diet. For example, *Nasutitermes corniger* and *Reticulitermes flavipes*, respectively, fix 10–100 and $<0.1 \ \mu g \ N \ g^{-1}$ body weight day^{-1}. The proportion of nitrogen generated by nitrogen fixation and uricolysis that is assimilated by the insect and the identity of the nitrogen compounds transferred are unknown. [*See* BIOLOGICAL NITROGEN FIXATION.]

Spirochetes represent a significant fraction of the microbial biomass in the termite paunch. They include pillotinas, such as *Hollandina*, *Pillotina*, *Diplocalyx*, and *Clevelandina*, and many others remain to be identified. Most spirochetes adhere to the insect gut wall, fragments of ingesta or protists via apical attachment sites, often of considerable structural complexity. Some of the spirochete–protist associations are specific; the diversity of protists involved and variation in structure of the attachment sites suggest that the associations have evolved independently many times. For example, the surface of the hypermastigote *Mixotricha darwiniensis* bears uniformly distributed projections, with spirochetes attached to their posterior surface (Fig. 3B). The rhythmic movements of the spirochetes serve to propel the protist forward, with the protist's flagella acting as a steering device. Lynne Margulis has proposed that cilia (and, hence, all microtubule-derived organelles in eukaryotes) have evolved from protist–spirochete associations analogous to those in *M. darwiniensis*. [*See* SPIROCHETES.]

C. Blattaria (Cockroaches) and Orthoptera (Grasshoppers, Crickets, Locusts, etc.)

1. Omnivores and Herbivores

Omnivorous and herbivorous Blattaria and Orthoptera possess a substantial gut microbiota, especially in the hindgut. For example, approximately 100 species of bacteria have been isolated from the proximal hindgut of the American cockroach *Periplaneta americana*, and they include both obligate anaerobes (e.g., *Clostridium*, *Fusobacterium*) at densities of 10^{10} cells ml^{-1} gut volume and facultative anaerobes (e.g., *Klebsiella*, *Citrobacter*, *Enterobacter*) at 10^8 cells ml^{-1}. The microorganisms in the house cricket *Acheta domesticus* and desert locust *Schistocerca gregaria* are exclusively facultative anaerobes, with *Klebsiella*, *Yersinsia*, *Bacteroides*, and *Fusobacterium* abundant in *A. domesticus* and Enterobacteriaceae, especially *Enterobacter agglomerans*, dominant in *S. gregaria*.

The hindgut microbiota appear to be of little significance to their insect host. *Schistocerca gregaria* and *A. domesticus* raised from the egg under sterile conditions had similar growth and development rates to control insects, and antibiotics that selectively eliminated the gut microbiota of *P. americana* (leaving its mycetocyte symbionts intact; see Section II.A) did not affect the fecundity of adult females. However, studies on insects lacking the gut microbiota indicated that (a) microorganisms in *S. gregaria* "protect" the insect from gut pathogens, especially gregarine protists; (b) microorganisms in *P. americana* are required for normal differentiation of the absorptive surfaces in the alimentary tract of young larvae, and bacteria-free larvae with impaired gut morphology may grow 10–20% more slowly than insects with a normal microbiota; and (c) the digestive tract of axenic *A. domesticus* has low carbohydrase activity, resulting in reduced efficiency with which the insect utilizes food; the axenic insects, however, can compensate for this by increased food intake.

2. The Gut Microbiota and Pheromone Production

It has been suggested that microorganisms in the crop of the migratory locust *Locusta migratoria* degrade ingested lignin and that one product 5-ethyl guaiacol (2-methoxyl phenol) is released via the feces. This compound, colloquially known as locustrol, supposedly acts as a pheromone, causing the gregarization of the locust. However, the putative lignolytic microorganisms have not been isolated or identified, and, if this hypothesis is valid, "locustrol" is probably just one of many inputs determining locust morphology and behavior.

3. The Woodroaches

A small minority of cockroaches feed on wood, and, by analogy with the xylophagous termites (Section III.B), they might be expected to require their gut microbiota. Two woodroaches have been investigated. The hindgut microbiota in *Panesthia cribatus* (family Blaberidae) comprises bacteria, none of which are cellulolytic; *Panesthia*, like higher termites, produces intrinsic cellulases. *Cryptocercus punctulatus* (family Cryptocercidae), by contrast, has up to 25 species of obligately anaerobic, flagellate protists, closely related (and often congeneric) with protists in lower termites (e.g., species of *Oxymonas* (a metamonad) and *Trichonympha* (a hypermastigote)). Some of the protists are cellulolytic. The protists in *Cryptocercus* differ from the termite symbionts in that they reproduce sexually, producing oxygen-resistant cysts prior to each insect molt. The protists are expelled from the hindgut and are reingested by the insect after molting. Linked to the complex colonial life-style and proctodeal trophyllaxis in lower termites (Section III.B), the protists in termites rarely reproduce sexually and never encyst.

It is widely accepted that the protists in *Cryptocercus* and lower termites have been inherited from the common ancestor of the Isoptera and Blatteria, possibly in the Carboniferous or Permian periods. Alternatively, *Cryptocercus* may have acquired the symbionts from termites living in the same habitat, sometime during the Mesozoic era.

D. Heteroptera

1. Reduviidae

Reduviids of the subfamily Triatominae feed exclusively on vertebrate blood, but unlike the hematophagous insects with mycetocyte symbionts described in Section II.C, the triatomid symbionts are located in the lumen of the midgut.

The midgut microbiota of triatomids is diverse and variable. In particular, the microorganisms in laboratory-reared stocks may differ greatly from those in insects from the natural environment. Isolates include *Pseudomonas* species, *Streptococcus*, *Mycobacterium*, *Corynebacterium*, and actinomycetes, especially of the genus *Nocardia*. *Nocardia rhodnii* is occasionally described as the sole or dominant symbiont of triatomids, but it is absent from some species of *Triatoma*, and laboratory cultures of *Rhodnius prolixus* can be raised with other *Nocardia* species or *Mycobacterium* as sole symbionts.

Most triatomids raised from surface-sterilized eggs die as larvae, usually in the fourth larval instar. The microbiota is believed to provide B vitamins (see Section II.C). Several isolates of *Nocardia rhodnii* synthesize "significant" amounts of folic acid, thiamine, and pantothenic acid, and developmental arrest of axenic insects is prevented by injection of B vitamins into the insect or their inclusion in the diet. However, when symbiont-free *Rhodnius prolixus* are infected with mutant *N. rhodnii* deficient in B vitamin synthesis, they develop normally, suggesting that the microbiota may provide other compounds that "spare" the insect's requirement for B vitamins.

2. Phytophagous Heteroptera

Many phytophagous "bugs," including some pentatomids (stink bugs), scutellerids (shield bugs), pyrrhocorids (fire bugs), coreids (squash bugs), and lygaeids (seed-sucking bugs) bear microorganisms in their midgut caeca. None of the microorganisms have been cultured, but their morphology at the light microscope level suggests that they include many taxa of bacteria. The microorganisms are apparently of value to the insects. Aseptically raised larvae of the pentamonid *Eurydema* grow and develop very slowly, and bacteria-free *Coreus marginata* die in the second larval instar.

E. **Diptera**

1. Tephritidae

Bacteria have been isolated from the alimentary tract of 40–50 species of tephritid flies. The microorganisms attain very high densities in the midgut caeca of larvae and, for some species, in unpaired evaginations of the foregut or hindgut (cephalic organ and rectal organ, respectively) of the adult female. For many years, it has been believed that the microbiota is dominated by a single species, required by the insect for normal development. The bacteria in some flies were supposedly important pathogens of the plants on which the flies feed. For example, the bacteria in the olive fruit fly *Dacus oleae* was identified as *Pseudomonas syringae* pathovar *savastonoi*, the agent of olive knot gall, and *Pseudomonas melophthora*, the agent of apple rot, was reported in the apple maggot *Rhagoletis pomonella*.

Modern studies indicate that the relationship between tephritids and their gut microbiota is far less specific. Neither *Ps. syringae* nor *Ps. melophthora* can be isolated from their putative insect hosts, and the microorganisms in the alimentary tracts of these species include many taxa, especially members of the Enterobacteriaceae, also widespread in the environment (e.g., *Klebsiella oxytoca, Enterobacter agglomerans*). These microorganisms appear to be of no significance to the insects, which develop normally under aseptic conditions.

2. Tipulidae

Larval tipulids harbor dense populations of microorganisms in evaginations at the proximal end of the hindgut. Bacteria, especially filamentous facultative anaerobes, predominate but ciliate and flagellate protists are also present. The microorganisms in *Tipula abdominalis* are morphologically similar to the microbiota contaminating detritus on which the insect feeds, suggesting that the gut bacteria are acquired by ingestion. Buchner proposed that the microbiota degrade cellulose and that the association is comparable to that in lower termites, but cellulolytic forms have not been isolated from tipulids, and the hindgut of *T. abdominalis* lacks cellulase activity.

F. **Coleoptera**

1. The Lamellicorn Beetles: Lucanidae and Scarabaeidae

The alimentary tract of Lucanidae (stag beetles) and some Scarabaeidae, especially the chafers, have substantial evaginations or dilations containing dense populations of bacteria. These structures have been interpreted as fermentation chambers, analogous to the paunch of lower termites (Section III.B) in which cellulolytic microorganisms degrade cellulose to SCFAs. However, there have been no experimental investigations of the associations in lucanids, and in the sole scarab beetle investigated, the rhinoceros beetle *Oryctes nasicornis*, cellulose is degraded primarily to methane (of no nutritional value to the insect).

2. Scolytidae

Bark beetles of the family Scolytidae bear various bacteria in their alimentary tract and yeasts, including species of *Candida* and *Hansenula* in an unknown location. These microorganisms can detoxify monoterpene hydrocarbons, defensive compounds in the trees on which the beetles feed, by various oxidative reactions. The significance of these capabilities to the insect is uncertain, because the beetles have intrinsic monoterpene detoxification systems. Potentially more important is the influence of microbial metabolism on bark beetle behavior, for bark beetles utilize various oxidation products of monoterpenes as pheromones. For example, the aggregation pheromone of *Ips paraconfusus* contains *cis*-verbenol (derived from the tree's α-pinene) and ipsenol and ipsdienol (from the tree's myrcene). *Ba-*

cillus cereus isolated from the gut of *I. paraconfusus* can oxidize α-pinene to *cis*-verbenol, and antibiotic treatment of *I. paraconfusus* abolishes ipsenol and ipsdienol production. However, the link between microorganisms and pheromone production is complex, as illustrated by the following observations:

Aseptically reared *I. paraconfusus* produces ipsenol and ipsdienol (in contradiction of the earlier antibiotic studies).

Yeasts in *Ips typographus* metabolize tree monoterpenes to antagonists of this beetle's aggregation pheromone.

The production of trans-verbenol, a component of the aggregation pheromone of *Dendronoctus ponderosae,* is increased sixfold in axenic individuals of this species.

Bibliography

Buchner, P. (1965). "Endosymbioses of Animals with Plant Microorganisms." Wiley, London.

Cruden, D. L., and Markovetz, A. J. (1987). *Ann. Rev. Microbiol.* **41,** 617–643.

Douglas, A. E. (1989). *Biol. Rev.* **64,** 409–434.

Martin, M. M. (1991). *Phil. Trans.* **333,** 281–288.

Smith, D. C., and Douglas, A. E. (1987). "The Biology of Symbiosis." 320 pp. Edward Arnold, London.

T

Taxonomic Methods

Claude Bollet and Ph. de Micco
Hôpital Salvator

I. Classifications
II. Numerical Taxonomy
III. Chemotaxonomy
IV. Molecular Taxonomy
V. Applications

Glossary

Nomenclature Rules that verify the legitimacy of the attribution of a name

Phylogeny Study of evolution and ancestry of an organism or both

Taxonomy Theory and practice of classification of individuals into groups

Systematics Study of the diversity and relationships among organisms

THE GOAL OF taxonomic classification is to assign an identity to a living or inanimate object. This attribution is first aimed at summarizing all of the properties assigned to a group of individuals so as to predict the properties of a new member of the group without having to completely explore it: The description of the model makes it possible to infer that of the individual. This predictive aspect of classification is very useful for passive users, such as clinicians, of bacterial classification for whom the assigning of a bacterium to the *Escherchia coli* group makes it possible to deduce all of the ecological, epidemiological, and therapeutic characteristics of this group.

In addition, classification is a prerequisite for the identification of new individuals since attribution to a group can only be made if the group has already been described: Identification is not possible without prior classification.

Finally, classification can serve to construct theories on the evolution of a group of individuals: The systematics of living beings can be used to study their phylogenesis, meaning the stages of evolution that are the origin of the individual today.

The diversity of these utilizations explains why classification must satisfy several criteria.

I. Classifications

A. Introduction

Classification must first follow a profound knowledge of the group in question. Otherwise, its predictive value would insufficiently fulfill the requirements of its users. A classification relying on a small number of characters that are considered "significant" by its authors would, in the end, reflect a reduced quantity of information and, thus, not be efficient in predicting individual characteristics. Because the choice of "significant" characters is essentially subjective, any taxonomist could question them and modify this unstable classification. The description of species would therefore rely on the idiosyncrasy of each taxonomist, such as Véron wrote. For example, this idiosyncrasy altered the description of species in the *Enterobacter–Erwinia* group, which were in turn described according to their biochemical characters and their phytopathogenicity to such an extent that this group is currently a real taxonomic wasteland where any rigorous study would demolish a section of the structure built in the 1950s.

The complexity of the classification should reflect the heterogeneity of the group being studied. For this reason, the former classification of *Acinetobacter* into *Acinetobacter lwoffi* and *Acinetobacter anitratus* was little used because it was too basic and nonpredictive of a particular pathogenicity or ecological biotope. Thanks to the judicious choice of reference strains, Bouvet and Grimont increased the number of bacterial groups from 2 to 12. This has made it possible, in most cases, to discern the origin of a bacterium and its ability to resist an antibiotic treatment. For example, the isolation of *Acinetobacter baumannii* in a blood culture, most often nosocomial and very resistant to antibiotics, does not have the same clinical and epidemiological significance as that of *Acinetobacter johnsonii*, which is commonly isolated from the skin and mucous membrane and is usually more sensitive to antibiotics.

Classification must appear to be stable for users. If the groups are rearranged too often, they give the user the impression that the classification is false, thus encouraging the user to be satisfied with a higher hierarchic level. In this respect, the description of the aforementioned *Acinetobacter* is an example of a well-accepted evolution of a classification because it immediately seems capable of resolving most identification problems: Few strains escape assignment to a species. On the contrary, the classification of *Pseudomonas* clearly suffers from the progressive description of new groups. The increasingly complex nature of the fluorescent pseudomonads (RNA Palleroni group I) makes it difficult to assign a pathological meaning to strains isolated in humans. The result is that we are now witnessing a multiplication of clinical case descriptions caused by these strains, but for lack of knowledge of their clinical significance, we are obliged to consider them as contaminants.

Classification should clearly be built upon solid definition criteria of the hierarchic levels. If the bacterial species is now well defined, it is not the same for the bacterial genus and even less the case for higher hierarchic levels. The general surveys made up to the 1980s vanished after the appearance of study techniques of parts preserved in genetic heritage that mixed morphologies, respiratory types, and wall structures—notions that supported the classic taxonomic structure. As R. G. E. Murray wrote, "the best is yet to be."

B. Special Purpose Classifications

Microbiologists and clinicians have often constructed classifications to fit their purposes. Numerous bacterial species have been described by characters chosen dogmatically for what one wished to obtain from the classification. The term nomenspecies was proposed for these taxa defined for medical or practical purposes. Medical bacteriologists have therefore tried to set up classifications that take bacterial pathogenicity into account. The result is a pathogenicity gradient that ranges from "nonpathogenic" *E. coli* strains to enteroinvasive, enterotoxigenic, or enteropathogenic *E. coli* strains, then to *Sigella,* which are distinguished by the serotypes to which one assigns a virulence or a variable aggressiveness. However, all of these microorganisms are taxonomically *E. coli,* more or less adapted to their medium and more or less able to cause damage to their host depending on the genes lost or the extrachromosomic characters acquired. Only their different clinical significance can justify the use of different vernacular names. The same is true for the distinction between *Bacillus cereus* and *Bacillus thurengiensis,* which cannot be taxonomically differentiated: Some strains contain crystalline formations that render them entomopathogenic, named *B. thurengiensis,* and others do not, *B. cereus.*

These classifications continue to exist to everyone's satisfaction. Such is the case with the subdivision of the *Mycobacterium tuberculosis* group, as it is true that one should be able to distinguish a pulmonary tuberculosis from a *Mycobacterium bovis–bacille* de Calmette et Guérin infection in an immunodepressed patient. This confusion and assimilation must nevertheless be kept in mind to recognize the emergence of new pathogens in immunodepressed patients.

Phytopathologists have favored the notion of pathovar, further weakening it with an oriented choice of strains. These taxa, made up of individuals adapted to their ecological biotope, have been called genospecies by Véron. The existence of these pathovars should now be discussed following the objective data from the study of the genetic heritage of bacteria.

Ravin proposed the use of the term genospecies to define species composed of individuals descending from the same ancestor and having the ability to recombine by transferring part of their genetic material. Véron noted that this definition was only applicable to a few bacteria with proven genetic transfer.

Moreover, the recent discovery of numerous *in vivo* transfers of plasmid-mediated properties limits the interest of this definition. It would therefore be better to use the expression genomic species when a species has been described by quantitative molecular hybridization.

C. Natural Phenetic Classifications

Bacterial groups were defined until the end of the 1950s by successive dichotomic keys that relied on the results of tests on strains that had been part of the definition of the species. The order of characters to be studied was defined in an arbitrary manner. Any individual that did not fit the key (e.g., by mutation) was not well oriented and therefore impossible to classify.

The principles of phenetic taxonomy were defined by the French botanist Michel Adanson in 1763. He understood that the natural similarities between individuals would be better studied by taking into account all of their characters and arranging individuals into groups "which cannot be arbitrary since they are not founded on one or several uncertain characteristics but on all of the parts."

P. H. A. Sneath was the first, in 1957, to introduce the concept of adansonian classification in microbiology, called numerical taxonomy, which he and R. R. Sokai defined in 1973 as "the grouping by numerical methods of taxonomic units into taxa on the basis of their characteristics." This involves studying all of the physiological and structural characteristics of bacteria, meaning those that correspond to appearance (phenotype; Grk. *phaïno*, I show) and heritage (genotype; Grk. *genos*, birth). Their "natural" character opposes them to the special purpose classifications cited in the first section. They make no reference to evolution or a possible common ancestor. Individuals are objectively arranged into polythetic groups or phenon, taking into account all of their characters. The relationships between individuals are expressed on a dendrogram. The high number of characters studied makes it possible to aggregate individuals that are atypical for certain characters: Certain individuals are outside of the definition of the group to the point that they cannot participate in this definition but can be included given all of their properties. The higher the quantity of information, meaning a higher number of studied characters, the higher the predictive value of an identification. In reality, this is tempered by the genetic support of characters, as discussed in Section II. Phenotypic classification methods have been used to analyze sets of inanimate beings (e.g., archeological furniture). Evolutionist taxonomists, who consider that taxa must be monophyletic (coming from a common ancestor), are against this classification of inanimate beings.

Monothetic classification, gathering individual nuclei of individuals possessing certain common properties, end up as fragile structures and are no longer used.

D. Natural Phylogenetic Classifications

Certain microbiologists have developed natural classifications that aim to reflect the phylogenetic evolution of a bacterial group. These cladist taxonomists (Grk. *klados*, branch) try to retrace the evolution that resulted in the emergence of known microorganisms. Individuals are clustered depending on a set of characters presumed to be inherited from a common ancestor. Cladism is based on an analysis of all of the characters during comparison of related taxa in order to distinguish those that are ancestral (plesiomorphic) from those that are derived (apomorphic). To find the connections in the phylogenesis, one follows the distribution of shared derived characters (synapomorphic). Apomorphic characters can only be shared by descendants of an ancestor in whom this character had appeared for the first time.

It is the common possession of derived characters that proves the ascending commonalty of a given group of species.

Group relationships are shown in what is called a cladogram, a series of dichotomies representing the successive separations of the various descendants.

E. Evaluation and Choice of a Classification

Phylogenetic classifications can only be verified by the construction of another classification that relies on other characters inherited in common and subsequent comparison of the cladograms. As Sneath wrote, "if both cladograms are constructed with the same characters, they will be congruent, without validating the characters studied. If the characters are different, the cladograms will be different."

Cladistic analysis was welcome for groups with numerous phenetic characters and for which the classification up to that point had been insufficient.

In these cases, the cladograms revealed that many of the taxa admitted were in fact polythetic. The study of the "genus" *Pseudomonas* brought out this problem. Clarification of the taxonomic position of this group will come from the genome study by consensual techniques and not from a different analysis of the characters expressed.

Phenetic classifications are always open to verification by their users: They rely on measurements that can be repeated outside of their original laboratory and that can give rise to a mathematical analysis. These techniques will be developed in Section II. They alone make it possible to integrate new individuals as they are discovered by modifying the definition of existing groups. They alone make it really possible to identify unknown individuals by measuring their proximity to known groups. They can now integrate similarity measurements of ribosomal RNA sequences, permitting them to reflect the phylogeny of the group being studied.

F. Definition of Species

The groups that the taxonomist assembles into genera are called species: The species is the basic unit of diversity in nature.

The increase in knowledge of physiology and molecular biology in the 1930s made the emergence of the concept of biological species possible: "a species is a reproducing community of populations (reproductively isolated from other communities) that occupies a particular place in nature. Members of a species are programmed in their DNA code to maintain a unique common mass of genes." (E. Mayr)

In bacteriology, morphologic similarity is of little value: All Enterobacteriaceae are gram-negative bacilli with peritrichous flagella; all *Clostridium* are endospore-forming, gram-positive bacilli. The criteria for isolated reproductive communities are difficult to use given the absence of true sexual reproduction in prokaryotes. Moreover, numerous gene-transfer phenomena remain poorly understood, making it impossible to use the previous definition of biological species. Microbiologists are confronted with great diversity: Prokaryotes appeared 3 or 4 billion yr ago and *Homo sapiens* 300,000 yr ago. Generation times are very different: 25 yr for humans (or approximately 10,000 generations since the beginning) and 20 min for *E. coli* in a liquid medium that could go through 10,000 generations in <6 mo. In addition, plasmid transfers and mutations are very frequent, explaining the adaptation of a given ecologic situation. In the absence of a particular concept of species for prokaryotes, the bacterial species is currently defined in the following manner: A species includes strains having DNA/DNA homologies expressed in percentages of hybridization superior or equal to 70% and presenting a stability of formed hybrids (measured by thermal stability of reassociated duplexes) inferior to 5°C. These techniques will be studied later.

II. Numerical Taxonomy

A. Introduction

Phenetic classification must use strict rules concerning the choice of strains, coding methods, and calculation techniques. If not, the results would be questionable.

B. Strain Selection

The choice of strains to be studied is essential. It is now thought that it is indispensable to incorporate type strains of the species being studied with type strains of neighboring species.

The minimum size of a batch of strains has been the object of much controversy. Two researchers defined a reasonable inferior limit as 60 strains. It is true that significant studies have been made with fewer strains, but they involved very homogenous bacterial groups, rare strains, or homogenous populations in restricted biotopes.

The most important studies in the literature have used batches of approximately 600 strains. Large computers (equipped with the appropriate software) connected to large plotters can analyze batches of >1000 strains and plot the dendrogram, but the study of dendrograms with >300 strains is visually difficult and requires a cutting of the dendrogram to isolate a few groups (Fig. 4).

The number of strains to be studied varies with the heterogeneity of the group. The choice of strains from different biotopes avoids redundancy and thus keeps the batch down to a reasonable size. Identical strains must gradually be eliminated from the batch. Strain purity must be verified each time the strain is studied.

C. Test Selection

Quantitative characters are rare in microbiology (bacterial cell measurements, colony size, reaction rate). In addition, such measurements are difficult to standardize.

Qualitative characters are more commonly used, which poses problems for coding the results. Table I shows the principal tests used in numerical taxonomy.

According to Adanson's principles, numerical taxonomy requires the study of the greatest possible number of characteristics of the microorganism. We now know that some characters are coded by a longer chromosomal sequence. Some characters depend on a single enzyme (catalase, tryptophanase, sucrose utilization), others depend on a single system (lactose operon, which carries three cistrons), and others depend on a polycistronic gene (histidine synthesis is coded by 10 adjacent cistrons; the presence of an endospore is due to the presence of approximately 50 operons).

The correspondence between the character and size of genetic information is usually unknown. Moreover, some characters are coded by extrachromosomic elements, as in lactose and/or rhamnose fermentation by *Serratia marcescens*. It is therefore customary to consider that all of the characters studied have the same taxonomic significance. The bias is minimized by increasing the number of studied characters. However, this increase is not unlimited: In addition to the limitations imposed by today's computers in microbiology laboratories, there is the problem of redundant characters, meaning those coded by the same genes. It is possible to avoid this redundancy if one knows the chemical components that have identical metabolism.

Those who use numerical taxonomy have noted that beyond 200–300 characters the quantity of explored information no longer increases.

Table I Biochemical and Cultural Tests Used in Numerical Taxonomy of Bacteria

Macroscopic aspect	Colony size	Q		Biochemical tests	Fermentation	L	(10)
	R/S aspect	M			Auxanogram	L	(11)
	Colony characteristics	M	(1)		Metabolites	L	(12)
	Corroding colony	L	(2)		Nitrate reduction	L	(6)
	Nonadhesive colony	L			Nitrites reduction	L	(6)
	Fluorescence	L			Enzymes	L	(13)
Microscopic aspect	Micromorphology	M	(3)	Inhibition tests	KCN	L	
	Size	Q			Antibiotics	L	(14)
	Gram-staining reaction	M	(4)		Heavy metals	L	
	Acid-alcohol fastness	L		Serology		M	
	Flagella	M		Phages		M	
	Motility	L	(5)	Bacteriocines		M	
	Spore	L	(6)	Chemotaxonomy	Cell wall	M	(15)
	Spore position	M			Metabolites	M	(16)
	Pleiomorphism	L			Proteines	M	(17)
	Pseudofilaments	L		Genetics	G+C content	Q	
Cultural characteristics	Temperature	M	(7)		DNA/DNA relatedness	Q	
	NaCl	M			DNA/RNA relatedness	Q	
	O₂ requirement	M	(8)		DNA sequences	Q	
	Growth factors	L	(9)		RNA sequences	Q	

(1) In charcoal form (*Clostridium difficile*), in candle-flame form (*Campylobacter*), with prominent center (certain *Bacteroides*), radiating (*Bacillus*), ...

(2) *Bacteroides ureolyticus, Eikenella corrodens*, ...

(3) Bacilli, cocci, spirochetes, vibrios, ...

(4) There are gram-variable bacteria (e.g., *Gardnerella vaginalis*) and bacteria losing the Gram stain in aged cultures (e.g., *Bacillus*).

(5) The temperature must be noted.

(6) Must be accompanied by the description of medium used.

(7) 4°, 10°, 20°, 37°, 42°, and 55°C are usually chosen.

(8) The following oxygen sensitivity thresholds are usually chosen: 0.1% (extremely oxygen-sensitive bacteria), 1% (strictly anaerobic), 5% (microaerophilic), and 20% (aerobic).

(9) Vitamins, coenzymes, divalent cations, ...

(10) Acid production from carbohydrates. Not useful in certain cases (*Serratia*).

(11) Utilization of carbohydrates, amino acids, organic acids, . . . , as sole source of carbon and energy. Must be accompanied by the description of minimum medium used.

(12) Acetoin production (Voges–Proskauer reaction), indole production, H₂S production, ...

(13) Must be accompanied by the description of substrates: arginine, ornithine, lysine, starch, various Tweeen, fatty acids, ...

(14) Susceptibility is most often considered as a positive response (existence of targets or receptors).

(15) Menaquinones, teichoic acids, mycolic acids, ...

(16) C3 to C8 acids (anaerobes), ...

(17) From cytoplasm, from outer membrane, esterases, ...

L, logical (Y/N); M, multistate;

Q, quantitative.

The introduction of the study of antibiotic susceptibility is under discussion. This susceptibility is very frequently used by traditional microbiologists—for example, for the classification of strict anaerobic gram-negative bacteria (*Bacteroides, Fusobacterium*). It is known that this susceptibility is often coded by plasmids and is therefore vulnerable to transfers that do not follow taxonomic classifications: A numerical taxonomy of *Straphylococcus* isolated from the pathologic products of hospitalized intensive care unit patients and only using the susceptibility to antibiotics would reflect the dynamics of the bacterial population and not their taxonomic heterogeneity. However, the capacity to express certain extrachromosomic resistance characters could be under the domination of chromosomic genes, which would justify their study. Because there are no certainties on this subject, most authors advise against incorporating susceptibility to antibiotics in a taxonomic study.

D. Reproducibility

Taxonomists can fall into certain pitfalls in their work: Bacterial strains can lose synthesis or degradation capacity through plasmid loss after successive subcultures, and they can also mutate. Moreover, interpretation of tests is sometimes difficult, particularly for the estimation of weakly positive results, which may depend critically on the exact experimental condition. In addition, taxonomic studies often involve several thousand results and it is unreasonable to exclude coding or data acquisition errors because important studies are most often multicentric.

Each of these problems can be minimized if precautions are taken: double reading and double execution of tests if the study is being simultaneously performed in several laboratories, utilization of a series of positive and negative controls for the auxanograms, and, especially, verification of stability of results on repeating them. Data-acquisition problems can be avoided by only noting the positive results, but always at the price of a careful verification of the phenotypes.

Sneath proposed a practical rule that would be to perform the study of 10% of the phenotypes twice in order to evaluate individual test variance, called S_i^2:

$$S_i^2 : \frac{n}{2t},$$

where n is the number of phenotypes with discrepancies in the test, and t is the total number of strains.

The global estimate of error is

$$S^2 = \tfrac{1}{N}(S_A^2 + S_B^2 + \ldots + S_N^2),$$

where N is the total number of tests, and S_A^2, $S_B^2, \ldots, S_N^2$ correspond to the individual test variances for tests $A, B, \ldots, N$.

The probability of error for an individual test is

$$P_i = \tfrac{1}{2}[1 - \sqrt{(1 - 4S_i^2)}].$$

Sneath recommended revising the test if there is an error probability superior to 10%. He noted that numerous classic biochemical tests (oxidase, gelatinase, urease, etc.) often entered into this category. One can also convince oneself that the effect of these errors is reduced by the number of characters studied: This has gradually been verified during successive verifications of large files.

E. Data Coding

Logical coding 1 and 0 (corresponding to + and −) is the most commonly used: 1 signifies the presence of a character and 0 the absence. This method is easy to use for nutritional characters or proteins (Table II).

It can be adapted to multistate character coding such as increasing temperatures or colony pigmentation by attributing to each character as many rows of coding as the number of possible states. Table III shows the coding techniques used for several multistate characters.

It can be noted that certain characters are translated by a limited list of possible phenotypes: Colony pigmentation cannot be multicolored. Other phenotypes can have several positive responses such as, for example, the study of temperature increase tolerances.

This greater importance given to multistate char-

Table II   Logical 1/0 Coding

Character	Presence	Absence
Motility	1	0
Urease	1	0
Indole production	1	0
Voges–Proskauer reaction	1	0
Gelatin liquefaction	1	0

Table III   Multistate Characters Coding

Character	State	Coding
Colony coloration	Red	1000
	Yellow	0100
	White	0010
	Gray	0001
	Colorless	0000
Colony size	<1 mm	10000
	<2 mm	11000
	<4 mm	11100
	<6 mm	11110
	>6 mm	11111
Growth temperatures	10°C	100000
	20°C	010000
	30°C	001000
	37°C	000100
	44°C	000010
	55°C	000001
	10°C $\leftrightarrow$ 37°C	111100
	44° $\leftrightarrow$ 55°C	000011

acters can be minimized by eliminating the constantly positive or negative characters before calculation, by not creating too many different classes, and, of course, by increasing the number of characters to be studied.

The susceptibility to antibiotics can be coded like multistate characters by defining the classes of minimum inhibitory concentrations or growth inhibition diameters or by simple 1/0 coding. Sensitivity to an antibiotic is considered as a positive test response. Bacteria possess enzymatic structures or mechanisms that allow the antibiotic to exert its inhibiting action.

In microbiology, only a few automatic devices are capable of reading results and sending the information to a computer. Most studies therefore include an important phase for acquisition of the results. Some programs, such as Taxan (Sea Grant College, University of Maryland), require acquisition of results in the form of a chain of characters made up of as many 0's or 1's as there are characters being studied. This method is very difficult to apply when the number of characters is greater than a few dozen: Acquisition is uncertain and data verification is impracticable. A technique was therefore developed that consists of entering only the rows of positive tests (Table IV). This method makes for easy acquisition and verification of the phenotypes. The computer takes care of transforming the positive rows into 1/0 character chains. [*See* IDENTIFICATION OF BACTERIA, COMPUTERIZED.]

Table IV   Entering the Results of 100 Tests

Logical 1/0 coding:
100011001110000100000000011010001110000101001001 01
1000100101101011000010010001001000010100101010 01

Entering only the rows of the positive tests:
1 5 6 9 10 11 16 26 27 29 33 34 35 40 42 45 48 50 51 55 58 60 61 63 65 66 70 73 76 80 83 88 90 93 95 97 100

F. Calculation of Distances

The computer measures the taxonomic distances or similarities between all of the individuals taken in twos. Dissimilarities are the complement of similarities, and dissimilarities are treated as distances. For example, 90% similarity = 10% dissimilarity, equivalent to a taxonomic distance of 10%. It is necessary to furnish the computer with the mathematical formula for calculation of these distances. Following the work of the mathematician Jaccard, numerous coefficients have been proposed. Table V displays a few of the most commonly used in bacterial taxonomy.

The most frequently used coefficient in microbiology is that of Jaccard–Sneath, which does not take negative similarities into account.

The distance calculation for each pair of individuals makes it possible to draw up the distance ma-

Table V   Some Coefficients Used in Numerical Taxonomy

Coefficient	Abbreviation	Formula
Jaccard	S_J	$\dfrac{a}{a + c}$
Czekanowski–*Dice*	S_{CD}	$\dfrac{2a}{2a + c}$
Kendall, Sokal, Michener (simple matching)	S_{KSM}	$\dfrac{a + b}{n}$
Rogers–Tanimoto	S_{RT}	$\dfrac{a + b}{n + c}$
Kulczynski	S_{K2}	$\dfrac{a}{2}\left(\dfrac{1}{n_i} + \dfrac{1}{n_j}\right)$
Ochiaï	S_O	$\dfrac{a}{\sqrt{n_i \cdot n_j}}$

a, number of positive matches; b, number of negative matches; c, number of discrepancies; n, number of tests; n_i, number of positive results for the phenotype i.

trix. It is a square table composed of as many lines and columns as there are individuals. This matrix is symmetrical and only has 0 in its first diagonal row (row i strain compared to itself). For n strains, it contains $n(n - 1)/2$ distances, or 499,500 distances for a 1000-strain file.

G. Aggregation Techniques

Each n line in the distance matrix can be considered as a vector whose coordinates are represented in a system of orthogonal axes constructed in an n-dimensional space. Their representation in a two-dimensional space is therefore not directly possible. Two mathematical processes are used.

1. Principal components analysis consists of projecting the cluster of n points into lower dimen-

sional space, usually a two-dimensional or three-dimensional space with maximum conservation of the inertia of the cloud. The resulting groups will vary with the projection technique. In addition, their limits are subjective, which thus limits any analysis for classification. Principal components analysis is seldom used except for population studies or to represent relationships between species and different characters (Fig. 1).

2. Hierarchical clustering is achieved by successive aggregations by first grouping the most similar individuals, i.e., those having the least distance between them. Then, step by step, the clusters and isolated individuals are aggregated into groups of higher hierarchic levels until all of the phenotypes are grouped. The distance between the last strain and/or the last groups is by definition the 100% hierarchic level.

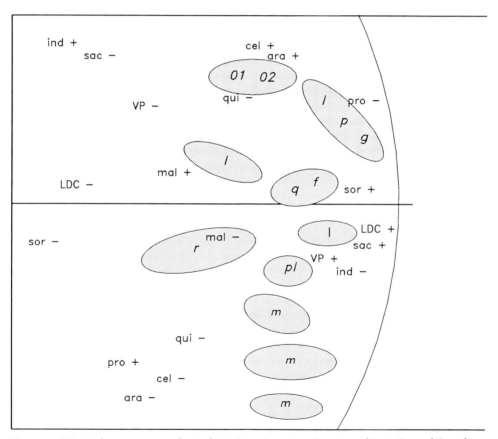

Figure 1 Principal component analysis of 720 *Serratia* strains. Species and tests. Jaccard-Sneath coefficient. Species: 01, *S. odorifera* 1 biotype; 02, *S. odorifera* 2 biotype; *f*, *S. ficaria*; *g*, *S. grimesii*; l, *S. liquefaciens*; *m*, *S. marcescens*; *p*, *S. proteamaculans*; *pl*, *S., plymuthica*; *q*, *S. quinovora*; *r*, *S. rubidaea*. Tests: ara, arabinose utilization; cel, cellobiose utilization; ind, indole production; LDC, lysin decarboxylase; mal, malate utilization; pro, prodigiosin production; qui, quinate utilization; sac, sucrose utilization, sor, sorbitol utilization; VP, acetoin production. [From Bollet, C., *et al.* (1988). *Ann. Inst. Pasteur/Microbiol.* **139**, 337–349.]

187

Several aggregation methods differ in the manner of calculating the distance between an isolated individual and a defined cluster. The single-linkage method consists of considering that the distance between an isolated individual and a group is the distance separating this individual from the nearest individual in the group. One aggregates an *i* cluster and a *jk* cluster according to the shortest distance (*i–j*) or (*i–k*):

$$D(i, jk) = \min D(i, j), D(i, k).$$

The effect of this method is to cause what is called the phenomenon of chaining: Strains that are closest to cluster are successively aggregated without having the constitution of homogeneous clusters. If there are too many atypical strains, one obtains a stairway or comb dendrogram and the study is uninterpretable (Fig. 2). It can only be used for batches with a distribution of homogeneous populations that are each very different. It has the advantage of being the easiest to program and does not require large computers.

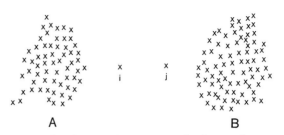

Figure 2 The chaining phenomenon. The distance between clusters A and B is reduced to the distance between individuals *i* and *j*.

The complete linkage clustering method produces different results as it does not cause chaining. It incorporates the individuals or aggregates the clusters with the least similarity between the individual and any member of the cluster.

The centroid sorting/clustering method diminishes the calculated distance by a factor, which is as proportionally as small as the distance to the new cluster.

The unweighted pair group method with averages

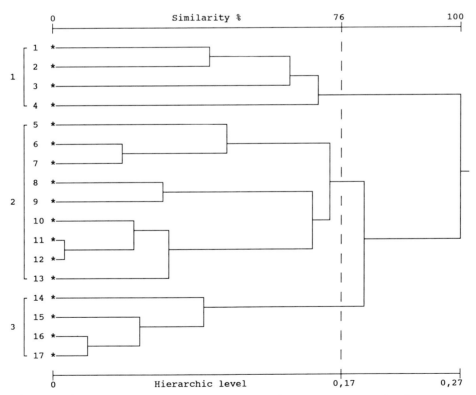

Figure 3 Cluster diagram of 17 strains of *Serratia rubidaea*. (1) Nonpigmented B2 biotype; (2) pigmented B1 biotype; (3) pigmented B3 biotype. [From Bollet, C., *et al.* (1989). *Trop. Agric.* **66**, 342–344.]

(UPGMA) is by far the most commonly used method. In this technique, an individual is aggregated to a group at a taxonomic level corresponding to its mean distance to all of the phenotypes that make up the group. The distance (*D*) between two clusters is calculated by taking into account the size (*M*) of each cluster:

$$D(i, jk) = [Mj \cdot D(i, j) + Mk \cdot D(i, k)]/(Mj + Mk).$$

The aggregation of two groups requires calculation of all of the distances of individuals in both groups taken in pairs and calculation of the mean distance. Mathematical calculations have shown that this technique is the most appropriate for representation of taxonomic structures. Its inconvenience is that it requires powerful computers because all of the matrix distances are recalculated at each stage of the process. To study a file with 100 phenotypes, approximately 150,000 calculations of distances between phenotypes are necessary, and for a file with 1000 phenotypes, 150,000,000 calculations.

H. Representation

Two diagrams are used for representation: the dendrogram and the similarity matrix.

(1) The dendrogram shows the successive arrangement of clusters and gives an idea of the affinity of the computed groups (Fig. 3). To plot it, one must know the intersection points of the diagram. Programs are complex and require large plotters when the number of phenotypes is high. The dendrogram may in turn be simplified by grouping phenotypes together at predetermined levels of similarity (Fig. 4). A simple calculation gives the hierarchic level (percentage of similarity) depending on the distance between phenotypes.

(2) Representation by similarity matrix is a less complex method. It consists of rearranging the distances in the matrix into several groups and attributing each group with a shade of gray or a more or less dark ASCII character, with the result giving a visual idea of the breakdown of clusters (Fig. 5).

It is generally admitted that bacterial species are individualized at similarity percentages of 80–85% using the Jaccard–Sneath coefficient and aggregation by the UPGMA. This approximate definition is acceptable for homogeneous and well-known bacterial groups (Enterobacteriaceae). It is not acceptable for less-known groups such as *Pseudomonas*. In any

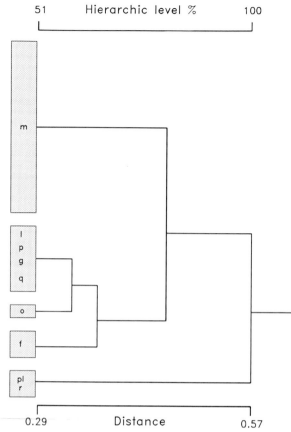

Figure 4 Hierarchic clustering of 720 *Serratia* strains. Jaccard–Sneath coefficient, clustering by unweighted pair-group method with averages. Cutoff of the dendrogram at hierarchic level 51 % (dissimilarity 0.29). Species: f, *S. ficaria*; g, *S. grimesii*; l, *S. liquefaciens*; m, *S. marcescens*; o, *S. odorifera*; p, *S. proteamaculans*; pl, *S. plymuthica*; q, *S. quinovora*; r, *S. rubidaea*. [From Bollet, C. (1988). *Ann. Inst. Pasteur/Microbiol.* **139**, 337–349.]

case, it is absolutely necessary to verify it with chemotaxonomy or DNA/DNA hybridization studies.

For descriptive purposes, it is necessary to choose a dendrogram cutoff point to define individualized groups and isolated phenotypes. This choice is frequently arbitrary. In practice, several cutoff levels are chosen and then one attempts to evaluate the rationality of each of these choices. This method usually enables observers to form a consensus.

The problem in choosing a cutoff level is all the more difficult when the batch is vast and heterogeneous. One must avoid cutoffs that are too high, which oversimplify the classification, or those that are too low, which render it unusable. A good compromise has been obtained in some studies by using Véron's acuteness coefficient: The dendrogram can

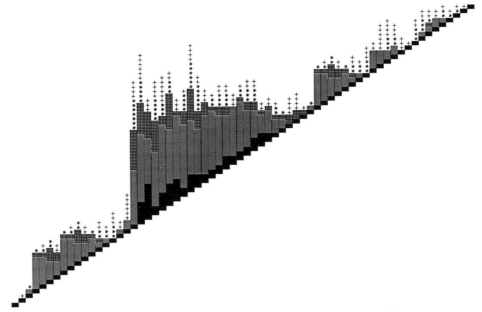

Figure 5 Similarity matrix; 66 strains of *Bacillus* from a hospital environment (five phenons). Entries in shading symbols: Range of similarity: blank, S < 0.70; +, 0.70 ≤ S < 0.80; *, 0.80 ≤ S < 0.85; ▊, 0.85 ≤ S < 0.90; ▉, 0.90 ≤ S < 0.95; ■, 0.95 ≤ S < 1.00.

be cut at any level of similarity between 0 and 100%. At each cutoff level, the added information or acuteness coefficient is measured: It takes into account the number of regrouped phenotypes, the cutoff level, and the difference between the cutoff level and the last aggregation. For each hierarchic level h_i, it is possible to plot a cutoff of the dendrogram. Each existent class g at this level can be characterized by three parameters:

1. n: number of phenotypes in this class;
2. h: hierarchic level at which class g is individualized;
3. d: distance between cutoff levels hi and hg. This parameter measures the degree of separation of class g from the other classes.

The acuteness coefficient for class g is represented by the ratio $(ng/hg)dg$. The sum of the acuteness coefficients for the various classes is the overall acuteness coefficient.

One obtains a curve with the maximum showing the most interesting cutoff level. Individualized clusters below this level are species and phenetic groups (Fig. 6).

This acuteness coefficient would seem to be of interest for very homogeneous batches (short taxo-

nomic distances) or when the centroid sorting/clustering method is used.

III. Chemotaxonomy

A. Protein Analysis

1. Sequences

Proteins are molecules that are well preserved during evolution, particularly those that are necessary for bacteria physiology (enzymes, ferredoxins, cytochromes). Cytochromes are electron transporting hemoproteins: They belong to four classes depending on the nature of their prosthetic group. Two cytochrome study techniques are used: the spectrometric dosage (cytochrome pattern) and the primary (amino acid sequence) or tertiary (X-ray diffraction) sequence study of these molecules. This technique was used to reveal great similarities between certain nonsulphur purple photosynthetic bacteria (i.e., *Rhodopseudomonas capsulatus*), the nonsulphur respiring bacterium *Paracoccus denitrificans*, and the mitochondria of eukaryotic organisms. This could lead one to think that *P. dentrificans* descends from *R. capsulatus* through loss of photosynthesis capacity and that these bacteria

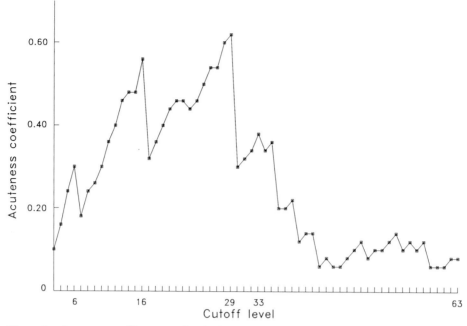

Figure 6   Acuteness coefficient in a batch of 64 anaerobes (clinical isolates). Best cutoff level: 29, acuteness:0.62, 10 phenons. [From Bollet, C., *et al.* (1988). *Pathol. Biol.* **36**, 203–208.]

could be close to endosymbionts, which were the origin of mitochondria. Taxonomic applications are limited because the distances between similar species are often very large.

Numerous sequencing studies of other proteins have been performed, usually with the aim to establish phylogenetic relationships by using proteins with different evolution speeds.

2. Electrophoretic Profiles

Proteins have their own electric charge and therefore migrate when placed in an electric field. The most commonly used method in microbiological taxonomy is the study of the banding of patterns of dodecyl sulfate–extracts in polyacrylamide gels.

Polyacrylamide gel is the result of polymerization of acrylamide by bisacrylamide. Polymerization is triggered by the addition of ammonium peroxydisulphate and accelerated with the addition of N,N,N',N'-tetramethyl-ethylenediamine. Because oxygen inhibits polymerization, one must degasify before the mixture is made. Such a gel can be considered as a porous support. Pore size is nearly the same as that of the protein molecules: It diminishes when the concentration of acrylamide is increased. Polyacrylamide gel acts as a screen. The bacteria are

broken, usually by sonication. The proteins are dissociated in a manner that frees their polypeptidic subunits. SDS, an ionic detergent, is used as the dissociating agent. The protein mixture is denatured by heating at 100°C with an excess of SDS and 2-mercaptoethanol, which break the protein disulfide bridges. Under these conditions, the polypeptides bind a constant quantity of SDS: approximately 1.4 g of SDS per gram of polypeptide. The polypeptide charge becomes insignificant compared to the negative charges brought by the SDS. The migration of polyacrylamide gel then depends only on the size of the polypeptides. This technique makes it possible both to analyze the polypeptidic composition of the mixture and to determine the molecular mass of these polypeptides as compared to markers with a known mass. Indeed, a linear relationship exists between the $\log_{10}$ of the peptide molecular mass and its relative mobility (R_f), which is estimated by the distance of migration in the gel.

A vertical electrophoresis device is used. The polyacrylamide gel is made up of two parts: The lower part is the separation or running gel containing 8–12% polyacrylamide, made with a Tris base buffer (0.4 M, pH 8.8). After polymerization, a second gel (concentration or stacking gel, 4% polyacrylamide, Tris base buffer 0.15 M, pH 6.8) is poured onto the surface of the separation gel. The proteins will be concentrated in a thin band during the migration through the wide-pore concentration gel. They are then revealed by Coomassie blue staining. This is known as the Laemmli system.

It is possible to study fractions containing total or membrane proteins. When the composition of the culture medium is perfectly constant, the protein composition of a given bacterial strain is constant. It is therefore possible to compare two bacterial strains by studying their protein pattern: This technique is very commonly used as an epidemiological marker and can be used as an identification technique (Fig. 9).

It is also possible to compare patterns after densitometric analysis. For this, a laser densitometer connected to a computer is customarily used: It is necessary to standardize the profiles by reference to proteins with known migration speeds. Measurements should be compared to those obtained with one or several standard strains. The computer makes it possible to directly use the measured values with the previously described numerical taxonomy methods. Several comparative studies have shown that the results are in excellent agreement with those obtained by DNA/DNA hybridization. However, it is difficult to obtain good interlaboratory reproducibility: It is necessary to use standardized procedures (growth conditions, electrophoresis technique, and reactives).

Using Western blotting, one can reveal proteins possessing a particular epitope. This technique is not applied in taxonomy.

B. Enzymes

Instead of studying all cellular proteins, it is possible to compare one or several enzymes (zymogram). Enzymatic activity is revealed by staining in the presence of a specific substrate. This technique is easier to standardize than the SDS-PAGE. It is especially applied to esterases and dehydrogenases. It gives approximately the same results as DNA/DNA hybridization. It has been proposed for bacterial identification.

C. Peptidoglycan

Peptidoglycan (murein) is present in all bacteria except mycoplasmas and archeobacteria. It is composed of amino–sugar backbones bearing tetrapep-

tide chains with a diamino acid in position 3. These chains are linked directly or by a bridge of one to six amino acids. The structure of the amino–sugar backbones is mostly preserved, but the composition of the tetrapeptide chains and the bridge show great variability, particularly in the gram- of gram-positive bacteria including antinomycetes and lactobacilli. Peptidoglycan is principally used for the taxonomy study of the taxonomy of actinomycetes. The most frequently studied structures are the peptidoglycan (structure and amino acid composition), acyl type of heteropolysaccharide muramic acid, and the composition of sugars fixed to the cell wall. The type of diaminopimelic acid (DAP) is usually studied by cellulose thin-layer chromatography of whole-cell hydrolysates and ninhydrin staining. This technique permits separation of meso-DAP, hydroxy-DAP, and the stereo isomeres LL-DAP and DL-DAP. This technique has also been used for *Bacillus* differentiation. In gram-negative bacteria, the composition of peptidoglycan is too homogeneous to be of taxonomic interest.

The study of heteropolysaccharide muramic acid acyl type is also interesting in actinomycetes: In some of them (and in some related taxa), the acetyl group of *N*-acetylmuramic acid is substituted by a glycolyl group. This *N*-glycolylmuranic acid has been found in *Aureobacterium, Microbacterium, Mycobacterium, Nocardia,* and *Rhodococcus* and in species belonging to the Micromonosporaceae and Actinoplanaceae families. This research is carried out by naphtalindol staining.

The study of cell wall sugars (sugar patterns) is also of interest in actinomycetes (rhamnose, ribose, fucose, xylose, arabinose, mannose, glucose, galactose, madurose). This is usually done by thin-layer chromatography and anilinphtalate coloration.

D. Cytoplasmic Membrane

1. Quinones

Quinones are components of the cytoplasmic membrane. Two types of quinones are found in bacteria: ubiquinones and menaquinones. Numerous studies have shown that these quinones can be used for taxonomic study of gram-positive bacteria, especially Actinomycetales, in which they can distinguish four groups. These menaquinones are usually written in the form MK-n(H$_m$), where n is the total number of isoprene units and m the number of saturated isoprene units. As with most other lipids, it is possible to recover them from bacterial cells by chloroform/ethanol extraction. The various types of menaquinones can be separated by reverse-phase thin-layer chromatography or high-performance liquid chromatography.

2. Determination of Phospholipid Types

Phospholipids are components of the cytoplasmic membrane of all living cells (except for archaebacteria where they are replaced by glycerolethers). These two classes are extracted by organic solvents and separated by thin-layer chromatography. Several studies have shown that determination of phospholipid types divide Actinomycetales into five groups.

3. Fatty Acids

There are two large groups of fatty acids (FAs): branched FAs and unbranched FAs. Branched FAs are principally found in gram-positive bacteria. They can be separated from polar lipids in the cytoplasmic membrane by esterification and analysis by gas or thin-layer chromatography. Precise identification is delicate. In standardized growth conditions, it is possible to obtain analyzable patterns by the habitual numerical techniques for obtaining classifications. In Actinomycetales, these patterns have shown a good correlation with the phylogeny of this group.

E. Outer Membrane: Mycolic Acids

Mycolic acids (2-alkyl 3-hydroxl long-chain fatty acids) are the principal component of the lipidic external membrane of *Mycobacterium*, *Corynebacterium*, and *Nocardia*. They can be differentiated by the length of their chain: "True" mycolic acids isolated from *Mycobacterium* are chains of 60–80 C-atoms, whereas mycolic acids with 50 and 30 C-atoms are found in *Nocardia* and *Corynebacterium*. These genera can be separated, depending on the length of their mycolic acids, in the form of methyl esters by silica gel thin-layer chromatography.

F. End-Products of Metabolism

End-products of metabolism are revealed by classic biochemical tests, such as the Voges–Proskauer reaction or the methyl red test. They can be studied quantitatively by gas liquid chromatography. These techniques, which can be computerized, are useful for the taxonomic study of fermentation, especially in anaerobic bacteria.

IV. Molecular Taxonomy

A. Chromosomal DNA

1. Mole % G+C Content

Researchers have demonstrated that the contents in purine (G, guanine; A, adenine) and pyrimidine (C, cytosine; T, thymine) bases vary from one organism to another but are constant inside a given species. G+C content is given by the following formula:

$$\text{mole \% G+C content} = (G + C) \times 100/(A + T + G + C).$$

It can be measured by four different methods.

1. Acid-hydrolyzing and separation of bases by paper chromatography, and then eluting and quantifying the bases. This reference method is no longer used.

2. Determination of the density of the DNA in cesium chloride density gradient by isopycnic ultracentrifugation. Under proper operating conditions, DNA density is proportional to G+C content, given the molecular weight difference between the two pairs of nucleotides: 654.4 for G+C, 653.4 for A+T. There can be numerous technical problems due to DNA contamination (polysaccharides, pigments) or to excessive DNA fragmentation.

3. Determination of the melting temperature (T_m). Progressive heating produces a denaturation of the DNA molecule, meaning separation of the two strands due to the rupture of the hydrogen bonds linking each pair of bases. Denaturation provokes an absorbance increase to 260 nm of the solution (hyperchromicity of denatured DNA). The exact temperature for separation of the DNA strands varies with the composition of DNA in bases. Destruction of the three hydrogen bonds that exist between G and C occurs at a higher temperature than that required for destruction of the two bonds between A and T. Under proper operating conditions, denaturation takes place at a temperature that is proportionally higher to the amount of G+C contained in the DNA molecule. The T_m is reached when approximately 50% of the sequences have become single-stranded. The T_m increases with the ionic concen-

tration and decreases with the addition of formamide or dimethysulfoxide.

The T_m is determined graphically on the curve of optic density at 260 nm.

Owen's formula is used, corresponding to a $0.1 \times$ SSC buffer:

$$G+C \text{ content} = (2.08 \times T_m) - 106.4.$$

Escherichia coli K12 may be used as reference DNA ($T_m = 75.5$, G+C content = 50.6).

4. It is possible to determine G+C content by high-performance liquid chromography.

These methods make it possible to estimate G+C content with a margin of error of <1%. They could fail if bacterial strains harbored very large plasmids with a G+C content that was very different from chromosomal DNA, but this has never been encountered. However, it is possible to detect this phenomenon because it produces a bump on the optic density curve. One can also first separate chromosome and plasmids by ultracentrifugation or electrophorese on agarose gel electrophoresis.

In bacteria, the G+C content is greatly variable, ranging between 25 (certain species of *Clostridium*) and 70% (certain species of *Pseudomonas*). According to one researcher, the variation in the same species is <2.5%. However, bacteria with identical G+C content could greatly differ in the primary structure of their chromosomal DNA. Therefore, it is the differences in G+C that are of primary significance, because this proves a taxonomic difference whereas the identity of G+C does not prove taxonomic identity (though, of course, it is consistent with taxonomic identity).

The taxonomic applications from the determination of G+C content have been very numerous. For example, they have made it possible to exclude the species *morganii* (G+C content = 50%) from the genus *Proteus* (G+C content = 39–40%) and to transfer it to the genus *Morganella* and to integrate the genus *Yersinia* (G+C content = 45.8–46.8%) in the Enterobacteriaceae (G+C content = 45–58%). Nevertheless, several authors have recently shown by other more precise techniques that bacteria with quite different G+C contents could belong to the same genus.

2. DNA/DNA Hybridization

Quantitative measurement of DNA/DNA hybridization is based on the formation of hybrid duplexes by base pairing: The DNA homology value measures the degree of genetic relationship of the two strains.

The first measurements were obtained by incorporating a heavy base (5-bromouracil) or a heavy isotope (^{15}N). After mixing the labeled and unlabeled DNA, an ultracentrifugation on cesium chloride was performed, which separated the heteroduplexes (intermediate density) and homologous duplexes (with a higher or lower density). This technique was delicate.

There are now numerous measurement techniques. They differ in the renaturation phase—it is usually liquid but some authors have suggested agarose gel or nitrocellulose membrane techniques—and the separation technique for single-stranded or double-stranded DNA.

a. Hybridization on Nitrocellulose Membrane

Hybridization on nitrocellulose membrane requires the following operations:

- DNA extraction and radioactive labeling (^{32}P or ^{3}H) of one of them (e.g., DNA A, written as A*).
- Binding of nonradioactive DNA A on nitrocellulose filter after heat denaturation (100°C, 5′).
- Fragmentation of DNA A* and B (sonication, French press) to encourage collisions between heterologous strands.
- Heat denaturation of A* and B.
- Hybridization: there is a competition between DNA A* and DNA B for the DNA A attached to the filter. Approximately 1500 times more B than A* is brought together so the labeled strands will have little chance of reassociation. After a 16-hr incubation at a temperature of (T_m − 25)°C, the radioactivity of the filter is compared to that obtained for the 100% controls (DNA B is replaced by A) and 0% (mixture only containing A*).

b. The Hydroxylapatite Technique

Hydroxylapatite is a calcium phosphate that enables separation of single-stranded (denatured) DNA from double-stranded DNA. The DNA are bound on hydroxylapatite and can then be differentially eluted: With a phosphate buffer molarity inferior to 0.20 M, only single-stranded DNA is eluted; >0.20 M, all of the single- or double-stranded DNA is extracted.

DNA A, A*, and B are incubated at (T_m − 25)°C for 16 hr, then sent through a hydroxylapatite column maintained at $(T_m - 25)°C$ and conditioned with a 0.14 M phosphate buffer. After several washes, the nonhybridized DNA is eluted. When the wash buffer is not radioactive, the duplexes are eluted with a 0.3 M phosphate buffer. The radioactivity of the eluted fractions is compared. This technique can be performed using the Brenner batch procedure where hydroxylapatite washing is obtained by repeated centrifugation.

c. The S1 Nuclease Technique

The S1 endonuclease is extracted from *Aspergillus oryzae*. It specifically shears single-stranded DNA. The remaining double-stranded DNA is separated from the nucleotides by trichloracetic acid precipitation and absorption on diethylaminoethanol (DE-AE)cellulose. By comparing the radioactivity of the filters having absorbed the hybridization mixture after nuclease or without nuclease action, one can deduce the percentage of hybridization. The DNA must first be dialyzed because the S1 nuclease is inhibited by SDS. Because the activity of the nuclease depends on DNA concentration, it is necessary to add a heterologous DNA carrier (salmon sperm, calf thymus).

This technique makes it possible to eliminate the single-stranded "hair pins" (Fig. 7). This is of interest when two chromosomes are of unequal size, which is frequently observed. This can be due to additions or deletions. In both cases, hair pins form during hybridization but are eliminated by the S1 nuclease treatment.

d. The Optic Technique

This technique consists of measuring the speed of hybrid renaturation. It is rapid and does not require the use of radioactive isotopes but only one sample can be studied at a time. In addition, it is difficult to study the thermal stability of duplexes.

3. Hybrid Stability

Two types of hybrids can form: very specific hybrids corresponding to very complementary sequences and fragile hybrids when two single-stranded fragments are face to face with a few nucleic bases arranged in complementary sequences.

Hybrid stability (and therefore their specificity) can be measured by studying hybrid denaturation kinetics as the temperature is progressively increased. $T_{m(e)}$ (thermal elution point) is the melting temperature of the hybrid, meaning the temperature at which 50% of the radioactivity can be eluted from

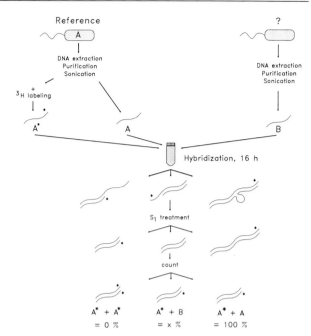

Figure 7 Principle of hybridization by the nuclease S1 technique. There is a competition between DNA A* and DNA B for DNA A. Approximately 1500 times more B than A* are mixed so that the labeled strands will have little opportunity to reassociate. Nuclease S1 treatment makes it possible to eliminate the hair pins.

the duplex. The difference in thermic stability $\Delta T_{m(e)}$ between the $T_{m(e)}$ of a heteroduplex and that of a homoduplex is the degree of stability of a heteroduplex. Numerous studies have shown that the thermic stability is correlated to the percentage of unpaired base pairs (1% of unpaired bases for a $\Delta T_{(me)}$ of 1.6°C).

4. Interpretation

The three techniques—hydroxylapatite, cellulose binding, and optic—produce equivalent results. The S1 nuclease method gives slightly higher values (10–15%). It is therefore necessary to specify the method that is used.

The relatedness below 20% (hydroxylapatite) or 30% (S1 nuclease) correspond to weakly genetically linked microorganisms.

Interpretation should take DNA homology values and thermic stability into account. A DNA homology superior to 70% and good hybrid stability ($\Delta Tm(e) <= 5°C$) are the current criteria for belonging to a common species.

In the uncertainty zone (30–70%), the hypothesis can be made that the bacteria belong to two species of the same genus. $T_{m(e)}$ is the most useful in this zone.

B. RNA

1. DNA/RNA Hybridization

DNA–ribosomal RNA (rRNA) hybridization techniques are comparable to those described for DNA-DNA hybrids. The technique is usually performed on nitrocellulose membrane. The total rRNA or only one of the two fractions, 16S or 23S, is isolated from the ribosomes, labeled with a radioactive isotope and hybridized with the single-stranded DNA bound to the nitrocellulose membrane. The relationship between the two rRNA is measured by the difference between the thermal stability of the hybrids.

Using this method, one can find a homology superior to 90% among all of the Enterobacteriaceae species. The five genetic groups in the *Pseudomonas* genus were described by DNA/rRNA hybridization.

2. 5S and 16S RNA Sequencing

The coding sequences for rRNA are among the most conserved molecules in the course of evolution. Their study makes it possible to not only know the relationships between species as in the case of DNA/DNa hybridizations, but also between genera and groups of genera.

Ribosomal RNA sequencing was applied for the first time in the 1970s in Urbana, Illinois, United States. The method consists in establishing, for each organism to be studied, a list of oligonucleotides derived from 16S rRNA after hydrolysis by T1 nuclease, extract from *A. oryzae*. This method, known as oligonucleotide cataloging, has been greatly used since that time. 16S rRNA was chosen because it is more stable than 5S RNA and easier to sequence than 23S RNA (1500 nucleotides instead of 2300). The principle of the technique is the following: *In vivo* ^{32}P-labeling during the growth phase, RNA extraction, 16S rRNA separation by PAGE, digestion by the T1 nuclease, which cuts between the 3′-guanylic acid and the 5′ hydroxyl group of the adjacent nucleotide. The residues are made up of 15–20 nucleotides and contain only guanine in the 3′ end position. These oligonucleotides are separated by bidimensional electrophoresis: first dimension on cellulose acetate at pH 3.5 and second dimension after transfer on DEAE–cellulose. In these conditions, spots form three or four series of wedge-shaped patterns. Each of these patterns corresponds

to a uracil number. Inside the patterns, spot position only depends on the composition of adenine and cytosine. The spots can be eluted and studied by digestion by various ribonucleases and migration on DEAE cellulose. Finally, a catalog is drawn up with the list of oligonucleotides encountered and the number of moles of each. Comparison of the two organisms is made by calculating a similarity index. The previously described numerical methods then make it possible to establish a classification and plot a dendrogram.

Ribosomal RNA sequencing techniques have been developed more recently. Primers that are complementary to the conserved sequences of the 16S rRNA are used. The primers are end-labeled by the T4 polynucleotide kinase and γ-^{32}P-ATP. They are hybridized with the rRNA. The elongation reaction is performed in the presence of avian myeloblastosis virus reverse transcriptase. Five aliquots are prepared, with four containing a mixture of deoxynucleotide triphosphate (dATP, dCTP, dGTP, dTTP) plus a dideoxynucleotide triphosphate (ddATP, ddGTP, ddCTP, or ddTTP). The fifth tube only contains deoxynucleotides and is used as control.

The elongation products are electrophoresced on polyacrylamide/7 M urea sequencing gels. The divergence between two sequences is usually measured by the number of nucleotide substitutions. Transitions are differentiated from transversions and deletions or additions are counted as transversions. The positions where nucleotides cannot be measured with certainty are not taken into account in the calculation. The K_{nuc} values are calculated (taking into account the probability of reverse mutation during the evolution):

$$K_{nuc} = -(1/2) \; ln \; [(1 - 2P - Q\sqrt{1 - 2Q})],$$

where P and Q are, respectively, transition-difference and transversion-difference proportions. It is then possible to plot a phylogenetic tree: A panel of experts recently recommended the Saitou and Nei neighbor-joining method (Fig. 8).

Sequencing methods after amplification by polymerase chain reaction have recently been developed. Primers corresponding to the preserved parts of 16S rRNA for its amplification are used. This technique makes it possible to sequence 16S rRNA and therefore to know the taxonomic position of noncultivable microorganisms.

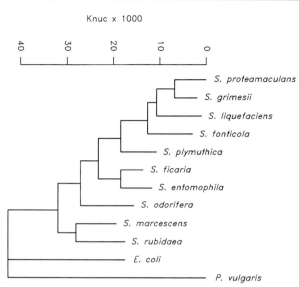

Knuc x 1000

40   30   20   10   0

- S. proteamaculans
- S. grimesii
- S. liquefaciens
- S. fonticola
- S. plymuthica
- S. ficaria
- S. entomophila
- S. odorifera
- S. marcescens
- S. rubidaea
- E. coli
- P. vulgaris

Figure 8 Nucleotide sequences of 16S ribosomal RNA from 10 *Serratia* species. Dendrogram by neighbor-joining method. The distance between two species (K_{nuc} value) is the sum of the horizontal branch lengths (in K_{nuc} units). [From Dauga, C., et al. (1990). *Res. Microbiol.* **141**, 1139–1149.]

V. Applications

A. Identification

1. Dichotomic Keys

Classically, bacterial identification calls for a certain number of tests concerning morphology, cultural and physiologic characters, experimental pathogenic ability, etc. These tests were performed one by one or in reduced groups (such as the "Pasteur gallery" for Enterobacteriaceae identification), according to successive dichotomic elimination inspired by the experience of microbiologists.

Identification is made in successive stages, the answer to the test of row N brings about a conditional linking on the test of row $N + 1$. One error in the reading of one of the first tests is enough to make all identification impossible. Dichotomic keys were replaced by diagnostic tables in the 1960s and 1970s. They are still found in student data files to simplify their training.

2. Diagnostous Tables

Diagnostic tables are matrices that contain the answers to tests on a batch of taxa. It is usually accepted that they must not contain empty cases (undocumented). Results are now most often expressed according to "Bergey's Manual of Systematic Bacteriology":

$+$, 90% or more of strains are positive
$-$, 90% or more of strains are negative
d, 11–89% of strains are positive
v, strain instability (not equavalent to d)
D, different reactions in different taxa

An unknown strain is submitted to the tests and the results are compared to the answers for each taxon. It is easy to understand that this process is all the more difficult when there are many tests and taxa.

Lapage's work has made it possible to make comparisons by computer so that no possiblity can be overlooked. This work led to modern numerical identification.

3. Numerical Identification

Numerical identification uses a data matrix: For each taxon, it contains the frequencies of positive answers to different tests. To avoid exclusions, 1 and 0 frequencies are not attributed—the values 0.01 and 0.99 are preferred.

When an unknown microorganism presents a positive result, the frequency value in the data matrix is retained. If the result is negative, the probability is obtained by subtracting the positive probability of the unit.

In 1973, Lapage defined probability products as the product of positivity frequencies for the various species. It reflected the atypia of phenotypes.

Two researchers have proposed the percent relative likelihood (PRL): This is the ratio of probability products of the unknown against a given species to the sum of probability products of the unknown against, any species in the matrix. This coefficient expresses the discrimination quality but does not make it possible to define threshold values. It is therefore preferable to use the standardized Lapage identification score (SSC): This is the ratio of the probability product of the unknown against a given species to the sum of probability products of the unknown against all the taxa in the matrix. It is generally multiplied by 100 with the following acceptable thresholds:

$>$99.9%, excellent identification
$>$99%, very good identification
$>$90%, good identification

>80%, acceptable identification
<80%, unacceptable identification.

Table VI presents an example of numerical identification using these parameters.

a. Probabilty Products

X compared to taxon 1: $0.80 \times (1 - 0.95) \times 0.05 = 0.002000$

X compared to taxon 2: $0.99 \times (1 - 0.01) \times 0.95 = 0.931095$

Total: 0.933095.

b. PRL

X compared to taxon 1: $0.002000/0.931095 = 0.0021480085$

X compared to taxon 2: $0931095/0.931095 = 1$

c. SSC

X compared to taxon 1: $0.002000/0.933095 = 0.0021434$

X compared to taxon 2: $0.931095/0.933095 = 0.9978566$

X belongs to taxon 2 with an identification score of 99.7% (very good identification). The sum of the probability products in the species are 0.93. This can also be expressed by saying that 93% of the taxon 2 strains present this phenotype.

4. Modal Frequency

Modal frequency is the ratio of probability products of the unknown against a given species to the probability product of the most typical phenotype of that species against it's own species. This index is independent of taxon variability. It is not affected by the presence of similar species in the matrix, but its value depends on the number of tests studied.

5. T Coefficient

In 1987, researchers proposed to weigh modal frequency with a significance threshold S that took into account the number of atypical results. This T coefficient varied from 1 (the most typical profile) to almost 0 (the least typical profile). These authors suggested the following thresholds:

>0.75, excellent identification
>0.50, very good identification
>0.25, good identification

6. Diagnostic Ability Coefficient

The Descamps and Véron diagnostic ability coefficient is derived from the Bayes theorem and measures the quantity of information furnished by each test.

A bacterial strain X could belong to N classes (N possible diagnostics). Let $P(A)$ be the probability that strain X belongs to class A. Let $P(C+/A)$ (read P from $C+$ if A) be the conditional probability so that a class A strain will have a C character. For each character mode ($C+$ or $C-$), the deviation between *a priori* and *a posteriori* distributions can be measured by various mathematical functions: $f(P(A), P(A/C))$. The interest of a character is given by the average of the deviations calculated for each mode weighed by the frequency of these modes. This average or diagnostic ability coefficient is expressed as follows:

$$CCD(C) = P(C+) \cdot f(P(A), P(A/C+)) + P(C-) \cdot f(P(A), P(A/C-))$$

This takes into account the frequency of each species in the batch and the data matrix positivity percentages. It can also be calculated by assigning an equal frequency to each species.

7. Vigor and Pattern

P. H. A. Sneath's work on vigor and pattern have profoundly influenced the thinking of taxonomists. Schematically, we must not only take into account genetic capacities, meaning the possession of metabolic circuits and their underlying genetic information, but also the manner in which this information is expressed. Weak strains can thus have the same aptitudes as strong strains, without having to express them when they are explored by usual cultural

Table VI Numerical Identification of Unknown Individual X

	Data matrix		
Taxon	Catalase	Oxidase	β-galactosidase
1	0.80	0.95	0.05
2	0.99	0.01	0.95

Phenotype (test response) for unknown individual X

Name	Catalase	Oxidase	β-galactosidase
X	+	−	+

or biochemical tests. Such weak strains can therefore be assigned a "pattern" or response profile for identical tests.

This conception perfectly accounts for the adaptation to their host of certain *E. coli* strains that have kept their "pattern" but lost their "vigor" during their evolution to *Shigella*.

It also enables us to explain the adaptation of *Hafnia alvei* to beer yeast fermenters: The selection pressure exerted by this very rich medium forces the evolution of *H. alvei* to *Obesumbacterium proteus*, which cannot ferment most of the substrates used by *H. alvei*.

This work led to a new conception of identification systems, assigning more importance to characters that are present than to absent characters, thus favoring the "pattern" to the "vigor." This sort of identification system, developed by P. A. D. Grimont, is used at the French reference center for Enterobacteriaceae at the Institut Pasteur in Paris.

B. Epidemiological Markers

The fresh outbreak of hospitalized immunodepressed patients in infectious disease or surgical units has demonstrated the need to control infection in hospitals. Therefore, the role of a microbiology laboratory is no longer limited to identification of the species of microorganism and its susceptibility to antibiotics. Its activity also includes evaluation of the incidence of germs that usually cause nosocomial infections, such as *P. aeruginosa*, *Serratia marcescens*, *Acinetobacter baumannii*, *Staphylococcus epidermidis*, etc. In the presence of a series of upper respiratory tract infections by *S. marcescens*, the microbiologist should have a sufficient number of reliable microbiologic clues at his or her disposal to distinguish the origin of the contamination: direct contact between patients or between patients and patients' care personnel, use of contaminated antiseptics, and even autoinfection or contamination of aqueous solutions intended for parenteral nutrition. [*See* HOSPITAL EPIDEMIOLOGY.]

By confronting the variable characteristics, called epidemiological markers, inside a bacterial species, one can suspect the identity or nonidentity of several strains and then discover the origin of the contamination, the transmission cycle, and the hospital reservoir, meaning the epidemiologic significance of the bacteria isolated from pathologic samples.

Nowadays, the most commonly used epidemiological marker is biotyping or auxanotyping. It consists of studying the most variable biochemical characters inside a species. It has the advantage of using basic bacteriologic methods that can be performed on a routine basis. It does, however, require enough time to verify the absence of biotype variation. Among the systems that have already been tested, we can mention the subdivision of *S. marcescens* into 19 biotypes (study of eight organic compounds by auxanogram and three cultural characters). Certain biotypes are ubiquitous (A3, A4); others are more particularly encountered in hospitalized patients. Biotypes A1 and A2/6 are rare in human pathology. Most biotypes are correlated to serotypes, which would tend to show the importance of their genomic basis.

The antigenic typing (serotyping) of *Klebsiella* reveals the capsular K antigen (Neufeld raction). There are 77 different capsular types. Some of them are found in *Klebsiella mobilis* (*Enterobacter aerogenes*). The sudy of the O (somatic) and H (flagella) antigenic factors of *S. marcescens* has revealed the existence of at least 26 H factors and 22 O factors. Serotype studies are generally very long and difficult to use in a hospital laboratory, with the exception, however, of *P. aeruginosa* (16 serotypes).

Phage-typing is a stable character because it is linked to the specific receptors of bacteriophages situated on the bacterial cell wall. One must have a batch of reference bacteriophages as well as "local" phages, obtained, for example, from effluent, to obtain results specific to the environment being studied. Reverse phage-typing techniques have been proposed for characterizing nontypable strains, using the action of nontypable strain culture supernatant on propagating strains. One of the best phage-typing applications concerns *Salmonella enterica* subsp. *enterica* serovar Typhi and other *Salmonella* with Vi antigen (Hirschfeldii and Dublin serovars): They use a series of phages adapted to the Craigie and Yen Vi-II phage. [*See* BACTERIOPHAGES.]

Numerous bacteria secrete bacteriocins, specific antibiotic protein substances for other bacteria, more often active against stains inside the same species. The most common is pyocinotyping. Other bacteriocynotype techniques reveal defective phages (type R pyocines, marcescines).

The study of susceptibility to antibiotics (antibiotyping), antiseptics, or heavy metals is easier to perform in a clinical laboratory. Differences in strain resistance profiles isolated during an epidemic are nevertheless frequent, by acquisition or loss of plasmids or by molecular rearrangement during bacterial division. This, for example, is the case for *Staphylococcus epidermidis.*

The detection of extracellular enzymes (zymotyping) such as esterases or β-lactamases is very reproducible but is little used: the bidimensional electrophoresis of esterases has been applied to *Yersina* and made it possible to discover specific profiles of certain *Yersina enterocolitica.*

Some virulence factors have been used with success, such as slime production in *Staphylococcus.* It is associated with adherence properties to plastic materials: It can be used to differentiate pathogens and contaminants and can also be used as an epidemiologic marker.

The genetic study of colony modifications is currently one of the rare markers available for *Candida albicans* and similar species.

The extraction of plasmid DNA, followed by an estimation of plasmid size by electrophoresis on agarose gel (plasmid profiles), only makes it possible to presume their identity. The plasmid profiles obtained by enzymatic digestion are very useful. It is in this manner that the existence of cryptic plasmids of identical size but of different nature in *P. aeruginosa* were discovered. Nevertheless, there are strains without plasmids.

The enzymatic restriction of chromosomic DNA following electrophoresis in agarose gel (chromosome restriction) is only of interest in rare cases in which strictly identical restriction profiles can be revealed. It is very much in use for phage (phage taxonomy) or *Rotavirus* study, and can be applied to *Chlamydia* or *Mycoplasma,* given the reduced size of the chromosome. The number of bands is too high in most other cases for interpretation.

The electrophoresis of membrane proteins in polyacrylamide gel after denaturation by SDS (SDS-PAGE) is, on the other hand, a technique that is very commonly used (Fig. 9). It is particularly useful for verification of strain identity. It can also be the object of a numerical analysis after standardization, using a scanning densitometer linked to a computer.

The *in vivo* labeling by ^{3}H-penicillin makes it possible to reveal, by electrophoresis and fluorography, the penicillin binding proteins. These proteins usually seem to be specific to speices (*Enterococcus*).

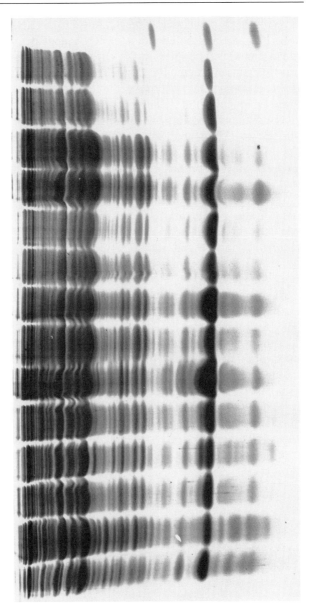

Figure 9   SDS-PAGE of 14 epidemic strains of 0:12 serotype *Pseudomonas aeruginosa.* Strains display identical patterns.

They could sometimes be used as epidemiological markers (*Staphylococcus*).

DNA/DNA hybridization with a chromosomic probe obtained by random cloning (chromosomal probe fingerprinting) has made it possible to differentiate the strains of *S. enterica* subsp. *enterica* serovar Dublin presenting identical chromosome restriction profiles. In addition, it suggested a clonal dissemination of serovar Enteritidis.

Ribosomal RNA gene restriction patterns have

recently been developed. The RNA operons (RNA 16S, 23S, 5S and a few transfer RNA genes) are present in several copies on the chromosome. Several restriction fragments carrying rRNA gene sequences can then be observed (except in slow-growing mycobacteria). The conserved character of rRNA makes it possible to use a single probe, generally *E. coli* ribosomal 16S + 23S RNA. The first results show excellent reproducibility. Inside a species, the various profiles correspond to hybrid thermal stability differences ranging between 2.5° and 5.5°C: In the case of the genus *Serratia*, the *S. marcescens* or *Serratia odorifera* biotypes can be perfectly distinguished. In other genera, such as *Brucella*, the size and number of coding fragments for rRNA are constant and correspond to the genus or to the species.

Restriction length polymorphism of polymerase chain reaction amplified fragments is a recent promising technique.

If the traditional epidemiological markers are still used, often with success, progress in molecular biology techniques has made new markers available to biologists. Their use requires the microbiologists to have additional training in the basic techniques and special equipment used in molecular biology, but their utilization is often less demanding than certain traditional techniques such as serotyping or phage-typing.

Bibliography

Austin, B., and Priest, F. (1986). "Modern Bacterial Taxonomy." Van Nostrand Reinhold, Wokingham, Berkshire, England.

Bollet, C. (1992). Taxonomie bactérienne. *In* "Manuel de Bactériologie Clinique" (J. Freney, F. Renaud, W. Hansen, and C. Bollet, eds.). pp. 7–28, Elsevier, Paris.

Brenner, D. J. (1985). Taxonomy, classification, and nomenclature. *In* "Manual of Clinical Microbiology" (E. H. Lennette, A. Balows, W. J. Hausler, and H. J. Shadomy, eds.), pp. 1–7, American Society for Microbiology, Washington, D.C.

Grimont, P. A. D. (1988). *Can. J. Microbiol.* **34,** 541–546.

Mayr, E. (1982). "The Growth of Biological Thought: Diversity, Evolution, and Inheritance." Belknap Press, Cambridge, Massachusetts.

Murray, R. G. E. (1984). The higher taxa, or, a place for everything . . . ? *In* "Bergey's Manual of Systematic Bacteriology," Vol. 1 (N. R. Krieg and J. G. Holt, eds.), pp. 31–34. The Williams & Wilkins Co., Baltimore, Maryland.

Sneath, P. H. A. (1984). Numerical taxonomy. *In* "Bergey's Manual of Systematic Bacteriology," Vol. 1 (N. R. Krieg and J. G. Holt, eds.), pp. 5–7. The Williams & Wilkins Co., Baltimore, Maryland.

Stackebrandt, E., and Goodfellow, M. (1991). "Nucleic Acid Techniques in Bacterial Systematics." Wiley, Chichester, England.

Staley, J. T., and Krieg, N. R. (1984). Classification of procaryotic organisms: An overview. *In* "Bergey's Manual of Systematic Bacteriology," Vol. 1 (N. R. Krieg and J. G. Holt, eds.), pp. 1–4. The Williams & Wilkins Co., Baltimore, Maryland.

Véron, M. (1982). Systématique bactérienne. *In* "Bactériologie Médicale" (L. Le Minor and M. Véron, eds.), pp. 113–145. Flammarion, Paris.

Temperature Control

Terje Sørhaug

Agricultural University of Norway

I. Temperature Ranges for Growth of
   Microorganisms
II. Temperature Control
III. Sublethal Injury at High and Low
   Temperatures
IV. Thermal Inactivation

Glossary

Indicator organism Microorganism that on detection gives a quantitative representation of a group of organisms, often of fecal contamination or pathogens in foods

Poikilotherm "Cold-blooded," with body temperature that approximately follows that of the surroundings

Resuscitation Procedure that restores the ability of an organism to grow and develop after sublethal injury

Stress Condition that may inflict injury on a microorganism

Sublethal injury Temporary loss in a microorganism of tolerance for specific conditions

Thermization Mild thermal process (e.g., 63°C, 10–20 sec)

Thermoduric Ability of a microorganism to survive (e.g., pasteurization)

MODERN FOOD MANUFACTURE depends heavily on temperature control to supply safe products of high quality. Thus, often pathogens and spoilage microorganisms are thermally killed or their activities are limited to acceptable levels by refrigeration or freezing. Many traditional fermentations (e.g., production of cheeses and beer) imply strict adherence to temperature schemes that have been further refined in todays industry.

Microorganisms commonly grow over temperature ranges of 30–40°C. Thermophilic organisms grow >50°C, and psychrotrophs may develop at 5° to <0°C. Foodborne representatives of all groups are found and psychrotrophic bacteria, yeasts, and molds selectively grow at refrigeration temperatures. Thermal processing at 70–80°C for 10–20 sec is sufficient to kill many vegetative microbial cells in environments of high-water activity. Destruction of some types of bacterial spores require heating >120°C for 15–20 min. Microbes exposed to high or low temperatures may suffer sublethal injury. These stressed organisms recover and resume growth under favorable conditions, sometimes in foods. Selective, analytical procedures may not allow injured organisms to recover for detection.

I. Temperature Ranges for Growth of Microorganisms

A. Minimum, Optimum, and Maximum Temperature

Growth of microorganisms is the combined expression of a complex system of biochemical and physicochemical events. These include at least energy metabolism, biosynthesis of monomers and polymers, assembly of enzyme complexes, ribosomes, genetic material, membranes and walls, cell division, and turnover of discarded intracellular components. If any one of these processes stops, growth of the organism will soon come to a halt.

Microorganisms often grow within temperature ranges of 30–40°C. Figure 1a shows an ordinary variation of growth rates with temperature for an organism. At the low temperature end for growth, the minimum temperature, membrane disfunction, and changes of hydrophobic protein interactions may be important factors to explain the limit to growth. The following near logarithmic increase of growth rate with rise in temperature is found for many microorganisms. The maximum growth rate at the top of the curve defines optimum temperature. At this point, destructive processes in the cells such as injury to nucleic acids and membranes and

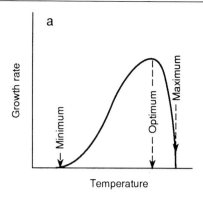

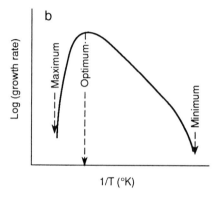

Figure 1 Temperature characteristics for a microorganism. (a) Variation of growth rate with temperature. Cardinal temperatures: minimum, optimum, and maximum. (b) Arrhenius plot. Details in text.

protein denaturation will dominate at only a small increase in temperature. The upper limit of growth is called the maximum temperature. The curve and its minimum, optimum, and maximum points, the cardinal temperatures, are characteristic for the organism, but deviations may occur if the growth conditions are changed.

B. Arrhenius Plot

Arrhenius developed an equation that expresses the relationship between chemical reactions and temperature:

$$k = Ae^{-E/RT}$$
$$\log k = -(E/2.303R)(1/T) + \log A,$$

where   = specific reaction rate constant at a given temperature, R = Boltzmann's constant (gas constant), E = energy of activation, A = frequency factor (Arrhenius factor), and T = °K. The validity of the equation is apparent when the plot of log k

versus 1/T, the Arrhenius plot, yields a straight line. The slope of this line is −E/2.303R. The activation energy can thus be calculated when experimental values for log k versus 1/T have been obtained.

Arrhenius plots have been made for the growth rate–temperature relationship for many bacteria. However, these plots may not conform to simple kinetics. A portion of the curve corresponding to the near log : linear part in Fig. 1a often gives an apparent straight line in the Arrhenius plot. Approaching the minimum and the maximum temperatures, the curve deviates from linearity and becomes almost vertical at both ends (Fig. 1b).

A simple linear relationship for part of the Arrhenius plot for an organism implies that one reaction/ function in the cell is limiting for growth in that temperature range. Although this would be attractive to explain growth kinetics, Arrhenius plots have seldom been used to describe details of the growth rate mechanism. Detailed analysis of Arrhenius plots of bacterial growth rates may also reveal that the apparent linear part of the curve is slightly curved or multiphasic. Finally, the corresponding temperature range is rather narrow, which implies that the precision of the E calculated from the slope will be low.

Nevertheless, Arrhenius plots can be useful for comparing the overall and greater differences between microorganisms. Organisms with preference for low-temperature growth can easily be distinguished from those growing at higher temperatures, and temperature mutants can be compared with their wild types.

Other mathematical models aimed to fit experimental data have been proposed [e.g., the Ratkowsky plot (square root of growth versus temperature], but their relationship to the biochemical functions remains to be seen.

C. Temperature Classification: Psychrophiles, Psychrotrophs, Mesophiles, Thermophiles

Growth of microorganisms has been reported from less than −10°C up to 250°C. The high-temperature extreme was claimed for bacteria inhabiting hot ocean floors near volcanic vents at pressures of 250 atmospheres, but the claims were subsequently challenged. In any case, the latter organisms are hardly of concern for food and fermentation microbiology, but among the microbes growing at <0°C, several are candidates for food spoilage in refrigerated storage. Temperature range specialization among microorganisms has prompted the proposals of various classification schemes both for scientific purposes and for more pragmatic reasons. The most common terms in use are psychrophiles, psychrotrophs, mesophiles, and thermophiles. Three of these definitions are based on minimum, optimum, and maximum temperature ranges (Table I).

The term psychrotroph is used mainly in food microbiology for organisms growing under refrigerated conditions. A common definition is that psychrotrophs produce colonies on agar media within 10 days at 7°C. A psychrotrophic count may thus include both psychrophiles and mesophiles. [*See* Low-Temperature Environments.]

In applied microbiology and screening work, isolates of microbes are seldom characterized strictly according to the definitions as psychrophiles, mesophiles, or thermophiles. As indicated earlier, organisms growing at low temperatures are often distinguished by a psychrotrophic count. Mesophiles are often referred to a total count, i.e., growth on agar media at 30–32°C in 2–3 days. Thermophiles are often reported as those organisms forming colonies at 55°C. [*See* Thermophilic Microorganisms.]

D. Critical Points on Maintaining Constant Temperature for Growth of Microorganisms

Establishing the growth rate : temperature curves requires measurements at a number of temperatures, each of which must be very exact. This is particularly evident between optimum and maximum temperatures, when a difference of only a few degrees greatly changes growth rates. Water baths for tubes and flasks allow good control, but often practical obstacles limit the distribution of temperatures. Temperature-gradient incubators can be

made, but the exactness of measurements should be carefully checked when in use.

Temperature fluctuations in air incubators are difficult to handle. Variations often occur within the incubator itself or may be caused by frequent opening of doors and when large numbers of plates or flasks are deposited. When stacking plates, the capacity of the incubator and the access of air between plates should be occasionally checked with thermometers (i.e., correct thermometers!).

Similar considerations apply when killing microorganisms in autoclaves or otherwise in the laboratory but also in food processing.

II. Temperature Control

A. How to Avoid or Select Microbial Activities

This section will briefly survey how temperature is used purposely by humans to avoid or regulate microbial activities.

Low-temperature storage of foods is now common in many parts of the world. Freezing and refrigeration, often developed into cold distribution chains, have become indispensable means to maintain the freshness of foods while diminishing the effects of spoilage microorganisms and preventing health hazards by pathogens. The convenience of extended storage periods is important both for the industry and for the consumer.

Freezing foods to −20°C or below effectively stops microbial growth both because of the low temperature and the low a_w, but microorganisms may survive and develop after thawing. Important factors that influence the degree or extent of injury and death by refrigeration and freezing are the rate of temperature change during chilling and thawing, the temperature differential rather than the actual temperatures, cryoprotectants, and the growth phase of cells. Inactivation by freezing occurs in two phases: first, there may be a rather quick decrease in viable numbers during freezing and then a slower change during storage. The action of lipases from microbes has been observed in some frozen foods rich in fats.

Refrigeration of foods is used alone or in combination with a previous heat treatment. Fresh fish is preferably kept on ice; fresh meat products, raw milk, and pasteurized milk and cream are stored at 0–5°C. Foods that have been further processed and often packed, but without (commercial) sterility,

Table I Temperature Classification of Microorganisms

	Temperatures (°C)		
	Minimum	Optimum	Maximum
Psychrophiles	<0	<15	20
Mesophiles	5–25	25–40	40–50
Thermophiles	35–45	45–65	>60

may be kept at somewhat higher temperatures, but lower temperatures are strongly recommended to avoid potential health hazards.

While bringing many advantages to the industry and the consumer, refrigeration of foods has also given new and interesting challenges to the microbiologists. Although refrigeration may have been thought of by some as putting a lid on microbial activities, the reality is rather to change priorities from mesophilic and thermophilic organisms more toward psychrotrophs. [*See* REFRIGERATED FOODS.]

Cold storage of raw milk with subsequent and alternative thermal treatments, followed again by low temperature storage, is an interesting scenario for presentation and consideration of advantages and problems associated with temperature control. Raw milk that has not been stored or heat-treated contains a wide variety of microorganisms. The percentage ranges of the organisms vary with the total microbial load. Mesophiles account for a relatively large fraction in high-quality milk. Pathogens may occur. The milk at this stage may contain only a moderate count of psychrotrophic bacteria. In areas with a complete cold chain, the milk is collected in farm tanks with cooling and later transported quickly to the dairy for further cold storage in silos. If this period extends to several days of cold storage, the psychrotrophic gram-negative rods will outgrow other microbes until complete domination. Among this group, *Pseudomonas* spp., particularly fluorescent pseudomonads, are most frequently detected, but strains of other genera are also regularly found: *Acinetobacter, Flavobacterium, Athrobacter, Alcaligenes, Aeromonas,* and *Enterobacter.*

In the dairy, the milk is pasteurized (e.g., 72°C, 15 sec) before distribution or further processing. Thermization, a mild heat treatment (e.g., 63°C, 10–20 sec), of the milk may also precede silo storage to safeguard against growth of psychrotrophs; pasteurization follows later. In either case, the majority of the psychrotrophs are killed. Some thermoduric bacteria survive pasteurization, namely, strains of *Micrococcus, Microbacterium, Alcaligenes, Arthrobacter, Corynebacterium,* and *Streptococcus* and spores of *Bacillus* and *Clostridium.*

Processed sweet milk for liquid consumption is kept in the cold chain, and among the surviving bacteria only some species of *Bacillus* have psychrotrophic strains that can develop and spoil the milk. Proteinases produced by the bacilli contribute to sweet curdling of the milk. Sweet curdling and bitter

taints occur mainly in the summer and are less of a problem when refrigeration is more strictly maintained.

When pasteurized milk is used for cheese production, relatively high temperatures prevail during processing (ca. 28–57°C) and ripening (ca. 3–27°C), varying with the type of cheese. The higher temperatures in combination with the other conditions in certain cheeses (pH, E_h, a_w, salt content) may allow growth of clostridia. Thus, Swiss-type cheeses and sometimes Gouda may suffer from development of poor flavor and excessive gas formation (late blowing) due to butyric acid fermentation of lactate by *Clostridium tyrobutyricum.*

The spores of *Bacillus* and *Clostridium* survive pasteurization of milk, but the heat treatment probably also contributes to activation of the spores, encouraging for later germination and outgrowth. This side effect is another good illustration of the duality that must be considered when temperature is used to control microbes.

When clostridia spoil cheese, the quantitative aspect is interesting. Less than 100 spores/100 ml milk is sufficient to cause problems. Detection of these small numbers is a challenge in analytical microbiology, involving destruction of vegetative cells by heat (70–80°C, 10 min) with simultaneous activation of the spores followed by most probable number or filtration methods. However, the ultimate goal is to get rid of the clostridial spores. Again heat treatment is called for after removal by centrifugation (bactofugation) or ultrafiltration (Bacto-Catch) of >95% of the spores from the bulk milk. The fraction enriched with spores is subjected to ultra-high temperature treatment (e.g., 145°C, 3–5 sec). At least these are two ways to challenge the problem; others involve the use of nitrate, lysozyme, or nisin.

Pasteurization, as we remember, killed the psychrotrophic gram-negative rods, those organisms growing most efficiently in the cold stored milk. Nevertheless, there are two more ways in which these bacteria may interfere with the quality of milk and dairy products.

Postpasteurization contamination (PPC) will occur. Of course, a few organisms will get access from air, water, equipment, etc., to the milk before cartons and bottles are closed, and cheese production is not an aseptic process. Constant attention to minimize the risk of PPC must be part of the hygiene program in a dairy. However, if microbes including pathogens get into the system in rather large numbers through leakages in lines, heat exchangers, or otherwise, the situation can soon become very serious. Some pathogens can grow at <5°C, i.e., strains of *Listeria monocytogenes, Yersinia enterocolitica, Aeromonas hydrophila*, enteropathogenic *Escherichia coli*, and *Clostridium botulinum* type E. Conditions in milk at 4°C may not favor growth of several of these species, but *L. monocytogenes* has been implicated in some cases, as indicated earlier.

Finally, proteinases, lipases, and phospholipases produced by psychrotrophic, gram-negative rods present a spoilage problem. The enzymes from *Pseudomonas* are mainly produced when the bacterial numbers reach 10^6–10^8 colony-forming units (CFU)/ml. Such dense floras are hardly compatible with high-quality milk, but, if the enzymes have been produced in milk remaining in poorly constructed or washed lines or other equipment, they may later be flushed into the bulk milk. The enzyme yield in a good culture can easily reach 100,000 ng/ml, and only 1 ng/ml (proteinase) may be sufficient to spoil milk. The proteinases and lipases can be produced at low temperature, and they have reasonably high activity at refrigeration temperatures; they release bitter peptides and free fatty acids, respectively, which both contribute to quality losses in dairy products. Destabilization of casein micelles and milk fat globules may also result from the action of the proteinases and lipases/phospholipases, respectively. The unfortunate fact is that most of these enzymes are not inactivated by any of our heat treatment processes. Very good hygiene, close control of temperature during storage, and thermization to reduce the bacterial load are important measures to fight the psychrotrophs and their enzymes [*See* DAIRY PRODUCTS.]

Unlike milk that is pasteurized and still considered fresh, meats, poultry, and fish are no longer "fresh" commodities after a heat treatment. While in cold storage (0–5°C), representatives of the initial psychrotrophic flora will therefore be definitive spoilers of these foods. Muscle tissues of all these foods when properly cut and handled can be obtained in nearly sterile condition. Contamination will thus be from skin, hide, hooves, nostrils, intestinal contents, water, and processing equipment for meats and poultry and from skin, gills, intestinal contents, water (ice), and processing equipment for fish. The tendency for all these foods during aerobic cold storage is an increasing domination by *Pseudomonas* on meats and poultry and *Pseudomonas* and *Shewanella putrefaciens* (formerly *Pseudomonas* type III/IV then *Alteromonas putrefaciens*) on

fish, but other bacteria also occur: *Acinetobacter, Aeromonas, Alcaligenes, Moraxella, Shewanella* and specific contaminants of the different foods.

Psychrotrophic counts of fish from temperate or cold waters are often higher than the respective total counts at higher temperatures (e.g., 30°C, 2 days), a situation quite different from that of meats and poultry. That many bacteria from these types of fish are relatively psychrophilic is a good example of the ecological control exerted in those vast areas of our oceans, lakes, and rivers, where temperature ranges at least from −1° to 5°C. Such bacteria will be natural inhabitants of skin, gills, and intestines of the resident fish and, thus, end up in our markets. We appreciate then the common practice of storing fish on ice. It is impossible to completely avoid contamination of the fish filets with the initial flora during handling, which is even worse with fish than with meats and poultry.

Contamination of tropical fish will often be characterized by gram-positive bacteria of the coryneform and micrococcus groups. Because these organisms are less psychrotrophic, if at all, fish stored in ice can have a very long shelf life (20–30 days).

Aerobic spoilage of meats, poultry, and fish becomes apparent as unpleasant odors, slime formation, and discoloration from metabolism of available low-molecular weight compounds: carbohydrates, lactate, amino acids, nucleotides, trimethylamine oxide (for fish), etc. Contrary to the situation in milk, proteinases and lipases from psychrotrophic bacteria, if they appear, are suggested to contribute to a secondary stage of spoilage with possible consequences for texture and rancidity. [*See* MEAT AND MEAT PRODUCTS.]

B. Combination Systems

When temperature is used to control microbial activities, other variables always influence the outcome (process) to a greater or lesser extent. This comes from the obvious fact that the microbes are part of a biological system with certain characteristics of pH, a_w, E_h, gas atmosphere, nutrient composition, salt content, antimicrobial content, etc.

Often combinations of factors that regulate microbe action are taken for granted because they represent common knowledge. Food processors generally accept that microorganisms are more easily killed by thermal processing when the pH is low (<4.5), and many foods can be stored at higher temperatures when the a_w has been lowered by drying, addition of salts or sugars, or combinations of these parameters. The combinations of parameters theoretically possible for inhibition or inactivation of microorganisms in processing of foods are indeed numerous. Naturally restrictive ranges (e.g., pH of milk and meats) simplify considerably, but still there are many alternatives to evaluate in each case.

Attempts to systematize the relationships between processes and parameters and the relative importance of parameters for a given process led to the hurdles concept illustrated in Fig. 2. For a specific process, there are one or more primary factors or hurdles that contribute most to prevent microbial growth and other activities; additional hurdles are also common. Pasteurization of milk is based on heat as the main hurdle, but subsequent refrigerated storage is decisive for extended shelf life. Fresh meats are preserved by chilling, but the advent of modified atmosphere packaging (oxygen removal— introduction of other gases) has contributed to even more extended shelf life. Application of preservatives is an additional hurdle of interest for meat products at low temperature (e.g., sorbate on chickens). Systematic use of hurdles can contribute to less-severe heat treatments and reduced addition of preservatives, and the safety of a process can be increased by additional hurdles. Different factors may act additively or with synergistic effects; thus, the complexity and potential benefits of exploiting several inhibitory principles in a process must be evaluated carefully.

Food fermentations are also combination systems with temperature as one of the factors. These complex processes depend on fine regulation of microbial growth and enzymatic conversions that imply temperature control to select some organisms and enzyme activities at the expense of others. The great majority of food fermentations have traditional roots, and the temperature during ripening was that of the storage environment [e.g., a cellar (cheeses, wine, beer), open air (cucumbers, olives, sausages), silos (forage)]. For some of these productions, elevated temperatures are applied during initial processing (cheeses, beer, sausages), which allows only the more resistant—and desirable—organisms (or enzymes) to survive. Modern food fermentations often demand strict regulation of temperature to obtain the correct flavor and texture of the product. As an example, the fine balance of *Streptococcus thermophilus* and *Lactobacillus delbrueckii* subsp. *bulgaricus* necessary for the desirable flavor of yogurt can only be accomplished at 42°C. [*See* BEER; WINE.]

Parameters \ Processes	Heating	Chilling	Freezing	Freeze drying	Drying	Curing	Salting	Sugar addition	Acidification	Fermentation	Smoking	Oxygen removal	IMF (f)	Radiation
F (a)	X(c)	+	+	+	+	+	o	+	o	+	+	+	+	o
t (b)	+(d)	X	X	o	o	+	+	o	+	+	+	+	+	o
a_w	+	+	X	X	X	X	X	X	o	X	+	o	X	o
pH	+	+	o	o	+	+	+	X	X	X	+	+	+	o
E_h	+	+	+	+	o	+	+	+	+	+	+	X	+	o
Preservatives	+	+	o	o	+	X	+	+	+	X	X	+	X	o
Competitive flora	o(e)	o	o	o	o	+	o	o	+	X	o	+	+	o
Radiation	+	o	o	o	o	o	o	o	o	o	o	o	o	X

Figure 2 Processes used in food preservation and parameters or hurdles they are based on. (a) High temperature, (b) low temperature, (c) main hurdle, (d) additional hurdle, (e) generally not important for this process, (f) intermediate moisture foods. [From Leistner, L. (1978). Hurdle effect and energy saving. *In* "Food Quality and Nutrition" (W. K. Downey, ed.), pp. 553–557. Applied Science Publishers, England.]

C. Mechanisms of Adjustment to Low and High Temperatures

The temperature limits for microorganisms are directly related to those of the activities and structures of the cells because, as poikilotherms, they have no shield against the surrounding temperature fluctuations. The further consequence is that any microorganism has enzymes, biochemical systems and structures more or less specialized for growth, or at least for survival, over a certain, often prevailing, temperature range. Dynamic systems for functional changes or adjustments with temperature fluctuations are also operative.

Thermophilic microbes have regimes of thermoresistant proteins seldom found in psychrophiles. A temperature shift-up induces qualitative and quantitative changes, which may be quite drastic when the final temperature is above the normal growth range. The fatty acids of the membrane phospholipids are mainly saturated; thus, the lipids are in a functional liquid-crystalline state in temperature ranges for thermophiles. However, at low temperature, for thermophiles, transistion to the gel state for membrane lipids may be a growth-limiting factor. [*See* THERMOPHILIC MICROORGANISMS.]

In addition to the specialization of components and structures for the high-temperature of life of thermophiles, microorganisms in general—in fact all living cells—appear to produce large amounts of so-called heat-shock proteins upon exposure to temperatures above normal growth conditions. Other inducers of the response include sulfhydryl reagents, ethanol, hydrogen peroxide, amino acid analogs, puromycin, nalidixic acid, and alkaline conditions. Regulation of the response includes polypeptide signal(s) produced by regulatory gene(s) (*rpoH*) that control the expression of the heat-shock protein genes. The induction results in the increased synthesis of a group of proteins, 12–24 in bacteria, some of which are involved in protecting the organism at the elevated temperature. Mutants without the proteins die under similar conditions. However, some or all of the heat-shock proteins appear to have functions in the cells at all temperatures: for cellular growth, correct assembly of bacteriophage and DNA replication.

The limiting conditions for psychrophiles at low temperatures have received considerable attention. General evidence with proteins at low temperatures (<5°C) show weakening of hydrophobic bonds, which may contribute to slight conformational changes, thus interfering with enzyme activities, regulatory functions in allosteric systems, and the assembly of complexes such as ribosomes. Sensitivity of DNA and DNA replication has also been implicated. However, sufficient evidence is not available to claim that these changes alone limit growth in general. Quite to the contrary, more protein is synthesized at 0°C in psychrophilic bacteria, apparently to compensate for reduced enzyme activities, and low temperature has not been proven to stop energy supply in psychrophiles. [*See* LOW-TEMPERATURE ENVIRONMENTS.]

With lipids, the situation is somewhat different because substantial information has been gained to suggest that changes from a liquid-crystalline state of membrane lipids to a gel state at low temperatures influence solute (nutrient) uptake in a decisive way. Accordingly, lowering the melting point of membrane lipids should be compatible with the maintenance of functionality (homeostasis) at low temperature. Comparison of psychrophilic bacteria with mesophiles or exposure of an organism to low temperature (0°) have revealed certain tendencies for changes in membrane lipid composition, as follows: Desaturation or fatty acids occurs as well as *de novo* synthesis of unsaturated fatty acids; shorter fatty acids may be synthesized; anteiso branched-chain fatty acids increase and cyclopropane acids decrease; rearrangements of fatty acids to the *sn*-1 long–*sn*-2 short configuration of *sn*-glycerol-3-phosphates also contribute to low melting points or, rather, melting ranges. Whether or not changes of composition must accompany a temperature decrease has been challenged by observations of mesophiles (e.g., *E. coli*) with fatty acids suitable for low temperature growth after incubation at mesophilic temperatures.

A recent addition to the repertoire of melting point depressing mechanisms is the transition of trans-unsaturated fatty acids into the cis configuration in *Vibrio* and other organisms. Both remodeling of phospholipids and the trans/cis-alterations may be reaction systems for rapid adaption to low temperatures.

Other functions, although not necessarily essential, may support growth and survival at low temperatures. The optimum temperature for chemotactic response with *Pseudomonas fluorescens* has been observed below the optimum for growth, and chemotaxis was strong at 0°C. Exopolysaccharide synthesis, visible as slime, is stimulated at low temperatures. For microaerophilic organisms (e.g., *Leuconostoc mesenteroides* and certain lactococci), capsules may protect against the increased oxygen solubility at low temperatures but also against threats like antibiotic action and dessication.

III. Sublethal Injury at High and Low Temperatures

A. Definitions

A variety of environmental changes from "normal" conditions for a microorganism are stressful and may inflict injury to the organism. The injury may be lethal (irreversible) so that the cells will not be able to produce progeny under any conditions or after any resuscitation procedure known to us. In contrast, sublethal injury is reversible and can be repaired under favorable conditions. The actual damages to the cells are seldom observed, but the overall consequences of the reversible injury are shown by the disparity of viable cell numbers (colony counts) in nonselective (base) and selective (test) media. The definition of reversible injury may thus be expressed as temporary loss of tolerance for specified conditions. This operative definition is illustrated in Fig. 3.

In the first phase in Fig. 3, a pure culture of an organism grows normally in a liquid medium, and plate counts with a test medium comes out with numbers close to those in a base medium. During phase two, cells are injured by exposure to a higher

temperature, growth stops, and some cells die as indicated by the decrease in counts on base plate medium. However, although sublethally injured and uninjured cells form colonies on the base plate medium, the test medium does not allow the injured cells to repair; thus, only the uninjured cells form colonies. The difference between the counts represents injured cells. When the stress condition is terminated (phase three) the injured cells in the culture gradually repair, provided the culture medium is rich. The steep rise of the curve sometimes observed in this phase indicates that repair is taking place and not regular growth. The difference between the base medium and the test medium may, for instance, be a higher salt concentration in the test medium.

Sublethal injury to vegetative cells can be produced by a variety of stressful conditions: modestly higher temperatures, lower temperatures including freezing–thawing, drying–rehydration, high osmotic conditions, low pH, salts, preservatives, sanitizers, depletion of essential ions, etc.

Differential plate count procedures to reveal that cells have been injured actually introduce additional or secondary stresses so that the cells injured reversibly in the first place (primary stress) do not repair and reproduce. Because reversible injury and lack of ability to form colonies have often been related to microbiological analysis and the use of selective media, differential conditions (secondary stresses) regularly seen are the following: increased salt concentration, minimal media, deoxycholate, lauryl sulfate, bile salts, detergents, crystal violet, brilliant green, other dyes, azide, antibiotics, etc.

Secondary stress may also occur during dilution of samples, thus Mg^{2+}-containing diluents are recommended to counteract this stress. Similarly a cold shock during dilution may kill some bacteria.

Sublethal injury has been observed in nearly all the foodborne pathogenic bacteria, in several indicator bacteria, spoilage bacteria, lactic acid bacteria, yeasts, and molds. Bacterial spores also experience injury, but generally more severe treatments are required. Drying and freezing are seldom injurious to bacterial spores. Mechanisms of injury vary between vegetative cells and spores.

B. Implications for Death or Survival of Microbes in Applied Microbiology and Research

Injury and repair in microorganisms have been studied thoroughly with foodborne pathogens and indi-

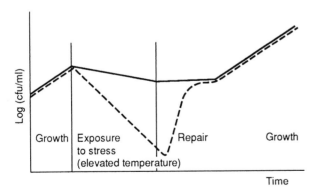

Figure 3 Model to illustrate injury and repair in a microorganism. Unbroken line, plate counts on base agar; broken line, plate counts on test agar. Further details in text. [Adapted from Busta, F. F. (1978). *Adv. Appl. Microbiol.* **23**, 195–201.]

cator organisms. Spoilage organisms have been given some attention, and starter organisms even less. The importance of injury in foodborne pathogens and indicator organisms is manifested in two ways. On the one hand, injured pathogens may repair and develop to the stage of food poisoning; on the other hand, injured organisms able to recover in a food may not be detected in traditional microbiological analysis. Substantial efforts have been devoted to remedy this situation by developing analytical procedures that will detect both uninjured and reversibly injured organisms.

Primary stress to microorganisms in foods may occur at various stages: in the fresh, raw material (refrigeration, chlorine in water, etc.), during thermal and other processing and during storage (refrigeration, modified atmospheres, etc.). Injured cells may recover or the damage may be exposed to secondary stresses in the food resulting in death. For the food processor, synergistic effects of several stresses are desirable (cf. hurdles concept) for better preservation and extended shelf life.

Samples for analysis are often stored transiently (frozen, refrigerated) with possible changes to the flora. Preparation of samples (thawing, mixing, heating) and dilutions (type of diluent, temperature) may contribute further stress. Even pouring plates has a potential for stress, particularly if the melted agar medium has not been cooled to 45°C. For pathogens and indicator organisms, the customary selective media, although suitable for uninjured cells, are harmful for injured cells. Therefore, in microbiological analysis, great care should in general be exercised with samples to avoid secondary stresses resulting in accumulated injury and subsequent death of organisms not necessarily experienced in the food.

Solutions to the problems with selective media focus mainly on preincubation procedures with nonselective media to allow the recovery of injured cells. Incubation on agar media are preferred over liquid incubation to avoid interference by multiplication of uninjured cells. Thus, the diluted sample is poured (or spread) in a nonselective medium, and after a 1–4-hr resuscitation period, the appropriate selective medium (single or double strength) is poured on top before the final incubation. Markedly higher counts are regularly obtained this way. Good alternatives are filter techniques with transfers between nonselective and selective media and the use of replica techniques.

Specific and important agents for injury are hy-drogen peroxide and superoxide radical, which, however, are effectively destroyed by the addition of catalase or pyruvate to the media.

Vegetative bacterial cells exposed to different kinds of stress may suffer damage at several sites: cytoplasmic membrane injury with leakage of intracellular material, degradation of DNA, ribosomes, and RNA, protein denaturation, and enzyme inactivation. With heating or low temperatures, the extent of injury will vary according to the severity of the treatment, and one or rather several of the sites are influenced.

Spores with other structures have different or additional mechanisms of injury in the spore and during activation, germination, and outgrowth.

IV. Thermal Inactivation

A. Classical Death Kinetics, Survivor Curve, and Thermal Processing

Experiments with many different organisms, particularly with bacterial spores, in various thermal processes have shown that the number of surviving microorganisms decrease exponentially with time:

$$N = N_0 e^{-kt},$$

where N_0 = the initial number of viable organisms, N = the number of survivors at time t, and k = the specific reaction rate for death of the organism. Transferred to logarithms, of base 10, it becomes

$$\log N/N_0 = -kt/2.303.$$

A plot of log N versus time is often called the survivor curve (Fig. 4). The slope is $-k/2.303$, and the intercept on the ordinate is log N_0. The decimal reduction time, D, is the time for a 10-fold decrease in the number of survivors, i.e., the time to kill through one log cycle (Fig. 4). A D value is given for a specific temperature (e.g., $D_{121°C}$). The D values decrease exponentially with increasing temperature; the slope of this line is called the z value, i.e., the number of degrees Celsius for a 10-fold change in D value. Typical examples of D values are the following:

Psychrotrophic pseudomonads, $D_{70°C}$ = 0.001–0.5 sec
Psychrotrophic sporeformers, $D_{110°C}$ = 0.4–8 min

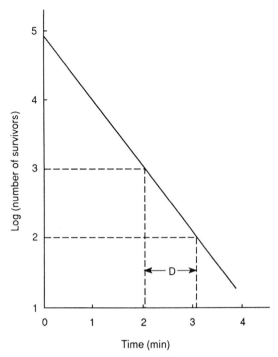

Figure 4 Thermal death of a microorganism. Model of survivor curve. Further details in text.

B. coagulans, acid condition, $D_{121°C} = 0.05$ min

B. stearothermophilus, low-acid food, $D_{121°C} = 4$ min

C. botulinum, low-acid food, $D_{121°C} = 0.21$ min

C. sporogenes, low-acid food, $D_{121°C} = 0.15–1$ min

From these data, it is apparent that strains of *Pseudomonas,* the predominant psychrotroph in milk, will be killed by pasteurization at 72°C for 15 sec, whereas psychrotrophic sporeformers will survive.

In canning of nonacid foods, a very extended thermal treatment would be needed for complete inactivation of *B. stearothermophilus* spores. However, spoilage by this organism is not a great concern except if the cans are cooled too slowly or stored at excessively high temperatures.

The safety requirement for a canning process with low-acid foods is to inactivate all *C. botulinum* spores because it is the most heat-resistant food-poisoning organism. This is accomplished by a so-called 12D process, which kills *C. botulinum* through 12 log cycles. An initial *C. botulinum* load of 10^4 per unit will thus be reduced to 10^{-8} per unit, or 1 survivor in 100 million units. With sealed containers and a processing temperature of 121°C, the 12D process will require $12 \times 0.21 = 2.5$ min, which in practice is rounded off to 3 min. The same calculation applied to spores of a mesophilic spoilage bacterium (e.g., *C. sporogenes*), gives a processing requirement of 12 min. However, occasional spoilage of a can is more acceptable than food poisoning, so a less severe treatment may be chosen. The preceding calculations give the so-called F_0 values, which are the total lethal effects expressed as minutes at 121°C. Different thermal processes may, thus, be compared by calculating the integrated lethality and then transferred to the equivalent value at 121°C (250°F). *F* values include contributions from heating, holding, and cooling periods.

Spores from different bacteria vary greatly in heat resistance. Also, psychrotrophic spore-formers have much less heat-resistant spores than mesophiles, and still less than spores of thermophiles. Less dramatic but still of considerable importance are the variations within strains or species caused by other factors. Lowering the pH decreases heat resistance markedly. Growth at high temperatures results in spores of increased resistance, but this may not be reflected in the cells after germination. Water activity, salts, preservatives, etc., may variously influence heat resistance. With spores, these considerations are complexed by impacts during activation, germination, and outgrowth.

Among the early regulations for food safety was pasteurization of milk to kill *Mycobacterium tuberculosis* and *Mycobacterium bovis* at 61°C for 30 min. However, it was the heat resistance of *Coxiella burnettii,* the etiological agent of Q-fever, that led to the requirement for 30 min at 63°C commonly reported for pasteurization of milk. High-temperature short-time (HTST) treatment is often referred to as 72°C for 15 sec, but equivalent combinations are 89°C, 1 sec; 90°C, 0.5 sec; 94°C, 0.1 sec; 96°C, 0.05 sec; and 100°C, 0.01 sec. Thermal processing of milk is also considered in Section II.A. Somewhat higher temperatures are required for cream and ice cream mixes. Other products may not demand as severe treatments because of lower or higher pH or other factors; thus, whole eggs are pasteurized at 60°C for 3.5 min. In comparison, cooking procedures have similar requirements (e.g., the minimum internal temperature of roast beef is 63°C with a holding time of 5 min). Fruit juices of low pH may need rather severe treatments to kill yeasts and molds. [*See* FOODS, QUALITY CONTROL.]

B. Deviations from Log : Linear Relationships of Survivor Curves

When D values are used, they often present simplifications of reality. The line in Fig. 4 represents the simple log : linear relationship between survivors and time, but partial deviations from linearity also occur for the death kinetics for spores. An initial increase of viable organisms may be the result of spore activation; a shoulder has been explained as a balance between activation and inactivation. However, assuming that multiple hits on sensitive sites in the spores are necessary for inactivation, a shoulder may also result. Heat-up time and clumping have likewise been mentioned as factors contributing to shoulders.

Concave survivor curves have mainly been associated with variations in heat resistance of individual organisms in a population. However, other explanations have been advanced. Extended tails as well as biphasic curves have been observed.

The log : linear relationship for thermal death with time has also been assumed for vegetative cells of microorganisms. However, even greater deviations are commonly observed than for spores. Differences among cells during heating have been suggested as important factors to explain this (cf. also Section II.C).

Bibliography

Gould, G. W. (1989). Heat-induced injury and inactivation. *In* "Mechanisms of Action of Food Preservation Procedures" (G. W. Gould, ed.), pp. 11–42. Elsevier Applied Science, London.

Gould, G. W., and Jones, M. V. (1989). Combination and synergistic effects. *In* "Mechanisms of Action of Food Preservation Procedures" (G. W. Gould, ed.), pp. 401–421. Elsevier Applied Science, London.

Greer, G. G. (1989). Red meats, poultry, and fish. *In* "Enzymes of Psychrotrophs in Raw Food" (R. C. McKellar, ed.), pp. 267–292. CRC Press, Boca Raton, Florida.

Gray, R. J. H., and Sørhaug, T. (1983). Response, regulation and utilization of microbial activities at low temperature. *In* "Food Microbiology" (A. H. Rose, ed.). pp. 1–45. Academic Press, London.

Herbert, R. A. (1989). Microbial growth at low temperatures. *In* "Mechanisms of Action of Food Preservation Procedures" (G. W. Gould, ed.), pp. 71–96. Elsevier Applied Science, London.

Palumbo, S. A. (1989). Injury in emerging foodborne pathogens and their detection. *In* "Injured Index and Pathogenic Bacteria" (B. Ray, ed.), pp. 115–131. CRC Press, Boca Raton, Florida.

Ratkowsky, D. A., Lowry, R. K., McMeekin, T. A., Stokes, A. N., and Chandler, R. E. (1983). *J. Bacteriol.* **154,** 1222–1226.

Ray, B. (1989). Enumeration of injured indicator bacteria from foods. *In* "Injured Index and Pathogenic Bacteria" (B. Ray, ed.), pp. 9–54. CRC Press, Boca Raton, Florida.

Skura, B. J. (1989). Standards, control, and future outlook. *In* "Enzymes of Psychrotrophs in Raw Food" (R. C. McKellar, ed.), pp. 245–263. CRC Press, Boca Raton, Florida.

Sørhaug, T., and Stepaniak, L. (1991). Microbial enzymes in the spoilage of milk and dairy products. *In* "Food Enzymology," Vol. I (P. F. Fox, ed.), pp. 169–218. Elsevier Applied Science, London.

Stepaniak, L., and Sørhaug, T. (1989). Biochemical classification. *In* "Enzymes of Psychrotrophs in Raw Food" (R. C. McKellar, ed.), pp. 35–55. CRC Press, Boca Raton, Florida.

Suhren, G. (1989). Producer organisms. *In* "Enzymes of Psychrotrophs in Raw Food" (R. C., McKellar, ed.), pp. 3–34. CRC Press, Boca Raton, Florida.

Witter, L. D. (1983). Elevated temperature inactivation. *In* "Food Microbiology" (A. H. Rose, ed.). pp. 47–75. Academic Press, London.

Tetanus and Botulism

David L. Smalley

University of Tennessee, Memphis

Glossary

Botulism Neuroparalytic illness, typically a foodborne illness, caused by a toxin produced by *Clostridium botulinum*

Infant botulism Foodborne illness most commonly seen in infants younger than 1 year; specific food products (e.g., honey) have been implicated as the source

Tetanospasmin Protoplasmic protein produced by *Clostridium tetani* that inhibits the release of acetylcholine

Tetanus Extremely dramatic neuroparalytic illness caused by a toxin, tetanospasmin, produced by *Clostridium tetani*

TETANUS is an acute disease induced by a bacterial neurotoxin produced by *Clostridium tetani*. It occurs when spores contaminate open wounds or sores. Soil and animal feces are frequently implicated as the source, and contamination may occur in very minor injuries. The disease may manifest itself in a period as short as 4 days but usually within a 2-week period from the introduction of the bacterial spores. The average period of incubation is 10 days but is dependent on the extent of the wound, vascularity, potential for anaerobiosis, and the infectious dose. No person-to-person transmission has been documented.

Botulism occurs in three forms. The most common illness of this type is food botulism or food poisoning. Canned or preserved foods contaminated with the spores of *Clostridium botulinum* produce a neurotoxin that causes the clinical manifestations.

Symptoms generally occur in 18 hr to 4 days. A second form of botulism is pediatric or infant botulism. In these cases, the spores contaminate food products (e.g., honey), are then ingested, and produce a culture of organisms in the gut of the infant. Toxin is produced, causing the clinical manifestations. A lesser known form is wound botulism. In these cases, the wound is contaminated with the spores, and the wound area becomes a cultural site for production of the neurotoxin as seen in the foodborne botulism cases.

1. Causative Agents

A. Clostridium tetani

Clostridia are anaerobic, gram-positive rods that are spore-forming and motile. *C. tetani* is an ubiquitous pathogenic organism in this genus that is found throughout the world, particularly in soil and in the feces of many domestic animals. *C. tetani* has been isolated from patients with tetanus and also from the fecal material of normal individuals. Although uncommonly cultured, *C. tetani* has unique biochemical and metabolic constituents. It does not ferment or use glucose but does produce indole. It liquefies gelatin, has terminal spores, and typically produces a significant toxin. *C. tetani* has a concentration of $G+C = 25\%$. Specific gas chromatograms with distinctive patterns have been reported by the Anaerobe Laboratory of Virginia Polytechnic Institute and State University.

C. tetani has several distinctive flagellar antigens, and all tetani species have a common somatic antigen. In addition, *C. tetani* produces a potent neurotoxin referred to as tetanospasmin. *C. tetani* is the causative agent of tetanus, which occurs when spores enter a wound via contaminated soil or foreign material. As the spores germinate, neurotoxin is produced, resulting in the clinical symptoms of tetanus.

B. Clostridium botulinum

Another pathogen in the genus *Clostridium* is *C. botulinum*. It is separated into several specific toxigenic types from A to G, all of which liquefy gelatin. *C. botulinum* types A, B, and F that are proteolytic have lipase activity, produce a neurotoxin, and usually hydrolyze esculin. *C. botulinum* type C ferments glucose, produces indole, and has lipase activity. Types B, E, and F that are nonproteolytic do not have lecithinase activity but do have lipase activity. Gas chromatography shows distinct patterns separating each of the types and is the most useful mechanism to identify specific types. Specific chromatograms for *C. botulinum* have also been reported by the Anaerobe Laboratory of Virginia Polytechnic Institute and State University.

II. Mode of Transmission– Disease State

Tetanus, which is an acute disease induced by the bacterial neurotoxin, occurs when spores contaminate open wounds or sores. Soil and animal feces are frequently implicated as the source of tetanus spores, and contamination may occur with very minor injuries. The disease may manifest within 4 days and usually within a 2 weeks period from initial contamination. The average period of incubation is 10 days but is dependent on the extent of the wound, vascularity, potential for anaerobiosis, and the infectious dose. No person-to-person transmission has been documented.

Because the organism is rarely recovered from the site of infection, diagnosis is usually based on clinical impression. The disease state is characterized by tonic muscular contractions involving the neck and trunk muscles. Initial symptoms involve unstimulated abdominal rigidity and muscle spasms in the area of the injury and the muscles in the jaw, thus referred to as lockjaw disease. The patient may be alert and fully conscious and may experience extreme pain when contractions occur. The mortality of tetanus can be as high as 50% and usually results from respiratory failure by muscular contraction interference.

Tetanus has been recognized throughout the world, although the incidence of disease is higher in developing countries. The lack of immunization programs in some underdeveloped countries may account for these higher incidences. For example, the

Centers for Disease Control reported in 1980 that only 95 cases of tetanus were identified in the United States, and among these patients, 72% were over the age of 50. During the years 1987 and 1988, 101 cases of tetanus were reported in the United States involving 35 states. Among these cases, 68% were over the age of 50. In contrast, approximately 621 cases annually were identified in India over a 14-year study period. In the same survey from India, the researchers reported that about one-half of the cases were associated with a wound entry and more than 20% were associated with otitis media.

Spores of *C. botulinum* are commonly found in soil and commonly contaminate food products such as fruits and vegetables. Because many of these types of foods are canned or preserved, the potential for contamination and proliferation of the organism is increased. Government regulation of food product preparation has decreased the incidence of botulism. Home-canned food products remain at high risk for botulism transmission. [*See* FOODS, QUALITY CONTROL.]

Antigenic types A, B, E, and F have been implicated as the predominant types of toxins causing botulism. Symptoms are noted 4 days after ingestion of contaminated food. The toxin produced by *C. botulinum* causes oculomotor problems such as blurred or double vision caused by the uncontrolled muscle spasms in the eye. Other complaints such as inability to swallow and speech difficulty are common. Unlike some other types of food poisoning due to toxin contamination, only mild gastrointestinal discomfort is reported. The patients may remain conscious until respiratory failure occurs because of muscular paralysis. The mortality rate is approximately 33% within 7 days of initial intoxication. Because the infection/toxicity is associated with foodborne transmission, no cases of person-to-person transmission have been reported. [*See* FOODBORNE ILLNESS.]

Pediatric botulism has also been reported in infants during the first few months of life. Honey contaminated with botulism spores has been implicated as a possible source of infection. In one study, 44% of the infants with type B illness had a history of honey ingestion. In a California investigation, spores were detected in 7.5% of the honey samples tested. It is a common practice for parents to saturate the bottle nipple or medications with honey to enhance ingestion by infants. As the spores germinate and toxins are produced, the infant becomes weak and may show signs of paralysis. Prolonged constipation followed by weakness, lethargy, and dysphagia are common. In two large studies of sudden infant death syndrome, 10 of 280 cases (3.5%) and 9 of 211 cases (4.2%) demonstrated the presence of either the toxin or spores in the infants' stools. The mortality of infant botulism is much lower than that in adults. Epidemiological studies report that as many as 250 cases may occur each year in the United States.

Wound botulism is another lesser known form of botulism and occurs after introduction of the bacterial spores into a wound. The organism proliferates within the wound, producing the same neurotoxin seen in foodborne botulism and causes the characteristic disease.

III. Pathogenesis–Toxin Effects

Tetanospasmin, the neurotoxin produced as a plasmid-encoded protoplasmic protein in *C. tetani,* is released after autolysis. The toxin has an affinity for central nervous system gangliosides and causes inhibition of motor neuron impulses. By blocking the motor impulses, muscle spasms are extended, affecting the flexor and extensor muscles. The toxin specifically blocks the release of acetylcholine, thus blocking the nerve inpulse. Although strain variations occur among the isolates of *C. tetani,* tetanospasmin has not been shown to vary antigenically or pharmacologically.

The toxins produced by *C. botulinum* are divided into seven types. Types A, B, E, and F are those typically seen in human disease. Types C and D have been reported in association with botulism in animals. Type G has not been implicated in human or animal botulism and has recently been reclassified as being produced by another Clostridium, *C. argentinense.*

The botulinum toxin is also a protoplasmic protein released by the organism during autolysis. It is resistant to low pH and is not destroyed by stomach acids. In the foodborne form of botulism, *in vitro* proliferation of the organism occurs, and the individual ingests toxin-contaminated food. In contrast, infant and wound botulism is characterized by the *in vivo* multiplication of the organism followed by toxin production.

The primary toxin effect is interference of the release of acetylcholine. The exact mechanism of action has not been clearly defined. A receptor mechanism associated with the neuromuscular junc-

tion most likely allows the penetration of the toxin into the cell, and then the toxin blocks the acetylcholine release.

IV. Preventive Measures– Treatment

The primary preventive measure in the United States is routine immunization with tetanus toxoid. Active immunization is initiated in children in a series of three injections, which is usually given with diphtheria toxoid and pertussis vaccine. Active immunization is now considered protective for a 10-year period.

Removal of foreign matter and thorough cleansing of the injury site is the most important initial step in treating a potential tetanus exposure. In serious wound cases where the 10-year period of immunization is exceeded or no record of active immunization is available, the use of intravenous human tetanus immune globulin should be administered as passive immunization. The passive immunity only lasts about 3 weeks, so adsorbed toxoid immunization is also initiated. In addition, antimicrobial therapy is recommended.

Because of the potential for respiratory failure in cases of tetanus, strict attention to the respiratory status of patients is necessary. Debridement of the wound to prevent continued microbial survival and toxin production is extremely important. In some cases, administration of tetanus immune globulin may be useful. The immune globulin has an affinity for the tetanospasmin. Neutralization occurs when immune globulin–tetanospasmin complexes are formed. Specific therapy for tetanus prophylaxis in routine wound management has been developed by the Immunization Practices Advisory Committee. If the patient has a history of receiving less than three doses of tetanus toxoid, clean minor wounds should be treated by administration of toxoid but no tetanus immune globulin should be given. In all other wounds, both are recommended. In individuals having received three or more doses of toxoid, clean minor wounds require no immunizations if the last dosage was received within the past 10 years. In other wounds, no immunizations are required if the individual has received toxoid within the past 5 years.

Because most cases of botulism are associated with contaminated food, secondary to home canning and preserving, preventive measures should include proper cooking times and temperatures to kill contaminating spores. Preventive measures are in place for commercial food preservation; however, surveillance by the Food and Drug Administration helps to ensure compliance. Education of the public is also important; individuals should not ingest food from atypical bulging cans and should not consume food products having an abnormal odor.

Diagnosis of foodborne botulism can be established by the demonstration of specific toxin in blood or feces or its presence in the suspected food. Intravenous or intramuscular administration of antitoxin is the usual treatment. A variety of polyvalent and monospecific antitoxins can be obtained from the Centers for Disease Control. Following the collection of samples for laboratory analysis, the food source should be autoclaved or incinerated to kill contaminating spores.

Infant botulism prevention is primarily restricted to the avoidance of honey for infants younger than 1 year. Treatment is usually supportive; antibiotics and antitoxins have been proven to be ineffective. Respiratory and nutritional support are the most critical treatments.

Preventive measures for wound botulism include cleaning the wound, debriding to minimize an anaerobic environment, supportive therapy, and if the toxin is identified, the administration of antitoxin.

V. Therapeutic Uses of Botulinum Toxin

Although the botulinum toxin is among the most lethal toxins known, therapeutic utilization of the toxin for a variety of disorders has been developed. The Food and Drug Administration recently approved the botulinum toxin as a therapeutic agent for patients with strabismus, blepharospasm, and other facial-nerve disorders. In strabismus and other ocular motility abnormalities, the toxin can be injected into extraocular muscles to weaken the muscle. This alternative to the conventional surgery has shown satisfactory improvement up to 5 years post-injection.

The toxin has also been successfully used in neurologic disorders associated with involuntary muscle contractions or spasms such as dystonia, a disorder characterized by spasmodic muscle contractions. Types of dystonia where the botulinum toxin may be useful include focal blepharospasm, spasmodic torticollis, oromandibular dystonia, laryngeal dystonia, task-specific dystonias, and segment or generalized tremor. A recent review article by Jankovic and Brin (1991) discusses specific applications and success according to the specific disorder.

Bibliography

Allen, S. D., and Baron, E. J. (1991). Clostridium. *In* "Manual of Clinical Microbiology" (A. Balows, W. J. Hausler, Jr., K. L. Herrmann, H. D. Isenberg, and H. J. Shadomy, eds.), pp. 505–521. American Society for Microbiology, Washington, D.C.

Arnon, S. S., Midura, M. F., Damus, K., Wood, R. M., and Chin, J. (1978). *Lancet* **i**, 1273–1276.

Benenson, A. S. (1990). Botulism and tetanus. *In* "Control of Communicable Diseases in Man," pp. 61–65, 430–434. American Public Health Association, Washington, D.C.

Dowell, V. R., Jr. (1984). *Rev. Infect. Dis.* **6** (suppl. 1), S202–S207.

Furste, W. A., Aquirre, A., and Knoepfler, D. J. (1989). Tetanus. *In* "Anaerobic Infections in Humans" (S. M. Finegold and W. L. George, eds.), pp. 611–627. Academic Press, New York.

Holdeman, L. V., and Moore, W. E. C. (1972). Clostridia. *In* "Anaerobe Laboratory Manual," pp. 67–89. Virginia Polytechnic Institute and State University, Blacksburg.

Jankovic, J. J., and Brin, M. F. (1991). *New Engl. J. Med.* **324**, 1186–1194.

Jawetz, E., Melnick, J. L., and Adelberg, E. A. (1987). Anaerobic sporeforming bacilli: The clostridia. *In* "Review of Medical Microbiology," pp. 208–211. Appleton & Lange, Norwalk, Connecticut.

Johnson, R. O., Clay, S. A., and Arnon, S. S. (1979). *Am. J. Dis. Child.* **133**, 586–593.

Morbidity and Mortality Weekly Report (1990). Tetanus—United States, 1987 and 1988, **39**, 37–41, U.S. Department of Health and Human Services, Atlanta, Georgia.

Thermophilic Microorganisms

S. Marvin Friedman

Hunter College of The City University of New York

Glossary

Archaebacterium Prokaryotic cell lacking a peptidoglycan cell wall and containing diether or tetraether lipids

Biotope Limited ecological region or niche in which the environment is suitable only for certain forms of life

Cardinal temperatures Minimum (T_{min}), optimum (T_{opt}), and maximum (T_{max}) growth temperatures

Hyperthermophile Microorganism growing optimally between 80° and 110°C and usually not below 60°C

Solfatara Hot, sulfur-rich environment

THERMOPHILIC MICROORGANISMS are unique cells that are endowed with the capability of growth at elevated temperatures. Although the upper temperature limit for life may not yet have been ascertained, we do know that certain hyperthermophilic archaebacteria can proliferate at 110°C in ocean depths where water remains liquid at this temperature. Thermophilic representatives are found among the algae, fungi, protozoa, cyanobacteria, eubacteria, and archaebacteria. These organisms inhabit a variety of ecological niches including thermal soils, desert sands, hot springs, solfataric fields, submarine hydrothermal systems, waters surrounding boiling outflows from geothermal power plants, and smoldering coal refuse piles.

I. Thermophilic Microbes and Their Habitats

Table I presents a list of representative thermophilic species for each phylogenetic group along with their maximum growth temperatures (T_{max}). It is important to note that these values are approximations since they will vary with the particular growth medium employed. For example, a lower T_{max} is observed for a synthetic medium, whereas higher growth temperatures are achieved when cells are cultured in rich media. Furthermore, variations in T_{max} have been recorded for different complex media.

Eukaryotic thermophiles are found among the algae, fungi, and protozoa. These organisms are confined to an upper temperature limit of about 61°C. The only thermophilic member of the algae is *Cyanidium caldarium*, an organism also capable of growth in extremely acidic environments. A Finnish company has developed a commercial facility for growing the thermophilic fungus *Paecilomyces varioti* on sulfite waste liquor, a by-product of the paper industry. The so-called Pekilo Process produces a single-cell protein that has proven to be an excellent supplement in animal feed and also shows promise for human consumption.

Thermophilic cyanobacteria are found in neutral to alkaline hot springs and in the fissures of desert rocks. The most thermophilic cyanobacterium, *Synechococcus lividus,* is capable of growth up to 74°C. This temperature is appreciably higher than that for any of the thermophilic eukaryotes, yet below that achieved by many thermophilic eubacteria and archaebacteria. *Mastigocladus laminosus* is the most thermophilic nitrogen-fixing cyanobacterium.

Thermophilic eubacteria can be classified into three distinct groups based on their cardinal growth temperatures. Facultative thermophiles have T_{max}'s between 50° and 65°C and also grow below 30°C. Examples include *Bacillus coagulans* and a few

Table I Representative Species of Thermophilic Microorganisms and Their Maximum Growth Temperatures

Group, genus, and species	T_{max} (°C)
I. Algae	
Cyanidium caldarium	56
II. Fungi	
Talaromyces thermophilus	57–60
Chaetomium thermophile	58–60
Aspergillus candidus	50–55
Paecilomyces spp.	55–60
Thermomyces ibadanensis	60–61
Dactylaria gallopava	50
III. Protozoa	
Cercosulcifer hemathensis	56
Vahlkampfia reichi	55
IV. Cyanobacteria	
Synechococcus lividus	74
Oscillatoria amphibia	57
Phormidium laminosum	57–60
Mastigocladus laminosus	63–64
V. Eubacteria	
Bacillus stearothermophilus	70–75
Bacillus coagulans	55–60
Bacillus acidocaldarius	70
Bacillus caldotenax	85
Bacillus caldolyticus	82
Clostridium thermocellum	70
Clostridium themosaccharolyticum	67
Thermomicrobium roseum	85
Thermoactinomyces vulgaris	70
Desulfovibrio thermophilus	85
Thermus thermophilus	85
Thermotoga maritima	90
VI. Archaebacteria	
Sulfolobus acidocaldarius	85
Acidianus infernus	95
Pyrobaculum islandicum	103
Pyrodictium occultum	110
Thermococcus celer	97
Pyrococcus furiosus	103
Archaeglobus fulgidus	95
Methanothermus sociabilis	97
Methanococcus jannaschii	86
Thermoplasma volcanium	67

strains of *Bacillus stearothermophilus*. Both of these soil microorganisms are among the most widely studied eubacterial thermophiles. *Bacillus stearothermophilus* is often isolated from "flat sour" spoilage of canned peas, string beans, and corn because spores of some strains are the most heat-resistant biological structures known. In this regard, spore suspensions of *B. stearothermophilus* have been used to check whether or not an autoclave has effectively sterilized its contents.

Obligate thermophiles will not grow below 40°–42°C and have an optimum growth temperature (T_{opt}) or 65°–70°C. Many strains of *B. stearothermophilus* belong to this class. Extreme (or caldoactive) thermophiles have a T_{max} >70°C, a T_{opt} >65°C, and a minimum (T_{min}) >40°C. *Thermus thermophilus,* a gram-negative organism isolated from a Japanese thermal spa, and *Bacillus caldolyticus* are representatives of this group. Thermotolerant bacteria, such as *Bacillus subtilis,* have T_{max}'s between 45° and 55°C and also grow below 30°C.

A new term, hyperthermophile, has been introduced recently to describe bacteria growing optimally at temperatures between 80° and 110°C. The T_{min} within this most thermophilic group is around 55°C. Most of the hyperthermophiles are archaebacteria, but a few eubacterial species are also known. It is of interest to note that the hyperthermophilic eubacterium *Thermatoga maritima* (T_{max} = 90°C) exhibits several unique properties. This organism possesses (1) a single set of ribosomal RNA (rRNA) genes, (2) a 30S ribosomal subunit that is insensitive to aminoglycoside antibiotics, (3) an elongation factor (EF-G) that is refractory to fusidic acid, and (4) unique lipids containing long-chain dicarboxylic acids. Based on its 16S rRNA sequence, *T. maritima* represents the deepest known branching in the eubacterial line of descent and also the most slowly evolving of eubacterial lineages. The only other deeply branched and slowly evolving eubacterial group, the green nonsulfur bacteria, is comprised predominantly of thermophilic species. Because it appears that slowly evolving bacterial lineages tend to retain ancestral characteristics, the original eubacteria may well have been thermophiles.

Archaebacteria can be distinguished from eubacteria in part by their lack of a peptidoglycan cell wall, possession of lipids that are isopranoid diethers or diglycerol tetraethers, and rRNA sequence characteristics. The archaebacterial domain contains two major branches. One lineage includes methanogens and extreme halophiles, while the other is comprised solely of hyperthermophilic sulfur metabolizers. Some hyperthermophiles are also found in the methanogen–halophile branch.

In terrestrial habitats, hyperthermophilic archaebacteria have been found in solfataric fields and hot springs. The marine biotopes for these organisms are shallow-water hydrothermal fields as well as deep, hot sediments, vents ("black smokers"), and volcanoes (seamounts). The nutritional modes for hyperthermophiles can vary from heterotrophy to obligate or facultative autotrophy. With the exceptions of *Sulfolobus, Metallosphaera, Acidianus,* and *Thermoplasma,* all hyperthermophilic archaebacteria inhabit anaerobic environments. It should be noted that *Thermoplasma,* a thermoacidophilic archaebacterium found in smoldering coal refuse piles and acid hot springs, is not strictly a hyperthermophile because it will only grow up to about 70°C. Because of its close relationship to hyperthermophiles, however, it is often listed with this group. [*See* ARCHAEBACTERIA (ARCHAEA).]

Based on its 16S rRNA sequence and unique type of RNA polymerase, the genus *Archaeoglobus* appears to form a separate branch intermediate between the methanogens and thermophilic sulfur metabolizers. Another hyperthermophile of particular interest is *Pyrodictium occultum* (Fig. 1), an anaerobic sulfur reducer, which grows up to 110°C and has a T_{opt} of 105°C. The disc-shaped cells of *Pyrodictium* are connected by a network of very thin, hollow fibers. This amazing organism has acclimated so well to its hot environment that it is not capable of growth below 82°C! Although *Pyrodictium* is among the most hyperthermophilic archaebacteria known to date, studies on the stability of amino acids and

nucleotides suggest that life could still be supported at temperatures as high as 150°C. Since the availability of liquid water appears to be an absolute requirement for the existence of life, we may still expect to find a new class of "superthermophiles" in hydrothermal environments where water is under considerable pressure. Figure 2 shows an electron micrograph of a novel hyperthermophile recently isolated from the Macdonald Seamount area. This organism bears no relationship (based on hybridization studies) to any of the known archaebacterial hyperthermophiles.

II. Intrinsic Stability of Macromolecules

A. Proteins

With a few exceptions (discussed in Section III), proteins isolated from thermophilic microorganisms are intrinsically more thermostable than their counterparts from mesophilic sources. Most of the investigations on stability have been carried out with enzymes, but the structural protein flagellin and the electron transport protein ferredoxin have also

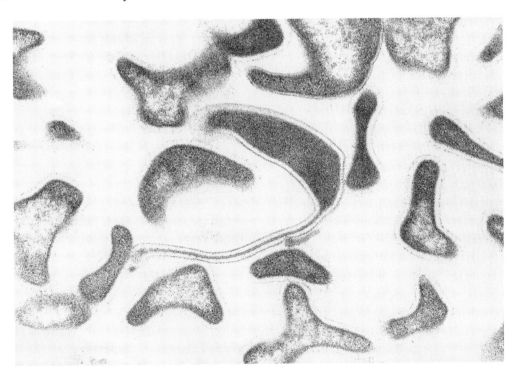

Figure 1 Electron micrograph of thin section of *Pyrodictium occultum*. Magnification: 43,500×. [Reprinted by permission from Stetter, K. O. *Nature* Vol. **300**, 258–259. Copyright © 1982 Macmillan Magazines Limited.]

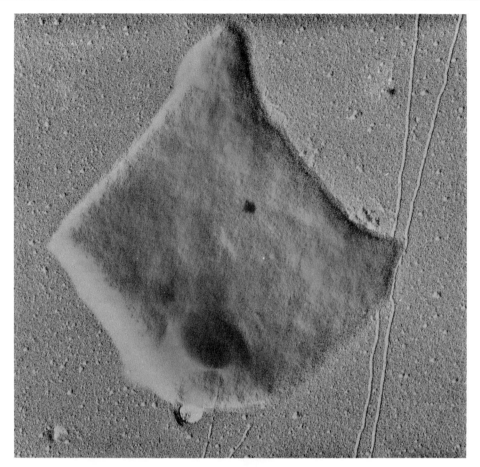

Figure 2   Electron micrograph of a platinum-shadowed novel hyperthermophile from slick sample MD 7. Magnification: 35,100×. [Reprinted by permission from Huber, R., Stoffers, P., Cheminee, J. L., Richnow, H. H., and Stetter, K. O. *Nature* Vol. **345**, 179–182. Copyright © 1990 Macmillan Magazines Limited.]

been studied. The thermophile enzymes that have been most extensively characterized are thermolysin, amylase, glyceraldehyde-3-phosphate dehydrogenase (GPDH), and lactate dehydrogenase (LDH). In the case of thermolysin and amylase, calcium ions have been found to play a major role in stabilizing this protease (see Section III).

Many investigators over a 25-yr period have attempted to understand the molecular mechanisms underlying intrinsic stability. Comparisons of amino acid sequences for homologous proteins isolated from thermophiles and distantly related mesophiles are difficult to interpret because the differences observed could result from evolutionary changes that are not related to the property of thermostability. This problem is minimized if one studies thermolabile and thermostable forms of the same enzyme synthesized by closely related species. The most direct approach, however, is to compare the amino acid sequence of a protein isolated from a temperature-sensitive (ts) or a temperature-resistant (T/r) mutant with that of the wild-type parent. Such studies have been carried out for the mutant mesophile enzymes phage T_4 lysozyme, tryptophan synthetase α-subunit, and kannamycin nucleotidyltransferase (KNTase). It can be argued that the mechanisms responsible for the stability of these enzymes, which are active across a temperature range of 30°–71°C, may not be the same as those for enzymes that are active at 75°–100°C. Nevertheless, the results of these studies show that single amino acid substitutions can greatly affect enzyme stability. Thus, replacement of Arg-96 in the wild-type phage T_4 lysozyme by His in a ts mutant results in a relatively thermolabile enzyme, although the three-dimensional structures of the two proteins are virtually identical. The presence of an imidazole at this position presumably destabilizes the hydrophobic core of the C-terminal lobe. Studies of mutants with single amino acid substitutions in the trytophan synthetase α-subunit indicate that these globular proteins are stabilized by hydrophobic residues buried in the interior of the molecule. Substitution of tyrosine for aspartate at position 80 in KNTase results in an appreciable increment in thermostability. An even more heat-stable KNTase contains the additional change of threonine to lysine at position 130. Both of these mutations occur in regions predicted to form α-helices. Since Lys-130 is on the protein surface, it can form an additional salt bridge with any neighboring negatively charged amino acid.

When compared to lobster muscle GPDH, three additional salt bridges are formed between subunit interfaces in the thermostable GPDH isolated from *B. stearothermophilus*. As discussed previously, this choice of organisms is far from ideal due to the extended evolutionary distance separating them. In searching for a more desirable system, it was found that the sequence of thermostable GPDH from *B. stearothermophilus* exhibits 87% identity to that of thermolabile GPDH from *B. coagulans*. A variable region that could contribute to the marked difference in thermostability between the two enzymes is the cyanogen bromide active-site peptide (residues 138–140a). A study of ferredoxins isolated from mesophilic and thermophilic clostridia suggests that stabilization of the thermophile enzymes can be achieved by a few extra salt bridges between external polar groups and by additional hydrogen bonds. Amino acid sequences of LDHs reveal a tendency for polar residues in the enzymes from psychrophilic and mesophilic bacilli to be exchanged for hydrophobic and charged residues in their thermophilic counterparts.

Separate studies based on the compilation of various parameters for selected proteins indicate a direct correlation between thermostability and (1) the hydrophobicity index coupled with the Arg/(Lys + Arg) ratio, (2) protein inflexibility, (3) the aliphatic index (relative volume of a protein occupied by aliphatic side-chains), and (4) solvent-accessible surface area. These proposals are of limited value due to the wide variations in the protein populations examined by each group.

A survey has been made of 70 amino acid sequences for mesophile and thermophile molecules representing six different protein families. The results show that the favored amino acid replacements (mesophile to thermophile) are Lys to Arg, Ser to Ala, Gly to Ala, Ser to Thr, Ile to Val, Lys to Ala,

Thr to Ala, Lys to Glu, Glu to Arg, and Asp to Arg. Further analysis indicates that the favored substitutions vary from protein to protein. Ser to Ala and Lys to Arg are frequent in LDH, whereas Gly to Ala and Ser to Ala are preferred by GPDH. Based on favored substitutions and secondary structural content of known tertiary structures, the overall conclusion is that decreased flexibility and increased hydrophobicity in α-helical regions are the main stabilizing influences in thermophile proteins. Only a few of these favored substitutions, however, are found in a newly presented amino acid sequence for GPDH from *T. maritima*. Because very little sequence data is currently available for hyperthermophile proteins, it remains to be seen if the mechanisms employed for achieving stability within this group are novel.

The results from a wide variety of studies show that the molecular basis for protein stability involves subtle rather than obvious effects. Furthermore, it appears that a universal mechanism cannot be formulated to explain all cases of stability. Rather, each protein will have to be analyzed on an individual basis. The availability of cloned mesophile and thermophile genes and of techniques for carrying out site-directed mutagenesis and for construction of chimeric enzymes will provide a productive approach to solve this intriguing problem. Indeed, oligonucleotide-directed mutagenesis has already been employed to replace one or both α-helical glycines in the N-terminal domain of λ-repressor with alanines. Substituting one of the best helix-forming residue (Ala) for a poor helix-forming residue (Gly) results in an increased melting temperature for the N-terminal domain of 3–6°C.

B. Ribosomes

A direct correlation exists between the T_{max} for eubacterial species and the thermostability of their ribosomes. Evidence for the heat stability of ribosomes isolated from the obligate thermophile *B. stearothermophilus* is based on two types of determinations. In the first case, thermal denaturation is monitored by measuring the hyperchromicity at 260 nm with increasing temperature. Such absorbance–temperature profiles of ribosomes and ribosomal subunits show that the thermophile 50S (melting point [T_m] = 80.2°C), 70S (T_m = 79.6°C), and 30S (T_m = 77.0°C) particles were all more heat stable than the *Escherichia coli* 50S (T_m = 72.5°C) and 30S (T_m = 66.0°C) particles. In addition, melting curves of 70S ribosomes from both organisms are biphasic, suggesting that dissociation into subunits may precede the final leg of the hyperchromic shift. A functional assay, measuring the residual capacity for polyuridylic acid-directed [^{14}C]phenylalanine incorporation at 37°C of ribosomal suspensions heated at 65°C for varying times, confirms the result of the physical measurements. Thus, after 5 min of preincubation at 65°C, 94% of the initial protein synthetic activity is retained by thermophile ribosomes, whereas this value is only 21% for *E. coli* ribosomes. When thermophile ribosomes are dissociated by lowering the magnesium concentration and subsequently reassociated at an elevated magnesium concentration, the 70S particles thus formed no longer display marked thermostability. The molecular basis for this interesting effect remains unknown, but likely candidates include conformational changes and/or the loss of some weakly bound component that is essential for maintaining stability. [*See* RIBO-SOMES.]

70S ribosomes isolated from the facultative thermophile *B. coagulans,* grown at 37° and 55°C, display the same thermostability (T_m = 74°C compared with 72°C for *E. coli* ribosomes under the same ionic conditions) and identical melting curves (T_m = 79°C) are also observed for ribosomes prepared from the obligate thermophile *B. stearothermophilus,* grown at 40° and 65°C. Whereas the heat stability of other cellular components can vary with growth temperature, thermophiles synthesize relatively thermostable ribosomes at both extremes of the growth temperature range. Thus, a major component of the protein-synthesizing machinery would have no problem functioning even if cells were faced with a sharp upward shift in temperature.

Pancreatic ribonuclease, which has a marked preference for single-stranded RNA, releases more acid-soluble nucleotides from *E. coli* ribosomes than from *B. stearothermophilus* ribosomes at each temperature studied. These results could not be accounted for by differences in the availability of RNA on the surface of the ribosome since titration with acridine orange reveals that approximately 90% of the RNA phosphates in both thermophile and *E. coli* ribosomes are available. Further evidence for the greater secondary structure of thermophile rRNA *in situ* comes from circular dichroism spectra of thermophile and *E. coli* ribosomes. The spectra are similar in shape, with a large, positive dichroic band centered at 265 nm and a small, negative trough at 300 nm. Thermophile ribosomes, however, display a small but significant increase of ellipticity at 265 nm when compared to *E. coli* ribosomes. The more highly ordered conformation of rRNA in *B. stearothermophilus* ribosomes undoubtedly contributes to the heat stability of thermophile ribosomes.

Archaebacterial ribosomes exhibit a mosaic of both eubacterial and eukaryotic characteristics. Although they contain 16S and 23S RNA species, archaebacterial ribosomes are resistant to many inhibitors that bind to eubacterial 70S ribosomes and are sensitive to a few inhibitors of protein synthesis in eukaryotes. Ribosomes isolated from the thermoacidophilic archaebacterium *Sulfolobus* contain a 30S subunit that is richer in protein (52% by weight) than the corresponding eubacterial particle (31% protein by weight). The heavier mass is due to both a larger number and a greater average molecular weight of the *Sulfolobus* 30S ribosomal proteins. In addition, much higher polycation concentrations are required for the association of *Sulfolobus* subunits to form 70S ribosomes than is the case for eubacterial subunits. When preincubated at 65°C for 40 min in a standard buffer, *Sulfolobus* ribosomes retain only 68% of the initial protein synthetic capacity, whereas this value is 97% if the buffer is supplemented with 3 mM spermine. These results explain in part the obligatory requirement of cell-free protein-synthesizing systems from *Sulfolobus* for polymines or polycations (see Section III).

C. Nucleic Acids

Although the stabilizing effect of a high G+C (guanine plus cytosine) content in DNA can be evoked in the case of *Thermus* (67–70% G+C) and *Pyrodictium* (61% G+C), many thermophiles (including hyperthermophiles) do not contain unusual amounts of guanine and cytosine. For example, *Sulfolobus* (T_{max} = 85°C) has DNA with 37% G+C and *Pyrococcus* (Fig. 3) (T_{max} = 103°C) has DNA with 38% G+C. Such organisms must contain DNA-stabilizing factors to maintain the integrity of their DNA at high growth temperatures (see Section III).

While determining the major and minor base composition of the DNA from 17 thermophiles, it was noted that only two species (both moderate thermophiles) contain the minor base, 5-methylcytosine (m^{5}C). In contrast, about 60% of mesophilic bacteria have m^{5}C in their DNA. Because m^{5}C residues in single-stranded DNA are deaminated at 95°C at three times the rate of C residues, many thermophiles apparently avoid having m^{5}C in their DNA so

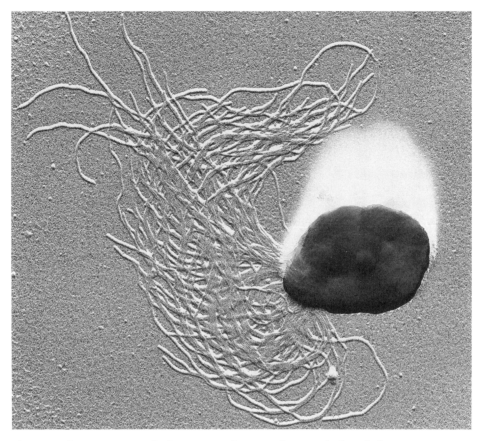

Figure 3 Electron micrograph of *Pyrococcus furiosus*, platinum-shadowed, showing a monopolar polytrichous flagellation. Magnification: 36,000×. [From Fiala, G., and Stetter, K. O. (1986). *Arch. Microbiol.* **145**, 56–61.]

as to escape from the mutagenic effects of $m^5C \rightarrow T$ transitions. It will be interesting to see if those thermophiles that do not contain m^5C in their DNA have either a high spontaneous mutation rate or a mismatch repair system, which is highly efficient for excising T residues at T·G mismatches. Several of the thermophiles were found to contain a novel base, 4-methylcytosine (m^4C), in their DNA. It is important to note that these m^4C residues are not susceptible to heat-induced deamination to T residues as are m^5C residues.

A novel topoisomerase, reverse gyrase, which produces adenosine triphosphate-dependent positive supercoiling of DNA, was originally isolated from *Sulfolobus*. Further screening showed that this unique enzyme is widely distributed in hyperthermophilic sulfidogens and methanogens. No reverse gyrase activity could be detected in mesophiles and moderate thermophiles. A close antigenic relationship exists among the reverse gyrase of *Sulfolobus* and those of other members of the Sulfolobales, but not with those of other hyperthermophilic organisms. Although positive supercoiling of DNA may eventually prove to be essential for life at extremely high temperatures, the rationale for this phenomenon is not clear at this time.

A correlation between the growth temperature of *T. thermophilus* and the T_m of transfer RNA (tRNA) has been observed. As the growth temperature is raised from 50° to 80°C, the T_m of extracted bulk tRNA increases progressively from 83.3° to 86.4°C. A linear relationship is found between the T_m and the mole fraction of 2-thioribothymidine (s^2T) replacing T at position 54 in the TψC loop. Further studies show that two purified major $tRNA^{Ile}$ species have the same nucleotide sequence, with the exception that $tRNA^{Ile}_{1a}$ contains s^2T at position 54, whereas $tRNA^{Ile}_{1b}$ contains T at the same site. The T_m of $tRNA^{Ile}_{1a}$ is 86.2°C, whereas that of $tRNA^{Ile}_{1b}$ is 83.3°C. Thus, the single replacement of an oxygen atom of T by a sulfur atom significantly contributes to the thermostability of this tRNA species. The conformational rigidity of s^2T is apparently due to the interaction between the bulky 2-thiocarbonyl group of the base and the 2'-OH of the ribose ring. Both of the thermophile tRNA[Ile] species contain 2'-O-methylguanosine at position 18 and 1-methyladenosine at position 58. These methylated bases, which are not found in the corresponding tRNA species from *E. coli,* may also play a role in stabilizing the thermophile tRNA.

The anticodon regions of six major tRNA species purified from *T. thermophilus* have been sequenced. The first letters of all anticodons are unmodified G or C. In the anticodons of three major tRNA species, the first letter (G or C) was found to be the same as the second letter (G or C). These anticodon structural studies can be correlated with the codons used most frequently in the structural gene for 3-isopropylmalate dehydrogenase. For this gene, >90% of amino acid codons contain either G or C in the third position. This is also true for the LDH gene from *Thermus caldophilus.* Thus, G·C base pairs containing three hydrogen bonds are formed between the first base in the anticodon and the third base in the codon. Such stable codon–anticodon interactions are probably utilized in this thermophile to ensure the accurate biosynthesis of proteins at elevated temperatures. When other cloned and sequenced thermophile genes from organisms that do not have high G+C contents in their DNA are examined, however, this type of nonrandom codon usage is not found.

D. Lipids

To survive at elevated temperatures, thermophiles must maintain the integrity of their plasma membranes. As we shall see, a variety of strategies have evolved to deal with this problem. Two basic explanations have been formulated to account for the relationships between growth temperature and lipid composition. The homeoviscous adaptation theory of Sinensky states that temperature-dependent changes in lipid composition are a means of maintaining constant lipid fluidity, thereby optimizing membrane function. Since there are several examples of bacterial lipids that do not maintain the exact level of fluidity throughout the growth temperature range, McElhaney advanced the alternative homeophasic theory. He argues that temperature-dependent changes in lipid composition are designed primarily to prevent the forma-

tion of nonbilayer phases, which would severely disrupt membrane organization and function.

Eubacteria contain predominantly C_{14}–C_{18} saturated and monounsaturated fatty acids that are joined to glycerol by ester linkage. Gram-positive eubacteria frequently have methyl-branched fatty acids as major components in which the methyl group is on the second (iso-branched) or third (anteiso-branched) carbon atom from the methyl terminus of the alkyl chain. The thermoacidophilic eubacterium *Bacillus acidocaldarius* contains iso- and anteiso-branched fatty acids, but the major components (50–90% of the total) are alicyclic ω-cyclohexyl fatty acids that possess a cyclohexyl ring on the terminal end of the hydrocarbon chain. The role, if any, of ω-cyclohexyl fatty acids in membrane stabilization or in providing an effective barrier to the acidic environment remains unclear.

The unique lipid compositions of several organisms place them in an intermediate position between thermophilic eubacteria and archaebacteria. In the extremely thermophilic gram-negative organism *Thermomicrobium roseum,* glycerolipids are replaced by a series of straight-chain or internally methyl-branched 1,2-diols of carbon numbers C_{18}–C_{23}. Fatty acids are ester-linked to the diols or amide-linked to amino sugars of polar head groups. *Thermomicrobium roseum* shares the archaebacterial characteristic of lacking a peptidoglycan cell wall, but it does not contain the typical glycerol ethers of archaebacteria (see later). The gram-negative eubacterial extreme thermophile *Thermodesulfotobacterium commune* has glycerol diethers that are similar to those of archaebacteria, but they contain iso- and anteiso-methyl-branched alkyl chains such as those found in many gram-positive eubacteria. Another unusual lipid adaptation is found in the thermophilic eubacterial clostridia, *Clostridium thermohydrosulfuricum* and *Clostridium thermosulfurogenes.* C_{30} dicarboxylic acids are formed by head-to-head condensation of iso-branched C_{15} fatty acids, possibly suggesting the presence of diglycerol tetraesters (analogous to archaebacterial diglycerol tetraethers). It should also be noted that C_{30}–C_{34} dicarboxylic fatty acids occur in *Thermotoga* and *Fervidobacterium* species.

In thermophilic eubacteria, increasing growth temperatures generally result in (1) increased fatty acid average chain length, (2) decreased content of unsaturated fatty acids, (3) increased proportion of straight- to branched-chain fatty acids, and (4) increased iso- to anteiso-branched chain fatty acids. All of these changes are known to increase the liquid-crystalline–gel phase transition temperature.

A general increase in glycolipid content is observed as one progresses from psychrophiles to moderate and extreme thermophiles. Growth temperature-dependent changes in sugar-containing lipids must play a role in stabilizing membranes at elevated temperatures. One possible mechanism is enhanced hydrogen bonding between the lipid bilayer surface and cell-wall components directly in contact with it.

Archaebacteria lack fatty acyl chains in their lipids. Instead, they contain phytanyl alkyl chains derived from C_{20} isopranoid alcohols, which have repeating methyl groups (four per phytanyl chain) and are joined to glycerol by ether linkage (Fig. 4A). The phytanyl chains can condense head-to-head to form a biphytanyl chain, which can be linked to two glycerol molecules yielding a tetraether structure (Fig. 4B). Because they apparently span the membrane, tetraethers would be expected to provide a stabilizing influence. Indeed, archaebacterial thermophiles with higher potential growth temperatures tend to have greater tetraether to diether lipid ratios than do thermophilic archaebacteria with lower T_{opt}'s. It is interesting to note that ether bonds are more resistant to chemical or enzymatic attack than are ester linkages, and this may have been a factor in the adaptation of archaebacteria to a variety of extreme environments.

The methyl groups of archaebacterial phytanyl chains can participate in an internal cyclization reaction to form cyclopentane rings. In aerobic archaebacterial thermophiles, a direct correlation has been noted between the number of cyclopentane rings per phytanyl chain and the T_{opt} of the organism.

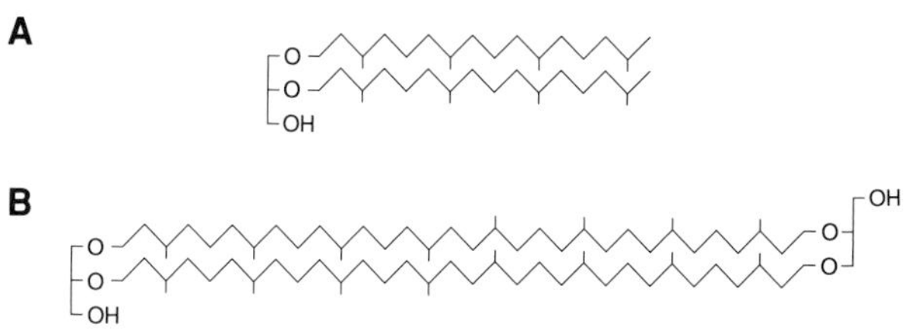

A

B

Figure 4 Structures of a phytanyl glycerol diether (A) and a dibiphytanyl diglycerol tetraether (B). [From Langworthy, T. A., and Pond, J. L. (1986). *In* "Thermophiles: General Molecular and Applied Microbiology" (T. D. Brock, ed.), pp. 107–135. John Wiley & Sons, New York.]

III. Stabilizing Factors

Although most enzymes isolated from thermophiles are intrinsically heat-stable, several important exceptions to this rule are known. Such proteins must therefore be protected *in vivo* from heat inactivation by their interaction with stabilizing factors. For example, calcium ions have been shown to play a major role in stabilizing the thermophile enzymes thermolysin and α-amylase as well as the cell envelope proteins of *T. thermophilus*.

Selected glycolytic enzymes (including GPDH) from the facultative thermophile *B. coagulans* have been shown to be as labile as their counterparts from mesophilic organisms regardless of whether the bacilli are grown at 37° or 55°C. Although increasing the concentration of inorganic ions *in vitro* provided thermoprotection to GDPH from *B. coagulans,* this mechanism could probably not operate *in vivo* because the inorganic salt concentration in *B. coagulans* is very similar to that in *E. coli*. Based on this and other evidence, it was concluded that a highly charged macromolecular environment induces stability in these proteins that would otherwise be heat-inactivated.

Analyses of enzymes isolated from the hyperthermophilic archaebacterium *Methanothermus fervidus* reveal relatively low thermostabilities. Elevating the growth temperature of *M. fervidus* results in increasing concentrations of potassium and the novel trianionic compound, cyclic 2,3-diphosphoglycerate (DPG). The potassium salt of DPG was subsequently found to be a potent stabilizer of both GPDH and malate dehydrogenase from *M. fervidus*. *Methanopyrus*, a newly discovered hyperthermophilic methanogen, grows at 110°C. This unique organism was found to contain unusually high concentrations (1.1 M) of DPG, thus lending support to the contention that this compound serves as a major stabilizer in hyperthermophilic methanogens. It will be interesting to see if the distribution of DPG is limited to methanogens or whether it will also be found in other thermophile groups.

Cell-free protein-synthesizing systems derived from the eubacterial extreme thermophile *T. thermophilus* and from the hyperthermophilic archaebacterium *Sulfolobus* are found to depend on supplementation with polyamines for activity at elevated temperatures. In the case of *T. thermophilus* extracts, tetraamines are most effective in stimulating protein synthesis at 60°C, whereas pentaamines are most effective at 70°C. These *in vitro* effects could be correlated with growth temperature-dependent changes in polyamine content. At higher growth temperatures, the cellular contents of pentaamines and hexaamines increase. Levels of a branched polyamine containing a quaternary nitrogen atom also increase when the thermophile is grown at a high temperature. Further studies with the *T. thermophilus* cell-free system show that spermine is necessary for the formation of a ternary complex (ribosome–messenger RNA–aminoacyl tRNA) and also for enhancing polypeptide chain elongation.

The polyamine composition of *Sulfolobus acidocaldarius* is also influenced by the growth temperature. Cells grown at 78°C have a higher tetraamine content than do cells grown at 70°C. In *Sulfolobus* extracts, spermine was found to protect ribosomes from thermal inactivation and also to promote association of subunits to form 70S particles. Thus, polyamines apparently exert their stimulatory effects at several different steps in the complex series of reactions comprising protein synthesis.

Polyamines can promote aggregation of DNA, increase its T_m, and induce the B to Z transition. One group reports that because pentaamines are more effective than spermine (a tetraamine) only in promoting DNA condensation at high temperatures, this may represent the true biological role. Other studies show that polyamines stabilize tRNAs and DNAs from both mesophilic organisms and *T. thermophilus*. Furthermore, the T_m's for tRNA and DNA are elevated dramatically in the presence of longer polyamines such as caldopentamine and caldohexamine.

Because the G+C content of many thermophile DNAs is not unusually high, stabilizing molecules must be present to prevent strand separation at elevated temperatures. In addition to polyamines discussed previously, DNA-binding proteins, which have been isolated from several thermophilic species, are probably also involved in this role. *Thermoplasma acidophilum*, a thermoacidophilic archaebacterium, contains a histonelike protein termed HTa. This protein, which binds to DNA and stabilizes it against thermal denaturation, displays some sequence homology to eukaryotic histones and some to the *E. coli* DNA-binding proteins NS1 (HU-1) and NS2 (HU-2).

DNA-binding proteins extracted from the thermoacidophilic arachaebacteria *Sulfolobus solfataricus* and *S. acidocaldarius* can be grouped into three molecular mass classes of 7, 8, and 10 kDa. The predominant 7-kDa protein from *S. solfataricus* (called Sso7d) has been purified to homogeneity. Its molecular weight is 7149 daltons and it contains 14 lysine residues out of 63 amino acids. Monomethylated lysine residues are located exclusively in the N-terminal and C-terminal regions of the molecule. Sso7d is strongly homologous to the 7-kDa proteins of *S. acidocaldarius*, with the highest homology to protein 7d.

Protein MCl from a thermophilic *Methanosarcina* is characterized by a high content of charged residues (27% basic residues and 15% acidic residues). This DNA-binding protein shows no homology to archaebacterial proteins HTa and 7d isolated from *T. thermophilus* and *S. acidocaldarius*, respectively.

DNA-binding protein II (BSb) from *B. stearothermophilus* has a monomeric relative molecular mass of 9716 and binds to 30S, 50S, and 70S ribosomal particles as well as to both single- and double-stranded oligodeoxyribonucleotides. BSb shows 58–59% homology to *E. coli* proteins NS1 and NS2 and a lower homology (32%) to *T. acidophilum* protein HTa. The secondary structure (45% α-helix) of BSb melts at about 68°C and the urea denaturation inflection point occurs at 3.5 M, whereas the comparable values for *E. coli* proteins NS1 and NS2 are 55°C and 2.5 M urea, respectively.

IV. Genetics of Thermophily

Thermophilic mutants of *Bacillus megaterium* growing at 55°C can be isolated from mesophilic strains at a frequency of 10^{-8}–10^{-7}, whereas mutants growing at 70°C are found with a frequency of 2×10^{-9}. It was thought that plasmids carry the thermophily determinants since treatment of the thermophilic variants with acridine orange cures them of the ability to grow at elevated temperatures. Based on electroporations into acceptor mesophilic strains and curing experiments, the same author recently reports that large plasmids found in obligately thermophilic strains of *B. stearothermophilus* code for temperature resistance. Some of these plasmids also carry markers for bacteriocin production. Two groups claim to have transformed *B. subtilis* to grow at 65–70°C successfully with DNA from *B. caldolyticus* or *B. stearothermophilus*, but this has not proven to be an easily reproducible system.

Using two strains of *B. subtilis* with different T_{max}'s (51° and 57°C), transformation of the ability to grow at 55°C was found to map between the *str* and *ery* genes at map position 10. Other workers isolated spontaneously occurring T/r mutants from mesophilic strains of *B. subtilis* and *Bacillus pumilus*. Retention of auxotrophic and antibiotic resistance markers in the thermophilic mutants attested to their mesophilic origin. Thermophilic mutant strains are recovered at a frequency of $<10^{-10}$ and are capable of growth at temperatures between 50° and 70°C. Transformation of genetic markers and the thermophilic trait between T/r mutants and mesophilic parents is achieved at frequencies of 10^{-3}–10^{-2} for single markers and about 10^{-7} for two unlinked markers and for the thermophilic trait. These findings suggest that the ability to grow at elevated temperatures is a phenotypic consequence of mutations in two unlinked genes. Further testing shows that all of the *B. subtilis* T/r mutants are resistant to nalidixic acid. When Nal^r mutants are isolated from the parental strain, however, their T_{max} is 56°C. Thus, resistance to nalidixic acid is sufficient to elevate the T_{max} from 52° to 56°C, but another and as-yet unidentified mutation is required for growth up to 70°C. When cell-free extracts prepared from the parental and T/r strains are examined, thermostabilities of the T/r protein pool and ribosomes are much greater than those in the mesophilic parent. In addition, spermidine stimulates protein synthesis at 55° and 65°C in the T/r extract, whereas this polyamine inhibits protein synthesis at each temperature tested in the extract from the mesophilic parent.

Thermotolerant nalidixic acid-resistant mutants of an *E. coli* strain carrying a tetracycline resistance transposon have been obtained by mutagenesis. Cotransductional analyses employing phage P1 indicate that the mutation resulting in the phenotype of growth at 48°C is an allele of the *gyrA* structural gene. Similar thermal inactivation kinetics are observed for ribosomes isolated from a thermotolerant (or T/r) mutant grown at both 37° and 48°C and from the parental strain grown at 37°C. Cell-free extracts prepared from the T/r mutant grown at 48°C exhibit a sharp increase in protein synthesis at 55°C, whereas this effect is not displayed by extracts from the T/r mutant or parental strains grown at 37°C. Furthermore, preincubation at 55°C enhances protein synthesis at 37°C up to 15-fold in an extract prepared from the T/r mutant grown at 48°C, whereas comparable values are 2.6- to 3.0-fold for extracts from the mutant and parental strains grown at 37°C. The mutated gyrase in certain Nal^r T/r mutants apparently affects DNA supercoiling in a manner that alters the pattern of gene expression. Further work is required to identify the specific gene products that enable cellular adaptation to elevated temperatures.

Although an understanding of the genetic basis for thermophily has been hampered in the past by the lack of an efficient genetic exchange system in thermophilic bacteria, important advances in this direction have been made during the past few years. Protoplast fusion has been successfully employed to construct isogenic mutant strains of *B. stearothermophilus* and for chromosomal mapping in this thermophile. In addition, generalized transduction by temperate and virulent bacteriophages has also been carried out in *B. stearothermophilus*, thus providing a second system for determining linkage groups and for ordering closely linked markers. Finally, an efficient protoplast transformation system has been developed for *B. stearothermophilus* using plasmids isolated from mesophilic and thermophilic bacteria.

The question of how many genes are involved in the conversion of mesophiles to thermophiles (or vice versa since both the ancestral eubacterium and archaebacterium are thought to have been thermophiles) is important to our understanding of bacterial evolution. Some evidence indicates that only two unlinked genes may be required, although mutations in one of these genes (*gyrA*) can cause a pleiotropic effect on gene expression. Nevertheless, it is difficult to reconcile these results with the fact that most proteins isolated from thermophiles are intrinsically heat-stable and therefore must have unique primary structures. On the other hand, we know that differences between thermolabile and thermostable proteins can often be very subtle. Temperature-tolerant extreme thermophiles that grow within an extended temperature span of 40°C or more (e.g., *B. stearothermophilus, Methanobacterium thermoautotrophicum, Thermoanaerobacter ethanolicus, C. thermohydrosulfuricum*) exhibit biphasic Arrhenius plots (growth temperature vs. doubling time). This data has been interpreted to indicate that these organisms possess two sets of key enzymes and that their synthesis is regulated by temperature. Indeed, two different ferredoxins, which vary in amino acid composition and in the temperature stability of the protein–(Fe-S)-cluster association, have been isolated from *C. thermohydrosulfuricum*. Temperature-regulated gene expression is an important concept that merits much more attention than it has been afforded in the past.

V. Biotechnological Applications of Thermophiles

Fermentations and other commercial processes utilizing thermophilic microorganisms offer several advantages over conventional systems. Although great care must be exercised to avoid contamination and overheating in most microbial fermentations, such precautions are hardly necessary in the case of a thermophilic fermentation. Elevated temperatures result in increased diffusion rates and solubilities of nongaseous substrates, thereby enhancing the efficiency of the fermentation. Enzymes produced from thermophiles not only catalyze reactions rapidly at elevated temperatures, but they are also much more stable than their counterparts from mesophiles. The latter property permits the storage of these products for relatively long time periods and also enables the researcher to manipulate the protein under experimental conditions that might easily inactivate a more labile molecule.

Thermostable enzymes are currently employed (1) as additives to detergents, (2) to produce natural sweeteners, and (3) in the manufacture of pharmaceuticals. Some examples are the xylose isomerase of *B. coagulans,* which is utilized in the production of high-fructose corn sugar at 60°C, the α-amylase from *Bacillus licheniformis,* which liquefies starch at temperatures of 95°–110°C, and the thermostable alkalophilic protease from *B. licheniformis,* which is most commonly used in detergent applications.

Thermophilic microorganisms could be useful in anaerobic waste-treatment systems leading to the production of methane, an energy-rich fuel. These processes require a combination of microbes with appropriate individual enzymatic capabilities. Thus, complex organic matter is first converted to soluble compounds (oligomers and monomers) by fermentative bacteria through the action of proteases, amylases, celluloses, etc. These monomers and oligomers are then fermented to volatile organic acids, H_2, and CO_2 by the hydrogen-producing acetogenic bacteria. Finally, CO_2-consuming methanogens carry out the reaction $4H_2 + HCO_3^- + H^+ \rightarrow CH_4 + 3H_2O$, whereas aceticlastic methanogens carry out the reaction $CH_3COO^- + H_2O \rightarrow CH_4 + HCO_3^-$. In laboratory scale studies that have been designed solely to produce methane as an alternate energy source, plant biomass (giant kelp and water hyacinths) and cattle manure have proven to be excellent sources of organic material for this anaerobic fermentation.

All of the ethanol produced microbially in the United States is from the fermentation of sugar by yeast. Alcohol is used primarily as a fuel additive ("gasohol" is the resultant mixture) for automobiles and in the production of ethylene. Ethylene, in turn, is employed in the production of both polymers (polyethylene) and monomers (ethylene glycol, vinyl chloride, and styrene). Several thermophilic bacteria have been investigated as candidates for a high-temperature ethanol fermentation. *Clostridium thermocellum,* an anaerobic thermophile, is capable of directly fermenting cellulose to yield ethanol. Thus, cellulosic biomass could in principle be used as the substrate. In practice, however, crude biomass is fermented more poorly than purified polysaccharides. The elevated incubation temperature of the fermentation facilitates recovery of the volatile (b.p. = 78°C) ethanol–water azeotrope. The use of *C. thermocellum* results in a mixed fermentation, with organic acids (acetate, lactate) as the major coproducts. A major problem has been the sensitivity of this anaerobe to the ethanol and acetate that it produces. A mutant strain that tolerates 5% ethanol has now been isolated, whereas growth of the parental strain is inhibited by 3% ethanol. Eventually, this well-studied system may be refined to the point where it will have commercial application.

Other chemicals that could be produced on a large scale by using thermophilic bacteria include amino acids (*B. coagulans*), antibiotics (*Thermoactinomyces*), lactic acid (*Clostridium, Thermoanaerobium,* and *B. coagulans*), acetic acid (*Clostridium* and *Acetogenium kivui*), and carotenoids (*Thermus aquaticus*).

Bacterial leaching refers to the process whereby microorganisms are used to remove metals from ores and mineral wastes. This technology has been industrially applied to leach copper from mine waste and to extract uranium from ore. The advantages of bacterial leaching over conventional smelting techniques are lower operating costs and the elimination of air pollutants such as sulfur dioxide. Leaching systems develop overheated areas due to exothermic oxidation reactions involving sulfides. Such hot areas are inhibitory to the growth of *Thiobacillus ferooxidans,* the mesophile most commonly used in metal leaching. This problem could be solved by the introduction of thermophilic bacteria. An additional advantage would be the acceleration of reaction rates and the concomitant reduction in recovery time. [*See* ORE LEACHING BY MICROBES.]

Because leaching processes take place at low pH values to avoid precipitation of solubilized metal ions, thermophilic candidates must be acidophiles and also capable of growing in the presence of elevated metal cation concentrations. Facultatively thermophilic strains of the eubacterium *T. ferooxidans* and the hyperthermophilic archaebacterium *Sulfolobus* have been employed in laboratory leaching studies. *Thiobacillus ferooxidans* is effective in leaching nickel from violarite (Ni_2FeS_4) and in forming soluble iron from the oxidation of pyrite. *Sulfolobus* can be used to leach molybdenum from molybdenite (MoS_2) and to extract copper from chalcopyrite ($CuFeS_2$). Both of these thermophiles are capable of solubilizing pyritic and organic sulfur from coal. *In situ* metal leaching may in the future serve as an alternative to conventional open pit or underground mining procedures.

The application of thermophilic microorganisms to industrial and biotechnological processes is still in its infancy. The use of *T. aquaticus* (*Taq*) DNA polymerase in the polymerase chain reaction carried out at elevated temperatures and of heat-stable proteins in the manufacture of semiconductor chips are two exciting current developments.

Bibliography

Brock, T. D. (1985). *Science* **230,** 132–138.
Brock, T. D. (ed.) (1986). "Thermophiles: General, Molecular and Applied Microbiology." John Wiley & Sons, New York.
Stetter, K. O., Fiala, G., Huber, R., and Segerer, A. (1990). *FEMS Microbiol. Rev.* **75,** 117–124.
Wiegel, J. (1990). *FEMS Microbiol. Rev.* **75,** 155–170.
Woese, C. R. (1987). *Microbiol. Rev.* **51,** 221–271.

Timber and Forest Products

David J. Dickinson and John F. Levy
Imperial College, London

Glossary

Habitat Place where organisms grow, develop, and find their nutrients, in association with other organisms

Hardwood Timber cut from trees belonging to the botanical group of plants Angiospermae

Heartwood Inner part of the wood of the trunk or branch of a tree

Moisture content Amount of water in a piece of wood, expressed as a percentage of the dry weight of that piece of wood, (% m.c.). Equilibrium moisture content (EMC) is the % m.c. in equilibrium with the moisture in the immediate vicinity. Fiber saturation point is the moisture content of the wood at which the cell walls are fully saturated with water while the cell lumina are empty.

Sapwood Outer part of the trunk or branch of a tree between the bark and the heartwood

Softwood Timber cut from trees belonging to the botanical group of plants Gymnospermae

Substrate Growth medium on which the organisms feed

Wood cell wall Structural element in plants, composed of layers: the primary wall, inside of which is a three-layered secondary wall with an outer layer (S1), a middle layer (S2), and an inner layer (S3)

DECAY AND DEGRADATION OF WOOD is part of nature's recycling process. When a tree or branch of a tree falls to the ground in a forest, the materials manufactured by the tree from the basic elements in the environment to produce its constitutent parts are broken down by a range of organisms to return those elements to the biosphere. These recycling processes are vital to life as we know it. Not only do they attempt to restore the balance of the basic elements in nature required for plant growth, but they also prevent the Earth's surface from becoming so littered with plant and animal remains that life as we know it would no longer be possible. It has been estimated that, without such decay processes, such littering would take little more than 20 years to achieve.

When wood is cut from a tree to be used as a material, such as timber, lumber or some other forest product, it passes through a sequence of events. These include the felling of the tree, cutting it into logs in the forest, the removal of the logs from the forest to the saw mill, storage of the logs prior to sawing, storage of sawn timber prior to seasoning, seasoning, further processing to a finished product, and the use of the finished product in service under a great variety of situations, some wet and some dry. Through all these major events, the timber is expected to remain sound, durable, stable in dimension, and free from all defects. It is expected to survive unharmed in exposed situations, such as posts and poles in ground contact, although nature has evolved processes for its destruction. This inevitably means further processing to ensure an induced durability for a long service life.

I. Wood Structure

Wood itself is a complex, anisotropic, polymeric, cellulosic composite. Its basic structure will depend on the tree species from which it was cut. Woods such as cedar, larch, pine, and spruce belong to the group of plants known to the botanist as *gymnosperms*, to the forester as *conifers*, and to the timber trade as *softwoods*. These woods are formed from two cell types (tracheids and parenchyma cells) and have some degree of similar structure among species. Woods such as beech, eucalyptus, mahogany, oak, and teak belong to the group of plants known to the botanist as *angiosperms,* to the forester as broadleaves, and to the timber trade as *hardwoods.* These woods are formed from four cell types (vessels and fibers, as well as tracheids and parenchyma) and, as a result, show a wider variation in basic structure among species. Softwoods and hardwoods not only differ in the cell types from which they are formed, but also in their chemical composition. Both hemicellulose and lignin constitution will depend on which of the two plant groups the wood belongs in. When wood becomes a habitat for organisms, therefore, the structure of that habitat will depend on the tree species from which the wood was cut.

In the living tree, the wood of the stem or trunk performs three main functions: mechanical strength, water conduction, and storage of food reserves. It provides mechanical strength to support the mass of branches and leaves forming the crown of the tree and to keep open the pathways of the water conducting system for the movement of water from the ground through the trunk and branches to the leaves. Transverse movement of water and materials between the inner and outer layers is via the rays. Here food reserves manufactured in the leaves are stored for future use. As the tree grows older and inceases in girth, the stored food reserves are moved to the outer layers. At the same time, the living parenchyma cells die and, in the process, produce materials that often color the center of the stem and are sometimes of a toxic nature. This may give the central portion of the stem a greater durability when the wood is used as timber. This part of the stem with no living cells is known as the *heartwood,* whereas the outer layers containing the stored food reserves in the living cells are known as the *sapwood.* Sapwood of all species is permeable to liquids to some degree, whereas heartwood of the same species may be impermeable.

II. Wood Degradation

A. Types of Degradation

Many factors can bring about the degradation of wood: fire; mechanical wear and tear (e.g., floors, lock gates); chemical action (e.g., dye houses, laboratory benches); weathering (a combination of chemical action and colonization by fungi and bac-

teria, sometimes followed by wasps stripping off the surface layers for nest building); biological activity by other plants, fungi, and bacteria; and biological activity by animals (e.g., birds, mammals, insects, termites, shipworm, and gribble).

The damage by biological factors may be casual (e.g., beaver and deer damage trees in the forest, woodpeckers and other birds damage trees and standing poles); for protection for softbodied organisms (e.g., shipworm); transitory (e.g., mice in buildings); the result of a search for stored food reserves in the sapwood (insects, termites, fungi, and bacteria); because of an association between colonizing fungi and insects (e.g., ambrosia beetles and deathwatch beetles); or due to the use of the constituent materials of the wood cell walls as food (e.g., fungi and bacteria).

B. Biological Factors

Although all types of degradation in service are important from time to time, the greatest destruction is caused annually by biological agents. These agents differ in different parts of the world and at different stages in the process of converting the wood in the tree to timber in service. Termites are the major factor in the tropics; shipworm, gribble, and marine fungi in the sea; and insects, fungi, and bacteria are widespread across the world. Microbial ecology is largely a matter involving the fungi and bacteria, but some mention will be made of the insects. [*See* BIODEGRADATION; ECOLOGY, MICROBIAL.]

C. Abiotic Factors

For fungi and bacteria to colonize and decay wood, water is an essential factor. Water is present in wood as free water in the cell lumina and as bound water in the cell walls. When the cell walls are fully wet but there is no free water in the cell lumina, the wood is said to be at the fiber saturation point. This point will vary between species, but is defined as the point at which the moisture content (m.c.) of the timber is about 30% of its oven dry weight. There must be enough water present in the wood to permit the products of extracellular enzyme activity to flow from and to the organism. If too little is present, the enzyme systems that decay the wood cannot work, whereas too much water can produce an anaerobic environment in which most wood rotting fungi cannot survive. [*See* ENZYMES, EXTRACELLULAR.]

Although water is by far the most important factor, temperature, related to geographical location, can also be important, particularly for insects.

Nutrients are also important. These are often provided during the early stages of colonization by the food reserves that had been stored in the sapwood of the tree before it was cut down. Once the wood-rotting fungi are established and have started to decay the wood, the breakdown products so formed are the food source for the later colonizers that are unable to degrade the cell walls themselves.

The abiotic factors that govern the biological activity of the organisms are water, nutrients, oxygen, and a suitable temperature.

III. Habitat

A. Habitat and Substrate

It has been said that, to the succession of microorganisms that colonize it, wood simply consists of a series of conveniently oriented holes surrounded by food. The holes are a means of travel for the organisms, from cell to cell through the wood, and also form a pathway for the movement of liquids and gases. The holes consist of the cell lumina, which are interconnected from cell to cell by localized thinner regions of the intervening double cell wall known as pits; only thin membranes separate the two cell lumina. Within the complex cellular structure of wood the microbial activity that leads to its destruction takes place. The wood is thus both habitat and substrate.

The basic structure of the habitat depends on the wood species from which it was cut and whether it is sapwood or heartwood. The suitability of the habitat for the establishment of wood-rotting organisms will largely depend on the moisture content. Naturally, durable heartwood may be subject to colonization without active cell wall breakdown by the usual wood-rotting fungi but, given time, an ecological system will evolve to bring about the decay of this modified habitat. Any chemicals impregnated into the wood, (e.g., wood preservatives) simply alter the habitat. Such treated wood is still capable of being colonized by some fungi and bacteria, which may be the start of an ecological succession capable of destroying the wood in the very long term.

IV. From Forest to Timber in Service

A. Ecosystems

At each stage of the processes involved in converting wood in the standing tree to timber in service, fungi and bacteria may infect the habitat. A microbial ecological succession involving many types of organism can develop. The early colonizers are scavenging for readily available nutrients, such as the stored food reserves of the sapwood. Many will be unable to penetrate cell walls or pit membranes. Some early colonizing bacteria can destroy the pit membranes, thereby opening up the structure to allow other organisms to penetrate more deeply. At the same time the destruction of the pit membranes will allow water and other liquids to flow from one cell lumen to another. It also enables gaseous diffusion of water vapor and oxygen to take place, which allows the habitat to dry and become less anaerobic. Such ecosystems may be specific to a particular stage in the conversion process and are worth considering before discussing what happens to timber in service, since they may affect the quality of the endproduct.

1. Standing Tree

Wood in the living tree consists of the living sapwood and the nonliving heartwood. In the former, the true host/parasite relationship can exist, and invading microorganisms may be controlled by the response of the living cells. In the heartwood, however, because of the absence of living tissue, there is no host response. Infection occurs through exposed sites caused by damage to the bark and outer layers of the roots, stems, or branches. Such damage is often kept in check by the host response in the sapwood, but causes decay in the unresponsive heartwood. The most common effect is butt or heart rot, which can give rise to the "hollow" standing tree. Control is largely a matter for the forester, since decayed and infected areas of logs will be rejected at the saw mill.

2. Felled Logs

Once the tree is felled and cut into logs, the host/parasite response of the sapwood ceases. The wood beneath the bark is now exposed at the cross-cut ends of the logs and at the wounds where branches were cut off. Both sapwood and heartwood are now open to infection from airborne spores and from organisms in the ground. Climate, season, and temperature may have an effect here. In those climates with a cold winter and warm or hot summer, the log has to be extracted from the forest during the cold season when the metabolic activity of the potential invading organism is low and colonization is slow. If left in the forest through the summer, metabolic activity may be high and the logs of certain species can be rendered too heavily infected to be of any use as sawn timber. Extraction has to be quick and the logs must be stacked without ground contact at the saw mill. Since the logs are wet and the sapwood is full of nutrients, fungal and bacterial spores will germinate on the wet wood and penetrate it. As the log dries out, the type of organism that can invade will change and an ecological sequence will develop. Deterioration is likely to occur if the logs are left in a stack with their cut ends exposed and unprotected.

In the sapwood, readily available carbohydrates will be the main source of nutrients. The earliest colonizers are bacteria and the primary molds. The former will include those bacteria capable of destroying the pit membranes, which enables the scavenging molds to penetrate more deeply into the wood beyond the surface layer of cells. The molds are closely followed by a group of fungi with pigmented hyphae known as the sapstain or blue-stain fungi. As the sapwood begins to dry out, the true wood-rotting fungi develop. They are of three basic types: soft rot, white rot, and brown rot. Each brings about a characteristic degradation of the wood cell wall—the micromorphology of decay. The soft rot fungi will develop first, to be followed by the white rot fungi at a slightly lower wood moisture content, and, as further drying occurs, finally the brown rot fungi.

In the heartwood, a similar succession of organisms will colonize. The pit membranes, following the deposition of materials during the process of heartwood formation, are no longer as easily destroyed by bacteria. This can prevent the penetration of scavenging molds beyond the surface layers, in the absence of stored food materials, the molds do not survive long. This will cause the ecological sequence of events to differ between heartwood and sapwood.

In addition, the wood moisture content in the standing tree may also differ between heartwood and sapwood. In certain pines, for example, the sapwood moisture content may be 120% of the dry weight of the wood whereas the heartwood moisture content of the same tree is as low as 35%. At the very

high moisture content, fungi will not develop, so logs are often stored in water or under continuous water spray to keep them safe from fungal decay. Under these circumstances, bacteria may still survive and be able to destroy the pit membranes of the sapwood. This can make the sapwood very permeable and enable all the sapwood to be penetrated by wood preservative chemicals relatively easily.

In the forest, the sapwood of felled logs may be so rapidly colonized by fungi that rapid penetration and egg-laying by the ambrosia beetles may follow before the log is removed. The larvae, on hatching from the eggs, will feed on the fungus until pupation, after which the adults will emerge from the log.

3. Logs in Storage

Water storage will prevent fungal activity in the log due to the high moisture content. If stored out of water, deterioration is likely to occur as drying proceeds. The logs must be rapidly converted into sawn timber, or the exposed wood at the cut ends and sides of the log must be treated with a fungicide and, when necessary, an insecticide. Like extraction from the forest, the time in storage should be short. In parts of the tropics it is very difficult to extract certain species before the sapwood is badly discolored by sapstain fungi. Only a very short time is required for this to occur. Even at the most hygenic saw mill yards, discoloration and decay can occur quickly if the climate and moisture content are suitable for fungal growth.

4. Converted Timber prior to Seasoning

Once the log has been sawn into timber, all surfaces are exposed. If it has been felled recently, the timber will have a high moisture content and is said to be in the "green" state. After drying, often called "seasoning," it is refered to as "dry" or "seasoned" timber, with a moisture content (% m.c.) given or implied. "Air-dry" timber will be below 20% m.c. In a centrally heated room, timber will be below 10% m.c. "Oven-dry" or "bone-dry" timber is at 0% m.c.

Drying is necessary to bring the moisture content of the wood into equilibrium with the atmospheric moisture content of the immediate area in which the timber will be in service as a finished product. The drying process must be fast enough to prevent deterioration due to fungal and insect penetration, but not so fast that it induces mechanical stresses in the wood that could cause distortion, collapse, or breakage.

If the wood must remain stacked for any length of time before drying, then chemical protection may be required to reduce the rate of degradation. After drying, the wood must still be protected from wetting by rain, snow, or dew, or from extreme desiccation of the surface layers by sun or wind. During such periods of storage, the wood may become a habitat for animal or insect penetration. Dry wood is free from microbial attack, but may be attacked by insects.

5. Converted Timber after Seasoning and in Service

If seasoned timber must be dried to a known moisture content, it must be maintained in that condition. This may not be easy to achieve. Many a load of timber has arrived on a construction site at the correct moisture content, only to be kept there for weeks or months. As a result, its m.c. increases and it is put into the building at too high an m.c., with consequent problems as it dries. Alternatively, it may be put into the construction at the right m.c. for its ultimate service life, but before the construction itself has dried, for example, wood floor set on a wet concrete scree.

If the ultimate service life is in ground contact as a pole or post, then the soil moisture will increase the m.c. of the bottom of the pole and render it liable to active fungal decay. The above-ground portion of the pole, however, is air-dry and only wetted occasionally on the surface by rain, snow, or dew, with a consequent change in the microbial ecology, and is only liable to some deterioration of the surface layers.

The wood, therefore, is in a situation of hazard. The degree of hazard will depend on where the timber is in service and whether it is likely to remain dry, intermittantly wet and dry, or with a permanant high m.c. Each hazard situation will produce a different series of abiotic factors for the habitat which the microorganisms will attempt to colonize.

B. Hazard Classes

In Europe, a new standard has been established defining a series of hazard classes. These classes have been defined with respect to the conditions of the timber in service, particularly the moisture content and the prevailing biological factors likely to be the cause of degradation under those conditions. Five classes are recognized.

1. Class 1: Aboveground, Covered

In this situation, the timber is protected from exposure to rain, snow, dew, and condensation. It is, therefore, considered to be permanently dry, with a moisture content less than 18%. Such a moisture content is well below the fiber saturation point, with no free water in the wood. It is, therefore, free from fungal and bacterial attack, but liable to attack by wood-boring beetles and termites if the temperature is right.

In Europe, the common furniture beetle (*Anobium punctatum*), the house longhorn beetle (*Hylotrupes bajalus*), and the powder post beetle (*Lyctus brunneus*) are important in different areas. Termites, both the dry-wood termites and the subterranean termites, are a very important and serious problem throughout the tropics and subtropics.

2. Class 2: Aboveground, Covered, Risk of Occasional Wetting

In this situation, particularly when condensation is likely to be a problem, the timber will, from time to time, be above 20% m.c. without being above the fiber saturation point. As a result, wood-boring insects and termites are likely to be more active and, in addition, quick-growing fungi can colonize when the wood is wet and, by their activity, provide a food source for the insects, termites, and other fungi. These will include those wood-decaying fungi capable of growing in such intermittant wet/dry situations. The brown rot fungi can often tolerate such conditions and are the most common wood-decay fungi present. In buildings accidental wetting due to leaks and poor maintenance gives rise to the brown rot fungi that cause wet rot and dry rot. A range of species is involved, but the most common in Europe are *Coniophora puteana* for wet rot and *Serpula lacrymans* for dry rot.

3. Class 3: Aboveground, Uncovered

In this hazard class, the uncovered timber is subject to the climate; rain, dew, or snow can bring about a moisture content above 20% for long periods of time. Although insects and the brown rots may still be a problem, the wetter conditions will produce an ecosystem in which the other group of wood-decaying basidiomycetes, the white rot fungi, can become established and decay external painted joinery or fence rails. The optimum moisture conditions for their growth, when the wood is neither too wet nor too dry, will support an ecological succession of microorganisms. Here the white rots are the climax organism, followed later by a colonization of microorganisms only capable of feeding on the breakdown products of the cell wall components left behind by the white rots. This occurs when the wood has remained wet for a long period of time and has developed a similar ecosystem to that in timber in contact with the ground.

The surfaces of such exposed timber may be wetted, but dried by sun and wind; a combination of surface colonization by quick-growing fungi feeding on the chemical breakdown products of the exposed wood surface by the sun's rays can bring about a discoloration and a very slow erosion of the surface over a long period of time.

Wood beneath painted surfaces, although protected from surface contact with liquid water by the paint film, may become wet in humid conditions by the diffusion of water vapor through the paint film. Alternatively, open joints that expose cross cut end grain can produce a situation in which water can penetrate from the cut end grain through the wood and build up a high moisture content beneath the paint film. Fungi then can colonize the wood surface through cracks in the surface finish. In the case of certain fungal species (e.g., *Aureobasidium pullulans*), colonization occurs by a direct penetration through the paint film and a build up of fungal growth beneath it that breaks up the paint film itself as fruiting bodies develop.

4. Class 4: In the Ground or Fresh Water

Contact with the ground or with fresh water will produce a permanent moisture content in the timber above 20%. If the timber is a post in the ground or a lake, the bottom part of the post will take up water until the equilibrium moisture content is reached. Free water in the cell lumina will rise by capillary action until it is in the upper part of the post above the ground or water surface. Here evaporation will take place, since water will be lost to the atmosphere as that part of the post attempts to set up an equilibrium moisture content with the relative humidity of the surrounding air. As water is lost by evaporation, more moves up to replace it; thus, the free water in the lower portion of the post is in constant movement. Any nutrients in the soil water will, therefore, also move upward and, unless intercepted and used by invading microorganisms, will accumulate above the groundline as the water evaporates from the post and leaves its dissolved solutes behind. The interface at the groundline or water surface, therefore,

will produce ideal conditions for microbial growth and development: water from below, with a supply of nutrients and oxygen from the air above, in a suitable habitat. In such a situation, a full ecological sequence of microorganisms has been shown to develop from the early colonizing bacteria to the final fungi scavenging for the breakdown products of wood decay. Within this sequence are the ascomycetes and fungi imperfecti that give rise to soft rot decay. These are a characteristic of hazard class 4.

Above the ground and water surfaces, the post is air-dry with intermittant wetting; under such conditions the sapstain fungi and those fungi that react with the products from the breakdown of the wood surface by sunlight invade, colonize, and develop. Essentially this is a situation characteristic of hazard class 3.

5. Class 5: In Salt Water

The timber is in permanent exposure to wetting by salt water and, therefore, permanently at a moisture content in excess of 20%. Like the post in the ground or freshwater, if part of the timber is above the high water mark, then it is in a hazard class 3 situation, and subjected to the types of ecosystems that can develop there.

In the sea water itself, there may be marine fungi capable of causing minor deterioration, but the major destruction of wood is by the group of organisms known as "marine borers." The chief of these are "shipworm" and "gribble." The former, of which there are many species, use their bivalve shells to burrow into the wood so their long soft body is protected by the presence of the surrounding wood. These organisms can cause great damage, particularly when associated with fungal decay. This was particularly true of wooden ships; it has been said that, in the Napoleonic Wars, the British Navy lost more ships to break up due to shipworm and rot than to enemy action. Perhaps the most famous was the loss of the H.M.S. Royal George when the bottom fell off while the ship was moored off Portsmouth Harbor.

Gribble damage is caused by organisms burrowing into the wood just below the surface. This damage is accentuated by the abrasive action of sand and gravel in the waves and surge of the sea, breaking off the thin surface layers above the gribble's burrowings. This can erode the surface to a great depth. A post with a 12 in. × 12 in. cross-section can be reduced to a 5 in. × 5 in. cross-section in the region of the post between high and low water levels well within 20 years of exposure.

These five hazard classes, each with its own characteristic ecosystem, are the main different types of hazard to which timber is subjected in Europe. Although other hazards may occur in other areas of the world, these classes also are represented worldwide.

In the tropics, the termites are the main organism responsible for the destruction of timber. Many different species are present, each with a number of basic types of life history. The ''ground'' or ''subterranean'' termites take advantage of the microbial ecology of wood decay to organize ''fungal gardens'' in their nests. The gardens are produced by the worker termites scavenging for timber and, when they find it, cutting off small pieces that they take back to their nests. Here they work the wood fragments into a three-dimensional labyrinth with wet soil. Fungi grow in and on the wet wood and sporulate. The fungal hyphae and spores, and probably the breakdown products of the fungal decay, then form the basic food on which newly hatched termites are nurtured. This portrays an interesting and specific microbial ecosystem vital to the development of the termite colony.

C. Microbial Ecology: The Organisms and Their Effect on Wood

The various ecosystems so far described all have been shown to be dependent on the moisture content of the timber, which will vary according to the type of exposure of the timber during its service life. It has also been pointed out that the habitat in which these ecosystems exist is dependent on the type of wood involved. Is it a softwood or a hardwood; is it sapwood or heartwood; from what tree species has it been cut? In addition, the four specific factors necessary for microorganisms to grow and develop—water, oxygen, nutrients, and a suitable temperature—can all show great variation. A combination of such variations can be a major factor in determining what ecosystem may develop.

Much work has been carried out to establish the sequence of events that occurs at or about the groundline of a timber post, the lower portion of which is buried in the ground. It has also been shown that the sequence of events that occurs at the groundline is very similar to that which occurs in simulated window joinery exposed to the weather out of ground contact. In both cases a succession of organisms occurs, some being very early colonizers and others taking weeks to be observed to have colonized the wood.

One great difficulty in attempting to find out what is happening inside the wood habitat is the fact that, as yet, there is no technique that can be used to make direct continuous observations without altering the structure of the habitat. Which organism colonizes first? What happens when a later colonizer meets up with an earlier colonizer? If attempts are made to isolate the organisms from the wood and culture them *in vitro* for identification, what was the state of the organism at the time the identification was made? Was it a young actively growing hypha; an old moribund hypha; or a reproductive spore of some sort? Can one be quite sure that every organism present in the wood at any one time has been isolated and cultured on the growth media available and, therefore, identified? In spite of these difficulties, a great deal of work by many people over the past 30 years has established a sequence of events.

In wood under ideal conditions for microbial development, the early colonizers are bacteria, often within a few hours of exposure. They are quickly followed by the fast growing molds and sapstain fungi. The former can only penetrate the wood through open pits and cell lumina, being incapable of destroying the cell wall itself. The sapstain fungi, however, are able to penetrate across cell walls from one cell lumina to another by a fine constriction of the hypha and a very narrow borehole through the walls. They are, thus, not as totally dependent on other organisms' destroying the pit membranes and opening up the structure to provide a pathway into the wood habitat as are the molds. The wood-rotting fungi appear some time later and are of three main groups: the ascomycetes and fungi imperfecti, which produce the type of decay known as soft rot, and the two groups of basidiomycetes, which produce the types of decay known as white rot and brown rot. The soft rot fungi usually appear first, to be followed by the white or brown rots. Once established, either basidiomycete group will suppress the soft rots and form the complex of organisms that can destroy the lignocellulose polymers that form the wood cell walls. As the cell walls are broken down, the final group of molds appears, which uses the breakdown products as a source of nutrients, although by themselves they cannot destroy the intact cell wall.

Such a complete sequence depends on the habitat being at the right moisture content and temperature.

If the habitat is too wet, the sequence may not proceed beyond the bacterial colonization which, in certain circumstances, may produce bacterial decay of the wood. As the habitat dries out, normal colonization by molds, sapstain fungi, and soft rot fungi can occur, but the moisture content may still be too high for the white and brown rots. In this case, the soft rot fungi become the climax organisms for wood decay and will be followed by the secondary molds. This situation is often found in the timbers in water cooling towers.

These microbial groups are composed of a range of organisms, often from widely separated genera. The grouping is independent of taxonomy and reflects the effect each group has on the structure of the wood habitat. Individual species are not necessarily confined to one group; *Phialophora fastigiata*, for example, will act as a sapstain fungus on initial colonization, but can, once established, act as a soft rot fungus and, under unusual culture conditions, has also been shown to cause a type of decay typical of the brown rots. Each of the groups has its own characteristic effect on the wood habitat and are worth considering separately.

1. Bacteria

The bacteria that colonize wood are of many types, belonging to many different genera. A high moisture content is normally required, above the fiber saturation point, so that sufficient free water is available in which the bacteria may move. Moving water in wood may well carry bacteria with it. Certainly they have been shown to move progressively through the open pathways in the wood structure. This broad group of organisms performs several functions.

1. Destruction of the pit membranes, thus opening up the structure of the wood to provide open pathways for the movement of liquids, gases, and microorganisms
2. Antagonism to other microorganisms, retarding their colonization
3. Synergism with other microorganisms to encourage and support their colonization
4. Nitrogen fixation progressively from the surface layers through the wood
5. Under specific circumstances, the destruction of the wood cell wall

Of these five functions, the destruction of the pit membranes is the most important since it permits the entry of the molds and sapstain fungi deep into the wood. The molds are therefore enabled to scavenge in depth for stored food reserves and degraded parenchyma cell contents in the rays, and mobilize these for the use of later colonizers that have been shown capable of saprophytism or parasitism of the earlier invaders. Antagonism and synergism may be important locally, but are not usually of major importance. Nitrogen fixation is only likely to be of considerable importance in restricted areas in the wood. Wood, especially heartwood, has a very low nitrogen content; any extra nitrogen produced by nitrogen-fixing bacteria may be of significant importance to initiate fungal or insect attack when the element has accumulated to a sufficiently high level. The wood-destroying potential of bacterial strains is of comparatively recent discovery and will be dealt with subsequently. [*See* BIOLOGICAL NITROGEN FIXATION.]

2. Primary Molds

The fungi that are the earliest colonizers of wood in service are usually the fast growing molds that scavenge for nitrogen, simple carbohydrates, and fats deposited in the rays of the sapwood. This group contains a wide variety of fungal species from many genera. Their life cycle is often short, and reproduction, particularly by asexual spores, is usually prolific. Both of these are very important criteria when intermittant wetting and drying occurs in timber in service. Entrance into the cut surface by growing mycelia, quickly followed by sporulation, allows the wood surface to dry and the mycelia to die by desiccation, leaving the spores ready to germinate when wetting recurs. The new mycelial growth can then penetrate a cell or two deeper from the surface before further desiccation occurs, followed later by spore germination which, repeated time after time, allows colonization to continue. Such mycelia and spores may provide a source of nutrients for other organisms.

These fungi, called primary molds, do not produce an enzyme system capable of breaking down the components of the wood cell wall. Their pathway into the wood must be opened for them by the action of other early colonizers, particularly the pit membrane-destroying bacteria.

3. Sapstain Fungi

This group is characterized by possessing dark-colored hyphae that can penetrate intervening cell walls from the lumen of one cell to the lumen of an adjacent cell by a very small bore hole in which the fungal hypha is greatly constricted. With the primary molds, they are early colonizers after the bacteria in sapwood, feeding on the stored food reserves in the sapwood rays. The dark-pigmented hyphae discolor the sapwood, which is normally a light color, turning it black, blue-black, or deep blue. This can occur quickly; timber felled in the cold season of the year must be removed from the forest very quickly and certainly well before the higher temperatures of spring and summer increase the rate of growth of the organisms. This is particularly important in the tropics and subtropics, where light colored hardwood species such as *Antiaris* and *Pycnanthus* are very difficult to extract unstained. Again these characteristics are not confined to a single species, but are common to a number of genera that may vary across the world.

4. Soft Rot Fungi

The soft rot fungi again represent a wide range of genera from the ascomycetes and fungi imperfecti. Their importance in causing decay of wood in water cooling towers was first discovered by Savory in the 1950s. These fungi have since been shown to be widespread across the world, wherever the conditions in the habitat are marginal for the wood rotting basidiomycetes because of excess moisture or treatment by wood preservatives, or for other reasons. They then become the major wood-rotting fungus and have caused great problems.

They are characteristically able to destroy the wood cell wall, confining themselves mainly to destruction of the middle layer of the secondary cell wall, the "S2 layer." Corbett, in the early 1960s, showed that a fungal hypha lying close to the inner layer of the cell wall produces a thin side branch that penetrates the cell wall at right angles to its surface. On reaching the S2 layer, this branch changes its penetration path by turning at right angles and along the grain within the S2 layer. At the turning point, a branch of the hypha develops that grows along the grain in the opposite direction; the original branch tip forms what is known as the "T" branch. Each of the two branch tips (the "proboscis hyphae") grow parallel to the helical orientation of the cellulose microfibrils in the S2 layer for a time. Elongation stops as the hyphae increase in girth. At the same time, a cavity with characteristic pointed ends is formed around each of the proboscis hyphae. Once the cavities have been formed, proboscis hyphae grow from the distal ends of the enlarged hyphae in their cavities following the orientation of the cellulose microfibrils. After a period of growth, the elongation of the hyphae stops again, the hyphae increase in girth, and new cavities with pointed ends develop. This stop/start growth and cavity formation can continue until the whole S2 layer is destroyed. Short side branches from the hypha in a cavity can give rise to further T branches, initiating successive chains of cavities parallel to those formed by the original T branch. [*See* HYPHA, FUNGAL.]

Once in the S2 layer, a T branch can grow into the S2 layer of an adjacent cell wall and the whole process can continue from cell wall to cell wall without penetrating a cell lumen, until the main structural component of all the cell walls in the infected timber is destroyed. The S1 and S3 layers of secondary cell wall remain, as does the primary wall–middle lamella complex.

5. White Rot Fungi

The white rot fungi comprise numerous genera belonging to the basidiomycetes and are characterized by their mode of destruction of the wood cell wall. The hyphae lie on the inner surface of the cell wall and, by the action of extracellular enzymes that are believed to be contained in the mucilaginous sheath surrounding the hyphae, destroy the lignin, cellulose, and hemicelluloses in direct contact. This lysis of the wall occurs along the line of hyphal contact, forming a groove or trough with a central ridge on which the hyphae rest. As the hyphae branch, the grooves begin to coalesce, thus eroding the wood cell wall from the inner surface. Degradation of both lignin and the carbohydrates can result in almost complete destruction of the wood, resulting in weight losses above 90% of the original wood dry weight. As decay proceeds, the wood becomes bleached or white in color.

6. Brown Rot Fungi

The brown rot fungi also belong to genera of the basidiomycetes, but have a different mode of destruction of the wood cell wall. They are characterized by their ability to degrade both the cellulose and hemicelluloses (the holocellulose) and leave the lignin essentially unaltered. The hyphae penetrate the cell lumina and lie on the inner surface of the cell wall. Both the hyphae and the cell wall surface appear largely unchanged, but the layers beneath the surface are converted to a brown amorphous mass—the residual lignin—giving the name "brown rot" to this type of decay. Much work and speculation is currently in progress worldwide to establish the mechanism by which the cell wall is degraded, often at a distance form the contact point of the fungal hypha with the inner surface of the cell wall. The enzyme system appears to involve a mediator capable of penetrating the cell wall structure and degrading the holocellulose despite the presence of the encrusting lignin. [*See* CELLULASES.]

7. Secondary Molds

The secondary molds are those fungi that appear to be incapable of attacking wood itself, although they possess an active cellulase enzyme system, as seen by their ability to break down ball-milled cellulose in agar culture. Like the jackals that feed on the carcass after the lion has made the kill, the secondary molds appear after the wood-rotting fungi have begun to destroy the cell wall and have broken through the lignin barrier to give access to the holocellulose and the products of cell wall breakdown. This cellulolytic food source may be a nutritional excess left behind by the wood-rotting species, or may present a source of competition between the two groups of organisms.

8. Wood-Destroying Bacteria

In the last 10 years, decay of wood by bacteria has been observed. Active work by Nilsson and his associates has characterized this effect on wood. Such decay was first seen in preservative-treated softwood stakes in vineyards and farms in New Zealand. On these sites, a regime of sprays and irrigation had produced a high moisture content in the wood that combined with the presence of the wood preservative, had prevented fungal colonization. This had given a range of bacteria unrestricted access to the habitat, free from competition for nutrients with the fast-growing fungi. Three basic modes of degradation have been observed and described, each with its own characteristic micromorphology.

These modes are (1) eroding, (2) tunneling, and (3) cavity forming. Detailed knowledge of the species involved, their life history and physiology, and the mechanisms that produce the mode of degradation of the cell wall characteristic of each group is still scanty. Much work is currently in progress to increase our knowledge of these organisms.

V. Durability of Timber

A. Natural Durability

The sapwood of all timbers is likely to be susceptible to fungal and insect attack because of the food reserves stored in the rays. *Lyctus* and *Hylotrupes* beetles will destroy the sapwood of species in which starch is the main stored product. Most fungi and bacteria associated with wood also will scavenge for this readily available source of nutrient. The heartwood of the tree is different. First, the stored food reserves are moved to the sapwood during the process of heartwood formation. Second, additional materials are often deposited in the cell walls and pit membranes as heartwood is formed. These materials may have a number of effects on the wood habitat. Additional lignification of the pit membranes may prevent their destruction by bacteria, so the wood structure is not opened. In some species, the materials deposited in the cell wall may be toxic to the usual colonizing microorganisms and thus inhibit the destruction of the wood structure by these organisms. Such heartwoods are said to show natural durability.

Such natural durability has an interesting impact on the organisms normally colonizing wood in ground contact. Many microbial species are absent from the wood and also from the soil in immediate contact with the wood. This clearly gives rise to a situation in which the organisms found in the soil in immediate contact with the wood are able to colonize the wood freely. Given time, an ecosystem is likely to evolve in which resistant strains of wood-rotting organisms are likely to proliferate and eventually bring about the destruction of this durable habitat.

B. Induced Durability

One means of inducing durability in wood is to keep it permanently dry. When wood in service is likely to be subject to deterioration by organisms, protection can be induced by the introduction of suitable chemicals into the wood structure. Such wood preservation has long been practiced in one form or another. For more than 150 years, the use of railways has necessitated the protection of the wooden sleepers or ties on which the rails are fixed, impregnating them with chemicals toxic or inhibitory to the wood-rotting organisms. Today a worldwide wood preservation industry attempts to keep wood in service in a sound durable condition for its entire projected service life.

1. Wood Preservation

For effective wood preservation to take place, the toxic chemical must penetrate deeply into the wood and, in particular, into the layers of the wood cell walls. A range of chemical systems are in current use, of which creosote has the longest service record, from the early days of the railway. Waterborne salts, such as copper chrome arsenate (CCA) and boron borate mixtures, and organic solvent systems all have given long service life to wood.

Once treated with a chemical wood preservative, timber becomes simply a new habitat. An additional abiotic factor, a toxin, has been added to it. Whereas the water availability tends to be the limiting factor for decay in untreated wood, the toxic chemical becomes the overriding abiotic factor determining the types of organism that can live in wet wood.

How the toxin works is another matter. The mere presence of the toxin in the habitat will prevent infection by the majority of the normal colonists. Others may grow through the habitat without any apparent inhibition but also without any deterioration of the treated cell walls until the organism has penetrated beyond the treated zone. The wood-rotting species among them can then decay the untreated wood. Under these circumstances, the chemical may be acting simply to block the sites in the cell wall at which the enzyme system of the invading fungus must bind to initiate decay. The soft rot fungi are examples of such organisms, and have been shown to colonize CCA treated sapwood of pine and birch. The former species showed no sign of decay, whereas treated birch fibers were destroyed by soft rot. The reason for this difference is the mode of action of the soft rot fungi and the micromorphology of the destruction of the S2 layer of the cell wall. In the pine sapwood, a permeable species, all the cell wall layers are well impregnated with the preservative, whereas the S2 layer of the fibers in the birch are largely untreated. As described earlier, once the fungus has penetrated into the S2 layer of one cell wall, it can move across the primary wall/middle lamella complex by means of a T branch into the S2 layer of the neighboring cell without penetrating a preservative treated cell lumen. Thus, the microdistribution of the preservative through the wall struc-

ture is a very important factor in successful wood preservation.

Since microorganisms, both fungal and bacterial, can survive in the new habitat of preservative-treated wood, it is highly probable that, given time, the treated wood will eventually be destroyed by the development of a suitable ecosystem.

Wet treated timber, from an ecological point of view, must be seen as a harsh environment for microorganisms in which few can grow and develop. It is, however, an accepted concept of microbial ecology that, in due time, an organism capable of using the resource of the modified habitat will do so. If none is present among the early colonizers, others capable of doing so will eventually arrive in the habitat. Many basidiomycete species are copper tolerant and can change a copper-containing chemical into a form such as copper oxalate, which is nontoxic. The fungus can subsequently decay the wood.

The degradation of wood-preservative chemicals is another microbial ecological system of interest to the wood preservation industry. In the years to come, such an ecological system could be exploited by the industry to bring about the conversion of toxic chemicals to nontoxic forms after the service life of the preservative-treated wood product has ended, so little or no toxic waste remains in the environment.

Probably the best and oldest example of the selection of a tolerant fungus in a preservative-treated habitat is the decay of poorly treated creosote-impregnated transmission poles by the basidiomycete *Lentinus lepideus*, which causes decay in the center of the pole and can do the same to railway ties.

VI. Conclusion

The decay of wood is the result of a microbial ecological system that uses the habitat in which it develops as a nutrient substrate through a succession of colonizing organisms. Each stage in the process of the use of wood, from tree to finished product in service, has its own characteristic ecosystems. If the finished product has been treated with a wood preservative to render it a durable commodity, it is likely to be only a matter of time before a suitable ecosystem evolves that will eventually lead to the destruction of the wood. In the case of preservative-treated wood, the art and science involved must ensure that this evolution does not occur within the specified life of the treated wood.

Bibliography

Bravery, A. F. (1971). *J. Inst. Wood Sci.* **5(30),** 13–19.

Carey, J. K. (1983). *In* "Biodeterioration" (T. A. Oxley and S. Barr, eds.), pp. 13–25. John Wiley & Sons, New York.

Corbett, N. H. (1965). *J. Inst. Wood Sci.* **3(14),** 18–29.

Dickinson, D. J. (1991). Wood preservation: The biological challenge. *In* "Chemistry of Wood Preservation" (R. Thompson, ed.). Royal Society of Chemistry, London.

Dickinson, D. J., and Levy, J. F. (1990). The microbial ecology of timber and forest products. *In* "Methods in Microbiology, Techniques in Microbial Ecology" (R. Grigorova and J. R. Morris, eds.), Vol. 27, pp. 479–496. Academic Press, London.

Harvey, P. J., and Palmer, J. M. (1989). *Spectrum, Brit. Sci. News* **217(1),** 8–11.

Levy, J. F. (1987). *Phil. Trans. Roy. Soc. London* **A321,** 423–435.

Levy, J. F. (1990). Fungal degradation of wood. *In* "Cellulose, Sources, and Exploitation" (J. F. Kennedy, G. O. Phillips, and P. A. Williams, eds.), pp. 397–407. Horwood, London.

Nilsson, T., and Daniel, G. F. (1982). *Proc. 16th Conv. Deutsche Gesellschaft Holzforsch.*

T Lymphocytes

Douglas R. Green
La Jolla Institute for Allergy and Immunology

Glossary

Antigen presenting cells (APC) Class II MHC$^+$ cells that are capable of taking up exogenous antigens, proteolytically cleaving them into antigenic fragments so that they associate with the class II MHC molecules; further, the cells must be capable of stimulating T cells bearing receptors specific for the ligand presented on the APC cell surface (e.g., macrophages, B cells, dendritic cells, and the Langerhans cells of the skin)

Antigen processing and presentation Action of proteolytically cleaving an antigen (either endogenous or exogenous) such that fragments become associated with molecules encoded by the Major Histocompatibility Complex (antigen processing); these are then transported to the cell surface, where they can be recognized by T-cell receptors (antigen presentation)

Class I MHC molecules Molecules encoded by the polymorphic class I genes of the major histocompatibility complex (MHC) (α-chain) and the monomorphic β2-microglobulin gene (β-chain); class I MHC molecules are found on essentially all nucleated cells of the body; the α-chain folds to form a groove that binds antigenic peptides. The complex of class I MHC molecule and antigenic peptide stimulates CD8+ T cells bearing the appropriate T-cell receptors

Class II MHC molecules Molecules encoded by the polymorphic class II genes of the major histocompatibility complex (MHC) (α- and β-chains); class II MHC molecules are found on relatively few cell types (versus class I MHC), most of which can act as antigen presenting cells; α- and β-chains are presumed to interact to form a groove that binds antigenic peptides; the complex of class II MHC molecule and antigenic peptide stimulates CD4$^+$ T cells bearing the appropriate T-cell receptors

Clonal selection Centrally important principle by which antigen specific immune responses occur; lymphocytes bear antigen receptors such that any two clones of cells bear receptors that are specific for different ligands, while all of the progeny of a cell bear (essentially) the same receptors; antigen (or an antigen fragment presented on an MHC molecule) induces the proliferation of a lymphocyte, resulting in the expansion of cells that are specific for that antigen

Cytokine Any of a number of different growth and differentiation factors secreted by lymphocytes, macrophages, keratinocytes, stromal cells, and other cell types; cytokines are important for cell–cell communication in the development and function of the immune system

Cytotoxic T cell T cells, often CD8$^+$, that are capable of lysing target cells that express the cell-surface ligand for which the cytotoxic T cell is specific; although most cytotoxic T cells are CD8$^+$, not all CD8$^+$ T cells are cytotoxic, since cytotoxic T cells must undergo a maturation process following initial activation in order to gain the ability to kill targets

Helper T cell T cells, usually CD4$^+$, that produce cytokines in response to presentation of the appropriate ligand by an antigen presenting cell; the cytokines stimulate proliferation of the activated helper T cell and also activate other cells to perform a variety of immunologic functions

Major histocompatibility complex Gene complex encoding several different classes of molecules, two of which (class I and class II) perform crucial functions for T-cell activation; each of the several class I and class II loci are highly polymorphic, (i.e., there are many different alleles for each locus), and heterozygous alleles are expressed codominantly; encoded molecules bind antigen fragments and present them at the cell surface to T cells (The name of this complex derives from the ability of allelic differences to dictate rapid rejection of organ grafts)

T-cell receptor (TCR) complex Collection of associated molecules found exclusively on the surface of T cells; two of the chains, usually TCRα and β (or γ and δ on a subset of T cells), include regions of high variability such that different clones of T cells bear different receptors; these variable regions dictate the specificity of the T cell for its specific ligand, composed of an antigenic fragment bound to an MHC molecule. The other components of the complex are the CD3 molecules (CD3γ, δ, ε, ζ, and η); these are involved in assembly and signal transduction of the complex

T-cell receptor ligand Ligand recognized by the T-cell receptor, which is composed of a proteolytically cleaved fragment of a protein antigen, bound into a groove on a cell-surface molecule encoded by polymorphic genes of the Major Histocompatibility Complex

T LYMPHOCYTES, or T cells, are a class of lymphocytes characterized by the presence of a structure, the T-cell receptor (TCR) complex, on the cell surface.[1] All T cells arise from pluripotent stem cells in the bone marrow, and most must then complete their maturation in the thymus (hence "T"). Like B cells, T cells are antigen-specific, but the ligand recognized by the cell surface T cell receptor is a complex one, composed of a combination of an antigen fragment and a host-encoded molecule. It is the interaction between this complex ligand and the TCR that forms the basis for the initiation of most antigen-specific immune reactions, regardless of the ultimate effector mechanism employed in the response.

---

[1] The other two major classes of lymphocytes are the B cells, which bear immunoglobulin on their surfaces, and the large granular lymphocytes, which bear neither surface immunoglobulin nor TCR.

Here we will consider how T cells perform their various functions in the immune system, and how these functions are triggered by the antigen-specific interaction of the TCR with its complex ligand. We will then examine the T cell-dependent strategies for controlling parasite[2] infection, and how different strategies may favor the host or the parasite. Finally, we will consider the pathological consequences of the interactions between T cells and some organisms.

I. What T Cells Do

T cells have two major functions that account for the majority of the effects T cells have on other cells in the immune system. These functions are roughly distributed among (and define) two functional subsets of T cells: the helper T cells and the cytotoxic T cells. The unifying concept behind the functions of these subsets is this: An antigen-specific activation event involving the interaction of the TCR with its ligand on another cell triggers the T cell to release mediators that have short-range effects. In the case of helper T cells, the released mediators are cytokines, while in the case of cytotoxic T cells the mediators deliver a lethal "hit," which destroys the cell bearing the T-cell receptor ligand. The process of antigen-specific T-cell activation is considered in more detail below.

These two subsets of T cells, helper and cytotoxic, are correlated with the expression of two cell surface molecules, CD4 and CD8, respectively.[3] These molecules are involved in the activation of T cells (see later) and can be readily identified through the use of fluorescence-labeled antibodies. Mature T cells bear one or the other of these molecules, but not both. Thus, helper T lymphocytes are usually

---

[2] We use the term parasite in its generic sense, to mean any invading organisms, including viruses, bacteria, and eukaryotic parasites, pathogenic and nonpathogenic.

[3] Throughout the 1970s and much of the 1980s, a third class of T cell was often included in discussions such as these: the suppressor T cells. Suppressor T cells, like helper T cells, regulate immune responses, but the functions are opposed; suppressor T cells inhibit immunity. It is now considered most likely that these do not represent a functional subset but, rather, a functional *feature* of the other subsets. For example, cytokines released from a helper T cell may inhibit rather than help a particular type of immune response. This is discussed in more detail in a later section. Other suppressive mechanisms also exist, but again they are unlikely to represent the activities of a uniquely specialized functional subset.

CD4$^+$, whereas cytotoxic T cells are usually CD8$^+$. The converse, however, does *not* necessarily apply; i.e., cells that are CD8$^+$ are not necessarily cytotoxic T cells. The reason for this is simple: Cells must often undergo a maturation process (often following primary activation) before they become functional. Therefore, simply assessing the percentages of CD4$^+$ versus CD8$^+$ cells does not provide useful information regarding the nature of the immune response occurring.[4]

When a helper T cell is activated, it releases cytokines, and the cytokines that are released by a particular T cell will determine the functions of that cell. This is shown schematically in Fig. 1A. Cytokines are growth and differentiation factors made by and affecting a number of different cell types, and some of those made by T cells are listed in Table I. Each cytokine reacts with cell surface receptors specific for the given cytokine, and the expression of these receptors determines which cells can react to the cytokine. Most cytokines have a short range of action, owing to their rapid degradation and the small amounts made by a single cell. This ensures that immune responses will be localized to sites containing antigen. [*See* CYTOKINES IN BACTERIAL AND PARASITIC DISEASES.]

One of the most important types of cytokine are those with T-cell growth factor activity [e.g., interleukin-2 (IL2), interleukin-4 (IL4)]. Activated helper T cells express receptors for these cytokines and respond to them by proliferating. Thus, a helper T cell produces a growth factor and its receptor upon activation and, thus, is stimulated to proliferate, a process called autocrine growth. This autocrine stimulation results in expansion of helper T cells that have contacted their specific ligand [which includes antigen (see later)], and, thus, antigen participates in the selection and expansion of helper T cells that can respond to it. The role of such clonal selection in the initiation and recall of specific immune responses is considered in more detail in the next section.

In addition to the T-cell growth factor activities,

---

[4] The assessment of the ratios of CD4$^+$ to CD8$^+$ T cells (CD4 : CD8 ratios) in various disease states formed the basis for much research in the early 1980s and overinterpretation of the results was common. Thus, an increase in CD8$^+$ T cells versus CD4$^+$ T cells (usually in the peripheral blood) was often taken to mean that helper T-cell activity was being blocked by excessive suppressor T cells (which are often CD8$^+$), an interpretation that cannot be supported (see Footnote 3). However, the practice of such assessment was not without value, because it led directly to the discovery that loss of CD4$^+$ T cells is associated with AIDS.

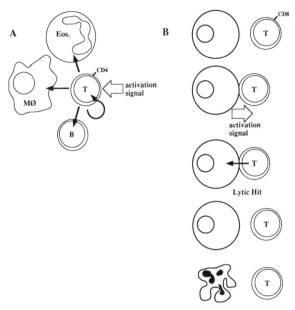

Figure 1 The functions of T cells. (A) Functional activity of helper T cells. In response to an activation signal, CD4$^+$ helper T cells (T) release cytokines that can have effects on a number of other cell types, including B cells (B), macrophages (MΦ), and eosinophils (Eos.), among others, depending on which cytokines are produced. In addition, the activated helper T cell releases cytokines that stimulate its own growth. (B) Functional activity of cytotoxic T cells. Unlike some of the functions of helper T cells, the activity of CD8$^+$ T cells requires that the cells first mature through an activation process. The mature cycotoxic cell will then specifically interact with target cells (e.g., virally infected cells) and destroy them. Contact with the specific target cell activates the cytotoxic T cell, when then delivers a lytic signal (the lytic "hit") to the target cell and then detaches. The target cell subsequently dies while the cytotoxic T cell goes off in search of more targets.

cytokines perform a number of other functions, including activation of B cells, macrophages, and eosinophils. Several of these additional functions are listed in Table I; however, this information should be treated with some caution for several reasons. First, the involvement of a cytokine in a particular function is not meant to imply that the cytokine acts in solo—for example, the activation of B cells requires other signals in addition to cytokines. Therefore, a given cytokine will not necessarily replace the T cell for the performance of the function. Second, several cytokines may have similar functions but may operate differently, depending, for example, on the expression of different cytokine receptors on the affected cells. Finally, many of the functions listed in the table are based on *in vitro* studies

Table I Some T-Cell Derived Cytokines[a]

Factor	TH distr	Chemical char.	Biological actions	Other names
IL2	TH1	15.5 kDa	Promotes growth of T cells; induces p55 subunit of IL2 receptor; NK cell activation; B cell growth and differentiation	T-cell growth factor, costimulator
IL3	TH1, TH2	30–40kDa	Mast cell growth; formation of neutrophils, macrophages, megakaryocytes, and mast cells from precursors	Colony-forming unit–stimulating activity, pan-specific hematopoietin, multi-CSF
IL4	TH2	15–16 kDa, pI 6.7 (human)	Promotes T-cell growth, B cell production of IgE and (with IL5) IgA, and mast cell growth factor	B-cell growth factor, B cell stimulating factor-1, T-cell growth factor-2, mast cell growth factor-2
IL5	TH2	12.4 kDa core; exists as 45–50-kDa dimer	Growth and differentiation of eosinophils; B-cell proliferation and IgA production (with IL4), and thymocyte stimulation	T-cell replacing factor, B cell growth factor II, eosinophil differentiation factor, IgA-enhancing Factor
IL6	TH2	19–26 kDa, pI 6.2, 6.4	B cell differentiation and IgG production; promotes growth of hybridomas; supports growth of T cells and thymocytes (with IL2); induces synthesis of acute phase proteins from hepatocytes	B cell stimulating factor-2, interferon-β_2 hepatocyte growth factor
IL9		16 kDa core, exists as 40-kDa dimer	Stimulation of T cells and mast cells	p40
IL10	TH2	17 kDa core, exists as 34–40 kDa dimer	Inhibition of antigen presentation to TH1 cells; stimulation of mast cells and thymocytes	Cytokine secretion inhibitory factor
TNFα	TH1, TH2	17 kDa	Mediates endotoxic shock; induces cachexia (wasting); stimulates fibroblasts and melanocytes; activates macrophages and neutrophils	Cachectin
TNFβ	TH1	17 kDa, 30% homology with TNFα	Tumor cytotoxicity; B-cell activation under some conditions; many other activities shared with TNFα	Lymphotoxin
INF-γ	TH1	Sensitive to low pH, 17 kDa	Activates macrophages; induces or increases class I and class II MHC on many cells; stimulates some B cells; enhances IgG$_{2a}$ production; inhibits IgG$_1$ and IgE; inhibits proliferation of TH2 cells; antiviral activity	Immune interferon
TGFβ	TH1	12.5 kDa, exists as 25-kDa dimer, activated at ph 2	Inhibits proliferation of T cells and B cells; IgA switch factor	
GM-CSF	TH1, TH2	13.5 kDa core, 18–22 kDa	Induces proliferation of macrophage and granulocyte colonies from bone marrow; potentiates neutrophil responses	

[a] The helper T-cell subsets associated with production of each cytokine (TH distr.) are indicated as TH1, TH2, or both, based mostly on the cytokine profiles observed for murine T cells.

CSF, colony-stimulating factor; IFN, interferon; IL, interleukin; GM-CSF, granulocyte–macrophage colony-stimulating factor; MHC, major histocompatibility complex; NK, natural killer; TGF, transforming growth factor; TNF, tumor necrosis factor.

and indicate a *potential* for the cytokine to perform a given function that it might not do under physiologic conditions.

A second mode of action of T cells is that of cytotoxic T cells. As shown schematically in Fig. 1B, cytotoxic T cells interact with target host cells and deliver a lethal hit to the cell. This function is important in those cases in which the host cell is infected with an intracellular parasite; destruction of the cell results in the effective destruction of the parasite. As with helper T cells, the activation of a cytotoxic T cell to perform its function is antigen-specific; i.e., a given cytotoxic T cell will destroy host cells infected with one type of parasite, but not another. Following destruction of the target cell, the cytotoxic T cell remains viable and can continue its "seek out and destroy" mission. The mechanistic basis for the cytotoxic function of these T cells is not fully understood but probably includes molecules involved in the formation of membrane pores in the target cells (perforin) as well as other molecules that appear to induce a form of cellular "suicide" in the targets.

Before considering how the functions of T cells drive immune responses to parasitic organisms, we will first backtrack to see how these functions are triggered in an antigen-specific manner.

II. How Antigen Stimulates T Cells

In general, T cells discriminate between proteins (or structures on proteins) that serve as antigens for an immune response. This specificity is dictated by the TCR on the cell surface; all the TCRs on a T cell and its progeny (collectively, a clone) are the same, and they differ from those found on a different clone of T cells.[5]

The TCR is composed of two peptide chains, most commonly TCRα and TCRβ. (A subpopulation of T cells bearing different chains, TCRγ and TCRδ, perform functions that remain somewhat obscure, and they will not be further considered in this article.) All of the specificity of a T cell is determined by the receptor formed by these two chains. The TCR chains are present on the surface of T cells as part of a complex with several other invariant molecules, collectively called CD3. While the TCR molecules are responsible for specific recognition, the CD3 molecules transduce the signal to the interior of the cell, leading to cell activation.

The ligand recognized by the TCR on any T cell is complex, composed of an antigen fragment held in a groove of one of a specialized group of molecules. The latter molecules are encoded by the genes of the major histocompatibility complex (MHC)[6] and fall into two types: the class I and the class II MHC molecules. Class I MHC molecules are expressed on all nucleated cells of the body, while class II MHC molecules are found on only a few cell types, including macrophages and B cells. Both classes of molecules are specialized to bind and present antigenic fragments to the TCR on a T cell.[7] If the ligand composed of the MHC molecule and an antigenic fragment effectively interacts with the TCR, then this "fit" provides a signal to the T cell to become activated. While other interactions of cell–surface molecules are also involved in the T-cell activation process, the specificity of the cell–cell interaction is entirely controlled by the binding of TCR to the MHC+antigen ligand.

Thus, MHC molecules bind fragments of antigen, forming a potential ligand for a TCR. Where do these fragments of antigen come from? The short answer is this: Fragments of antigen are produced inside of cells by proteolytic degradation and this "antigen processing" provides peptides that can then associate with MHC molecules. However, the source of the antigen depends on whether the fragments are

---

[5] This remarkable diversity is generated by an equally remarkable genetic mechanism. In the pre-T cell (and all other non-T cells), the genes responsible for encoding the TCR molecules are found in collections of gene segments. During the differentiation of the T cell, one member from each group of gene segments is fused with a member from another group, to eventually form a TCR gene. A very large number of different T-cell receptor genes is generated via this random process, and this number is sufficiently large that it ensures that different clones of T cells will bear different T-cell receptors. The process of gene rearrangement also forms the basis for the generation of diversity in immunoglobulin genes.

[6] The molecules responsible for "presenting" antigen to the TCR are highly polymorphic; i.e., they differ between individuals (owing to a large number of alleles at each locus). For this reason, and because of their roles in stimulating T cells, these molecules are the most responsible for determining whether a tissue graft will be accepted or rejected (histocompatibility). It is, perhaps, an historical accident that this function of the MHC molecules was discovered long before their role in antigen presentation.

[7] Each T cell is selected during development to ensure that the TCR on its surface will recognize "self" MHC plus some foreign antigenic fragment. Thus, the TCR discriminates between different MHC molecules as well as different antigenic fragments.

presented with class I MHC or class II MHC. That is, antigens can come from outside of cells (exogenous antigen) or may be produced inside the cell itself (endogenous antigen); an example of the latter is an antigen produced by an intracellular parasite. Exogenous antigens are processed and their fragments are presented only on class II MHC molecules. Endogenous antigens are processed and their fragments are presented mainly on class I HMC molecules.

Recall that all cells bear class I MHC molecules; therefore, any cell that has become infected will display fragments of the parasite's antigens on the class I MHC molecules on the surface of the infected cell. Uninfected cells in the vicinity will not display this class I MHC+antigen combination, because the parasite antigen will not be produced endogenously in such cells. Parasite antigens released into the surrounding area (exogenous antigens) may, however, be picked up by antigen-presenting cells, i.e., any cell that expresses class II MHC molecules and is capable of presenting them to T cells. (Recall that only a few cell types express class II MHC molecules.) The exogenous antigen is taken up by the antigen-presenting cell into endocytic vesicles and proteolytically cleaved, and the fragments then associate with class II MHC molecules. These then move to the surface. The pathways of antigen processing and MHC association are shown schematically in Figure 2.

Thus, a cell displaying parasite antigen fragments on a class I MHC molecule is probably infected by the parasite, whereas a cell displaying parasite antigen fragments on a class II MHC molecule need only be in the vicinity of the parasite.[8]

III. Directing an Immune Response

The important distinctions between endogenous and exogenous antigens and between class I and class II MHC molecules guide the immune system into making an appropriate response. This is because different T-cell subsets recognize antigen fragments with different classes of MHC. That is, CD8+ T cells recognize antigen fragments on class I MHC mole-

cules, whereas CD4+ T cells recognize antigen fragments on class II MHC molecules. This is because the CD4 and CD8 molecules participate in T-cell activation and bind to nonpolymorphic (invariant) parts of class II and class I MHC molecules, respectively.

The net result is summarized in Table II. Recall that CD4$^+$ T cells tend to act as helper T cells, whereas cytotoxic T cells are CD8$^+$. Therefore, cytotoxic T cells will tend to kill only infected cells, because these are the cells displaying antigen fragments on class I MHC (and because the CD8$^+$ T cells only recognize antigen on class I MHC). Antigen-presenting cells, which present exogenously derived antigen on class II MHC molecules, will induce helper T cells (because these CD4$^+$ T cells recognize antigen on class II MHC) but will not serve as targets for cytotoxic T cells (because the antigen is not on class I MHC in these cells), unless, of course, they are also infected (see Footnote 8).

In the next section, we consider how these directed interactions summarized in Table II lead to immune responses of various types. Stimulation of helper T cells to release different cytokines, and thereby direct different immune reponses, is an important key to understanding the functions of the immune system.

IV. Putting Together an Immune Response: Antigen Presentation, T-Cell Subset Function, Cytokines, and Immune Effector Mechanisms

We will consider three different types of cellular interaction that focus on the roles played by T lymphocytes in immune responses: macrophage–T cell, T cell–T cell, and T cell–B cell. In each case, we will see how the principles outlined earlier apply to give a response. The types of responses we will discuss depend on T cells, as do most antigen-specific responses by the immune system. However, the reader should be aware that many immune defense mechanisms can function independently of T cells (e.g., T independent antibody responses, inflammatory responses, natural killer cell function) and such responses are extremely important in host defense.

The induction phases of all of the T-dependent immune responses proceed along essentially similar

---

[8] If an antigen-presenting cell is actually *infected* with a parasite, then the antigenic fragments are presented on both class I and class II MHC molecules on the surface of the cell.

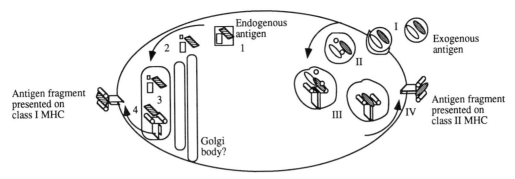

Figure 2 Two pathways of antigen processing and presentation. The processing and presentation of endogenous antigen on class I major histocompatibility complex (MHC) molecules and exogenous antigen on class II MHC molecules is shown. Endogenous antigen (1) present in the cytoplasm (because, for example, it is synthesized by intracellular parasites) is proteolytically cleaved (2) and transported (probably) to the Golgi body (3), where some of the fragments associate with class I MHC molecules (3). The complex is then transported to the cell surface (4). Exogenous antigen (I) is taken into the cell by phagocytosis or pinocytosis, and in the endocytic vesicle it is degraded into antigen fragments (II). Some of these antigen fragments then associate with class II MHC molecules within the vesicle (III). The vesicle fuses with the cell membrane, so that the complex is now presented on the cell surface (IV).

lines (Fig. 3). Most antigens that enter the body are likely to be engulfed by phagocytic cells present in the tissues. Many of these phagocytic cells express class II MHC molecules and can function as antigen-presenting cells. Thus, most immune responses are initiated when a phagocytic antigen-presenting cell takes up an antigen, processes it, and presents it on surface class II MHC molecules.

At this point, nothing happens until a CD4$^+$ T cell that bears the appropriate TCR contacts this antigen-presenting cell. That is, the T cell must bear a TCR that is specific for one of the antigen fragments presented in one of the class II MHC molecules on the cell surface. If such a T cell makes contact, it will be activated.

Recall that activation of a CD4$^+$ T cell results in two early events: the production of a T-cell growth factor (e.g., IL2) and the expression of a receptor for this cytokine (e.g., IL2R). The binding of the growth factor to its receptor now directs the cell to proliferate, thus expanding the numbers of CD4$^+$ T cells

that bear the TCR that is specific for the particular combination of class II MHC molecule and antigen fragment that triggered the response in the first place. This expanded pool of antigen-specific CD4$^+$ T cells is now available to act as helper T cells for different types of immune responses (Fig. 3).

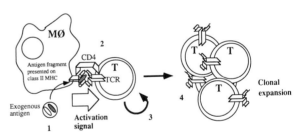

Figure 3 Activation of a CD4$^+$ T helper cell. The process begins (1) when an exogenous antigen is taken up by a macrophage (mø) or similar antigen-presenting cell and cleaved into fragments, and a fragment associates with a class II major histocompatibility complex (MHC) molecule (see Fig. 2). A T lymphocyte that bears a T-cell receptor (TCR) specific for the combination of class II MHC plus antigen fragment contacts this ligand, and the CD4 molecule on the T-cell surface binds to a class II MHC molecule as well (2). This produces an activation signal for the T cell, such that the cell expresses growth factor and a growth factor receptor (3), leading in turn to proliferation of the T cell (4). The result is an expansion of T cells bearing receptors capable of recognizing the antigenic fragments (on class II MHC molecules) derived from the original antigen. This expanded T-cell clone is now prepared to participate in immune responses (Figs. 4–6).

Table II How Antigen Location Influences T-Cell Function

Antigen	Major histocompatibility complex	T cell	Function
Exogenous	Class II	CD4$^+$	Helper
Endogenous	Class I	CD8$^+$	Cytotoxic

The precise conditions under which the expansion of the antigen-specific T cell occurred ultimately determines what additional cytokines this T cell and its progeny will produce, if and when they re-encounter the antigen fragment on the class II MHC molecule (e.g., on another antigen-presenting cell). While few of the conditions that affect this process have been defined, it is likely that these include factors produced by the antigen-presenting cell at the time of T-cell stimulation. Other factors probably include initial concentration of antigen, the anatomical site at which the response is occurring, and nonspecific effects of parasite components (e.g., bacterial cell wall components), among others. Profiles of cytokines that have been identified in such expanded helper T cells are listed in Table I. Note that these cytokines cluster into two groups: those expressed in type 1 helper T cells (Th1) and those expressed in type 2 helper T cells (Th2). There are certainly other types of helper T cells as well, but the Th1/Th2 designation, based on cytokine profiles, is both useful and accurate for a first approximation.

The result to this point is the expansion of a population of CD4+ T cells, bearing TCR specific for fragments of the original antigen (complexed with specific class II MHC molecules), which produce certain cytokines upon subsequent presentation of the antigen (Fig. 3). If the subsequent antigen presentation is by a macrophage (Fig. 4), and if the induced T cells produce cytokines capable of macrophage recruitment and activation (e.g., interferon γ, granulocyte–macrophage colony-stimulating factor), the result will be a delayed-type hypersensitive (DTH) response. The phagocytosis and oxidative bursts produced by the activated macrophages often represent an effective response against a parasite. Prolonged stimulation of the T cell and production of such cytokines eventually induce collagen formation and fibroblast proliferation, effectively "walling off" the affected area.

Alternatively, the T cells may be of a type that they produce other cytokines upon subsequent activation. For example, the T cell might release IL-5, a cytokine that is effective for inducing expansion of eosinophils, which in turn will serve as effector cells in the response. Other interactions between T-cell cytokines and myeloid effector cells (e.g., basophils, mast cells, neutrophils) lead to responses that differ with respect to the ultimate effector mechanisms employed. Nevertheless, the role of the activated T cell in such interactions remains essentially the same (Fig. 4).

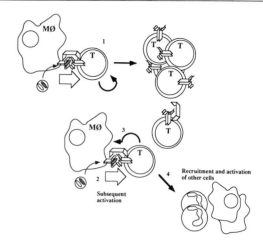

Figure 4   T cell-mediated immunity. The activation and clonal expansion of a helper T cell proceeds as in Fig. 3 (1). The expanded population of antigen-specific T cells now confronts the specific ligand on the surface of an antigen-presenting cell (2), and the T cells produce cytokines in response to the stimulus (3). Depending on the cytokines produced, this might result in activation of the antigen-presenting cell (3) and/or recruitment and activation of other cell types (4). The result is a cell-mediated immune response.

CD4+ T cells that recognize antigen fragments presented on class II MHC molecules are also involved in the development of an antibody response (Fig. 5). As with the preceding responses, the CD4+ helper T cells usually first contact the specific ligand on the surface of a macrophage or other ubiquitous antigen-presenting cell. And, again, this leads to expansion of the helper cells and acquisition of the capacity to produce a profile of cytokines upon restimulation. In this case, however, the secondary stimulation (antigen presentation) comes from an antigen-specific B cell. (Recall that B cells are class II MHC+.) B cells bear specific receptors on their surfaces (surface immunoglobulin) capable of directly binding determinants on an intact antigen.[9] Upon binding, the antigen is taken into the cell and processed, and the antigen fragments are then re-expressed on the surface on class II MHC molecules. Because the B cell binds to its specific antigen before processing it, the presence of the antigen

---

[9] As with TCR, the surface immunoglobulins are all identical on a given B cell and its progeny but are different from those of another B cell. While the genes encoding TCR and immunoglobulins are different, the generation of diversity proceeds by analogous mechanisms of gene segment rearrangement in the developing cell.

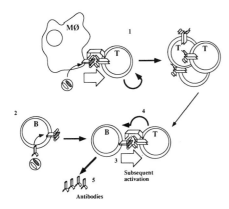

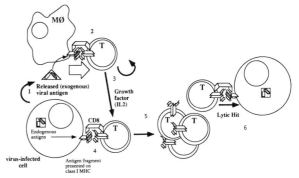

Figure 5 Induction of a T cell-dependent antibody response. Again, the process begins as in Fig. 3, when an antigen-presenting cell activates an antigen-specific helper T cell, leading to clonal expansion of the T cell (1). Meanwhile, a B cell that bears surface immunoglobulin molecules specific for the same antigen takes up and processes the antigen, which then becomes associated with class II major histocompatibility complex (MHC) molecules on the B cell surface (2). (The processing and presentation of antigen by the B cell is similar to that of other antigen-presenting cells, except that the uptake of antigen tends to be antigen-specific.) This B cell then presents the specific antigen fragment on class II MHC to the expanded antigen-specific T cells (3), which in turn produce cytokines that stimulate the B cell (4). The B cell then produces and secretes antibodies (5) (which, because the B cell makes only one specificity of immunoglobulin, is specific for the original antigen).

Figure 6 Activation of a CD8$^+$ cytotoxic T cell. This process begins when a cell becomes infected by a virus or other intracellular parasite (or expresses novel antigens for any other reason). Some of the antigen is released from this cell (1) and is taken up, processed, and presented on class II major histocompatibility complex (MHC) molecules by an antigen-presenting cell (2). Antigen-specific helper T cells respond to this presented antigen as in Fig. 3, including the production of lymphokines that act as T-cell growth factors such as interleukin-2 (IL2) (3). Meanwhile, the virally infected cell processes and presents the endogenous viral antigens on class I MHC molecules on its surface, and this stimulates CD8$^+$ T cells that bear T-cell receptors that are specific for this ligand (4). However, the activated CD8$^+$ cell cannot proceed without the growth factor provided by the helper T cell in step 3. In the presence of this "help" (growth factor), the CD8$^+$ cell proliferates and matures into cytotoxic T cells (5). Upon contact with a cell expressing viral antigen fragments on class I MHC molecules (i.e., a virally infected cell), the mature cytotoxic T cell is activated to deliver its lethal "hit," thus destroying the infected cell (6).

fragment on the class II MHC molecule can be thought of as a signal, identifying B cells that are specific for the given antigen. The T cells recognize the presented antigen fragments and release cytokines, which, in turn, stimulate the B cell (which has been partially activated through contact with antigen) to produce their specific antibodies. Different profiles of cytokines induce the production of different types of antibody (i.e., different functional classes), which then can bind to the specific antigen and perform a number of different effector functions. [*See* B Cells.]

All of the immune responses discussed so far involve CD4$^+$ T helper cells, and, indeed, the critical roles played by these cells are underscored by the devastating effects of losing CD4$^+$ T cell function in AIDS. Nevertheless, CD8$^+$ T cells also perform important functions as cytotoxic T cells, but their activation again requires the participation of a CD4$^+$ helper T cell. The stimulation of a cytotoxic T-cell response is illustrated in Fig. 6. Antigens released from an infected cell are picked up by an antigen-

presenting cell, processed, and presented on class II MHC molecules on the cell surface. This leads to the activation, expansion, and maturation of CD4$^+$ helper T cells as described earlier. Meanwhile, the antigens produced *inside* the infected cell are processed and presented on class I MHC molecules on the surface of the infected cell. These are recognized by CD8$^+$ T cells bearing the TCR specific for this combination of antigen fragment and class I MHC molecule. The CD8$^+$ T cells are activated to express growth factor receptors but do not produce growth factors. Cytokines (such as IL2) produced by the activated CD4$^+$ T cells bind to the receptors on the CD8$^+$ T cells and stimulate these cells to proliferate and mature. This maturation involves activation of genes involved in cytotoxic effector functions. Subsequently, when the TCR on the mature cell recognizes its ligand on the surface of an infected cell, it delivers its cytotoxic hit and destroys the target (see Fig. 1B). [*See* Acquired Immunodeficiency Syndrome (AIDS).]

These represent a few of the ways in which T cells recognize and respond to parasitic organisms. In each case, the basic principles of antigen processing and presentation, TCR recognition, T-cell activation, and maturation of T-cell function are essentially the same.

V. Cytokine Feedback and the "Canalization" of Immune Responses

The various types of T cell-directed immune responses discussed in the preceding section do not all proceed simultaneously; rather, one or a few types will dominate the response. This is due to a regulatory feedback mechanism whereby some of the cytokines that induce one type of response will inhibit another.

For example, interferon γ is a T cell-derived cytokine that plays important roles in macrophage activation and directing the production of certain antibody isotypes. At the same time, however, interferon γ also inhibits the proliferation of those helper T cells that are primed to produce the cytokines IL4, IL5, and IL10 (i.e., Th2 cells; see Table I). Conversely, the cytokine IL10 acts upon antigen-presenting cells in such a way that they now fail to stimulate helper T cells that produce IL2 and interferon γ (i.e., Th1 cells; see Table I). Thus, interferon γ and IL10 each regulate cells producing the other cytokine, while having no effect on their own helper cell type. [*See* INTERFERON; INTERLEUKINS.]

Because of this cytokine feedback, T cells tend to be limited in a particular immune response as to which cytokines they can produce. This, in turn, channels the reponse in defined directions. Thus, responses dominated by Th1 cells may involve DTH, cytotoxic T cells, and the production of IgG_{2a} antibodies, whereas those dominated by Th2 cells may involve eosinophil activation and the production of IgG_1 and IgE antibodies.

Clearly, then, this has important consequences in infectious disease. In mice, administration of killed *Brucella abortus* channels responses toward the Th1 type, whereas the presence of the nematode *Nippostrongylus* favors TH2-type responses. In humans, the two types of response to *Mycobacterium leprae* correlate with the dominant helper T cell type involved: Granulomatous responses are associated with Th1-type T cells (as determined by cytokine profile), and lepromatous responses are associated with the presence of Th2 cells. Similarly, susceptibility or resistance to *Leishmania* in mice is controlled by the profile of cytokines produced by T cells, and this is subject to experimental manipulation (see Fig. 7).

As mentioned in Footnote 3, cytokine regulation is an important mechanism of immune control and probably accounts for many of the phenomena previously ascribed to suppressor T cells. Nevertheless, other mechanisms of immune regulation also exist, and these may also play important roles in the control of immune responses.

VI. Going Too Far: Pathologic Consequences of T-Cell Responses to Parasites

It is often the case that the disease associated with an infection is caused by the immune response to the parasite rather than a direct assault by the infecting organism. Thus, tissue destruction by cytokines (e.g., tumor necrosis factor), by activated myeloid cells (e.g., macrophages), or as a direct consequence

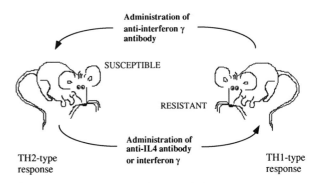

Figure 7 Interconversion of susceptibility and resistance to *Leishmania* infections in mice. Different strains of mice are genetically susceptible or resistant to infection with *Leishmania* parasites. However, this status can be experimentally altered by agents that affect the balance between helper T-cell subsets. Thus, resistant animals produce helper T cells of the Th1 type, i.e., that express interleukin-2 (IL2) and interferon γ, and this state of resistance can be produced in susceptible mice by administration of agents that promote Th1 cells and inhibit Th2 cells (such as interferon γ or anti-IL4 antibodies). Alternatively, susceptible mice produce helper T cells of the Th2 type, i.e., that express IL4, IL5, and IL10; therefore, normally resistant mice can be made susceptible by administration of agents that favor the generation of Th2 cells (such as anti-interferon γ antibody).

of a specific immune response (e.g., cytotoxic T cells) can result indirectly in pathology. While not all immunopathology is a consequence of overactive T lymphocyte function, much of it is, and it is likely that overstimulation of T cells probably can benefit the parasite by ultimately impairing host defense.

One example of such pathology is the result of a granulomatous reaction. Granulomas are collections of macrophages and epithelial cells, often surrounded by fibrosis, and result from the cellular response to persistent irritating stimulation. As already discussed, persistent stimulation of T cells by antigen can produce such foci, and this is the cause of the granuloma formation in infectious diseases such as leprosy and schistosomiasis. In the latter case, the antigen derives from the eggs that may deposit in tissues such as the liver (and, because they cannot develop, are otherwise harmless). The resulting granulomas, however, obstruct blood flow and are the cause of the pathology.

Overstimulation of T lymphocytes is perhaps best understood for those parasites that express "superantigens," including Staphylococcus enterotoxins, *Streptococcus* M antigen, toxic shock syndrome toxin, and the Mls and related antigens encoded by the mouse Moloney viruses. Superantigens are at least partly defined by the following phenomenon: The majority of T cells bearing certain TCRβ chain V regions respond to a given superantigen.[10] The result is that a much larger proportion of T cells becomes activated than would normally respond to an antigen (or the parasite). In experimental studies, animals without superantigen-responsive T cells (i.e., the T cells bearing the responsive V regions were removed) have been shown to be resistant to the pathological effects of superantigen administration; therefore, at least some of the toxic effects associated with such superantigens are due to this T-cell overactivity.

One of the least understood consequences of the interactions between infectious disease and T-cell function is the problem of autoimmune disease. No satisfactory theory has been proposed to account for how an infection or the immune response to it can later result in autoimmune disfunction, yet epidemiological evidence has implicated infections (and genetic predisposition) in a wide range of autoimmune diseases. The roles played by T cells in such diseases have been established in animal models and related studies in human disease, and by the observation that several MHC alleles are strongly associated with predisposition for autoimmunity. It is likely, therefore, that a greater understanding of T-cell response to infection will pave the way to understanding autoimmunity and, in turn, provide deeper insights into T-cell function in general.

Bibliography

Carbone, F. R., and Bevan, M. J. (1989). Major histocompatibility complex control of T cell recognition. *In* "Fundamental Immunology," 2nd ed. (W. Paul, ed.), p. 451. Raven, New York.

Davis, M. M. (1989). T cell antigen receptor genes. *In* "Molecular Immunology," (B. D. Hames and D. M. Glover, eds.), p. 61. IRL, Oxford.

Davis, M. M., and Bjorkman, P. (1988). *Nature (London)* **334,** 395.

Gajewski, T. F., Schell, S. R., Nau, G., and Fitch, F. W. (1989). *Immunol. Rev.* **111,** 79.

Golub, E. S., and Green, D. R. (1991). "Immunology: A Synthesis." Sinauer Press, Sunderland, Massachusetts.

Herman, A., Kappler, J. W., Marrack, P., and Pullen, A. M. (1991). Superantigens: Mechanism of T cell stimulation and role in immune responses. *Annu. Rev. Immunol.* **9,** 745.

Locksley, R. M., and Scott, P. (1991). Helper T-cell subsets in mouse leishmaniasis: Induction, expansion and effector function. *In* "Immunoparasitology Today" (C. Ash and R. B. Gallagher, eds.), p. 58. Elsevier Trends Journals, Cambridge.

Marrack, P., and Kappler, J. (1987). *Science* **238,** 1073.

Mosmann, T. R., and Coffman, R. L. (1989). *Annu. Rev. Immunol.* **7,** 145.

Playfair, J. H. L. (ed.) (1989). *Curr. Opinion Immunol.* **1,** 425.

Sell, S. (1987). "Immunology, Immunopathology, and Immunity," Ch. 24. Elsevier, New York.

Townsend, A., and Bodmer, H. (1989). *Annu. Rev. Immunol.* **7,** 601.

Ullman, K. S., Northrop, J. P., Verweij, C. L., and Crabtree, G. R. (1990). *Annu. Rev. Immunol.* **8,** 421.

Unanue, E. R. (1989). Macrophages, antigen-presenting cells, and the phenomenon of antigen handling and presentation. *In* "Fundamental Immunology," 2nd ed. (W. Paul, ed.), p. 95. Raven, New York.

Werdelin, O., Mouritsen, S., Petersen, B. L., Sette, A., and Buus, S. (1988). *Immunol. Rev.* **106,** 181.

---

[10] This is in contrast with the usual situation in which very few T cells with a given Vβ region will respond to a given antigen.

Toxoplasmosis

James L. Fishback

University of Kansas School of Medicine

I. History
II. Epidemiology and Life Cycle
III. Infection and Disease in Cats
IV. Infection and Disease in Humans
V. Prevention and Treatment

Glossary

Bradyzoite Slowly dividing tissue form of *Toxoplasma* found within the characteristic tissue cyst, from which they may be released by proteolytic enzymes or unknown immune mechanisms; once released, they either proliferate rapidly as tachyzoites, are killed by immune mechanisms, or form new "daughter cysts"; tissue cysts are the source of recrudescent disease in the immunocompromised host since they persist for many years, perhaps for life

Chemoprophylaxis Use of drugs, usually sulfa derivatives, to prevent the active replication of *Toxoplasma* during periods of inadequate immunity, as occurs during immunosuppression for cancer or organ transplant, or in the case of AIDS

Computed axial tomography Radiological technique that images a patient with X-rays from several different angles and then uses a computer to reconstruct the images in a series of axial "slices" through the patient's body; often, the patient is imaged twice, both before and after the use of radio-opaque contrast material, given intravenously; these are known as pre- and postcontrast films; contrast material concentrates in areas of increased blood flow or inflammation, such as an abscess or tumor

Oocyst Environmentally resistant form of *Toxoplasma* produced by the sexual mating of gametes within the intestine of cats; unsporulated oocysts are excreted in cat feces, which then sporulate within 48–72 hr in the presence of the proper combination of heat, oxygen, and mois-ture; sporulated oocysts remain infectious in soil for many months, transmitting the infection to birds, rodents, herbivores, and sometimes humans

Premunition Type of immunity where the host animal remains infected with the organism yet has good protection against re-infection and symptomatic disease; also referred to as infection-immunity or concomitant immunity

Recrudescent disease Relapse of latent toxoplasmosis, which exists within tissue cysts in many organs of the body, but in greatest concentration in the brain and muscles; this is the form of the disease most commonly seen in patients with severe immunosuppression, such as AIDS patients; the vast majority of people and animals become immune after their initial acute infection and never suffer any ill effects from the encysted *Toxoplasma* bradyzoites

Sporulation Process in Sporozoa by which a biologically active organism is produced by the division of a zygote; in the specific case of *Toxoplasma*, the oocyst sporulates in the environment to produce two sporocysts, which are still confined with the oocyst wall, and contains four sporozoites each, for a total of eight

Tachyzoite Rapidly dividing tissue form of *Toxoplasma*; some authors have referred to this stage as the trophozoite (feeding form); occurs only when the organism is actively proliferating, as in acute infection or relapsing disease; at all other times after the initial acquisition of immunity to acute infection, the organism remains latent within the host as bradyzoites in tissue cysts

TOXOPLASMA GONDII is an obligate intracellular organism belonging to the kingdom Protista, phylum Sporozoa (Apicomplexa), class Sporozoea, subclass Coccidia, order Eucoccidida, suborder Eimeriina, family Eimeriidae. It is the only member of the genus *Toxoplasma*, and *gondii* is the only

species. The organism can infect virtually any
animal, and probably any cell type, making it one of
the most common infections of mammals and birds
worldwide. The definitive host is the cat, both wild
and domestic, because the sexual life cycle of *Toxoplasma* occurs only within the intestinal epithelium
of cats. Therefore, all other affected species are considered intermediate hosts and harbor only the
asexual replicating tissue forms, known as tachyzoites and bradyzoites. Infection in intermediate
hosts is acquired by ingesting *Toxoplasma* oocysts,
which are excreted in cat feces, or by eating raw or
undercooked meat of an animal chronically infected
with *Toxoplasma*. Congenital infection is also possible, in both animals and humans. Curiously, toxoplasmosis is at once a common infection and a rare
disease since there is a high prevalence of subclini-
cal or inapparent infection in most host species.
Toxoplasmosis has recently gained notoriety in human medicine because relapsing or recrudescent
toxoplasmosis is a common cause of encephalitis
(brain infection) in patients with the acquired immunodeficiency syndrome, or AIDS. Immunosuppressive viruses and drugs have thus fundamentally altered the historic parasite–host relationship that
Toxoplasma has had with most hosts over evolutionary time.

I. History

A. Discovery

The organism was discovered in 1908 in a colony of
sick laboratory rodents by Nicolle and Manceaux
while they were working at the Pasteur Institute of
Tunis. The native North African rodents they were
working on, *Ctenodactylus gondi,* were often captured in the wild for experimental use in the laboratory, thus accounting for the species name. The genus name is derived from the Greek *Toxon,* meaning
bow or arc. This designation refers to the crescent or
banana shape of the organism, easily seen once liberated from the host cell. *Toxoplasma* was also observed in a sick laboratory rabbit in the same year by
Splendore in Sao Paulo, Brazil. The disease was not
recognized as a cause of human illness until 1923,
when it was observed in the enucleated eye of a
congenitally infected infant by Janku, in Czechoslovakia. The first autopsy cases of congenital and
acute disseminated toxoplasmosis began to be described in the late 1930s.

B. Defining Human and Animal Disease

Since the initial discovery period, symptomatic disease has been amply documented in humans and
many other species. As witnessed by the epidemics
within laboratory animals already described, *Toxoplasma* is capable of causing significant localized
outbreaks of illness. Epidemic illness has been described in sheep, resulting in abortions. Epidemics
have also occurred in zoo animals, particularly Australian marsupials, many of which died as a result of
infection. Even a few human epidemics have been
described, which are discussed further in Section II.
However, by calling undue attention to such epidemics, we would be in danger of missing the point
that most infections, in humans and animals, are
completely asymptomatic.

Indeed, seroepidemiologic surveys, starting in the
1950s, have documented that a large percentage of
the adult population, approaching 50% in some areas
of the United States, are chronically infected with
Toxoplasma. Few of these people would have visited a physician complaining of disease, and even
fewer would actually have been diagnosed as having
toxoplasmosis, because most physicians would
probably not think of toxoplasmosis and order the
proper serologic tests. Only a handful of seropositive individuals can remember any kind of illness
with fever and swelling of lymph glands in the neck,
two of the most common manifestations of acute
toxoplasmosis.

In humans, congenital disease has received much
attention in the medical literature. *Toxoplasma* is
known to cause blindness and mental retardation
when congenitally transmitted to the fetus during
acute infection of the mother. Some of the older
literature suggests that toxoplasmosis may be a
cause of chronic abortions, or that it may be transmitted in successive pregnancies. While a small
number of cases of transmission of *Toxoplasma* to
the fetus during chronic infection of the mother have
been published, this event is considered extremely
rare by most authorities. For practical purposes, the
disease can only be transmitted to the fetus during
acute infection. Mothers with AIDS or severe immunosuppression are a possible exception. The infection rate of the fetus is about 30–50% in acutely
infected mothers, as determined from a large number of cases in Paris, France, and documented by
Hohlfeld and his associates. The number of cases of
congenital toxoplasmosis in the United States is
probably between about 500 and 10,000 annually.

The total annual preventable cost of congenital toxoplasmosis in the United States has been estimated by Roberts and Frenkel at between 368 million and 8.7 billion dollars. The range is wide because of the extreme difficulty in knowing the exact number of cases and differences in various methods for estimating preventable costs. Of course, emotional costs of an infected infant are incalculable.

Recrudescent disease in humans was first noted by Arias-Stella in 1954. A number of such cases have since been reported, coinciding with the common use of immunosuppressive drugs for cancer and other diseases. These drugs interrupt or corrupt the normal immune relationship of premunition that *Toxoplasma* has with most hosts, acquired over evolutionary time. After 1981, viral immunosuppression by human immunodeficiency virus also led to many reports of relapsing toxoplasmosis in AIDS patients. [*See* IMMUNE SUPPRESSION; ACQUIRED IMMUNODEFICIENCY SYNDROME (AIDS).]

C. The Definitive Host

Research by Dubey and Frenkel at the University of Kansas in 1969 led to the discovery of the complete life cycle within the cat intestine and the description of the *Toxoplasma* oocyst. This discovery helped to explain some of the outbreaks of *Toxoplasma* that were not associated with carnivorism, since the oocyst remains viable in the environment for many months, where it may be ingested by susceptible hosts, such as humans, birds, rodents, and herbivores. Subsequent studies have shown that many species of wild cats are susceptible to *Toxoplasma* and that all excrete the characteristic oocysts in their feces shortly after oral infection with bradyzoites.

D. Molecular Biology

Several genes of *Toxoplasma* have now been cloned, although research on *Toxoplasma* genetics can still be considered to be in its early stages. Development of drug-resistant mutants and tissue culture cloning methods by Pfefferkorn and Pfefferkorn has been an important breakthrough in the study of *Toxoplasma* genetics, because these drug-resistant mutants provide readily identifiable genetic markers to which other genes can be mapped. *Toxoplasma*, so far, appears rather pedestrian in its molecular biology. For example, no evidence indicates elaborate gene shuffling, as seen in the genes coding for the variable surface glycoproteins of trypanosomes. Indeed, the cell surface of *Toxoplasma* appears to be occupied by a rather constant protein population, with one, designated P-30, accounting for >5% of the total protein produced by the tachyzoite stage of the organism. However, the different stages of *Toxoplasma* (bradyzoite, tachyzoite, and sporozoites within the oocyst) all appear to have distinct stage-specific antigens. Infection of a host cell with *Toxoplasma* also leads to the production of several distinct antigens that are secreted by the host cell. Several of these excretory antigens have been characterized and cloned and may eventually prove useful in diagnosis and possibly vaccination.

E. Vaccine Research

There has been considerable interest over the years in a *Toxoplasma* vaccine that could be used to protect human or animal fetuses from congenital infection. However, research with standard types of killed vaccines showed that they had absolutely no efficacy. The first experimentally successful *Toxoplasma* vaccine developed was a temperature-sensitive mutant developed by Frenkel and his associates and designated *ts*-4 (*ts* denotes temperature-sensitive). Its efficacy was indicated by experiments in which mice and hamsters resisted 1000 times the lethal dose. In addition, vaccinated animals survived without subsequent chronic infection, because the *ts*-4 strain did not form bradyzoites. This stood in contrast to infection with natural strains of *Toxoplasma* that were regularly followed by chronic infection with bradyzoites.

Some of the cloned antigens of *Toxoplasma* have also been tested as vaccines, such as the P-30 antigen and P-28, one of the excretory antigens. They offer only moderate protection against challenge with a wild strain. Indeed, P-30 actually increased mouse mortality in early studies. Because killed vaccines do not work, one would expect that it would be difficult for single recombinant proteins to achieve the protection afforded by the *ts*-4 live vaccine. Perhaps if a mixture of antigens were used, or a new type of adjuvant were formulated, then recombinant vaccines might have better immunogenicity. This has recently been reported using the P-30 antigen, and immunogenicity appears increased. Recombinant antigens do have potential advantages over live vaccines.

First, they are easily stabilized for widespread distribution to potential clients. Second, they could possibly be administered to immunosuppressed patients with some hope of boosting immunity, and without concomitant risk of killing or debilitating the patient by undue replication of the vaccine strain.

There has also been some interest in vaccines that would prevent oocyst shedding by the definitive host, the common house cat. It was known that wild-type *Toxoplasma* strains would lose the ability to form oocysts after rapid passage (twice weekly) as tachyzoites in mice. These strains seemed to be good candidates for a cat vaccine that would not result in the shedding of oocysts. Unfortunately, cats challenged after vaccination with these strains still shed oocysts.

One relatively simple method that provided excellent immunity (80–85%) against oocyst shedding was to infect a kitten with a wild strain of *Toxoplasma* and give anti-*Toxoplasma* chemoprophylaxis for a period of days after immunization. Both 0.02% monensin or appropriate doses of sulfadiazine and pyrimethamine were successful in preventing oocysts shedding after immunization.

Although impractical as a vaccine, these results suggested that formation of oocysts was not absolutely necessary for the development of resistance to oocyst shedding. Subsequently, chemically mutagenized clones of *Toxoplasma* were screened for lack of oocyst shedding in cats. The T-263 strain successfully immunized 33 of 37 cats (84%). Unfortunately, the T-263 strain currently suffers from the same basic disadvantage as the *ts*-4 vaccine, because it is a live vaccine. Better means of stabilizing *Toxoplasma* are being sought.

II. Epidemiology and Life Cycle

The relationship between the host animal and *Toxoplasma* is a rather cordial one, with the organism causing little or no symptomatic disease. Tachyzoite and bradyzoite stages are found in all animals with systemic toxoplasmosis, including cats. However, it is the distinctive enteroepithelial stages, which occur only in the cat intestine, that distinguish felines as the definitive host for *Toxoplasma* (Fig. 1). The life cycle starts with a cat eating a rodent or bird that is chronically infected with *Toxoplasma,* which contains numerous bradyzoites in tissue cysts (Fig. 2). The bradyzoites initiate a rapid cycle, with the various stages simultaneously maturing with the gut epi-

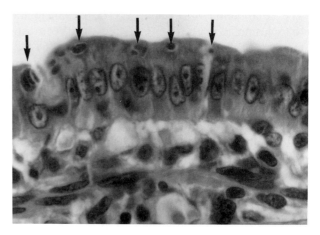

Figure 1 Photomicrograph of the enteric (gut) epithelium of a kitten infected orally 8 days previously with *Toxoplasma* bradyzoites. The enteroepithelial forms can be seen in the superficial aspect of the epithelial cell (arrows), above the cell's oval nucleus. A schizont, or dividing form, is seen at the far left. Five distinct stages have been recognized, which occur sequentially prior to the development of male and female gametes. The histology is quite complex, and the stages may be difficult or impossible to find in weaned kittens, or in the case of light infection. The male and female gametes mate to form the oocyst, which is excreted in the cat's feces. 400×.

thelium, which is eventually sloughed near the tip of the villus. Male and female gametes are formed from the enteroepithelial types, which mate to produce the oocyst. Within 3–4 days after oral infection with tissue cysts, cats will begin shedding many thousands of unsporulated oocysts daily in their feces. The shedding of oocysts generally continues for 10–20 days. The time from ingestion of an infected animal by a cat and the shedding of oocysts varies with the number of tissue cysts ingested and with whether or not an acutely infected mouse was ingested (containing tachyzoites only) (Fig. 2).

The excreted oocysts sporulate in moist soil after 24–48 hr and are then infectious to other animals. Deposits of these oocysts have been recovered from soil in Kansas and Costa Rica and have been shown to be infectious for at least 18 mo. Intermediate hosts containing tissue cysts, especially birds, spread the infection widely when they become prey for cats, which then shed oocysts to become a source of local contagion for additional intermediate hosts, including sheep, goats, dogs, swine, and humans. This transmission cycle is most apparent in the farm environment (Fig. 3). It is important to note that humans can be infected by both oocysts and tissue cysts.

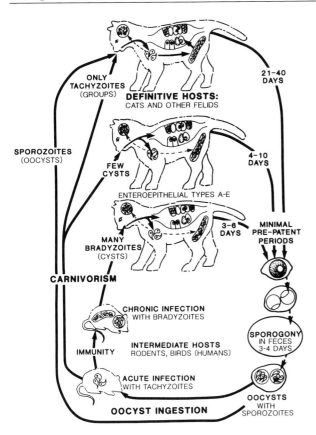

Figure 2 Life cycle of *Toxoplasma gondii*. This schematic diagram depicts the gut enteroepithelial life cycle in cats and the tissue cycle in the many intermediate hosts (represented in the drawing by mice). The rapidly proliferating tachyzoites are characteristic of the acute infection and are depicted as half-moon-shaped organisms, as shown in the mouse in the lower left-hand corner. For simplicity, tissue cysts containing the slowly growing bradyzoites are shown as round cysts in the brain and elongated cysts in the muscle, although they can occur in other tissues. Three cats are shown to trace the fate of different inocula. Carnivorism of a chronically infected animal containing some *Toxoplasma* tissue cysts leads to the enteroepithelial cycle in the cat intestine and prepatent period (before oocyst shedding) of 3–10 days, depending on the size of the infecting dose (bottom and center cats). Most cats also develop tissue infections in addition to the gut cycle, and this is depicted by the drawings of tachyzoites and bradyzoites in tissue cysts within each cat shown. However, if cats ingest mice containing only tachyzoites (in the case of acute infection), or sporozoites from oocysts (top cat), a generalized infection with tachyzoites develops first, followed by bradyzoites, which then initiate the enteroepithelial cycle. The prepatent period between ingestion of the acutely infected animal and the shedding of oocysts is then lengthened to between 21 and 40 days. [Reprinted with permission from Freyre, A., Dubey, J. P., Smith, D. D., and Frenkel, J. K., (1989). Oocyst-induced *Toxoplasma gondii* infections in cats, *J. Parasitol.* **75**, 750–755.]

Animals such as Australian marsupials evolved in a location that was originally devoid of cats (cats first arrived in Australia with the First Fleet of Captain Cook) and, thus, were not selected for resistance to *Toxoplasma*. In addition to these Australian marsupials, certain tree-dwelling New World monkeys are also highly vulnerable to toxoplasmosis and often get sick and die when they acquire an acute infection. In the latter case, there is only a relative lack of cats (and cat feces), because the monkeys rarely come down to the ground where they might be exposed, and certainly cats rarely defecate in trees. In the wild, populations of these monkeys demonstrably have a very low seroprevalence rate, although ground-dwelling animals in the same location show a high prevalence of infection. Epidemics of toxoplasmosis may thus occur in zoos that house such exotic species. The zoo cycle depends on exhibited and/or stray cats, which may contaminate cages or grain stores with their feces (and oocysts). Food contaminated with cat feces may then be fed to kangaroos or wallabies, which would then sicken and possibly die. Similarly, if the brooms used to clean the cages of wild cats were inadvertently used to clean the cages of marsupials or New World monkeys, they too could become ill.

Human epidemics are rare, but a few have been described, usually traceable to oocysts or consumption of *Toxoplasma*-infected meat that was consumed raw or rare. The incubation period in all published epidemics varied from 5 to 18 days following exposure. Oocyst exposure occurred in several different ways. In one epidemic in Atlanta, an indoor riding stable frequented by cats was implicated. It was postulated that the patients became infected from cat feces aerosolized by the trotting horses. Other outbreaks have been documented in children who often have the habit of eating soil, which could easily be contaminated with cat feces. An additional epidemic was documented in a group of United States soldiers training in the Panamanian jungle, all of whom apparently became infected when they ingested oocysts from a water source contaminated by cat feces. The epidemics traceable to raw or undercooked meat have involved hamburger, venison, and lamb chops. [*See* WATERBORNE DISEASES; FOODBORNE ILLNESS.]

Again, the vast majority of infections are completely asymptomatic. The history of scientific inquiry into toxoplasmosis initially overlooked this fact, concentrating instead on the distinctive pathology found in congenitally infected infants and in

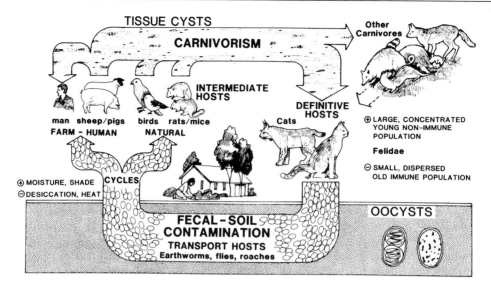

Figure 3 Transmission of *Toxoplasma* to humans and animals. All infections can ultimately be traced to oocysts excreted by cats. Oocyst contamination of soil is depicted by the round to oval symbols in the fecal–soil contamination arrow at the bottom of the drawing. At the lower right is depicted an unsporulated oocyst, and a sporulated oocyst to its immediate left, with its eight sporozoites contained within two sporocysts. The natural cycle in birds and rodents (which are easily caught by cats) is separated from the farm–human cycle, in which humans may become infected by eating the undercooked or raw meat of sheep or swine. Environmental effects on viability of oocysts in the soil is shown on the left of the drawing, whereas the effect of the cat population on the number of oocysts excreted is shown on the right. Infection of the fetus takes place when a woman, sheep, goat, or pig first contracts the infection during pregnancy (not depicted). [Reprinted with permission from Freyre, A., Dubey, J. P., Smith, D. D., Frenkel, J. K., and Erron, J. L. A., 1981, Transmission of *T. gondii, Am J. Epidemiol.* **113**, 254–269.]

those rare individuals who died of acute disseminated disease. Only through a series of serologic surveys in the last 30–40 years have we discovered how widespread *Toxoplasma* infection is among humans and animals. In France, and in many parts of Central and South America, seropositivity rates in late adulthood approximate 100%. Several factors may be operative to explain these statistical data. In France, and throughout Europe, consumption of raw or rare meat is common ("steak tartare"). This habit, acquired in adulthood, predisposes the person to infection by the bradyzoite stage of *Toxoplasma*. Beef appears relatively safe compared to pork or lamb. In tropical climates, oocysts may be more important to transmission than rare meat, because they can persist in the soil for long periods in moist climate, and stray cats are not usually controlled well. Epidemiological studies are still in progress in various locations worldwide to determine which factors are most important in *Toxoplasma* transmission. [*See* MEAT AND MEAT PRODUCTS.]

III. Infection and Disease in Cats

In adult cats, acute infection is usually mild or asymptomatic and rarely causes owners to bring their cats in for treatment, similar in this respect to humans. However, in unweaned kittens or very old cats, death may result. Newborn kittens infected experimentally with *Toxoplasma* tissue cysts have a uniformly fatal clinical course. Weanling kittens usually survive acute experimental infections, but up to 30% die suddenly between 16 and 30 days postinfection. Subacute encephalitis is nearly always found at autopsy. When cats weighed >1.5 kg, experimental infection caused no deaths. Thus, age is probably the most important factor in clinical outcome, perhaps related to maturation of the immune system. Natural disease in adult cats may present as poor appetite, diarrhea, pneumonia, fever, or encephalitis.

A serologic survey of 1000 adult cats (>3 mo of age) in the United States yielded a *Toxoplasma* seropositivity rate of approximately 40%. Cats in Central America and other tropical climates have a somewhat higher rate of seropositivity. This is probably because oocyst contamination of soil is more widespread in Central America than it is in the United States, and more rodents and birds thus become infected.

Once a cat has been infected with *Toxoplasma* and sheds oocysts, it is usually resistant to further oocyst shedding upon subsequent re-infection with *Toxoplasma*. Although resistance to oocyst shedding after primary infection occurs in >90% of cats, serologic evidence of past infection is no guarantee that the cat will not shed oocysts again. Immunosuppression, from whatever cause, presumably modifies the gut immune response and may allow oocyst shedding to recur in cats that were previously resistant.

Domestic cats have experienced an epidemic of retroviral illnesses, similar in many respects to the human experience with AIDS. Both feline immunodeficiency virus (FIV) and feline leukemia virus (FeLV) can have profound effects on the feline immune system. Experiments with FeLV-infected cats have indicated that relapse of chronic toxoplasmosis, with myocarditis, encephalitis, and pneumonitis, occurs in such cats, just as in humans infected with the AIDS virus. What effects FIV will have on chronic toxoplasmosis in cats is still unclear, but one would expect similar effects.

IV. Infection and Disease in Humans

In humans, infection with *Toxoplasma* is rarely a clinical problem. Most people do not notice the mild symptoms of acute infection, and immunity is acquired within a few days or weeks. Following acute infection of immunocompetent individuals, the organism generally persists into a chronic latent stage. However, pregnant women who become acutely infected during pregnancy may transmit the infection to their unborn child, regardless of whether or not they experience symptoms. The disease can then prove devastating to the immunologically immature fetus, where it proliferates unchecked in the brain and eye, resulting in severe mental retardation and blindness.

Cerebral toxoplasmosis has been recognized as a common infectious complication of AIDS, occurring in 5–25% of patients, depending on geographical location. This alludes to the geographic variance in seropositivity rates mentioned in Section II. Curiously, not all *Toxoplasma* seropositive AIDS patients have a relapse of their chronic *Toxoplasma* infection. Apparently, some immunity against toxoplasmosis is preserved, with only about 50% of *Toxoplasma* seropositive AIDS patients ever experiencing symptoms traceable to toxoplasmosis during life. Cerebral relapse may present clinically as a severe headache, or it may mimic a brain tumor, causing focal paralysis. Classically, a computed axial tomography scan of the patient will show "ring-enhancing" lesions (Fig. 4).

Tachyzoites are the rapidly dividing tissue forms that cause cell death and tissue necrosis if not controlled by a cellular immune response. An adequate immune response by the animal slows the proliferation of the tachyzoites and results in the formation of slow-growing bradyzoites in tissue cysts. These encysted forms of *Toxoplasma* probably persist for the life of the animal, rupturing only occasionally and causing no ill effects in normal animals, because the released organisms are rapidly destroyed by immune mechanisms. The normal immune response

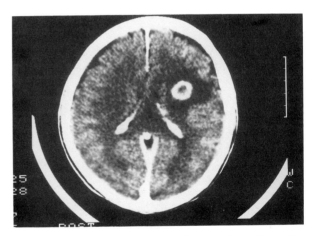

Figure 4 Computed axial tomography scan of a patient with cerebral toxoplasmosis and AIDS. The ring enhancement is caused by inflammation in the periphery of the enlarging necrotic lesion, which concentrates the radio-opaque contrast material injected by the radiologist in the characteristic ring pattern. These lesions may continue to enlarge unless checked by anti-*Toxoplasma* drug therapy, which can eventually allow the resolution of the lesion. 200 ×.

primarily depends on so-called cellular immunity from the action of T cells. The precise immune mechanisms that regulate the encystment of *Toxoplasma* are still not known. Gamma-interferon probably plays a major role, as determined by studies of the depletion of gamma-interferon in mice by monoclonal antibodies. Other mediators are probably also involved, including one that has been shown to be specific for *Toxoplasma,* and does not affect immunity to other organisms, unlike gamma-interferon.

After the animal has recovered from the acute infection with tachyzoites, and all organisms have encysted, then it can be said that a state of infection-immunity (premunition) exists in the animal, with tissue cysts rupturing periodically and releasing bradyzoites. Some bradyzoites may live long enough to enter adjacent host cells and form new ''daughter'' cysts, especially if there is some impairment in the host immune system, as is the case in AIDS or immunosuppression for cancer.

Figures 5–10 show the relationship between the host immune system and *Toxoplasma.* These figures present the pathology of the evolution of a ring-enhancing lesion in an AIDS patient with relapse of chronic toxoplasmosis. The photomicrographs actually represent a composite view of several different patients, because patients are often treated or partially treated with anti-*Toxoplasma* medications before coming to autopsy, which alters the histologic

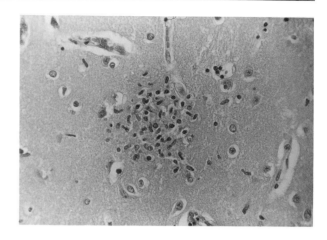

Figure 6 Immunity to *Toxoplasma.* A scar within the brain of an immunocompetent patient, formed by proliferating glial cells, typical of what one finds in an autopsy patient who was seropositive for *Toxoplasma* during life but who never had evidence of active disease. The scar can be thought of as the tombstone of a previously ruptured *Toxoplasma* tissue cyst, where all of the bradyzoites have been destroyed by an effective host immune response. Occasionally, one finds evidence of a residual *Toxoplasma* antigen in an immunoperoxidase reaction, but generally no organisms are seen. 200 ×.

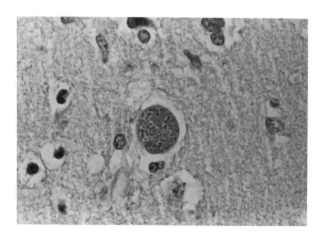

Figure 5 Immunity to *Toxoplasma.* An intact *Toxoplasma* tissue cyst within a human brain at autopsy. Such cysts are occasionally found in completely normal patients, although they are found in larger numbers in those who are immunosuppressed. The cyst contains 100–1000 bradyzoites and has evoked no inflammatory response from the host. 200 ×.

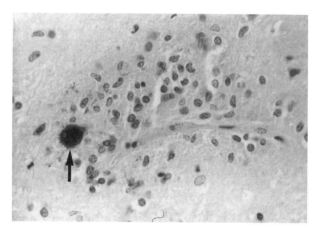

Figure 7 Immunity to *Toxoplasma.* A glial scar in an AIDS patient (immunoperoxidase reaction) with an intact tissue cyst [arrow] seen at the periphery of the scar. This can be interpreted histologically as evidence of imperfect immunity, because the normal result of intact infection-immunity (premunition) would be complete destruction of the released bradyzoites, as in Fig. 6. The defect in immunity could be the result of decreased numbers of *Toxoplasma*-specific T cells due to destruction by human immunodeficiency virus, lack of sufficient gamma-interferon, or as-yet-unknown mechanisms. 200 ×.

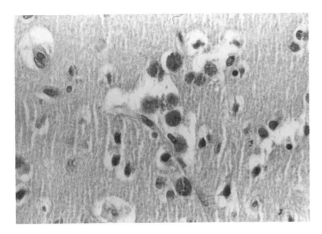

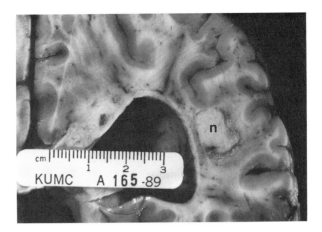

Figure 8 Immunity to *Toxoplasma*. A cluster of bradyzoites in tissue cysts in the brain of an AIDS patient. At the center of the figure, bradyzoites are seen breaking out of one cyst. These will soon begin to divide rapidly as tachyzoites in the absence of adequate immunity. 200 ×. This photomicrograph likely represents a very early stage of a ring-enhancing lesion, as seen in Fig. 4.

Figure 10 Immunity to *Toxoplasma*. This is a gross photograph of a ring-enhancing lesion in the brain of an AIDS patient at autopsy. The necrotic center (n) is usually quite soft. *Toxoplasma* tachyzoites and bradyzoites, plus inflammatory cells, are concentrated in the edge of the lesion. This accounts for the ring enhancement seen on a computed axial tomography scan postcontrast, as in Fig. 4. 1x.

appearance. By studying the sequence of photographs, one can see that tachyzoites are the destructive stage of *Toxoplasma* and are not found in an immune individual, because the host immune response would destroy them. On the other hand, bradyzoites can be found in the tissues of both immune and immunocompromised patients. The two stages can co-exist, depending on the immune status of the individual.

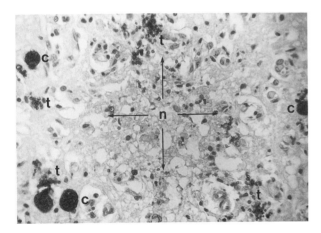

Figure 9 Immunity to *Toxoplasma*. This represents the further evolution of a ring-enhancing lesion in an AIDS patient. Note that central destruction or necrosis (n) of brain tissue has now occurred. Around the periphery of this expanding area of necrosis, one sees tissue cysts (c) and rapidly proliferating tachyzoites (t). The presence of both forms of *Toxoplasma* (bradyzoite and tachyzoite) suggests that there is at least a modicum of immunity, enough to cause the organism to encyst but probably not enough to keep it encysted for any appreciable length of time. 50 ×.

V. Prevention and Treatment

From examination of the *Toxoplasma* life cycle and transmission diagrams (Figs. 2 and 3), one can see that all infections ultimately derive from the oocyst. Environmental contamination with *Toxoplasma* oocysts thus constitutes a reservoir of infection that can be effected by public health measures. Key among these would be advocacy for good personal hygiene, feeding of *Toxoplasma*-free food (dry cat food) to pet cats, and control of stray cats. Parents may also wish to cover children's sandboxes in which local cats may defecate. Pregnant mothers should take the additional precautions of washing their hands after handling raw meat and cooking all meat well done to avoid infection from tissue cysts. It is not necessary to sell or euthanize a favorite cat because of fear of toxoplasmosis.

Another useful preventive measure takes advantage of the fact that oocysts must first sporulate at room temperature before becoming infectious, which takes roughly 24–48 hr. If kitty litter were to be changed daily and then disposed of, the risk of infection to the owner or other members of the household would be virtually eliminated.

Because cats develop good immunity to *Toxoplasma* and rarely shed oocysts upon re-infection, several vaccination schemes were developed, as outlined in Section I. The most practical is the recently described mutant strain, designated T-263, which immunizes about 84% of cats. This mutant apparently undergoes a partial enteric cycle, stopping short of oocyst production, and immunizes the cat against oocyst shedding upon challenge with bradyzoites of an oocyst-producing wild strain. This vaccine could prove useful for cat owners who find it difficult or impossible to adequately follow the suggested hygienic measures outlined earlier.

The combined use of the cat vaccine (T-263) and the animal vaccine (*ts*-4) could be envisioned, especially in the farm environment, as diagrammed in Fig. 3. Under these conditions, use of the *ts*-4 vaccine in sheep or pigs, coupled with use of the T-263 vaccine in farm kittens, could reduce or eliminate oocyst shedding by cats. The resultant decrease in the number of oocysts in the environment would eventually reduce the number of *Toxoplasma* infections suffered by ground-feeding livestock.

Treatment of toxoplasmosis, in humans and animals, still depends on use of sulfonamide drugs and pyrimethamine, a folic acid antagonist. Because of the high incidence of adverse reaction to sulfonamides, alternative drugs such as clindamycin are now being tried in AIDS patients, with some success. However, all of these drugs act mainly on the tachyzoite stage of *Toxoplasma*. If drugs could be found that would kill the bradyzoite stage of the organism, then it might be possible to achieve sterile immunity within the host. That is, all of the bradyzoites would be killed by the drug, leaving no reservoir of infection for reactivation of disease. The host would still retain good immunity against reinfection and symptomatic disease. The accomplishment of this goal would be very advantageous to that increasingly large group of immunosuppressed patients who are currently chronically infected with *T. gondii*.

Bibliography

Fishback, J. L., and Frenkel, J. K. (1991). *Sem. Vet. Med. Surg. (Small Animal)* **6**, 219–226.

Frenkel, J. K. (1990). *J. Am. Vet. Med. Assoc.* **196**, 240–248.

Frenkel, J. K. (1990). *J. Am. Vet. Med. Assoc.* **196**, 233–240.

Frenkel, J. K., and Escajadillo, A. (1987). *Am. J. Trop. Med. Hyg.* **36**, 517–522.

Frenkel, J. K., Pfefferkorn, E. R., Smith, D. D., and Fishback, J. L. (1991). *Am. J. Vet. Res.* **52**, 759–763.

Frenkel, J. K., and Roberts, T. (1990). *J. Am. Vet. Med. Assoc.* **196**, 249–256.

Hohlfeld, P., Daffos, F., Thulliez, P., Aufrant, C., Couvreur, J., MacAleese, J., Descombey, D., and Forestier, F. (1989). *J. Pediatr.* **115**, 765–769.

McLeod, R., Mack, D., and Brown, C. (1991). *Exp. Parasitol.* **72**, 109–121.

Pfefferkorn, L. L., and Pfefferkorn, E. R. (1980). *Exp. Parasitol.* **50**, 305–316.

Remington, J. S., and Desmonts, G. (1990). Toxoplasmosis. **In** "Infectious Diseases of the Fetus and Newborn Infant" 3rd Edition (J. S. Remington and J. O. Klein, eds.). pp. 89–195. W. B. Saunders, Philadelphia.

Transcriptional Regulation, Eukaryotes

Institute of Genetics, Bulgarian Academy of Sciences

I. Cis-Acting Regulatory Sequences and
   Trans-Acting Protein Factors
II. Chromatin and Transcription
III. Loop Organization and Transcription
IV. Concluding Remarks

Glossary

Chromatin The state of DNA in the eukaryotic
nucleus where linear DNA molecules are compacted by interaction with proteins, mainly histones

DNA supercoiling Topological states of the double helix caused by partial unwinding or over-
winding of the helix as compared to the thermodynamically favored B-form

Enhancer Region enhancing transcription from
a distance in a position- and orientation-
independent fashion

Histones Highly basic, evolutionary conservative,
low molecular mass proteins binding to DNA to
form chromatin structure

Loops Organization of linear DNA molecules into
topologically constrained domains by attachment of DNA sequences to components of the
nuclear matrix

Nuclear matrix Proteinaceous structure of the nucleus remaining after removal of the nuclear
membrane and extraction of the histones

Promoter Control region in the immediate vicinity of the transcription start site to which RNA
polymerase binds

Transcription factors Sequence-specific DNA-
binding proteins that activate or inhibit transcription by binding to elements in promoter or
enhancer regions

FOR QUITE A LONG TIME it was believed that
with respect to the organization and the regulation of
the expression of the genetic material what was true
of *Escherichia coli* was also true of man. With the
advent of recombinant DNA technology our knowledge of eukaryotic genes, of their structure and expression, has increased enormously. It became clear
that although some basic principles are shared
among prokaryotes and eukaryotes, there are also
significant differences in the organization and regulation of individual genes. Protein-coding genes in
eukaryotic cells are often discontinuous [i.e., coding
sequences (exons) are interrupted by noncoding sequences (introns)]. The primary product of transcription is an RNA copy of the entire length of the
gene that contains sequences well beyond the 5' and
3' end of the coding sequence itself. The primary
transcript is modified at its two ends ("capped" at
the 5' end and, as a rule, polyadenylated at the 3'
end) and then subjected to a complex splicing
process that leads to excision of the introns and
splicing of the exons with the formation of mature
messenger RNA (mRNA) molecules. These are then
transferred from the nucleus to the cytoplasm where
they serve as templates for protein synthesis.

Prokaryotes and eukaryotes are also distinct with
respect to the regulation of the expression of their
resident genes. Most, if not all, of the genes present
in a bacterial cell are potentially active (i.e., they are
either actually transcribed or can be quickly activated on changes of growth conditions). The situation in the nucleus of a eukaryotic cell is significantly
different. Only about 5–10% of the structural genes
are active at any given moment; some other genes
can be activated later in development or under the
influence of some environmental stimuli; the majority of genes are, however, permanently "silent" or
repressed and can never be activated in a certain cell

type. This selectivity of expression of certain sets of genes in some cells but not in others forms the molecular basis of development and differentiation, processes that are unique to eukaryotic organisms. The mechanisms that control gene inactivation in eukaryotes are remarkably efficient: They are estimated to be some 10^5-fold stronger than those effected by bacterial repressor proteins.

Transcriptional regulation in eukaryotic cells is exerted at at least three levels: (1) via interactions of trans-acting factors, usually proteins, with cis-acting regulatory elements (DNA sequences linked to a gene); (2) via organization of the nuclear DNA in chromatin; and (3) via topologically constrained loops formed by attachment of DNA to components of the nuclear matrix.

I. Cis-Acting Regulatory DNA-Sequences and Trans-Acting Protein Factors

A. Cis-Acting Regulatory Elements

Detailed studies of several prokaryotic promoters and the regulatory sequences associated with them, as well as of the proteins interacting with them to activate or repress transcription, have led to the following picture. Conserved sequences located 10–35 base pairs (bp) upstream from the transcription initiation site are required for binding of the RNA polymerase molecule. Additional sequences adjacent to the constitutive polymerase binding sites are involved in positive control that is exerted by binding of activator proteins to these sequences next to the polymerase molecule. For negative regulation, a repressor protein binds to control sequences overlapping the constitutive ones, interfering with RNA polymerase binding. Thus, the system controlling gene activity in prokaryotic cells consists of two basic interacting moieties: Short stretches of DNA present in cis to the gene and trans-acting factors of protein nature encoded somewhere else in the genome.

The answer to the question whether or not eukaryotic cells regulate gene activity in the same general fashion became possible with the advent of the techniques collectively called "reverse or surrogate" genetics. In the classical genetic approach one identifies a mutant cell or organism and then looks for the changes in gene structure or expression that lead to the mutant phenotype. In "reverse ge-

netic" approaches native cloned DNA sequences are first manipulated *in vitro* and then tested for transcriptional activity in some kind of an *in vitro* or *in vivo* transcriptional assay (Figs. 1 and 2).

The structure of eukaryotic promoters turned out to be similar to that of bacterial genes. While most bacterial promoters are able to bind RNA polymerase specifically, the eukaryotic enzymes cannot recognize their cognate promoters by themselves and do so via interaction with the trans-acting proteins that do possess sequence specificity of binding. There are several conserved oligonucleotide sequences, usually termed "boxes," present to about 100 bp from the start of transcription. These elements are required for accurate positioning of the transcription initiation site and for efficient initiation. The most frequently encountered elements include the **TATA** and **CCAAT** boxes and the **GGGCGG** motif. In addition to the common conserved boxes, the promoters of different genes contain other elements that have been recognized to participate in transcriptional regulation. The number of these elements can vary with the gene and can sometimes be quite impressive (for an example see Fig. 3). The mutual disposition of the elements can also be highly variable. Different elements or combinations of them determine the developmentally-regulated and tissue-specific expression, the inducibility of transcription under a variety of inner and outer stimuli, the cell-cycle specificity of expression, and so on. All these cis-acting sequences act, as in bacteria, via sequence-specific binding of some regulator, trans-acting proteins (see following discussion). Certain elements bind ubiquitous trans-acting factors, others bind tissue- or developmental-stage specific factors, while still others bind factors that lead to inducible transcription.

A surprising development was the discovery in eukaryotic genes of control elements activating transcription at a great distance (sometimes over distances of up to 10 kbp from the transcriptional start site). The study of these elements, called *enhancers,* unravelled some extraordinary properties: They can elevate transcription from both homologous and heterologous promoters with exceptional positional flexibility, independent of their orientation or position relative to the gene (5' or 3' of the gene). These properties serve as functional criteria to define enhancer elements. Thus, for instance, the upstream activating sequences (UAS) typical of yeast genes seem to behave like enhancers in that they function when their position or orientation with respect to the

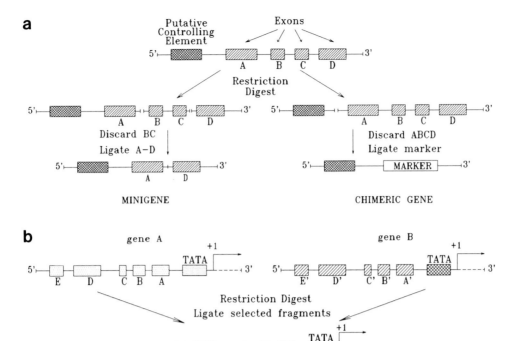

Figure 1 Construction of hybrid genes and hybrid promoters. In (a) the eukaryotic gene is schematically represented with its controlling elements in the 5' region and its separate exons designated A, B, C, and D. Alternative restriction digestions and ligation of the digestion products lead to the construction of minigenes or chimeric genes. Marker genes encode easily identifiable traits. [Adapted from Kelly, J. H., and Darlington, G. J. (1985). *Ann. Rev. Genet.* **19**, 273–296, with permission.] In (b) the transcription regulatory region is presented as a combination of several distinct gene-specific *cis*-elements (marked A to E in gene A and A' to E' in gene B). The generally occurring TATA box is also shown.

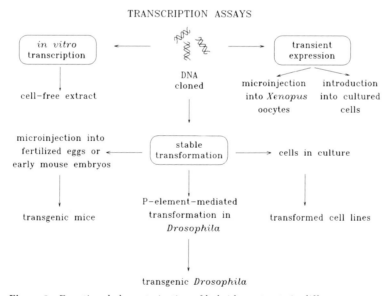

Figure 2 Functional characterization of hybrid constructs in different types of *in vitro* and *in vivo* transcription assays. Wild type or manipulated cloned DNA is introduced either into living cells or in cell-free extracts where it is transcribed. The products of transcription are analyzed either immediately (*in vitro* transcription) or following some time (a few hours in the case of oocyte injection, 1–2 days in the case of transient transfection of cells in culture, and weeks or months in the case of stable transformation) to assess the fidelity and level of the transcription process.

Figure 3 Schematic of human *hsp70* promoter. Promoter sequences from −1 to −188 are represented schematically by a line. The boxes represent the approximate locations of transcription factor-binding sites in the promoter. The factors that bind these regions are indicated above the boxes; the corresponding genetic elements are indicated below the boxes. Black and hatched boxes indicate protein-DNA interactions characterized by two different methods; *, sites where binding activity has been detected but not identified. [From Williams, G. T., McLanahan, T. K., and Morimoto, R. I. (1989). *Mol. Cell. Biol.* **9,** 2574–2587, with permission.]

TATA box has been altered. They are, however, not defined as enhancers as they do not function in positions downstream of the TATA box. Evidence of regulatory sequences located hundreds of nucleotides away from their promoters has been reported also for some *Escherichia coli* operons (*gal, ara, lac, deo,* etc.).

Enhancers have been characterized in a large number of viral and cellular genes (Fig. 4), and it has turned out that enhancers are highly complex, both structurally and functionally. They contain a large number of transcription elements, many of which overlap; some of them are functionally redundant. The properties of some enhancers observed *in vivo* have been reproduced to some extent *in vitro* using purified DNA templates and nuclear extracts. These studies suggested that enhancer function depends on interaction with trans-acting factors. Direct analysis using techniques that can probe DNA–protein binding *in vitro* has demonstrated that all sequences that were found by mutational analysis to be essential for function bind specific factors.

An intriguing feature of some enhancers is their striking cell-type specificity. Thus, for instance, the immunoglobulin heavy chain enhancer preferably mediates transcriptional activation in cells of lymphoid origin, in particular in cells of the B lineage. The tissue-specific expression is achieved via the concerted action of the respective enhancers and promoters because constitutively-active enhancers such as the SV40 enhancer mediate cell-specific transcription in combination with immunoglobulin promoters.

Despite the intense effort devoted to the study of enhancers, the mechanism of their action remains elusive. Some models imply an indirect effect via induction of DNA conformational or chromatin structural changes or via directing genes into compartments of the nucleus rich in polymerase and other components of the transcriptional machinery. Alternative models imply a direct involvement of enhancers and proteins specifically bound to them in the formation of transcriptional complexes. This can be accomplished either by sliding of the bound factors from their original binding site in the enhancer to the transcriptional start site, or by bringing the factors into direct contact with the transcriptional machinery by looping out the intervening DNA. The case for DNA looping has been experimentally demonstrated by several independent approaches including electron microscopic visualization of loops formed *in vitro* by two regulatory proteins linking two distant DNA sites.

Other cis-acting elements include the so-called *silencers* (elements that inhibit transcription), *terminators* (elements that determine the termination of transcription) and so on. They act by mechanisms analogous to those previously described for promoters and enhancers. Note that transcription termination can be used as a mechanism for control of gene expression (see, for example, Fig. 5 for a scheme demonstrating the regulation of the adenovirus major late transcription unit).

B. Trans-Acting Protein Factors

Significant progress has been made in the study of proteins that specifically bind to regulatory DNA elements, thus, conferring promoter selectivity to RNA polymerase. Various *in vitro* DNA binding assays allow detection and identification in crude extracts of sequence-specific DNA-binding factors. The DNA-binding properties of these low-abundance proteins can be used for their purification via DNA-affinity chromatography.

Cloning of the genes encoding trans-factors has been achieved via several alternative approaches

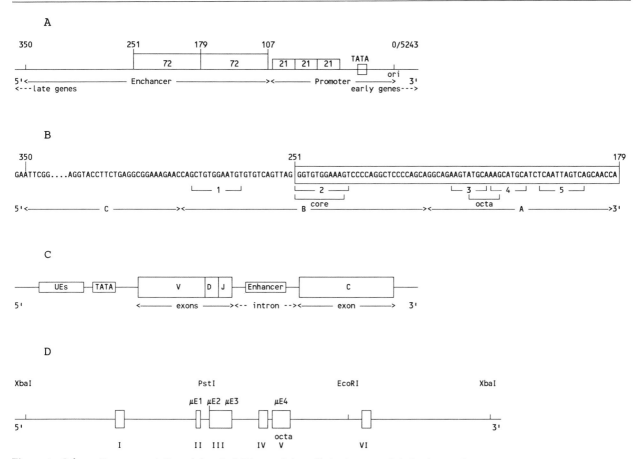

Figure 4 Schematic representation of the viral SV40 and the cellular immunoglobulin heavy chain (IgH) enhancers. (A) SV40 regulatory region. The enhancer (including the two 72 bp repeats from position 107 to 251 and sequences further upstream up to position 350), the promoter (including the TATA box and the three 21 bp repeats) and the origin of replication (ori) are shown. Broken arrows indicate the orientation of transcription of the early and late viral genes. (B) Sequence of the SV40 enhancer and the structural elements (brackets 1 to 5) responsible for activity. Only one of two 72 bp repeats is displayed (nucleotides 179–251, inside the rectangle). The enhancer is dissected into three functional units (domains A, B, and C indicated at the bottom), each of which can cooperate with the others or with itself, if duplicated, to enhance transcription. (C) IgH gene and its regulatory elements. Variable (V), diversity (D), joining (J), and constant (C) exons are rearranged for transcription in lymphoid cells. The promoter consists of a TATA box and upstream elements (UEs). The enhancer is situated in the intron between the rearranged VDJ and the C exons. (D) IgH enhancer within the 1 kb *XbaI* intron fragment. Boxes I to VI represent areas of DNA-protein interaction; μE1 to μE4 are occupied only in lymphoid B cells. [Adapted from Hatzopoulos, A. K., Schlokat, U., and Gruss, P. (1988). *In* "Frontiers in Molecular Biology. Transcription and Splicing" (Hames, B. D. and Glover, D. M., eds.), pp. 43–96. IRL Press, Ltd., Washington, DC, with permission.]

Figure 5 Regulation by termination in the major late transcription unit of adenovirus type 2. This unit encodes five families of mRNAs designated L1–L5. The major late promoter is active early as well as late in infection, but distinct mRNA populations exist at the two states of infection. Early in infection transcription terminates between L3 and L4, whereas late in infection it does not terminate until past the L5 polyadenylation site. Thus, termination precludes L4 and L5 expression early in infection. [Adapted from Friedman, D. A., Imperiale, M. J., and Adhya, S. L. (1987). *Ann. Rev. Gen.* **21,** 453–488, with permission.]

and has turned out to be important in several respects:

1. Expression of cloned wild-type and mutated genes in bacteria allows the production of the respective factor in quantities amenable to biochemical analysis.
2. Comparison of sequences at the DNA and protein levels helps in identifying important functional regions of the respective molecules.
3. The availability of cDNA clones allow structure-function studies by a parallel estimation of the *in vitro* DNA-binding properties of wild-types and mutant factors and of their transcriptional activities in transient expression assays. To that end, tissue culture cells are simultaneously transfected with the respective cDNA cloned in an expression vector and with a reporter gene carrying in its promoter binding sites for the factor.
4. The availability of genomic clones will make it possible to study their regulatory *cis*-acting elements and the trans-acting factors involved in the activation–repression processes.

This knowledge is essential to the understanding of the cascades controlling the switches in gene expression during successive steps of development and differentiation. The study of transcription factors (TFs) has shown that different factors are composed of several, usually separable domains, the most important of which are the DNA-binding and the trans-activation domains (Fig. 6).

1. DNA-Binding Domains

Specific DNA-binding regions of TFs have been localized to small autonomous protein domains containing 60–100 amino acids. The binding domains are necessary but not sufficient for trans-activating function. Although detailed structural information is not yet available, the primary sequences and biochemical properties of the domains studied thus far are suggestive of the existence of at least three basic structural motifs participating in the sequence-specific binding of the factors to their cognate sequences.

The *helix-turn-helix* (HTH) motif is a structure initially characterized in prokaryotic activator and repressor proteins. It contains two α-helices separated by a β-turn (Fig. 7); the geometry is highly conserved despite considerable sequence variability. The amino acids in the so-called "recognition helix" directly contact bases exposed in the major groove of DNA; the other α-helix lies across the major groove and makes nonspecific contacts to DNA. The prokaryotic HTH proteins bind as dimers to symmetric DNA sites.

In eukaryotes, the HTH motif was first identified as a conserved protein segment in a family of Drosophila proteins known to be involved in decision making in early development: They specify the identity and spatial arrangement of the individual body segments. Later, similar domains called "homeodomains" were found in many other eukaryotic organisms (Table I), ranging from yeast to man. In several cases, the homeodomain proteins have been shown to specifically bind defined DNA sequences, and

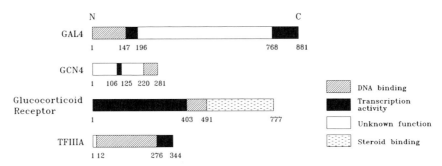

Figure 6 Functional domains of eukaryotic TFs. The location of amino acid sequences required for DNA binding, transcription activity, or steroid hormone binding is shown for several TFs. The functional domains are represented by boxes as indicated in the figure. The number of amino acid residues from the amino terminus of the respective proteins is indicated. [Adapted from Maniatis, T., Goodbourn, S., and Fischer, J. A. (1987). *Science.* **236**, 1237–1245. Copyright 1987 by the AAAs.]

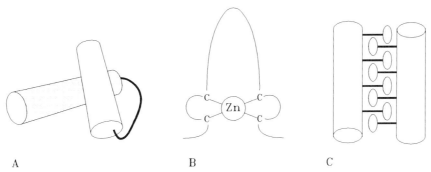

A   B   C

Figure 7   Schematic representation of the three structural motifs encountered in DNA-binding domains of eukaryotic TFs. (A) HTH (Adapted from Wharton, R., and Ptashne, M. (1986). *Trends Biochem. Sci.* **11,** 71–73, with permission). (B) Zn finger [From Struhl, K. (1989). *Trends Biochem. Sci.* **14,** 137–140, with permission]. (C) Leucine zipper [From Landschulz, W. H., Johnson, P. F., and McKnight, S. L. (1988). *Science.* **240,** 1759–1764. Copyright 1988 by the AAAs]. Alpha-Helices are represented by cylinders; conserved cysteine residues are shown by C; the interdigitating protrusions symbolize leucine side chains.

there have been some direct demonstrations that these proteins can be considered as bona fide TFs.

Conclusive evidence that HTH motifs are present in TFs other than the developmental regulators came from studies on three recently cloned mammalian proteins: The octamer-binding factors OCT-1 and OCT-2 and the pituitary-specific factor Pit-1; the HTH motif was also identified in proteins known to activate transcription through specific binding to cis-elements present in some enhancer sequences.

A second type of DNA-binding domain is the so-called "*Zn finger*" (Fig. 7). This motif was originally identified in TFIIIA, a TF that binds to the internal control region of the 5S RNA gene and activates RNA polymerase III transcription. Two types of Zn fingers have been described. The first one, exemplified by Sp1, consists of about 30 amino acid residues with two cysteine and two histidine residues that stabilize the domain by tetrahedrally coordinating a Zn^{2+} ion. The region of about 12 amino acids between the invariant cysteine-histidine pairs contain scattered basic residues and several conserved hydrophobic residues and projects out as a finger. The second class of Zn fingers, exemplified by the DNA-binding domain of steroid hormone receptors, uses two pairs of cysteines for coordination of Zn^{2+}. This is a distinct structure, as replacement of one cysteine pair by a histidine pair abolishes DNA-binding activity. The Zn finger-containing TFs display different DNA sequence specificities. This suggests that although the Zn fingers have an essential role in

structuring the binding domains, the determinants of binding specificity must lie elsewhere.

A new class of proteins has been identified that uses an alternative motif for DNA binding, the "*leucine zipper*" (Fig. 7). This protein class includes the mammalian enhancer binding protein C/EBP, the products of the oncogenes *fos* and *jun*, the yeast TF GCN4 involved in the general control of amino acid biosynthetic genes, etc. (see Table I). These factors share a bipartite region of primary sequence similarity consisting of a highly basic stretch of about 30 amino acid residues immediately followed by a region containing several leucine residues positioned at every seventh position. Display of the latter sequence on an idealized α-helix revealed notable amphipathy: One side of the hypothetical helix was predominantly composed of hydrophobic amino acids (particularly leucines), while the other side was composed of residues with charged or polar side chains. It was proposed that the leucine side chains extending from one α-helix interdigitate with those displayed from a similar α-helix of a second polypeptide, facilitating dimerization. Thus, the interaction of two polypeptide chains leads to the formation of a Y-shaped molecule: The stem of the Y corresponds to a coiled pair of α-helices and the bifurcating arms constitute the DNA contact surfaces. Because the bifurcation point closely approaches the center of the DNA target (which possesses a dyad symmetry), each arm is positioned so as to track along each half of the recognition site (*scissors-grip* model for DNA recognition).

Table I   Specific DNA-Binding Motifs in Transcription Factors and Proteins Controlling Development

Helix-turn-helix	Zn finger	Leucine zipper	S(T)PXX
Homeodomain proteins[a]:	Segmentation proteins[a]:	Yeast:	Homeodomain proteins[a]
Drosophila:	Kruppel, Hunchback, Snail,	GCN4	Segmentation proteins[a]
ANTP, FTZ, ENG, DEF	Serendipity		Steroid hormone receptors
Other species: human, mouse, rat,	Other developmental regulators:	Mammals:	Yeast transcription factors
Xenopus, chicken, earthworms	Aspergillus blrA gene product	jun, fos, myc oncoproteins	
Transcription factors:	Steroid hormone receptors	C/EBP, CREB, TFE3, OCT-2	Fos-gene products
Yeast:	Yeast transcription factors:	Plants:	RNA-polymerases
MATα2, MATa1	HAP, ADR1, PDR1, GAL4, SWI5,	HBP-1	Some histones
Mammals:	ARGRII		
OCT-1, -2, Pit-1, TFE3	Miscellaneous:		
	TFIIIA, Eryf1, Sp1, TDF		

[a] Homeodomain and segmentation proteins are involved in the control of early development in *Drosophila*.

Other DNA-binding domains have also been proposed, for example α-helices with high density of basic amino acids and the SPXX (serine-proline-X-X) motif (Table I). The primary structures of several cloned genes for mammalian TFs indicate that the number and the types of DNA binding domains are not restricted to the ones described here.

2. Transcriptional Activating Domains

The domains responsible for transcriptional activation are also relatively short stretches of the molecules (30–100 amino acids) and are separate from the DNA-binding domains (Fig. 6).

The first activation regions were identified and analyzed in detail in studies of the yeast TFs GAL4 and GCN4. These domains possess significant negative charge and can form amphipathic α-helical structures. Seemingly, the primary amino acid sequence is not of great importance. This view is substantiated by several independent lines of evidence. Thus, for instance, GAL4 has two separate acidic domains that, despite their lack of obvious sequence homology, are functionally redundant (i.e., each of them is active in the absence of the other one). When linked to heterologous DNA-binding domains, they can activate transcription from genes with a binding site for the heterologous domain. Active factors are readily formed by attaching random bits of the *E. coli* genome to the DNA-binding domain of GAL4, as long as the *E. coli* piece can form amphipathic α-helix, bearing negatively charged residues along one surface and hydrophobic residues along the other. Interestingly, addition of an yeast activating

domain to a repressor molecule can transform it into an activator.

Activating domains with similar properties were described in the glucocorticoid hormone receptor, the AP-1/Jun TF, etc. It is envisaged that acidic activation domains may act by interacting in a relatively nonspecific way with some of the general components of the transcriptional machinery, such as the TATA-binding TFIID or the polymerase itself.

More recently, additional types of activating domains have been identified, such as a glutamine-rich domain in Sp1 and a proline-rich domain in CTF/NF1. Still other types of structures of activating domains will undoubtly be identified in the future.

C. Families of TFs with Related DNA-Binding Specificities

Fractionation of nuclear extracts by DNA-affinity chromatography usually leads to protein preparations containing several proteins. Thus, for instance, a series of mammalian proteins encoded by the genes *c-jun, junB, junD, fos,* and *fra-1* (AP-1 family) specifically bind to oligonucleotides bearing the sequence TGACTCA. The shared binding specificity of AP-1 family members correlates with the conservation of their leucine zipper-associated binding domains. Formation of homo- and heterodimers between members of the family may create a variety of TFs with different functional properties to satisfy the needs for complex regulation of genes with AP-1 binding sites. An example of extreme complexity of regulation is the *fos* gene itself, whose expression may have different consequences to the cell depending on the particular stimulus of activation.

Another family of mammalian TFs recognizes the frequently encountered promoter element CCAAT. Several CCAAT-binding proteins are produced by differential splicing of the primary transcript from the CTF/NF1 gene. Other unrelated genes also code for CCAAT-binding proteins.

Studies on factors with similar DNA-binding specificities demonstrate that sequence specificity is not the only factor determining the promoter selectivity of a TF. An intriguing example is the octamer binding factors OCT-1 and OCT-2, which bind to their cognate site with apparently identical specificities *in vitro;* yet only one of these factors, OCT-2, can selectively activate transcription of immunoglobulin genes in lymphoid cells.

D. Regulation of TF Activities

A question that reasonably follows from the recognized cis-trans level of gene regulation is how TFs themselves are regulated. There are several possible ways for the cell to do so. (1) TFs can be regulated at all levels of their synthesis, starting from transcription of their genes down to the translation of the respective mRNA. Thus, the *fos* gene is transcriptionally regulated by its own protein product; alternatively, the yeast GCN is regulated at the translational level. (2) Some TFs are regulated by ligand binding (heme, galactose, steroid hormones, etc.). (3) A commonly used pathway of activation is via postsynthetical modifications of the protein molecule. A typical example is the increased phosphorylation of the heat shock TF (HSTF) in yeast, which correlates with increased transcription of heat shock genes; interestingly, the factor is phosphorylated when already bound to the respective promoters. (4) Protein–protein interactions can constitute an important regulatory pathway. Cases have been reported when, in the absence of inductive signal, the TF is found in the cytoplasm in an inactive association with another protein. The inductive stimulus causes the inactive complex to dissociate, the freed TF is translocated into the nucleus where it binds its appropriate response element and affects transcription. Thus, inactive steroid hormone receptors associate with the abundant cytoplasmic heat shock protein, hsp90, via the steroid-binding domain (Fig. 6). The factor NF-κB is regulated by association

with a specific cytoplasmic inhibitor, IκB; TPA, the phorbol esther inducing the NF-κB regulated genes, apparently acts through changes in IκB rather than in NF-κB.

Other examples of regulating TFs via protein–protein interactions include the known promiscuous activating action of the early adenovirus gene product E1A. This protein, not a sequence-specific binding protein by itself, activates a very broad range of viral and cellular promoters. Its action could be explained by: (1) its ability to bind any accessible DNA nonspecifically; (2) its ability to interact with a variety of cellular TFs; and (3) the presence, particularly in the N-terminal exon, of several clusters of negatively charged residues that could provide acidic activating surfaces suited to different cellular factors. Similar activity is described for another viral trans-activator, VP16, encoded by a herpes simplex virus gene. Another interesting example of a viral protein, trans-inhibiting transcription of specific viral genes is the large T antigen of simian virus 40; it acts through complex formation with the cellular TF AP-2, thereby inhibiting the binding of the latter to SV40 enhancer elements. [*See* TRANSCRIPTION, VIRAL.]

Thus, the picture emerging from studies of cis-elements and trans-factors and their interactions is exceedingly complex. Each gene has a particular combination of a number of positive and negative *cis*-elements that are uniquely arranged. These elements interact with one another, are often redundant, can be positioned at any point of the gene even at relatively large distances from the transcription initiation site, and sometimes operate in both orientations. Some elements in the regulatory gene regions are common to many genes transcribed by a particular type of polymerase. A large number of less common elements participate in cell- or developmental-specific transcriptional regulation; still other elements are involved in inducible expression in response to inner and outer stimuli (growth factors, hormones, heat shock, heavy metals, agents causing malignant transformation, etc.). These elements form specific binding sites for TFs that activate or repress transcription. Because the structure of a gene is invariable in every cell of an organism (with some notable exceptions), it is the abundance and activity of the DNA-binding factors in different cells that determine gene expression. The interplay amongst the numerous cis-acting elements and trans-factors ensures the subtle regulation of gene expression, characteristic of cells in a eukaryotic organism. The mechanisms by which TFs activate or repress initiation of transcription are still unknown.

II. Chromatin and Transcription

One of the fundamental differences between prokaryotic and eukaryotic cells lies in the degree of compactness and the structural organization of DNA.

Prokaryotes, like eukaryotes, face the problem of how to compact DNA in a usable form inside the cell. The degree of compaction is achieved by two distinct levels of condensation: The organization of DNA in independent supercoiled domains and condensation by DNA-binding proteins. The nature of these proteins is still not known but it is clear that they do not organize DNA in any repeating structure, at least not a structure comparable to the eukaryotic nucleosome (see following section). In addition, no ordered higher-order structures exist in bacteria.

A. Chromatin Structure

DNA in the eukaryotic cell is highly condensed. According to current estimates the degree of compaction is about 250-fold in the interphase nucleus and about 10,000-fold in mitotic chromosomes. This compaction is achieved at several successive levels through interactions with histones and nonhistone proteins with the formation of chromatin and loops (see also Section III). The fundamental repeating structure of chromatin is the *nucleosome:* The core of the nucleosome consists of 146 bp of DNA wrapped in 1.75 turns around a core of eight histone molecules (two molecules each of histones H2A, H2B, H3, and H4); core particles are connected to one another by linker DNA whose length varies from 15 to over 100 bp; the fifth histone class, the lysine-rich histone H1 binds to the linker DNA, sealing off two turns of DNA around the nucleosome (Fig. 8). The linear array of nucleosomes that appears as "beads-on-a-string" under the electron microscope is further compacted into higher-order structures (Fig. 8). The formation or maintenance of the higher-order structures requires histone H1. Several models have been proposed for these structures; one of them is illustrated in Fig. 8.

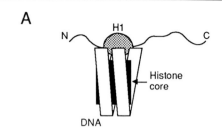

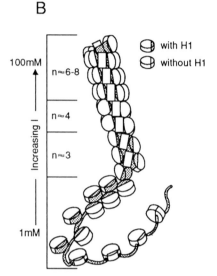

Figure 8 Schematic representation of chromatin structure. (A) Nucleosome: the DNA is wrapped around the histone octamer (core); the two turns of DNA are sealed off by the globular domain of histone H1 whose random-coiled N- and C-tails are also shown. (From Bradbury, E. M. (1983). *In* "Mobility and Recognition in Cell Biology" (H. Sund, and C. Veeger, eds.), pp. 173–194, Walter de Gruyter & Co., Berlin, New York, with permission.) (B) Idealized drawing of the higher-order structure of chromatin containing H1 with increasing ionic strength. The open zigzag of nucleosomes closes up to form helices with increasing numbers of nucleosomes per turn (n). When H1 is absent, no zigzags or definite higher-order structures are found. [From Thoma, F., Koller, T., and Klug, A. (1979). *J. Cell Biol.* **83**, 403–427, with permission of the Rockefeller University Press.]

B. Transcriptionally Active Chromatin

Many studies have been devoted to elucidation of the characteristics of transcriptionally active chromatin. Initially, these were performed on "fractionated" chromatin; however, artifact-prone fractionation procedures were circumvented by the use of total chromatin and cloned DNA sequences to probe chromatin structure of defined genes in their transcriptionally active or inactive state.

1. General Biochemical Characteristics

The first biochemical evidence that the structure of active gene chromatin differs from that of inactive genes came from nuclease digestion studies. Relatively large regions containing a particular active gene were found to possess an increased sensitivity to the enzyme DNase I. This property of active chromatin was interpreted as reflecting a more "open," accessible conformation. Active chromatin is enriched in acetylated forms of the core histones, in nonhistone high-mobility-group (HMG) proteins, in topoisomerase I; on the other hand, active chromatin contains hypomethylated DNA, is sometimes impoverished in histone H1. These and other biochemical characteristics of active chromatin have so far only indicated that chromatin structure of a particular gene correlates with its transcriptional activity. No causal relationships have, however, been established; it is not clear whether these structural features are a cause or a consequence of transcription.

2. DNase I Hypersensitive Sites and Nucleosome Phasing

There are two important aspects of chromatin structure which seem to be directly related to the cis-trans regulation described above. The first one concerns the so-called *DNase I hypersensitive (DH) sites,* initially reported in studies of *Drosophila* heat shock genes and SV40 viral chromatin. All DH sites share the characteristic of being nucleosome-free regions of the chromatin fiber, thus representing "open windows" that allow enhanced access of trans-factors to resident cis-elements. These accessible regions are operationally defined by their highly elevated sensitivity to nuclease cleavage or chemical modification (typically two orders of magnitude higher than that of other regions in bulk chromatin and an order of magnitude higher than the general sensitivity of the large active regions mentioned previously).

The sites usually encompass a unit length close to that of the nucleosomal repeat (approximately 200 bp) or several multiples thereof, reflecting the absence of canonical nucleosome(s). The fine structure of any given site exhibits multiple "hot" and "cold" spots, the "cold" protected ones reflecting the presence of bound trans-factors.

DH sites appear to be an essential feature of chromatin structure, being encountered in fungi, plants and animals. Evidently, there are functionally distinct classes of DH sites, as they are mapped to DNA sequences of different functions, such as promoters, enhancers, UAS, silencers, origins of replication, recombination elements, and sites within and around telomeres and centromeres.

Within a given cell type, DH sites in regulatory gene regions fall into two major categories: *Constitutive* and *inducible.* Constitutive sites exist independently of the actual status of transcription and are often present in promoters of potentially active genes. Inducible sites appear prior to transcription and often persist long after removal of the induction stimulus. Sites that appear transiently during embryogenesis are called *developmental* and those that exist only in certain cell types are *cell-* or *tissue-specific.* A typical pattern of DH sites for a gene with tissue-specific expression, the chick lysozyme gene, is presented in Fig. 9.

The hypersensitivity of a site is usually determined by its underlying or flanking DNA sequences. Mutational analysis of DH sites directly demonstrates that sequences representing binding sites for trans-factors are necessary for generating hypersensitivity. Furthermore, appearance of sites in inducible genes is accompanied by binding of factors to their cognate sequences. Direct evidence for this mechanism of establishment of DH sites is presented by order-of-addition experiments for *in vitro* reconstitution of sites onto some cloned DNA templates: Sites are formed only if nuclear extracts or purified TFs are present prior to or during nucleosome assembly. (Other studies indicate that nucleosome formation inhibits transcriptional initiation.)

The second aspect of chromatin structure related to the cis-trans regulation is the so-called *nucleosome positioning. In vivo* and *in vitro* studies recognized that nucleosomes are frequently positioned at preferred or even fixed positions along specific DNA sequences. Several striking examples exist in lower eukaryotes, yeast centromeric DNA, URA3 gene, and the PHO5 gene being among them. The biochemical determinants of the *nucleosome phasing* are still not known, but they might include factors such as bendability of the DNA sequence itself, local protein-DNA interactions, and existence of "boundary conditions" determined by the properties of nearby sequences (e.g., stable nucleosome-free regions).

It is clear that the disposition of particular DNA sequences relative to the histone octamer is of con-

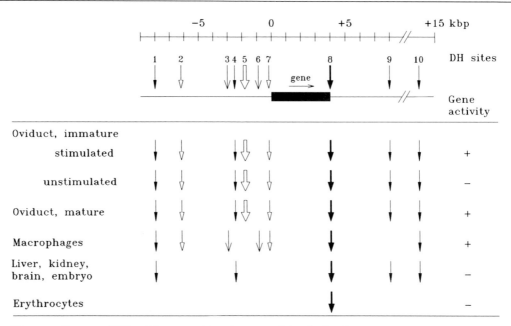

Figure 9   Patterns of DNase I-hypersensitive sites around the chicken lysozyme gene in various tissues and different functional states of the gene. A compilation of all DH sites observed is given in the upper row. Subsequent rows give the pattern observed in nuclei from the indicated tissues. The DH site at the 3' end of the coding sequence (position 8) is present in all cases; DH sites at positions 2 and 7 correlate with gene expression; site 5 is oviduct-specific hormone-dependent. Sites at positions 1, 3, 4, 6, 9, and 10 do not show any clear-cut correlations with activity. [Adapted from Elgin, S. C. R. (1988). *J. Biol. Chem.* **263,** 19259–19262, with permission.]

siderable importance regarding the potential for recognizing specific sequences by trans-factors: Sequences located in the linker are more readable than those in the core; and DNA on the surface of a core particle exposes half of its information—the half apposed to the histone octamer is quite likely unavailable for interaction with trans-factors.

So far, *phasing frames* seem to be constant whether or not a gene is destined to be expressed or not. Thus, for instance, the positioning of nucleosomes in the mouse β-globin gene is the same in expressing and nonexpressing cells; induction of expression in erythroleukemia cells is accompanied by removal (or modification) of four nucleosomes from the promoter regions without disturbing the adjacent phasing frames. Similar observations were reported for nucleosomes positioned in the mouse mammary tumor virus (MMTV) promoter and also for the promoter of the yeast acid phosphatase (PHO5) gene (Fig. 10). It appears that nucleosomes are positioned so as to favor the reversible interactions of resident DNA regulatory sequences with trans-factors; these interactions, in their turn, lead to the induction of nucleosome-free regions (DH sites) or their rever-

sion back to nucleosomes without disrupting flanking chromatin structure.

What happens actually with chromatin structure upon transcriptional activation of a gene? Detailed studies of chromatin structure of the 5' regulatory regions of three genes, the *Drosophila hsp26,* the yeast PHO5, and the MMTV long terminal repeat and its changes accompanying transcription, have led to the notion that mechanisms of transcriptional activation can be extremely versatile.

In some cases, all essential regulatory regions are positioned in nucleosome-free DNA and no displacement of nucleosomes is required for activation (hsp26). In other cases (PHO5), the initial responsive element is in nucleosome-free DNA but access to other *cis*-elements (e.g., the TATA box) necessary for the formation of the transcriptional initiation complex is provided by nucleosome displacement (Fig. 10). In still other cases (MMTV), the critical regulatory sequence can be incorporated in a nucleosome but is nonetheless accessible and capable of binding of trans-factors to start the process of transforming inactive chromatin structure into an active one (including nucleosome displacement).

A

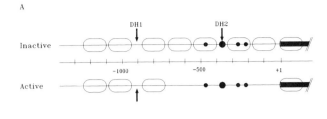

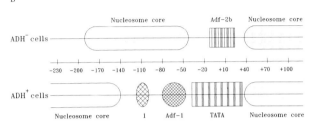

Figure 10 (A) Chromatin structure at the promoter region of yeast PHO5 gene in transcriptionally inactive and active state. Arrows denote the two DH sites; ovals indicate the positioned nucleosomes; solid circles represent four 19 bp UAS elements. The scale is in base pairs with +1 referring to the initiation codon of the PHO5 coding sequence represented by the solid bar. (Adapted from Almer, A., Rudolph, H., Hinnen, A., and Horz, W. (1986). *EMBO J.* **5,** 2689–2696, with permission.) (B) Chromatin structure at the distal promoter of *Drosophila Adh* gene in ADH⁻ and ADH⁺ cell types. Locations and boundaries of specific protein interactions are indicated aligned to the distance from the RNA initiation site. Nucleosome cores are shaded differently to show their apparently different structures that correlate with the different transcriptional activity of the gene. [From Ewel, A., Jackson, J. R., and Benyajati, C. (1990). *Nucl. Acids Res.* **18,** 1771–1781, with permission.]

Possibly, other activation mechanisms will be found.

III. Loop Organization and Transcription

As already mentioned, bacterial DNA is organized in independent supercoiled domains, in which it is torsionally strained. The bacterial chromosome forms about 40 topological domains of about 100 kbp each.

The concept of the loop organization of DNA in the eukaryotic nucleus was first advanced in physical studies of histone-depleted interphase nuclei: It was found that such nuclei behave in a manner analogous to closed circular DNA molecules, a fact that was interpreted to mean that DNA was somehow topologically constrained, probably as a conse-

quence of DNA attachment to some intranuclear structures such as the nuclear matrix. Later, the existence of constrained loops of DNA was directly proven by electron and fluorescence microscopy. In contrast to bacteria, DNA in the loops is not torsionally strained, probably as a result of the winding of DNA around the nucleosomal core. The role of the loop organization in the regulation of transcription is still a matter of debate, but a brief discussion of this issue is included here for completeness.

A. General Features of Loop Organization

The terms *SARs/MARs* (*scaffold attachment regions/matrix attachment regions*) were introduced to denote DNA sequences found attached to matrix structures. The major features of SARs are illustrated in Fig. 11. Scaffold attachment regions map to 300–1000 bp fragments and contain multiple sites of protein-DNA interactions. They are generally AT-rich and contain several recognizable nucleotide sequence motifs, including topo II cleavage consensus sequences.

Methodological difficulties have hampered the definitive identification of many of the constituents of

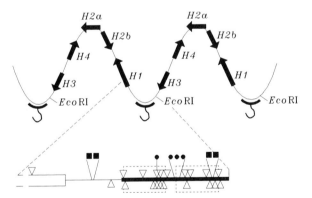

Figure 11 Structure of SARs in the repeat of *Drosophila* histone genes. Two repeats of the 5 kbp DNA fragments that contains all five histone genes are shown in the upper part of the figure. The 657 bp SAR occurs in the nontranscribed spacer between the H1 and H3 genes and is indicated by a bar containing a hook. Approximately 100 tandem histone gene repeats are present in the genome forming a series of small loops. Sequence motifs common to a number of SARs are indicated in the enlarged map. ▽ sequences with 70% homology to topo II consensus sequence; ●, the 10 bp A-box; ■, the 10 bp T-box. The domains encompassed by dotted lines contain protein-DNA complexes. [From Mirkovitch, J., Gasser, S. M., and Laemmli, U. K. (1987). *Phil. Trans. R. Soc. Lond. B.* **317,** 563–574, with permission.]

278

Transcriptional Regulation, Eukaryotes

the matrix. Functional nuclear proteins reported to be associated with the matrix include DNA polymerase α and β, DNA methylase, DNA primase, ribonuclease H, hormone receptors, and topo II. Whether or not RNA polymerases are an integral component of this structure is still uncertain.

The major structural components of the matrix include the lamins (the proteins forming the so-called *lamina,* the proteinaceous skeletal structure situated on the nucleosolic side of the inner nuclear membrane), the pore complex glycoprotein gp-188, actin, vimentin; topo II is also discussed as a structural component analogous to the lamins. Especially interesting is the *repressor-activator binding protein (RAP-1).* This is an abundant TF with multiple binding sites in the yeast genome, including numerous UASs and also the cis-acting silencers that flank the silent mating-type loci. Repressor-activator binding protein fractionates efficiently with the nuclear matrix; affinity-purified RAP-1 is capable of *in vitro* reconstitution of loops at the silent mating-type locus HMLα via silencer–silencer or silencer–promoter interactions. Thus, binding of TFs to *cis*-regulatory elements may define topologically sequestered loops capable of being regulated individually.

B. Topological Organization of Transcribed Sequences with Respect to Loop Structure

The issue of how the active gene sequences are organized relative to the attachment sites has attracted significant attention. The data are, however, extremely controversial, mainly as a result of experimental uncertainties inherent to the approach used. The claimed enrichment of transcribed sequence at the bases of the loops should be viewed with caution.

Several studies have indicated a loose inverse relation between loop size and level of transcription of the gene(s) contained within a loop. Thus, the actively transcribed histone genes are found in small 5 kbp loops (Fig. 11), while genes that are transcribed to give low-abundance mRNA are located in much larger loops as those present in the region of the *rosy* and *ace* loci in *Drosophila* DNA. In some instances, alternative SARs have been identified for one and the same gene that could be used differen-

tially (forming larger or smaller loops) depending on gene activity. This notion, however, still awaits its experimental proof.

In many instances, identified SARs lie close to promoter elements or to other regions that contain transcription regulatory sequences (Table II). This is, however, not a general rule as, for instance, sequences required for the tissue-specific expression of *Adh* and *Ftz* are not matrix-bound.

At present it is difficult to speculate on the functional significance of the proximity of SARs to regulatory sequences. The difficulty is caused by the fact that the mechanisms of action of the regulatory elements themselves are far from being clear. It has been proposed that enhancers can modulate chromatin structure, serving as the origin of the changes in the generalized DNase I sensitivity accompanying transcriptional activation; the changes are dampened as the altered structure propagates away from the enhancer. Active chromatin, once propagated to a distant promoter, could favor the binding of transcriptional factors or RNA polymerase, thus stimulating transcription at a distance.

A number of genes possess SARs at both the 5' and 3' flanking sequences; the flanking SARs might define a topologically sequestered functional unit. Examples of this have been reported. Thus, the MARs of the chicken lysozyme gene co-map with the boundary of the chromatin domain that, when active, is featured by a disrupted nucleosome structure and elevated nuclease sensitivity. The domain encompasses all nine DH sites believed to be involved in the tissue-specific and developmentally-regulated expression of the gene. Similarly, the DNA segment limited by two successive attachment sites in *Xenopus* rDNA corresponds exactly to the domain traversed by RNA polymerase I. On the other hand, SARs can also delineate "inactive" chromatin domains. Thus, SARs that co-map with silencing elements define a loop containing the efficiently repressed *Drosophila* mating-type genes in this locus.

That flanking SARs define independent functional units is clearly demonstrated in gene transformation experiments. It is well known that the transcriptional activity of genes that are randomly integrated into the genomes of transformed cells is in general unpredictable, as it depends on the particular site of insertion. The results from transfection experiments clearly show that the presence of SARs in the constructs allows a constitutively high-level expression

Table II   Proximity of SARs to Transcription Regulatory Sequences

Gene	SAR/MAR	Localization at or near regulatory elements
Drosophila Adh[a], *Ftz, Sgs-4*	5′	Cohabitation with an enhancer
Human β-globin		Cohabitation with a regulatory element
X. laevis rDNA	5′	Cohabitation with upstream regulatory elements including an enhancer
Mouse Ig kappa and lambda chains		Upstream of the tissue-specific enhancer of the JκCκ or J-Cμ domain
Drosophila hsp70	5′	Upstream of sequences necessary and sufficient for heat-shock response
Chicken lysozyme	5′	At about 2.8 kbp upstream of an enhancer and a binding site for NF1
Yeast silent mating type locus *HMLα*	*HML-E* *HML-I* α1/α2	*HML-E* and *HML-I* are regulatory sequences with ARS activity; α1/α2 is a promotor
dhfr in Chinese hamster ovary cells	5′	Upstream from the transcription start site, in a DNaseI-hypersensitive region

[a] For the *Adh* locus, from which two transcripts are made, two enhancer-like regulatory regions have been identified that map to the 5′ SAR fragment.

of the respective gene on integration, independent of its site.

Thus, it seems beyond doubt that the loops can define differentially regulatable domains. The nature of the changes that determine the actual transcriptional state of individual loops remains elusive. One actively discussed possibility is that DNA of different superhelical density might provide conditions for differential binding of regulatory molecules (e.g., histone H1 versus TFs). Indeed, many proteins are known to exhibit preferential binding to either supercoiled or relaxed DNA forms, histone H1 being among those preferring the supercoiled forms.

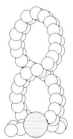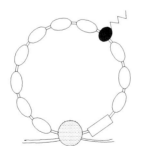

Figure 12   Major structural features of transcriptionally inactive and active genes. Inactive "repressed" domain (condensed loop) (left) and active "poised" domain (decondensed loop) (right). Chromosomal loops are depicted with their bases attached to components (dotted circle) of the nuclear matrix. Nucleosomes are depicted as open circles or ovals. The black circle indicates bound TF. [Modified after Gross, D. S., and Garrard, W. T. (1987). *Trends Biochem. Sci.* **12**, 293–297, with permission.]

IV. Concluding Remarks

From this discussion it becomes clear that the issue of transcriptional regulation in eukaryotic cells is an extremely complex one, much more complex than could be envisaged only a decade ago. Features of gene regulation inherited from the simple prokaryotic organisms are superimposed by features that evolved to satisfy the needs of development and differentiation. The three levels of regulation (cis-trans, chromatin, and loops) are intimately and multisidedly interrelated to give the picture of the subtle regulation of gene expression in a eukaryotic cell. A schematic drawing illustrating the major structural features of the transcriptionally active and inactive genes is presented in Fig. 12. It is evident that much more has to be done in order to get a clearer and much more thorough knowledge of the fascinating issue of transcriptional regulation in eukaryotic cells.

Bibliography

Elgin, S. C. R. (1988). *J. Biol. Chem.* **263**, 19259–19262.
Gasser, S. M., Amati, B. B., Cardenas, M. E., and Hofmann, J. F.-X. (1989). Studies on scaffold attachment sites and their relation to genome function. *In* "IRC Reviews in Cytology" (K. Jeon, ed.), Vol. 119, pp. 57–96. Academic Press, New York.
Gross, D. S., and Garrard, W. T. (1987). *Trends Biochem. Sci.* **12**, 293–297.

Gross, D. S., and Garrard, W. T. (1988). *Ann. Rev. Biochem.* **57,** 159–197.

Hatzopoulos, A. K., Schlokat, U., and Gruss, P. (1988). Enhancers and other *cis*-acting regulatory sequences. *In* ''Frontiers in Molecular Biology. Transcription and Splicing'' (B. D. Hames and D. M. Glover, eds.), pp. 43–96. IRL Press, Ltd., Washington, DC.

Maniatis, T., Goodbourn, S., and Fischer, J. A. (1987). *Science.* **236,** 1237–1245.

Mitchell, P. J., and Tjian, R. (1989). *Science.* **245,** 371–378.

Pederson, D. S., and Simpson, R. T. (1988). *ISI Atlas of Science: Biochemistry.* **1,** 155–160.

Ptashne, M. (1988). *Nature (London).* **335,** 683–689.

Simpson, R. T. (1986). *BioEssays.* **4,** 172–176.

Transcription Attenuation

Charles Yanofsky
Stanford University

Glossary

Antiterminator RNA hairpin structure that prevents formation of a terminator hairpin

Attenuator Segment of DNA that functions as a site of regulated transcription termination

Charged tRNA Transfer RNA bearing an appropriate amino acid, (e.g., Trp-tRNA)

Leader peptide coding region Short peptide coding region involved in regulation of transcription termination at an attenuator

RNA hairpin structure RNA structure composed of base-paired complementary sequences

Transcription attenuation Regulation of transcription termination at an attenuator

Transcription pausing Temporary interruption of RNA polymerase movement and transcript elongation

Transcription termination Cessation of transcript elongation and dissociation of polymerase from its DNA template

Translational attenuation Mechanism regulating availability of a ribosome-binding site for translation initiation

TRANSCRIPTION ATTENUATION is a general transcription regulatory strategy used by bacteria to control gene expression. Mechanisms of transcription attenuation have as a common feature modulation of transcription termination at a site, the attenuator, located before one or more structural genes of an operon. In most examples of transcription attenuation, a metabolite is sensed as the signal that determines whether or not transcription termination will occur at the attenuator. Thus, like repression and transcription activation, which modulate transcription initiation, attenuation exploits one of the fundamental features of RNA polymerase action, transcription termination, in regulating the levels of specific messenger RNAs (mRNAs). Perhaps the principal advantage of transcription attenuation as a regulatory strategy is that it permits the cell to use RNA sequences and structures as the basis for regulatory decisions. RNA structures generally are not relevant to transcription initiation decisions. Bacteria use transcription attenuation to regulate expression of operons concerned with the biosynthesis, utilization, breakdown, and transport of amino acids and the biosynthesis of purines and pyrimidines. Transcription attenuation is also used to regulate expression of operons conferring drug resistance and utilization of sugars. The varied forms of transcription attenuation, and the range of organisms in which it has been observed, establish the general utility of this regulatory strategy. In prokaryotes, transcription termination occurs by two general mechanisms, factor-dependent and factor-independent. Factor-dependent termination is mediated by the protein factor Rho. Factor-independent termination is a property of RNA polymerase itself when it encounters a specific hairpin structure in the transcript it is synthesizing. There are examples of transcription attenuation employing each form of transcription termination. Studies with eukaryotes have also revealed instances in which transcript elongation is regulated. Some eukaryotic examples have features resembling mechanisms of prokaryotic attenuation; however, whether elongation is blocked or there is true transcription termination is not yet certain.

I. Mechanism of Attenuation Control of Amino Acid Biosynthetic Operons

Numerous operons concerned with amino acid biosynthesis, in a variety of bacterial species, are regulated by a similar mechanism of transcription attenuation. The essential structural features responsible for this form of attenuation are diagrammed in Fig. 1. In each biosynthetic operon, the level of an appropriate charged transfer RNA (tRNA) is sensed as a signal of whether or not overall protein synthesis is being limited by a deficiency of that charged tRNA. If there is a deficiency, events occur that prevent transcription termination at the attenuator in the respective operon. This leads to the synthesis of full-length, functional mRNAs. Such transcripts increase the likelihood that the polypeptide products of the operon will be synthesized and will correct the deficiency. Lack of a specific charged tRNA is sensed during attempted translation of a short peptide coding region in the leader segment of the transcript of each operon (Fig. 1). Generally there are multiple regulatory codons for the respective amino acid in the coding region for the leader peptide; i.e., in the *his* operon there are seven contiguous histidine codons, in the *phe* operon there are also seven

phenylalanine codons, but they are not contiguous, and in the *ilv* operon there are threonine, leucine, and isoleucine regulatory codons. A deficiency of any one of the corresponding charged tRNAs reduces transcription termination at the respective attenuator. In the *trp* operon of *Escherichia coli* there are only two adjacent tryptophan codons in the leader peptide coding region. However, unlike most amino acid biosynthetic operons regulated by attenuation, the *trp* operon is regulated only modestly by attenuation; a six- to eightfold increase in operon expression is observed in response to a significant decrease in the intracellular concentration of charged tRNA^{Trp}. Attenuation control of the *trp* operon is thought to be modest because transcription initiation in this operon is tightly regulated by repression. Three overlapping segments of the transcripts of the leader regions of amino acid biosynthetic operons fold to form alternative RNA hairpin structures, called the pause, antiterminator, and terminator structures (Fig. 1). Which of these structures form is determined by several cellular events (see next section); transcription is terminated only when the terminator structure is formed. The *pheST* operon encoding the polypeptides of the phenylalanyl–tRNA synthetase of *E. coli* has a regulatory region with the features depicted in Fig. 1. It is regulated by charged tRNA^{Phe} in much the same manner as amino acid biosynthetic operons.

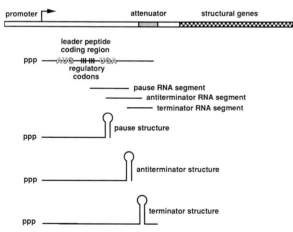

Figure 1 Transcription attenuation in amino acid biosynthetic operons. The structural features essential to transcription attenuation are shown. The leader peptide coding region contains codons for the regulating amino acid(s). The overlapping RNA segments that form the transcription pause, antiterminator, and terminator structures are shown. Below are transcripts that have just formed the pause, antiterminator, and terminator hairpin structures, respectively.

A. Attenuation Regulation of the trp Operon of *E. coli*

The key features of the two molecular modes responsible for attenuation control of amino acid biosynthetic operons are illustrated in the *trp* operon example in Fig. 2. When there is an adequate level of Trp-tRNA^{Trp} (tRNA^{Trp} charged with tryptophan), the leader peptide is synthesized and a transcription termination signal, the terminator, is formed. The existence of the terminator causes RNA polymerase to stop transcription at the attenuator. When there is a deficiency of Trp-tRNA, the ribosome attempting synthesis of the leader peptide stalls over one of the tryptophan codons. Stalling of the ribosome at this position promotes folding of the adjacent segment of the transcript to form the RNA hairpin structure, the antiterminator. The antiterminator and terminator are alternative RNA structures relying on a common sequence of nucleotides (Fig. 1); therefore, whenever the antiterminator forms it prevents formation

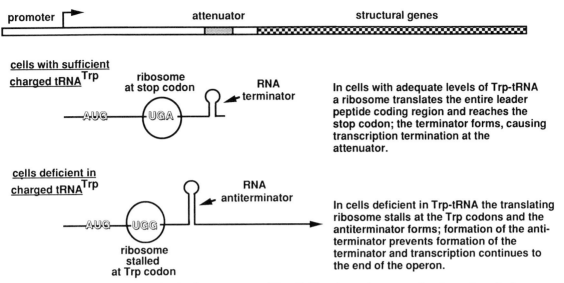

Figure 2 Transcription attenuation in the *trp* operon of *E. coli*. The alternative events that determine whether transcription terminates at the attenuator or continues into the structural genes of the operon are pictured. Ribosome stalling at a Trp codon or movement to the stop codon are the deciding events.

of the terminator and, hence, the transcribing polymerase molecule continues to the end of the operon.

In the leader segments of the transcripts of many amino acid biosynthetic operons, including the *trp* operon, there is a third essential RNA hairpin structure, called a transcription pause structure (Fig. 1). This structure plays the crucial role of ensuring synchronization of transcription of the leader segment of the operon with translation of the leader peptide coding region of the transcript. Synchronization is essential for attenuation to be an effective regulatory mechanism. If the transcribing polymerase molecule completed transcription of the leader segment of the operon before a ribosome began translating the leader peptide coding region, the terminator would invariably form and there would be no opportunity to regulate termination.

The transcription pause structure is a specific RNA hairpin (Fig. 1). When the transcribing RNA polymerase molecule forms this structure and pauses, a transcription delay is provided. This delay allows a ribosome to initiate synthesis of the leader peptide before transcription is resumed. As the translating ribosome proceeds along the transcript and approaches the paused polymerase, it activates the paused polymerase, presumably by disrupting the pause hairpin structure. Thereafter the polymerase continues to transcribe the leader region while the translating ribosome synthesizes the leader peptide. Either the antiterminator or terminator then

forms, depending on the availability of the regulating charged tRNA.

B. Basal Level Control of Attenuation in Amino Acid Biosynthetic Operons

An ancillary feature of attenuation control allows basal level expression of an operon to be set at a level somewhat elevated over what it would have been if there were no translation of the leader peptide coding region. In the *trp* operon of *E. coli*, for example, basal level expression results in fivefold higher expression—increased transcription readthrough at the attenuator—than occurs when there is no translation. This level of readthrough is only 15% of the level observed when there is no termination. Because basal level expression provides 5-fold readthrough, while starvation, as mentioned, provides a 6- to 8-fold increase in expression, the features of attenuation allow about a 30- to 40-fold overall range of transcription termination control at the attenuator.

Basal level control exploits many of the features involved in starvation control. However, one additional event plays an essential role—the rate of ribosome release from the leader peptide stop codon (Fig. 3). In the presence of adequate levels of all the charged tRNAs, a translating ribosome would reach the leader peptide stop codon. The translating ribosome, having completed synthesis of the leader pep-

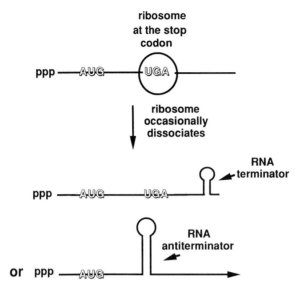

The ribosome translating the leader peptide coding region reaches the stop codon before the RNA terminator is formed.

On most operons the RNA terminator forms and terminates transcription. On a fraction of the operons, the ribosome dissociates and the transcript folds to form either the terminator or the antiterminator.When the terminator is formed, transcription stops. When the antiterminator forms transcription continues to the end of the operon. In the *trp* operon of *E. coli*, for example, ribosome release from the leader stop codon leads to five fold higher basal expression than is observed when there is no release.

Figure 3 Basal level control in amino acid biosynthetic operons. Following occasional ribosome release from the leader peptide stop codon, the transcript can fold to form either the terminator or the antiterminator, as illustrated. Basal level expression is higher than it would be if the ribosome did not release from the stop codon.

tide, would subsequently bind an appropriate release factor and dissociate from the transcript. Whenever the ribosome dissociates from the transcript before the attenuation decision has been made, the ribosome-free transcript would be free to fold and would form either the antiterminator or terminator. If the antiterminator formed, as it does during starvation control, the engaged polymerase would continue transcription through the attenuator and synthesize a full-length transcript of the operon. If the terminator formed, termination would occur. Basal level expression is believed to influence an organism's ability to recover quickly following a shift from a medium rich in an amino acid to a medium lacking that amino acid. [*See* RIBOSOMES.]

II. Attenuation Mechanisms Employing RNA-Binding Regulatory Proteins

RNA-binding regulatory proteins also modulate transcription termination at attenuators. The best understood examples of this form of attenuation control involve the *trp* operon of *Bacillus subtilis* and the *bgl* operon of *E. coli*. During transcription of these operons, a regulatory protein binds to RNA sequences and disrupts an RNA hairpin structure,

thereby controlling transcription termination. Binding of the respective regulatory protein has opposite effects in the *trp* and *bgl* operons. Other operons of *B. subtilis* appear to be regulated by attenuation mechanisms that are similar to those used to control the *trp* operon of *B. subtilis* and the *bgl* operon of *E. coli*.

A. Regulation of the trp Operon of B. subtilis

With the exception of *trpG*, all of the genes required for tryptophan biosynthesis in this bacterial species are arranged in the same order as that in *E. coli*, in a single operon. However, unlike *E. coli*, there are other aromatic genes at the end of the *trp* operon, and an aromatic operon precedes and overlaps the *trp* operon. The *trp* operon of *B. subtilis* is regulated by transcription attenuation (Fig. 4), but the mechanism is quite different from that utilized by enteric bacteria. Tryptophan rather than tRNA^{Trp} provides the regulatory signal, and translation of a leader peptide coding region is not the basis of signal sensation. Rather, a tryptophan-activated RNA-binding regulatory protein, not a translating ribosome, is responsible for conveying the sensed signal. The leader transcript of the *trp* operon of *B. subtilis* also can form alternative antiterminator and terminator

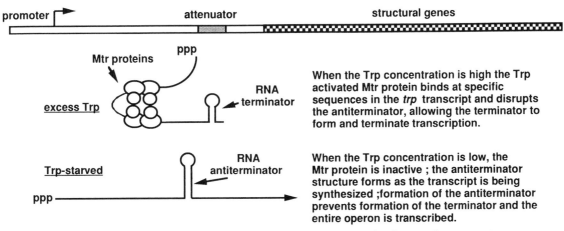

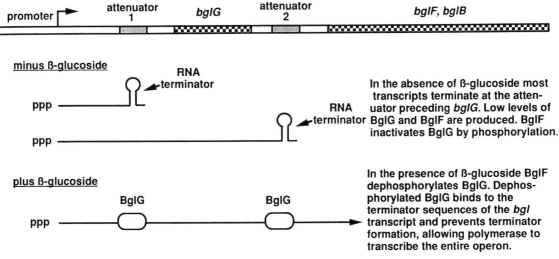

Figure 4 Transcription attenuation in the *trp* operon of *B. subtilis*. An RNA-binding regulatory protein, Mtr, when tryptophan-activated, binds to two specific sequences in the antiterminator segments of the *trp* operon transcript and disrupts the antiterminator. When the antiterminator is disrupted the terminator forms and stops transcription in the leader region. When cells are tryptophan-deficient, the Mtr protein is inactive, and the antiterminator is formed.

structures. The antiterminator and preceding RNA region have binding sites for the tryptophan-activated Mtr protein. When this regulatory protein is activated by tryptophan, it binds to the transcript and disrupts the antiterminator, allowing the terminator to form (Fig. 4). When the intracellular concentration of tryptophan is low, the Mtr protein is inactive, the antiterminator forms, and transcription termination at the attenuator is prevented. Expression of the *trp* operon of *B. subtilis* is regulated several hundred-fold in response to changes in the intracellular content of tryptophan. Part of this regulation is due to secondary effects on translation initiation at the start codon of *trpE*, the first structural gene of the operon.

B. Regulation of the bgl Operon of *E. coli*

The *bgl* operon consists of three genes that provide regulated use of β-glucosides as sole carbon sources (Fig. 5). *bglG* encodes a positive regulatory protein,

Figure 5 Transcription attenuation in the *bgl* operon of *E. coli*. Expression of the *bgl* operon is controlled by dual attenuators in response to the presence of a β-glucoside as a sole carbon source. In the absence of a β-glucoside, transcription is terminated at the two attenuators. In the presence of a β-glucoside, the BglG protein is activated by dephosphorylation. It binds to the transcript and prevents formation of the RNA terminators.

bglF encodes a β-glucoside transport protein that phosphorylates β-glucosides and transports them across the cytoplasmic membrane, and *bglB* encodes an enzyme that can dephosphorylate phosphorylated β-glucosides. In the absence of β-glucosides, most transcription is terminated at the factor-independent terminator preceding *bglB* (Fig. 5). Under these conditions, the expression level of *bglG* and *bglF* is low. The low level of BglF is sufficient to phosphorylate BglG, inactivating it as an RNA-binding protein. In the presence of β-glucosides, BglF dephosphorylates BglG. Dephosphorylated BglG binds to the *bgl* transcript at sequences overlapping the two RNA terminators and prevents formation of these terminators. This allows transcription to continue to the end of the operon.

III. Attenuation in Pyrimidine Biosynthetic Operons

The *pyrBI* and *pyrE* operons of enteric bacteria are regulated by a mechanism of transcription attenuation featuring uridine triphosphate (UTP)-dependent transcription pausing as the crucial event (Fig. 6). The transcripts of these operons contain important leader peptide or protein-coding regions. In the *pyrBI* operon, in a cell with a UTP deficiency, the transcribing polymerase pauses while attempting to synthesize the U-rich segment of the transcript. This provides an opportunity for a ribosome to bind to the leader transcript and begin synthesizing the leader peptide. The translating ribosome presumably approaches the paused polymerase, and thereafter polymerase and ribosome move synchronously on template and transcript, respectively. Because the leader peptide coding region overlaps the terminator, the translating ribosome would disrupt the terminator and thereby prevent transcription termination. In cells with high levels of UTP, the transcribing polymerase presumably does not pause sufficiently long to permit synchronization of polymerase and ribosome movement. Hence, the terminator would form and function before the translating ribosome would have an opportunity to approach this RNA structure.

In the *pyrE* operon, a coding region for a polypeptide of unknown function is located immediately upstream of the regulatory region for *pyrE*. Translation of this upstream coding region appears to play the same role as translation of the leader peptide coding region of the *pyrBI* operon. This upstream coding region terminates just before the terminator hairpin preceding *pyrE*, but the stop codon is sufficiently close to the terminator that it should be disrupted by the translating ribosome. Regulation of *pyrBI* and *pyrE* by attenuation is appreciable in response to changes in the UTP concentration.

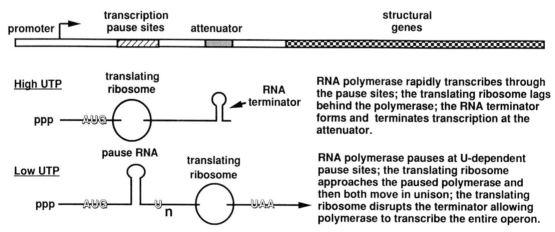

Figure 6 Transcription attenuation control of the *pyrBI* operon of *E. coli*. The alternative states of the *pyrBI* transcript are shown in the figure. In attenuation control in this operon, pyrimidine-dependent transcription pausing plays a crucial role. Pausing occurs when the transcribing polymerase encounters an AT-rich region on the template and the cell is deficient in UTP. Pausing provides a transcription delay during which a ribosome can bind to the transcript and begin translating the leader peptide coding region. When the translating ribosome disrupts the terminator before it has had an opportunity to function, termination is prevented.

IV. Attenuation Regulation of the *ampC* Operon of *Escherichia coli*

The *ampC* operon of *E. coli* encodes a β-lactamase that can degrade penicillin and related antibiotics. Expression of this operon is subject to growth rate regulation; i.e., the shorter the cell division time, the higher *ampC* expression. The *ampC* operon is regulated by factor-independent transcription attenuation (Fig. 7). The leader region of the *ampC* operon contains a simple ribosome sensing arrangement—a ribosome-binding site start codon immediately followed by a stop codon, with the start and stop codons adjacent to a transcription terminator (Fig. 7). It has been proposed that transcription termination at the *ampC* attenuator is regulated by the frequency of ribosome attachment at the leader region start codon. At low growth rates, when the ribosome content per cell is low, there is a low probability that a ribosome will bind at the start codon and disrupt the terminator. However, when the growth rate is high, and there are many more ribosomes per cell, the probability increases that a ribosome will bind at the start codon, disrupt the terminator, and permit transcription of the remainder of the operon.

Ribosome release at the leader stop codon may be a factor in *ampC* operon regulation, as it is in basal level control of amino acid biosynthetic operons. It is conceivable that ribosome binding to the *ampC* transcript is determined by the ribosome content per

cell, but the bound ribosome then moves to the stop codon, where it remains for a period sufficient to disrupt the terminator.

V. Attenuation Regulation of the Tryptophanase Operon of *Escherichia coli*

Escherichia coli can use tryptophan and many other amino acids as sole sources of carbon or nitrogen. Breakdown of tryptophan is accomplished by the polypeptide products of the *tna* operon. Like the promoters of many degradative operons, the promoter of the *tna* operon is subject to catabolite repression control; hence, the operon is transcriptionally off when a more suitable carbon source, such as glucose, is available. Transcription of the structural genes of the *tna* operon is induced by the presence of tryptophan (Fig. 8). The mechanism of tryptophan induction involves transcription antitermination at termination sites located in the leader region of the operon. Unlike the other examples of transcription attenuation described in this article, termination in the leader region of the *tna* operon occurs at multiple Rho factor-dependent termination sites. The presence of tryptophan apparently activates an antitermination mechanism that prevents transcription termination at these Rho factor-dependent termination sites. The leader region of the *tna* operon encodes a 24-residue peptide that has a single tryptophan residue. Translation of

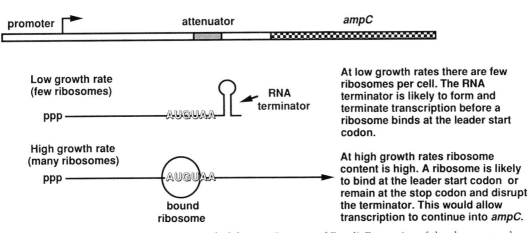

Figure 7 Transcription attenuation control of the *ampC* operon of *E. coli*. Expression of the chromosomal *ampC* operon of *E. coli* is growth rate-dependent. A change in the ribosome content per cell is sensed as the signal for whether or not to terminate transcription at the attenuator located just before *ampC*. The mechanism of regulation of attenuation is illustrated in the figure. Binding of a ribosome at a start codon located adjacent to the RNA terminator results in disruption of the terminator. This is the crucial event that regulates the frequency of termination.

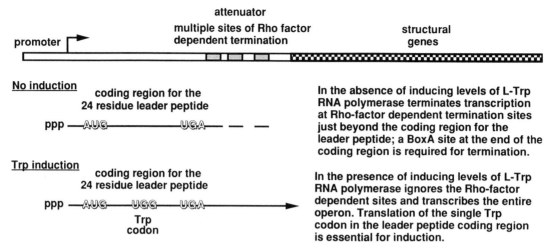

Figure 8   Transcription attenuation control of the *tnaAB* operon of *E. coli. Escherichia coli* can use tryptophan as a sole source of carbon or nitrogen. In the absence of a catabolite-repressing carbon source and in the presence of tryptophan, an antitermination mechanism is activated that prevents transcription termination in the leader region of the operon. In this example of attenuation, Rho factor-dependent termination is involved. Translation of a 24-codon leader coding region is essential to tryptophan induction. In addition, for induction to occur, the single Trp codon in the leader peptide coding sequence must be translated by tRNA^{Trp}. The mechanism of tryptophan induction is not known.

this coding region—particularly the tryptophan codon—is essential for tryptophan-induced antitermination. There is, in addition, a crucial recognition sequence, called BoxA, near the DNA segment specifying the end of the leader peptide coding region. Mutations altering this BoxA sequence lead to constitutive expression of the operon. The mechanism of tryptophan-induced transcription antitermination is not yet known. [*See* CATABOLITE REPRESSION.]

VI.   Conclusions

A.   Advantages of Transcription Attenuation as a Regulatory Mechanism

The examples described in this article illustrate the diverse transcription attenuation mechanisms that bacteria employ in regulating gene expression. A variety of molecular events modulate the frequency of transcription termination at attenuator sites located before structural genes of operons. Most mechanisms involve alteration of the structure of the transcript, usually promoting or disrupting an RNA hairpin, as the basis of influencing the frequency of transcription termination. The common feature of transcription attenuation mechanisms that distinguish them from regulatory mechanisms that control

transcription initiation is the use of RNA sequences and structures in the sensing of metabolic events. This allows the elongation and termination functions of RNA polymerase to be exploited in the control of gene expression. In addition, because a transcript can contain a short upstream coding region, events that affect ribosome structure and function can be used to influence attenuation decisions. The options of control of initiation, elongation, or termination of transcription permit organisms greater flexibility in regulating gene expression. Transcription attenuation studies in eukaryotes are at a very early stage. It remains to be seen what variations on the mechanisms uncovered in prokaryotes are operable in eukaryotes.

A second important feature of most mechanisms of transcription attenuation is that relatively little genetic information is needed for specific control. This fact suggests that attenuation may have evolved early as a regulatory strategy. In the amino acid biosynthetic operons of enteric bacteria, for example, only the region rich in regulatory codons—generally less than 10 codons in length—provides for a specific response to amino acid starvation. Even if the entire leader region is considered as essential information for attenuation, 200 bp are sufficient to permit an amino acid-specific response. Only the common cell components involved in transcription and translation are involved in many examples of attenuation control. By contrast, regulation by repression or activation involves unique regulatory proteins. Mechanisms of transcription attenuation reveal how specific regulation can be achieved by the combined action of common molecular events.

B. Translational Attenuation in Bacteria

A somewhat comparable translational regulatory mechanism exists—translational attenuation. In examples of translational attenuation, formation or disruption of a specific RNA structure determines whether or not a ribosome-binding site becomes available for initiation of translation. Like many examples of transcription attenuation, the transcripts of operons regulated by translational attenuation contain a short leader peptide coding region preceding the coding region that is translationally regulated. Translation of the upstream peptide coding region regulates formation of an RNA hairpin that blocks the ribosome-binding site for the downstream major coding region. For example, when a ribosome translating a leader peptide coding region is stalled by an antibiotic such as chloramphenicol, the ribosome-binding site of the adjacent polypeptide coding region is activated and becomes accessible to ribosomes. Translational attenuation is commonly used to regulate expression of plasmid operons encoding proteins that confer drug resistance in bacteria.

Bibliography

Chamberlin, M. J. (1985). Transcriptional regulatory factors and the structure of prokaryotic transcription units. *In* "The Proceedings of the Robert A. Welch Foundation Conferences on Chemical Research XXIX. Genetic Chemistry: The Molecular Basis of Heredity," pp. 347–356. The Robert A. Welch Foundation, Houston, Texas.

Landick, R., and Turnbough, C. (1992). Transcriptional attenuation. *In* "Transcriptional Regulation" (S. L. McKnight and K. R. Yamamoto, eds.). Cold Spring Harbor Laboratory, Cold Spring Harbor, New York (in press).

Landick, R., and Yanofsky, C. (1987). Transcription attenuation. *In* "*Escherichia coli* and *Salmonella typhimurium*: Cellular and Molecular Biology," Vol. 2 (F. C. Neidhardt, J. L. Ingraham, K. B. Low, B. Magasanik, M. Schaechter, and H. E. Umbarger, eds.), pp. 1276–1301, American Society for Microbiology, Washington, D.C.

Platt, T. (1986). *Annu. Rev. Biochem.* **55,** 339–372.

Yager, T. D., and von Hippel, P. H. (1987). Transcript elongation and termination in *Escherichia coli*. *In* "*Escherichia coli* and *Salmonella typhimurium*: Cellular and Molecular Biology," Vol. 2 (F. C. Neidhardt, J. L. Ingraham, K. B. Low, B. Magasanik, M. Schaechter, and H. E. Umbarger, eds.), pp. 1241–1275. American Society for Microbiology, Washington, D.C.

Yanofsky, C. (1988). *Trends Genet.* **3,** 356–360.

Yanofsky, C. (1988). *J. Biol. Chem.* **263,** 609–612.

Transcription, Viral

David S. Latchman
University College, London

Glossary

Enhancer Region of DNA that, although incapable of directing transcription itself, can enhance the ability of a promoter to do so
Genome DNA or RNA molecule of the virus that enters the infected cell
Long terminal repeat Element found at either end of the retroviral genome that contains the promoter/enhancer element driving viral transcription
Promoter Region of DNA essential for directing the transcription of a particular gene
Reverse transcription Process whereby RNA is converted to DNA by the enzyme reverse transcriptase
Splicing Production of a mature messenger RNA molecule from the initial RNA transcript by the removal of intervening sequences (introns)
Transcription Process whereby DNA is converted to RNA

TRANSCRIPTIONAL REGULATION of eukaryotic viruses is the process whereby these viruses control the expression of their proteins by regulating the transcription of the genes that encode them. Following infection of eukaryotic cells with these viruses, the virus must sequentially express different sets of proteins. First, these are the regulatory proteins; second, the proteins required for genome replication and metabolism; and, finally, structural proteins. This will allow the virus to replicate its genetic material, package it appropriately, and leave the cell. This sequential production of specific proteins is controlled by the sequential transcription of the genes encoding them. In turn, this sequential transcription depends on the interplay between cellular transcription factors present in the cell prior to viral infection and virally encoded factors that are either synthesized in the infected cells or enter with the incoming virion. The regulatory processes that occur in the small DNA viruses, the large DNA viruses, and the RNA viruses will be discussed in turn. [*See* TRANSCRIPTIONAL REGULATION, EUKARYOTES.]

I. Small DNA Viruses

A. Simian Virus 40

Among the small DNA viruses (genome size 5–10,000 bases), transcriptional regulation is best understood for simian virus 40 (SV40). This virus, which naturally infects monkey cells, has a genome of 5243 base pairs in size and encodes five proteins. Two of these, the large T and small t antigens, are produced immediately after entry of the virus into the cell, whereas the others encoding the coat proteins VP1, 2, and 3 are expressed subsequently. The arrangement of the genes encoding these proteins within the circular viral genome is illustrated in Fig. 1. The large T and small t antigens play a vital role both in the replication of the viral DNA and in the activation of the genes encoding the coat proteins. Therefore, they must be produced early in infection, allowing DNA replication, coat protein synthesis, and virion production to occur subsequently (Fig. 2).

The large T and small t antigens are encoded by a single gene within the virus, with the critical RNA product of this gene being spliced in two different ways to produce the messenger RNAs (mRNAs) encoding the two proteins. Therefore, this gene must be transcribed immediately following entry of the virus into the cell so that the large T and small t antigens can be made and fulfill their function of activating transcription of the other genes and DNA replication.

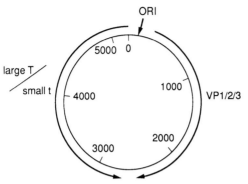

Figure 1 The SV40 genome showing the origin of DNA replication (ORI) and the transcription units encoding the large T and small t antigens and the coat proteins VP1, 2, and 3. The numbers indicate number of DNA base pairs.

This early transcription of the large T–small t gene depends on a promoter element located immediately upstream of the start site for transcription and an enhancer element located upstream of the promoter. The region of the viral genome containing these control elements also contains the origin of DNA replication and the late promoter driving transcription of the coat protein genes, which are transcribed late in infection, in the opposite direction from the gene encoding the large T and small t antigens (Fig. 1). [*See* DNA REPLICATION.]

The promoter and enhancer driving large T–small t expression contain binding sites for a number of different cellular transcription factors, which are present in the uninfected cell. In particular, the

SV40 promoter, in common with many cellular and viral genes, contains a so-called TATA box sequence approximately 30 bases upstream of the transcription start site. This TATA box will bind the cellular factor TFIID, which interacts with the RNA polymerase itself allowing transcription to occur. The rate of such transcription will be greatly increased by the binding of other cellular factors to their appropriate binding sites upstream of the TATA box within the promoter and enhancer. These factors will interact with TFIID activating it and greatly increasing transcription. The binding sites for the various factors within the promoter and enhancer are illustrated in Fig. 3 and the factors themselves are listed in Table I, together with the other viral gene promoters/enhancers that bind these factors.

Following entry of the virus into the cell, these multiple cellular transcription factors will bind to the viral promoter and enhancer, thereby activating transcription of the large T and small t antigen gene. Thus, the expression of these genes occurs early in infection without the need for viral protein synthesis allowing them to activate the later stages of the lytic cycle.

Once the large T antigen has been synthesized, it then binds to three binding sites (T1–T3) within the promoter region illustrated in Fig. 3. This results in the inhibition of large T–small t gene transcription since the bound large T antigen lies across the start site for early RNA production and, hence, interferes with the initiation of transcription. Moreover, the large T antigen also stimulates the replication of the viral DNA, which begins at the site where it binds, as well as the transcription of the viral coat protein genes, which occurs from the same control region but in the opposite direction of early gene transcription (Fig. 3).

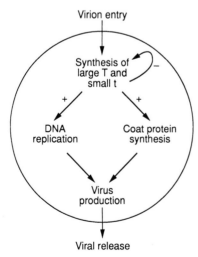

Figure 2 Life cycle of SV40 showing the stimulatory effect of the early antigens on DNA replication and coat protein production and the inhibitory effect on their own synthesis.

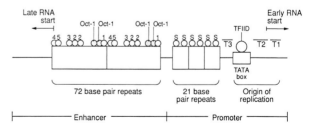

Figure 3 The SV40 promoter/enhancer region. The binding sites of the viral large T antigen (T1, T2, T3), and the cellular factors TFIID, Sp1 (S), Oct-1, and activating transcription factors 1–5 (1–5) are indicated.

Table I Cellular Factors Interacting with the SV40 Promoter/Enhancer

Factor	Binding site	Size	Other viral genes with binding sites for the factor
AP1	TT/GAGTCA	47 kDa	Polyoma enhancer, adenovirus E3 gene, human papilloma virus enhancer
AP2	CCCCAGGC	52 kDa	Polyoma enhancer, adenovirus major late promoter, BPV enhancer
AP3	GGGTGTGGAAAG	?	Polyoma enhancer
AP4	CAGCTGTGG	?	Polyoma enhancer
AP5	CTGTGGAATG	?	—
Oct-1	ATGCAAAT	90 kDa	Related sequence in HSV IE genes
Sp 1	GGGCGG	10,595 kDa	Adenovirus E1B, HSV IE genes, HSV tK gene, HIV LTR
TFIID	TATA box	120–140 kDa	Many genes

BPV, bovine papilloma virus; HIV, human immunodeficiency virus; HSV, herpes simplex virus.

Hence, the transcription of the SV40 genome occurs in two precisely defined stages with cellular factors stimulating early transcription of the large T–small t antigen gene while the proteins encoded by these genes then stimulate late gene transcription of the coat protein genes and DNA replication.

B. Other Small DNA Viruses

A similar two-stage pattern of gene transcription is also observed in the other small DNA viruses. In the mouse virus polyoma, for example, a single gene encoding the three regulatory proteins large T, middle T, and small t is transcribed first following infection with subsequent transcription of the genes encoding the viral coat proteins. As in the case of SV40, the enhancer element driving early gene transcription contains binding sites for many cellular transcription factors, several of which are identical to those that bind the SV40 enhancer.

Similarly, in the papillomaviruses such as human papilloma viruses types 16 and 18 (HPV 16 and 18), transcription of the early regulatory genes is controlled by cellular transcription factors that bind to a region of the genome known as the long control region, which also contains the origin of DNA replication. The subequent transcription of the late genes encoding viral structural proteins depends, as before, on the prior expression of the early regulatory proteins.

Interestingly, however, the regulatory elements of HPV 16 and 18, unlike those of SV40 or polyoma, display a strong cell type specificity in their activity. Thus, the long control region is considerably more active in driving early gene expression in epithelial cells than in any other cell types, whereas the late genes are only expressed in terminally differentiated keratinocytes within the upper layers of stratified epithelia. No cell type-specific transcription factors binding to the long control region have thus far been identified and, hence, the basis for this cell type-specific activity remains unclear.

II. Large DNA Viruses

A. Herpes Simplex Virus

Herpes simplex viruses (HSV) types 1 and 2 are large DNA viruses with a genome of approximately 150,000 bases that can infect human cells. The large size of these viruses is paralleled by their ability to produce over 70 proteins, far more than the number encoded by the small DNA viruses. This increased number of proteins is paralleled by a more complex temporal pattern of protein production. Thus, viral proteins are synthesized following infection in three temporal phases. The first of these results in the production of the five viral immediate-early proteins, which have a regulatory role. In particular, these immediate-early proteins induce the expression of the viral early proteins, which include the viral DNA polymerase and other enzymes of DNA replication. Subsequently, the action of the immediate-early proteins also results in the expression of the late proteins, which encode viral structural proteins and are only expressed following DNA replication (Fig. 4). [*See* HERPESVIRUSES.]

Like the large T–small t gene of SV40, the genes encoding the immediate-early proteins must be expressed immediately after the viral genome enters the cell. As in the SV40 case, the promoters of the genes encoding the immediate-early proteins contain multiple binding sites for cellular transcription

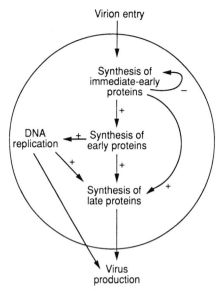

Figure 4 Life cycle of herpes simplex virus.

factors. In fact, the cellular factors that have been shown to bind to the HSV promoters such as TFIID, Sp1, and Oct-1 also bind to the SV40 promoter or enhancer. The binding sites for cellular transcription factors in the major immediate-early gene IE3 encoding Vmw175 or ICP4 are illustrated in Fig. 5.

Unlike SV40, however, HSV employs a unique mechanism for modulating the activity of one of these factors. Thus, the incoming HSV virion contains a protein known as Vmw65, which, in addition to being a structural component of the virion, also forms a complex with the cellular transcription factor Oct-1. In the absence of Vmw65, Oct-1 will bind only weakly to the TAATGARAT (where R is purine) binding site in the viral immediate-early promoters because this site is only distantly related to its normal octamer binding site (consensus ATGCAAATNA, where N is any base), which is found in cellular promoters and in the SV40 enhancer. In contrast, when complexed to Vmw65, the

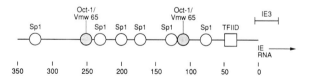

Figure 5 The herpes simplex virus immediate-early three-gene promoter. The binding sites for the IE3 protein itself, the cellular factors TFIID and Sp1 and the complex of the viral factor Vmw65, and the cellular factor Oct-1 are indicated.

complex binds to the TAATGARAT sequence with high affinity and strongly activates transcription.

Hence, in HSV, the problem of transcribing the first set of viral genes in the absence of viral proteins made in the infected cell is solved in a more complex manner than in SV40 by utilizing both cellular transcription factors and a viral protein contained in the incoming virion.

Interestingly, as in the papilloma viruses, the transcription of the immediate-early genes of HSV is cell type-specific. Thus, although high-level transcription occurs following infection of most cell types, such transcription is not observed following infection of neuronal cells. Hence, these cells do not support the viral lytic cycle illustrated in Fig. 4; instead, the virus establishes long-lived latent infections in which viral proteins are not synthesized.

In lytic infection with HSV, as with SV40, once the regulatory proteins are synthesized, they play multiple roles in the next stage of the lytic cycle. In particular, the protein encoded by the HSV immediate-early IE3 gene (Vmw175 or ICP4), like T antigen, binds to its own gene promoter, covering the start site of transcription and, hence, inhibiting its own transcription (Fig. 5). Moreover, this protein also plays an essential, positive role in stimulating transcription of the genes encoding the viral early and late genes, because mutant viruses that lack a functional ICP4 protein fail to transcribe either the early or late genes.

Unlike the repression of its own synthesis, however, gene activation by ICP4 apparently does not require direct binding of the protein to the early or late gene promoters. Thus, no specific sequences within the early or late promoters that mediate the response to ICP4 have been identified. Rather, the response to ICP4 apparently depends on the same sequences that are necessary for the low level of transcription of these genes that is seen in the absence of ICP4 and that are binding sites for cellular transcription factors. Hence, the activation of the viral early late genes by ICP4 appears to be mediated via its interaction with cellular transcription factors increasing the rate of gene transcription.

Interestingly, although both early and late promoters respond to ICP4, they can be distinguished by the binding sites they possess for these cellular transcription factors. Thus, while early promoters contain, in addition to the TATA box, binding sites for factors such as Sp1 or the CCAAT box binding factor, late promoters appear to contain only the TATA box, and it is through this binding site that the response to ICP4 is mediated. This relatively simple structure of the late promoters is in contrast, however, to their complex pattern of regulation. Unlike the early promoters, they require both DNA replication and the presence of the immediate-early proteins ICP22 and ICP27 in addition to ICP4 for full activation.

Thus, the three-phase transcription of the HSV genome is controlled by the different control elements in the promoters of the different gene classes. In addition to binding sites for cellular factors, the immediate-early promoters contain the TAAT-GARAT motif, which allows them to respond to the virion protein Vmw65. The early promoters contain binding sites for cellular transcription factors, allowing a response to ICP4. The late promoters have a much simpler structure, which leads to a requirement for other immediate-early proteins and an increased copy number following DNA replication for full activity to occur. The promoter structure of the different HSV gene promoters is summarized in Fig. 6.

B. Adenovirus

The action of the HSV ICP4 protein in activating subsequent stages of viral gene expression is paralleled in the adenoviruses by the immediate-early E1A protein, whose mechanism of action is understood in great detail. This protein, which is made in the cell immediately following infection, transactivates a number of other adenovirus genes, including the early genes E1B, E2, E3, and E4 as well as the major late promoter.

This transactivation does not depend on binding of E1A to these promoters. Thus, as with ICP4, there are no specific sequences in the induced promoters that confer responsiveness to E1A; rather, the promoter sequences necessary for such a response are the same as those required for basal transcription in the absence of E1A and are sites for the binding of cellular transcription factors. Moreover, unlike ICP4, which can bind to its own promoter in a sequence-specific manner, E1A apparently cannot bind directly to specific DNA sequences. Hence, as with the interaction of ICP4 with the HSV early and late promoters, the E1A protein stimulates gene expression by interacting with cellular factors bound to the promoter.

The cellular transcription factor with which E1A interacts has been identified as activating transcription factor-2. This cellular factor binds to sites within the promoters of E1A responsive genes, and the E1A protein then binds to it. The E1A protein, which is bound to the promoter indirectly in this way, then interacts with other components of the transcription machinery (such as TFIID and the RNA polymerase itself) to stimulate transcription (Fig. 7).

Hence, as in HSV, the interaction of cellular and viral transcription factors results in the appropriate, coordinate expression of the viral genome.

III. RNA Viruses

Among the RNA viruses, the regulation of transcription is best understood in the retroviruses. Despite their RNA genome, these viruses face the same problems in the regulation of gene transcription as the DNA viruses. Thus, following entry into the cell, the viral genome is converted into its DNA equivalent by the process of reverse transcription, and this DNA molecule then integrates into the host cell genome. Therefore, to produce the viral proteins, this DNA genome must be transcribed into RNA exactly as occurs for the DNA viruses. [*See* RETROVIRUSES.]

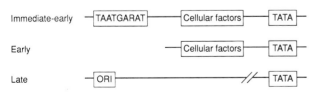

Figure 6 Structure of different classes of herpes simplex virus promoters. Note that although only the immediate-early promoters contain the TAATGARAT binding sites for the Oct-1–Vmw65 complex, both the immediate-early and early promoters have binding sites for other cellular factors and a TATA box. In contrast, the late promoters have only a TATA box and depend on DNA replication initiating at a replication origin (ORI) for their expression.

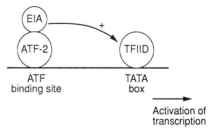

Figure 7 Activation of transcription by the adenovirus E1A protein, which binds to the cellular activation transcription factor-2 (ATF-2) and then stimulates the TFIID protein bound to the TATA box.

The various retroviruses that have been studied can be divided into two groups, namely the simpler viruses, which do not encode any transcriptional regulatory proteins of their own, and the more complex viruses, which do encode such regulatory proteins. These two groups will be considered in turn.

A. Simple Retroviruses

The simpler retroviruses are typified by the intensively studied Moloney murine leukemia virus (MoMuLV), which encodes only three proteins: the virion structural proteins gag and env and the pol protein, which reverse-transcribes the RNA genome. The single RNA molecule encoding these proteins is transcribed from a promoter/enhancer element contained within the long terminal repeat (LTR) sequence found at either end of the viral genome (Fig. 8).

This promoter/enhancer is highly active in most cell types and evidently does not depend on the activity of a viral regulatory protein, because no such protein is encoded in the viral genome. Rather, as in the SV40 early promoter, the LTR contains binding sites for a number of cellular factors that are present in all cell types and whose binding activates transcription. These factors include several such as NF1 that are known to bind to other promoters as well as three cellular factors (LVa, b, and c), which have not been previously identified.

A similar pattern of multiple binding sites for cellular factors is also seen in the LTR of the chicken virus Rous sarcoma virus (RSV), which, like MoMuLV, does not encode its own regulatory proteins. The factors that bind to the RSV promoter/enhancer are different from those that bind to that of MoMuLV, however, and include proteins that also bind to the enhancers of SV40 or polyoma.

In both RSV and MoMuLV, therefore, the binding of multiple cellular transcription factors to the

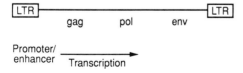

Figure 8 The Moloney murine leukemia virus genome showing the long terminal repeats (LTR) that contain the promoter/enhancer element and the positions of the genes encoding the structural proteins (gag and env) and the gene encoding the reverse transcriptase (pol).

LTR allows it to direct a high level of transcription in the absence of any viral protein. Because these cellular factors are present in all cell types, this high rate of transcription is observed in virtually all cells into which the LTR is introduced. Interestingly, however, the MoMuLV LTR cannot direct transcription in early embryonic cells, which, therefore, are not permissive for viral infection. This failure of viral transcription does not arise, however, from the lack of one of the positively acting cellular factors mentioned earlier. Rather, it is due to the presence of an inhibitory protein that is found only in early embryonic cells. This protein binds to a sequence in the MoMuLV LTR known as the CCAAT box and not only fails to activate transcription itself but also prevents the binding to this site of a positively acting factor, which is essential for the activation of transcription.

Hence, the regulation of the simple retroviruses is controlled by the interplay of positively and negatively acting cellular factors without any involvement of virally encoded regulatory proteins.

B. Complex Retroviruses

The complex retroviruses that do encode their own regulatory proteins are typified by the human immunodeficiency virus (HIV), which has been intensively studied due to its role as the causative agent of AIDS. As is the case for MoMuLV and RSV, the LTR of HIV contains binding sites for several cellular transcription factors. Several of these factors such as Sp1 (which also binds to SV40 and HSV promoters) are present in all cell types, whereas others such as NFκB and NFAT-1 are present in an active form only in T cells that have been stimulated with antigen or cytokines (Fig. 9). The binding of these factors together is sufficient to produce a low level of transcription from the HIV LTR, which is enhanced in activated T cells.

This low level of transcription results in the production of an mRNA, which is then multiply spliced to produce small RNAs encoding the HIV regulatory proteins Tat, Nef, and Rev. The Tat protein is a strong activator of HIV LTR-driven transcription, and its presence results in a great increase in the amount of HIV RNA that is made. Simultaneously, the Rev protein changes the splicing/transport pattern of this RNA so that its own production and that of Tat and Nef is inhibited and the production of

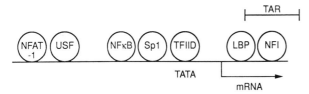

Figure 9 The human immunodeficiency virus long terminal repeat. The binding sites for the cellular factors nuclear factor 1 (NF1), leader binding protein (LBP), TFIID, Sp1, NFκB, upstream stimulatory factor (USF), and NFAT-1 are indicated together with the target site for activation by the viral Tat protein (TAR). mRNA, messenger RNA.

RNA encoding the viral structural proteins is favored. Hence, in this case, unlike SV40 or HSV, the viral regulatory proteins made in the early phase do not promote the use of a different promoter late in infection. Rather, they stimulate the same promoter as is used early in infection but alter the pattern of splicing so that new proteins are made (Fig. 10).

Interestingly, Tat increases transcription by binding not to DNA but to newly made nascent HIV RNA in a region at the 5′ end of the RNA molecule known as Tar (see Fig. 9). When bound in this position, the Tat protein can interact with cellular transcription factors bound to the DNA upstream of the transcription start site and stimulate transcription.

HIV is not unique in encoding viral regulatory proteins that interact with cellular transcription factors to modulate LTR-driven transcription. Thus, the human T cell leukemia virus HTLV-1 encodes the protein Tax, which greatly increases transcription from its LTR. In contrast to Tat, however, Tax does not bind to RNA but, rather, interacts with the NFκB protein, which binds to an upstream region of the HTLV-1 LTR, thereby increasing LTR-driven transcription. Hence, in this case, NFκB serves not only as a means of stimulating LTR activity in response to T cell activation (as occurs in HIV) but also as the response element for a viral transactivator protein.

Retroviruses such as HIV and HTLV-1, therefore, have a more complex pattern of transcriptional regulation than RSV or MoMuLV, involving both virally encoded transactivators and the response to cellular signals such as T cell activation. These complex gene regulation strategies are probably related to the ability of these viruses to produce long-lived latent infections as well as a full replicative cycle and are, therefore, central to our understanding of virally induced disease.

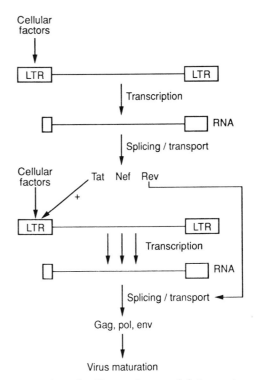

Figure 10 Life cycle of human immunodeficiency virus showing the early phase of viral infection, which yields RNA species encoding the regulatory proteins Tat, Nef, and Rev and the subsequent late phase in which viral transcription is stimulated by Tat while a change in RNA splicing/transport mediated by the Rev protein produces RNA encoding the viral structural proteins.

IV. Summary

In a wide variety of viruses, the initial phase of gene transcription following entry into the cell depends on the binding of cellular factors to the viral regulatory sequences. Even in the case of HSV, where transcription is stimulated by a virally encoded transactivator protein contained in the incoming virion, such activation occurs via a cellular transcription factor with which the virus protein forms a complex.

Similarly, although the later phases of viral transcription in the infected cell depend, in most cases, on virally encoded transactivator proteins made in the first phase of infection, these factors often act by binding to or stimulating cellular factors that can bind to sites in the viral control elements. Such a requirement for cellular factors in all phases of viral transcription exemplifies the parasitic role of viruses and their dependence on the cell for their gene expression and replication.

Bibliography

Cullen, B. R., and Greene, W. C. (1989). *Cell* **58,** 423–426.
Everett, R. D. (1987). *Anticancer Res.* **7,** 589–604.
Goding, C. R., and O'Hare, P. (1989). *Virology* **173,** 363–367.
Goodwin, G. H., Partington, G. A., and Perkins, N. D. (1990). Sequence specific DNA binding proteins involved in gene transcription. *In* "Chromosomes: Eukaryotic, Prokaryotic and Viral," Vol. 1 (K. W. Adolph, ed.), pp. 31–85. CRC Press, Boca Raton, Florida.
Jones, N. C., Rigby, P. W. J., and Ziff, E. B. (1988). *Genes Dev.* **2,** 267–281.
Serfling, E., Jason, M., and Schaffner, W. (1985). *Trends Genet.* **1,** 224–230.
Tooze, J. (1981). "Molecular Biology of Tumour Viruses DNA Tumour Viruses," revised 2nd ed. Cold Spring Harbor Laboratory, New York.

Transmissible Spongiform Encephalopathies

Stephen F. Dealler
Leeds General Infirmary

Richard W. Lacey
University of Leeds

Glossary

Astrocytosis Increase in the astrocytes of the brain tissue, usually associated with the death of neurons

Infective range Range of animals that can be infected from another with a spongiform encephalopathy

Infective unit Minimum amount of spongiform tissue required to infect an animal of the same species

Inversely related Increase in one factor is associated with a drop in another factor

Neuropil Nervous tissue close to the surface of the cerebrum and cerebellum

Passage Infection of an animal passing to another, and then this latter animal infecting another, i.e., the infection of an animal from another, usually of the same species

Spongiform encephalopathy Pathological condition of the brain in which vacuoles appear

TRANSMISSIBLE SPONGIFORM ENCEPHALOPATHIES (TSEs) are naturally occurring diseases of some mammals (but no other species) that can be transmitted to a limited number of others. They have a long asymptomatic incubation period, which leads to an inevitably fatal protracted neurological illness, and share histopathological changes in the brain. TSEs of different animal species appear to be transmitted by a very similar agent, which either enters the body from an external source or is created *de novo* in the body, and gives rise to no immunological response. The agent can be found in most of the tissues of the body during the course of the infection, albeit by complex means, but in the highest concentration in the nervous and lymphoid tissues. As the disease progresses, the amount of cerebral infective agent rises and is associated with the degeneration of neurons, astrocytosis, and a spongelike vacuolar change to the tissue.

The relationship among the TSEs of different species is only now becoming clearer. The illnesses that will be discussed in this article are those that have taken place without the intervention of humans, but it must be made clear that the spread of many of these conditions is probably due to human action (or lack of it). The TSE that has been most researched is scrapie (of sheep), and this will be considered first.

I. Scrapie

Scrapie is a naturally occurring disease of sheep found in many parts of the world (Fig. 1). Now

Figure 1 Distribution of scrapie. The possibility that scrapie was derived originally from Spain and was transmitted to sheep in northern Europe has been considered. Scrapie probably was present in the sheep exported from Britain after intense breeding. ▦, country reporting scrapie; ■, country where scrapie has been eradicated.

known for over 200 years, it possibly started in Spain and spread to the whole of western Europe. The export of strictly bred sheep from Britain in the nineteenth century is thought to be involved in the rapid spread to other countries. Work in Iceland has shown that the land on which infected sheep graze may retain the condition and infect later sheep, even if all scrapie-infected sheep are removed. Although sheep that were shipped to Australia and New Zealand could not have been tested for scrapie before they were sent, the illness does not seem to exist there; eradication procedures appear to have worked. Sheep that are imported into those countries are kept separately for some years before being allowed into contact with local sheep, and scrapie has appeared in sheep during this quarantine period. This makes it extremely unlikely that an infected animal will gain entrance into these countries. Some countries claim to have low numbers of cases (e.g., Germany), but limited outbreaks occur apparently randomly among unconnected flocks despite this. The incidence of scrapie in a flock appears to be related to the breed of the sheep, with some being relatively resistant to the illness (e.g., Scottish blackface) and others that are prone to it (e.g., Herdwick, Suffolk). Attempts have been made to eradicate scrapie from certain countries (e.g., United States) by slaughter of infected flocks. This, however, has been largely unsuccessful, and the occurrence of scrapie has been claimed to be increasing in the United Kingdom. Poor statistics on the prevalence of scrapie abound, and farmers not

recognizing the condition may merely slaughter the infected animal and fail to report it to an agricultural officer.

In 1936, researchers showed that scrapie could be transmitted to a healthy sheep by the intraocular inoculation of a homogenate of scrapie-infected brain. This experiment led to large amounts of research concerning the mode of transmission of TSEs.

Sheep inoculated with scrapie-infected tissue intracerebrally will have a short (possibly as low as 2 mo) incubation period, but on farms it is older sheep, usually >5 yr, that show signs of the disease. The sheep show increased irritability, excitability, and restlessness at the onset, giving rise to scratching, biting, rubbing the skin (hence, the name scrapie), patchy loss of wool, tremor (hence, the French name *tremblante*), loss of weight, weakness of the hindquarters, and, in some animals, impaired vision. The disease is always fatal. Only a small number of animals in a herd suffer from the clinical signs of scrapie, and experts have rarely seen 10 cases in a single herd. The natural mode of transmission between sheep is unclear. Experiments in which sheep with scrapie and those without it have been kept together on a farm have given rise to differing results, but goats have been shown to catch the disease from sheep in this form of experiment. Lambs of scrapie-infected sheep are more likely to develop the disease themselves later in life, but the reason for this is unclear. The infective agent is present in the membranes of the embryo but has been demonstrated neither in the colostrum and milk of the mother nor in the tissues of the newborn animal. Many cases of scrapie appeared following the accidental contamination of a louping ill vaccine; however, the mode of infection in most cases of scrapie seen on farms is unknown.

II. Transmissible Spongiform Encephalopathies of Humans

TSEs of humans are divided into specific clinical types, which may appear similar histopathologically but are either spread differently or have differing patterns of distribution and prevalence.

A. Creutzfeld–Jakob Disease

Creutzfeld–Jakob disease (CJD) was first described in 1920 and 1921 when it was known as "spastic pseudosclerosis" or "subacute spongiform encephalopathy." The illness exists throughout the world (Fig. 2) and is claimed to have a similar prevalence in each of the countries tested with an annual incidence of approximately one case per million of the population. This is almost certainly an underestimate, because histopathologists dislike carrying out necropsies on cases that may have died of CJD and many older people dying of a dementing illness do not have necropsies performed. There is an increased incidence among Libyan Jews (26 cases per million) and

spatial or temporal clusters in areas of Czechoslovakia, Hungary, England, the United States, and Chile. Cases are clustered in urban areas, but this can be accounted for by the increased population density. The average age in typical CJD is 56 yr, and only seven cases between the ages of 18 and 29 yr have been reported. Between 4 and 15% of cases have a familial connection with other cases. There is a slight excess of CJD in women.

Clinical prodromal symptoms start with changes in sleeping and eating patterns and progress over a few weeks into a clearly neurological syndrome. A rapid onset of neurological symptoms appears in 20% of cases, most commonly, vibrating muscular spasms, dementia, loss of higher brain function, and behavioral abnormalities. The disease progresses with deterioration in cerebral and cerebellar function to a condition in which most neurological activity is decreased, sensory and visual function decays, and the patient dies possibly after a decrease in lower motor neurological function and seizures. Ninety percent of the cases end in death within 1 yr of onset and a further 5% die in the following year. However, for 5% of the cases, fatality may take up to 10 yr, and in these cases neurological decay is relatively slow.

Diagnosis is by clinical assessment of patients with presenile dementia and by examination of electroencephalogram patterns, which characteristically show triphasic 1 cycle/sec activity or slow wave bursts with intermittent suppression (also found in animals with TSE). Enlargement of the lateral ven-

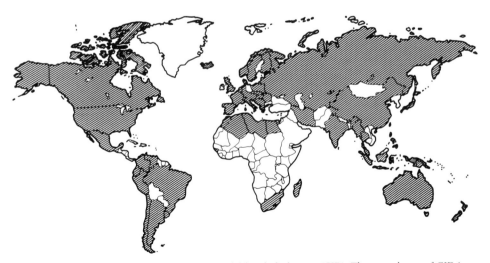

Figure 2 Worldwide distribution of Creutzfeld–Jakob disease (CJD). The prevalence of CJD in any country depends on many factors, including the willingness of pathologists to carry out necropsies.

tricles and an increase in IgG may be found, but these factors are of limited diagnostic value. Postmortem diagnosis is currently carried out by histological examination under the light microscope of cerebral tissue, although this is not always reliably diagnostic. Research techniques have been used to demonstrate CJD (and other TSEs). These may involve the electron microscope examination of brain tissue for scrapie-associated fibrils (SAFs), the staining of the tissue for prion protein antigens, or the intracerebral injection of tissue into animals, which will go on to die of the disease.

In some patients, the source of CJD has been claimed to be an infection transferred from other patients with the condition. For example, in one case, cerebral electrodes that had been sterilized with only alcohol and formalin vapor after use in a patient with CJD, were used in the brains of two young epileptic patients, both of whom contracted CJD after a short incubation. The transfer of CJD by corneal transplant in one patient, by cadaveric dura mater grafts in two patients, and by concentrated human growth hormone injections in eight more have been reported. Some cases in the literature seem too improbable for the low incidence in the community—for instance, the report of CJD in a neurosurgeon, in a mortuary attendant, in two men living 200 m apart and sharing a general practitioner, in a patient who had visited the Eastern Highlands of New Guinea (the kuru area) 10 mo previously, in three patients who had been operated on in the same neurosurgical unit within a period of 8 mo, in two people living together but not genetically linked, and in an individual marrying into an affected family (although the spouse did not suffer from CJD). Two husband and wife couples have died of CJD, as well as a lifelong vegetarian. The mode of disease transmission may be by personal contact, but only medical procedures have been described as to how this takes place. If the disease is transmitted from animal sources, many routes have been suggested but insubstantial evidence is available to prove them. The distribution of CJD in the world does not seem to be the same as that of scrapie in sheep, and human exposure to sheep is poorly associated with CJD. None of the animals that suffer from TSE except cows appear to be present in large numbers in all the countries where CJD is prevalent. Although their tissues are unlikely to be infective, pigs, which are generally slaughtered, are not consumed by Muslims and Jews, who also suffer from CJD.

B. Kuru

Kuru is a condition of the Fore tribe of the Okapa district of the Eastern Highland in Papua New Guinea in which a practice of ritual cannibalism of fellow tribesmen took place until around 1956. The disease affected mainly adult women and children of both sexes to give an annual disease-specific mortality of approximately 3%. Most deaths of women occurred from this disease, and some men who died from this disease were thought to have contracted it when young.

Kuru is caught by eating infected tissue. The brain of the dead tribal member was eaten by women and children and the muscle tissue by men. The possibility that this has also transmitted the disease to men but with a lower dose of infective agent and, hence, a longer incubation period has not been ruled out. The cohort of children born since 1956 has not suffered from kuru.

Clinically the disease is of a progressive cerebellar ataxia leading to uncoordinated movements, neurological weakness, palsies, and decay in brain cortical function. Most patients dying of kuru are not demented, and this is a major clinical difference between kuru and CJD. Patients with a longer incubation period appear to have a slower progression of symptoms, but generally death from intercurrent infection or medullary involvement takes place with an average clinical period of 12 mo.

C. Gerstmann–Straussler–Scheinker Disease

Gerstmann–Straussler–Scheinker disease (GSS) is an autosomally dominant condition rarely present in families. The disease is similar to CJD except that it has a more extended onset and duration, a tendency toward cerebellar ataxia as the initial predominant neurological sign, and a large number of amyloid plaques present among the spongiform encephalopathic changes of the brain. It has been transmitted to monkeys and rodents by intracerebral inoculation and to hamsters merely by the insertion of the human abnormal PrP gene from chromosome 20 into the hamster genome.

D. "Alper's Disease"

Alper's disease represents a group of very rare chronic progressive degenerative disorders of the central nervous system of infants and children. Histologically, this condition is similar to CJD and can be transmitted, like CJD, to hamsters easily but not to guinea pigs by intracerebral inoculation. Unlike CJD, however, there is also a fatty degeneration or cirrhosis of the liver.

III. Transmissible Mink Encephalopathy

Transmissible mink encephalopathy is an uncommon, fatal disease that occurs as outbreaks in ranch mink (*Mustela vision*). The condition was first reported in 1947 in Wisconsin and has also been reported in Canada and Finland with a similar pattern. Most of the mink on the farm die rapidly after a short encephalopathic period. The incubation in experimental situations is considered to be approximately 6 mo. Because mink are generally separate from each other when on farms (except when <3 mo old), and because little contact is made between them and external animals, the disease is thought to be derived from their food, which is contaminated with a TSE of another animal. Fighting and cannibalism among young mink is difficult to prevent, and this may be the reason why most of the animals on the ranch become infected. One outbreak in Stetsonville, Wisconsin, United States, followed the feeding of the mink with the meat and bone meal of a cow that died of a disease clinically similar to BSE. No sheep were included in their diet. The experimental feeding of food that contained scrapie infectious agent to mink has also given rise to the disease.

IV. Chronic Wasting Disease of Deer

Chronic wasting disease of deer is a TSE seen in 1978 in a mule deer herd and an adjacent herd of elk at Fort Collins, Colorado, United States. Both herds were captive. The disease shows typical spongiform change in the cerebral gray matter and can be transmitted to deer and ferrets by inoculation.

V. Bovine Spongiform Encephalopathy

Bovine spongiform encephalopathy (BSE), a condition seen generally in adult cattle of either sex, was first recognized in 1986 in the United Kingdom,

where it now infects >20% of milking herds. The numbers are highest in southern England, where >60 cases have been reported in a single herd but are generally spread throughout the British Isles, often as <3 cases per herd. It has been reported now in Oman, Switzerland, and France, but these cases are probably associated with the export of either infected animals or infected meat and bone meal for bovine food from the United Kingdom. The disease is thought to have been derived either from a change in the manufacture procedure of meat and bone meal (for bovine consumption) or from the inclusion of an uncommon bovine case of spongiform encephalopathy in bovine food in approximately 1981–1982. Claims have been made that this is not a new disease; in the past, although not histologically diagnosed, it has been seen in approximately 1 cow in 20,000–30,000. The rapid increase of the disease (500 cases reported per week in 1991) is probably due to the inclusion of undiagnosed cases of BSE in the meat and bone meal used for bovine food. This was stopped in the United Kingdom in July 1988, but the meal was simply exported to other countries by its manufacturers. BSE has been transmitted to cattle, pigs, and mice (both orally and by inoculation) and has been reported in the calf of a cow with BSE (born after oral infectious material was banned).

The long incubation period (presumed to be >2 yr and most commonly 4–5 yr) means that case numbers have continued to increase despite preventative measures. It is hoped that this incidence will reach a peak between 1993 and 1996 in the United Kingdom. Clinically, the cow initially appears alert but agitated, anxious, and apprehensive. As the disease progresses, however, the animal starts to take a wide base stance, the abdomen is drawn up, and the gait becomes abnormal and exaggerated and it gives rise to tumbling and skin wounds. Fine muscle contractions are seen involving small muscle groups over the surface of the neck and body with occasional larger muscular jerks. The animal loses weight and is taken to frenzied movements including aimless head butting.

The possibility that BSE may be infectious to humans was considered minimal in the United Kingdom until November 1989, when the feeding of bovine nervous tissue, lymphoid tissue, spleen, or gut to humans was banned. All animals that show signs of BSE in the United Kingdom must now be slaughtered and disposed of by incineration or burial. Beef in the United Kingdom would be expected to carry a low titer of the infective agent at the present time, but the large amounts eaten by humans and the long human life span make its safety unclear.

VI. Feline Spongiform Encephalopathy

Feline spongiform encephalopathy (FSE) is a condition that was reported in May 1990 in a 5-yr-old male Siamese cat and has since then been reported in many others in the United Kingdom. The epidemiology of FSE is unclear at present, but attempts to find previous cases among demented or neurologically degenerate cats from the past have been unsuccessful. We must therefore consider it to be a new disease. Histologically, it is similar to other TSEs. The worrying factor about this disease is that it was probably caught from feeding the cat on food that contained the agent of BSE. The owner of the original cat with FSE denied feeding it tinned cat food and insisted that it was fed fresh meat. The possibility that bovine meat contains adequate BSE to infect an animal of a different species orally is currently considered unlikely, but the outbreak of FSE suggests it to be possible. Attempts to transfer FSE to other species are currently taking place.

VII. Zoological Spongiform Encephalopathy

Zoological animal TSE has been reported since 1986 in an eland, a nyala, an oryx, a kudu, and a gemsbok in British zoos. An offspring of a mother kudu with the disease also developed it in 1990 and is considered to have caught it from the mother. It had not been given food thought to be infected. These animals became clinically unwell after the appearance of BSE on British farms, and they were probably infected either from the same source as the cattle or from BSE-contaminated foodstuff. No TSE in similar animals has been reported before and, hence, these must be considered new diseases.

VIII. Transmission of Spongiform Encephalopathies

Infected nervous tissue from some animals was injected, often intracerebrally, into others to find the range of infectivity of the agent (Table I). From this

Table I Range of Animals to Which TSEs Can Be Transmitted from Species with the Conditions[a,b]

Host	CJD	Scrapie	TME	Kuru	BSE
Human	+	NT	NT	NT	NT
Sheep	−	+	+	−	NT
Mink[c]	−	+	+	−	+
Cow	NT	+	−	NT	+
Chimpanzee	+	−	−	+	NT
Gibbon	−	−	−	+	NT
Monkeys[d]					
Capuchin	+	−	NT	+	NT
Marmoset	+	NT	NT	+	NT
Spider	+	+	−	+	NT
Squirrel	+	+	+	+	NT
Woolly	+	NT	NT	+	NT
Cynomolgus	−	+	NT	−	NT
Managabey	+	NT	NT	−	NT
Rhesus	−	−	+	+	NT
Pig-tailed	+	NT	NT	+	NT
Bonnet	NT	NT	NT	+	NT
African green	+	−	NT	−	NT
Baboon	+	NT	NT	NT	NT
Bush baby	+	NT	NT	−	NT
Patas	+	NT	NT	NT	NT
Stump-tailed	−	NT	+	−	NT
Talapoin	+	NT	NT	NT	NT
Goat	+	+	.+	−	NT
Ferret (albino)	?+	NT	+	NT	NT
Cat[c]	+	NT	−	−	+
Raccoon	NT	NT	+	NT	NT
Skunk	NT	NT	+	NT	NT
Mouse	+	+	−	−	+
Rat	−	+	NT	−	NT
Hamster (golden)	+	+	+	−	−
Gerbil	+	−	NT	−	NT
Vole	NT	+	NT	NT	NT
Guinea pig	+	−	+	−	NT
Rabbit	−	−	−	−	NT
Pig	NT	NT	NT	NT	+

[a] Relatively little work has been done on chronic wasting disease of deer, which can be transmitted to the ferret and to deer.

[b] Tables I, II, and III are acknowledged with thanks to the editor of the journal *Food Microbiology*.

[c] Infection of mink and cat with BSE is presumed from epidemiology.

[d] Some primates failed to develop the disease when inoculated with mouse-adapted scrapie.

+, Denotes transfer of infectious agent; −, denotes lack of transfer of infectious agent; BSE, bovine spongiform encephalopathy; CJD, Creutzfeld–Jakob disease; NT, not tested; TME, transmissible mink encephalopathy; TSE, transmissible spongiform encephalopathy.

it was clear that one animal strain of TSE could experimentally infect approximately 60% of the other species.

A. Dose Experiments

Dose experiments were carried out to understand the nature of the infective agent. Scrapie-infected brain was exposed to various agents (e.g., irradiation) and then injected in multiply diluted forms into the brain of an uninfected animal. In this way, the agent could be filtered. Multiple 10-fold dilutions were made of the infective material, and each of the dilutions were injected in a similar amount into an animal of the same species. The infective dose of the greatest dilution of agent that caused the animal to die (often after years of incubation) was defined as an infective unit (IU). The brain of an animal dying of TSE commonly contained between 10^6 and 10^{10} IU/g. The oral infectious dose of scrapie for a mouse was 4×10^4 IU, which represents between 10^{-2} and 10^{-5} g of infected brain tissue. In these experiments, researchers noticed that the incubation period was inversely related to the dose given to the animal and that an animal may be infected with a small dose of TSE but die of old age before clinical signs appear.

B. Effect of Host Passage on the Properties of the Infectious Agent

1. The infectious dose between species is usually higher than between animals of the same species (possibly a millionfold), but it is sometimes the same (e.g., scrapie doses for mink).

2. When a species has been infected with the TSE of a different species, it can then go on to infect a range of animals that the original species could not and with a different dose.

3. When a species has been infected, it can then infect additional animals of the same species with much lower doses of agent.

4. The histopathology of the disease in an animal infected from another species is not the same as it would be if infected from one of the same species.

5. The incubation period of an animal infected from another species is much longer than that of an animal infected from one of the same species.

To demonstrate these factors (Fig. 3), brain tissue of a sheep with scrapie would only need 1 IU to infect another sheep, but if mice were injected, a much larger dose would be needed, the incubation period would be relatively long, and a low percentage of the mice would be affected. If brain tissue from these scrapie-infected mice was inoculated into additional mice, the dose would be 1 IU (a very small amount), the incubation period would be much lower, a high percentage of the mice would become infected, and the histopathology would be the same as in further passages of the disease in mice (but different from the histopathology of the mouse infected from the sheep). The mice would also be able to infect a distinct range of animals other than sheep.

These factors are known as the species barrier (SB) and behave as if the agent is altered by passage through a species into a form that is more likely to infect that species. The insertion of the hamster PrP gene (*vide infra*) into the genes of a mouse removes the SB between the animals; i.e., when injected with scrapie from a hamster, such a mouse will develop scrapie as if it is a hamster. This has been explained by the possibility that the PrP protein is all or part of the infective agent and, as it is produced from the genes of the host animal, it has a different structure in different species.

C. Tissue Infectivity of Clinical Cases of Transmissible Spongiform Encephalopathies

TSE infectivity is present in most tissues tested, and the distributions vary among species (Table II). The

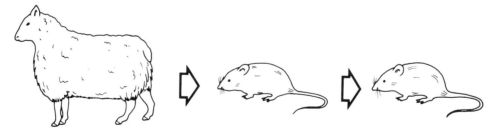

Figure 3 Species barrier. Transfer of scrapie from a sheep to a mouse is difficult in that it requires a high dose, has a low success rate, produces an illness with a long incubation period, and produces histopathological changes not found in mice infected from the same species. Transfer of scrapie from mouse to another mouse is relatively easy, however, in that it requires a low dose, has a high success rate, produces an illness with a relatively short incubation period, and produces histopathological changes similar to those found in other mice infected from the same species.

Table II Infectivity of Tissues from Transmissible
Spongiform Encephalopathy (TSE)-Infected Animals[a]

Tissue	Animal with TSE					
	Sheep	Goat	Mouse	Mink	Human	Cow
Brain	+	+	+	+	+	+
Spinal cord	+	+	+	+	+	
Peripheral nerve	+	+	+		+	
Eye						
Adrenal		+				
Lymph node	+	+				
Tonsil		+				
Salivary gland	−	+	+	+		
Spleen	+	+	+	+		
Gut	+	+	+	+		
Liver		+	+	+		
Kidney	−	−	+	+		
Bladder				+		
Pancreas		−				
Heart	−	−				
Lung	−	−	+	+		
Thyroid	−					
Thymus			+	+		
Testis		−				
Ovary		−				
Uterus			+			
Blood/serum	−	−	−	+	+	
Bone marrow		−	+	−		
Cerebrospinal fluid	−	+			+	
Urine		−		−	−	
Feces						
Saliva	−	−			−	
Milk	−					
Mammary gland	−					
Muscle	−	+		+		

[a] Scrapie-infected hamsters have been reported as also have infective muscular tissue.

+, Indicates that an attempt to transfer TSE with this tissue was successful; −, indicates that the attempt was not successful.

finding of infectivity in the buffy coat of blood has led to fears that CJD may be transferred by blood infusions, but there has been no report of this at this time. The finding that the scrapie agent was present in peripheral as well as central nervous tissue and in lymphoid tissue has given little surprise to the finding of TSE agent in muscles of goats, hamsters, mink, and possibly humans.

D. Tissue Infectivity during the Incubation Period

The animal is asymptomatic for a long period before the disease becomes clinically apparent. During this time, many of the tissues of the body are infectious but at a relatively low titer compared to the nervous

system during the symptomatic period (Table III). This titer is adequately high, however, to permit infection of other animals by intracerebral inoculation and possibly by parenteral or oral routes.

E. The Mode of Spread of Transmissible Spongiform Encephalopathies inside the Body

Research has shown that the agent will pass along peripheral nerves and hence may travel in this way from a site of absorption to the brain. Other research has shown it to be present in the buffy coat (probably the macrophages) of the blood. The exact mode of spread of TSE inside the body is unclear.

F. Immunity

Developed immunity against the infective agent has not been demonstrated. Apparently, no antibodies that react with it are produced, even in chronically infected animals. The possibility that this may permit multiple inoculations of sublethal doses of the agent to be effectively additive in their effect has been considered and is presumed by some researchers, but no specific proof of this has been shown.

Table III Stage of Incubation Period at Which Tissue
Infectivity was Found to Start

Tissue tested[b]	% Incubation period[a]								
	30	40	50	60	70	80	90	100[c]	110
Brain				m, g, k					
Pituitary					g				
Spinal cord		m		g, k					
Peripheral nerve					g, k				
Spleen	m	g			k				
Adrenal					g				
Lymph node	g, k								
Thymus	m, k								
Lung		m							
Liver/kidney						k		m	
Muscle									g
Gut	m, g							k	
Bone marrow						m			
Salivary gland				g		k			

[a] m, g, and k represent the incubation period percentages at which the tissues of mice, goats, and mink become infective, respectively.

[b] Some tissues (e.g., the spleen of mice) become infective before 30% of the incubation period.

[c] 100% of the incubation period coincides with the onset of clinical disease.

Rabbits may produce antibodies against PrP derived from sheep (*vide infra*).

IX. Resistance

A. Resistance of the Agent to Destruction

Chemical disinfectants (e.g., domestic bleach), weak acids, DNAase, RNAase, proteinases (including those found in the animal gut), ultraviolet light, ionizing radiation, heat (cooking temperatures), and chemicals that react with DNA (psoralens/ultraviolet light, hydroxylamine, zinc ions) all have little effect on the infectivity of the agent. High-temperature autoclaving (135°C for 18 min) decreases the infectivity dramatically, as does the use of 1 *M* NaOH, but neither will fully destroy the agent, as it has been found to remain infective after 360°C for 1 hr or even after incineration. Internment of infective tissue in the soil for 3 yr did not destroy the agent. Some phenols and proteases will decrease the infectivity of the agent but not to an adequate degree to be of value in disinfection.

B. Prevention of Transmissible Spongiform Encephalopathies

Nosocomial CJD should be prevented by prohibiting CJD, GSS, or Alper's disease patients (or those with obscure neurological conditions) from becoming blood or tissue donors, by the incineration or high-temperature autoclaving of all materials that came into contact with blood or postmortem tissue from such a patient, and by the disposal of *all* surgical instruments used for brain surgery on such a patient. The body should not be used for teaching anatomy or surgery. Correct action to be taken concerning BSE-infected herds is currently under intense discussion.

X. Histopathological Changes

Characteristic lesions under the light microscope consist of spongy changes in the neuropil nerve cells (Fig. 4) and astrocytes with nerve cell degeneration and astrocytosis. These changes generally take place in the gray matter of the cerebrum and cerebellum. The distribution of this histopathology may

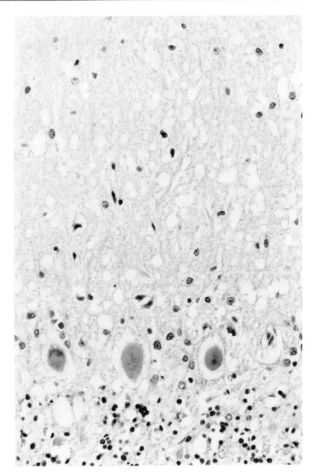

Figure 4 Histopathology of Creutzfeld–Jakob disease. Formalin-fixed and paraffin-embedded section of cerebellum stained with hematoxylin and eosin. This shows the multiple clear vacuoles (spongiform change) seen throughout the neuropil. [Photograph provided by Dr. Bridges, University of Leeds, United Kingdom.]

vary according to the strain of disease, as has been shown with inbred strains of mice. It may also vary with the site of inoculation—for example, if the infection reaches the brain through the optic nerve, then the spongy degeneration is clustered around the occipital lobe. Amyloid plaques may be seen between cells in some TSEs (e.g., kuru and hamster scrapie) and stained with Congo red. These contain PrP and have been shown to be infective. Electron microscopy shows twiglike structures 12–16 nm in width and 100–500 nm long, which are found only in TSEs and are now called scrapie-associated fibrils (SAFs).

XI. Nature of the Infective Agent for Transmissible Spongiform Encephalopathies

A. PrP: A Cellular Protein Found Altered in Cases of Transmissible Spongiform Encephalopathies

PrP is a sialoglycoprotein produced from a gene normally found in the genome of the infected animal. This gene, on chromosome 2 in mice (or chromosome 20 in humans), may be conserved between strains but altered between species. PrP in a normal brain has glycoside modifications and is held onto the membrane of the cell by covalent attachment to phospholipid. In scrapie, however, the PrP is present much more within the membrane and has a different structural form (PrP^{Sc}). This modification, whatever it may be, renders the protein resistant to heat and to most proteases, and when treated with protease K (a very powerful enzyme) it is split into particles of 27–30 kDa, the structures of which are unknown. PrP^{Sc} structure may vary with the strain of scrapie and may vary among animals of the same species. The PrP of patients with GSS is mutated so that proline is replaced by leucine at position 102. The PrP of CJD in the Libyan Jew and Czechoslovak clusters and some cases of familial CJD have a different alteration from that in GSS and various changes have been reported. Patients with CJD that is not familial do not generally have PrP mutations. When the gene for PrP in a patient who suffers from GSS (PrP^{GSS}) is inserted into a hamster genome, it spontaneously develops a TSE. PrP is therefore greatly involved in the infectivity and pathogenesis of the disease, but how it could be the infective agent (*vide infra*) is not clear. Inbred mouse strains with long and short incubation periods (known as *sinc* genes) for scrapie have been found to have differences in their PrP codons, which are claimed by some to be the *sinc* genes themselves.

B. Structure of the Transmissible Spongiform Encephalopathy Infective Agent

Conjecture is currently complex and, hence, there are many hypotheses:

1. *Virus* No specific particle has been isolated, but the transmissibility is similar to this form of agent.

2. *Prion* A protein infectious agent. Felt to be the PrP-altered form (PrP^{Sc}), which is resistant to destruction, would not cause antibody production in the infected animal and would vary according to the species infected. Purified PrP^{Sc} has been claimed to transmit scrapie, but experimental methods have been challenged. How a prion would transmit strains of disease without an independent genome is currently unclear. The recent transmission of GSS to hamsters by an artificial PrP^{GSS} gene has made this hypothesis more likely.

3. *Virino* Small fragment of DNA/RNA closely wrapped in protein. The SAFs that are only found in TSEs and that have PrP involved in their structure could be virinos. They are suggested to contain some genetic material (although none has been found), which could explain the different strains of disease. However, the agent is resistant to chemicals and ionizing radiation, which would be expected to destroy the nucleic acid.

XII. Future Prospects

The incidence of CJD is apparently unchanged over many years, although poor statistics are available to confirm this. The epidemic of BSE in the United Kingdom is expected to continue there for up to 10 years but for the incidence to decrease after 1993. It is, however, likely to spread to Europe and to other countries that imported infected British meat and bone meal. The precise risk that this may be followed after 5–25 years by a wave of CJD, FSE, and zoological spongiform encephalopathy, is unknown but the possibility should be considered.

Bibliography

Aiken, J. M., and Marsh, R. F. (1990). *Microbiol. Rev.* **54,** 242–246.

Brown, P., Cathala, F., Castaigne, P., and Gajdusek, D. C. (1986). *Ann. Neurol.* **20,** 597–602.

Casaccia, P., Ladogana, A., Xi, Y. G., and Pocchiari, M. (1989). *Arch. Virol.* **108,** 145–149.

Collinge, J., Owen, F., Poulter, M., Leach, M., Crow, T. J., Rossor, M. N., Hardy, J., Mullan, M. J., Janota, I., and Lantos, P. L. (1990). *Lancet* **336 ii,** 7–9.

Committee on Health Care Issues, American Neurological Association (1986). *Ann. Neurol.* **19,** 75–77.

Dealler, S. F., and Lacey, R. W. (1990). *Food Microbiol.* **7,** 253–279.

Fraser, H., Richie, L., Farquhar, C., Somerville, R., and Hunter, N. (1989). *Prog. Clin. Biol. Res.* **317,** 659–667.

Gabizon, R., McKinley, M. P., Groth, D., Westaway, D., DeArmond, S. J., Carlson, G. A., and Prusiner, S. B. (1989). *Prog. Clin. Biol. Res.* **317,** 583–600.

Gajdusek, D. C. (1990). Subacute spongiform encephalopathies: Transmissible cerebral amyloidoses caused by unconventional viruses. *In* "Virology," 2nd ed. (B. N. Fields, D. M. Knipe, *et al.,* eds.). Raven Press, New York.

Hadlow, W. J., Race, R. E., and Kennedy, R. C. (1987). *J. Virol.* **61,** 3235–3240.

Kimberlin, R. H. (ed.) (1977). "Slow Virus Diseases of Animals and Man." North-Holland, Amsterdam.

Kimberlin, R. H. (1982). *Nature (London)* **297,** 107–108.

Manuelidis, E. E., and Rorke, L. B. (1989). *Neurology* **39,** 615–621.

Narang, H. K., Asher, D. M., and Gajdusek, D. C. (1988). *Proc. Natl. Acad. Sci.* **85,** 3575–3579.

Oppenheimer, D. R. (ed.) (1983). "Scrapie Disease in Sheep." Academic Press, London.

Pattison, I. H. (1990). *Vet. Rec.* **Jan 20,** 68.

Prusiner, S. B. (1989). *Annu. Rev. Microbiol.* **43,** 345–374.

Race, R. E., Graham, K., Ernst, D., Caughey, B., and Chesebro, B. (1990). *J. Gen. Vir.* **71,** 493–497.

Westaway, D., Carlson, G. A., and Prusiner, S. B. (1989). *Trends Neurosci.* **12,** 221–227.

Wyatt, J. M., Pearson, G. R., Smerdon, T., Gruffydd-Jones, T. J., and Wells, G. A. H. (1990). *Vet. Rec.* **May 19,** 513.

Transposable Elements

Peter M. Bennett
University of Bristol

Glossary

Composite transposon Modular structure comprising a nontransposing sequence flanked by copies of the same IS element

Insertion sequence Small cryptic DNA element that can migrate from one genetic location to another, completely unrelated location

Inverted repeat Sequence that defines both ends of a transposable element and that is found as an inverted duplication

Target site Site at which a transposition event occurs (occurred)

Transposing bacteriophage Bacteriophage that replicates using a form of transposition

Transposition immunity Inhibition of transposition of one copy of a transposable element by a second copy of the same element on the target DNA

Transposon DNA element that can migrate from one genetic site to another, unrelated site and that encodes a function(s) other than that required for transposition

TRANSPOSABLE ELEMENTS are discrete DNA sequences that can move from one location on a DNA molecule to another location on the same or on a different DNA molecule. The process of transposition is, *ipso facto*, a recombination event in that it involves breaking and reforming phosphodiester bonds but does not require homology between the element and its sites of insertion. Accordingly, it does not depend on the homologous recombination system of the host cell; in practice, this means that transposition is *recA*-independent. Several different types of transposable elements are known, and transposition can occur by one of several different mechanisms. Transposable elements have been discovered in gram-negative and gram-positive bacteria, in archaebacteria, and in yeasts and are probably responsible for much of the macromolecular rearrangement of microbial genomes.

I. Introduction

A. General Structure of a Transposable Element

Two points of caution should be interpolated at the start. First, although transposition results in the insertion of one DNA sequence into another, in many instances, but not all, the transposed sequence is also retained at its original location; i.e., the event is not only recombinational but also replicative (see later). Second, not all transposition events involve discrete DNA elements [e.g., one-ended transposition mediates the transposition of a nested set of DNA sequences rather than one discrete sequence (see later)].

The majority of documented transposition events in microorganisms involve transposable elements. Each element is a defined structure that is preserved from one transposition event to the next. Most, but not all, elements terminate in short perfect, or near perfect inverted repeats (IRs), which act as specific recognition signals for the transposition enzymes. IR sequences differ from one transposable element to another and are usually 15–40 bp long. Each element usually encodes at least one protein that is needed for its own transposition, specifically, a transposase, which mediates the recombination events between the ends of the element and the target site. In addition, functions unrelated to transposition may be encoded.

B. Insertion Sequences and Composite Elements

The smallest elements capable of self-promoted transposition are the prokaryotic insertion sequences, or IS elements, which range in size from about 0.7–2 kb. Each encodes only one or possibly two functions necessary for control and execution of its own transposition. Transposons are larger structures that also accommodate functions unrelated to transposition and that confer a predictable phenotype on the host cell (e.g., resistance to an antibiotic), unlike IS elements that alter the cell phenotype at random by mutation. Indeed, it was this mutational activity that first drew attention to the existence of transposable elements in bacteria.

C. Transposons

Two types of transposons have been recognized: composite transposons and complex transposons. An element of the former type has a modular construction in which a central unique sequence, accommodating the genes not involved in transposition, is flanked by two copies of an IS element. These terminal elements provide the functions that mediate transposition of the composite structure. The terminal IS elements may be oriented as direct or, more commonly, as inverted repeats and usually retain the ability to transpose as independent IS elements.

Complex transposons have no obvious modular structure. Functions unconnected with transposition are encoded by genes that are part of the basic transposable element. The whole gene ensemble is flanked by short IRs, as seen for the majority of IS elements, and no component can transpose independently of the rest of the structure. In crude, comparative terms, a complex transposon is more analogous to an IS element than to a composite transposon. Indeed, pairs of some complex transposons can also function together to transpose DNA sequences flanked by them. This cooperative behavior seems to be a feature of many prokaryotic transposable elements and has been exploited in the use of mini-Mu to "clone" chromosomal genes *in vivo* (see later).

D. Transposing Bacteriophages

Another group of transposable elements found in bacteria are the transposing bacteriophages, typified

by bacteriophage μ (or Mu). This element was identified by J. Taylor in the early 1960s in a culture of *Escherichia coli* and was the first bacterial transposable element to be discovered, although this was not fully appreciated at the time. It was so called because Taylor deduced, correctly, that the increased numbers of auxotrophic mutants he found among populations of *E. coli* lysogenized by this temperate phage had been generated by phage integration into different sites on the bacterial chromosomes in individual cells, instead of into a single site, as is seen for bacteriophage lambda. Some of these integration events, because the sites of insertion were within genes, generated auxotrophic mutations. Hence, the phage was described as a mutator phage, now abbreviated to Mu. Transposing bacteriophages replicate by a form of transposition. They differ from other transposable elements in that they can exist independently of other DNA molecules and the host cell, i.e., as bacteriophage particles. [*See* BACTERIOPHAGES.]

E. Distribution of Transposable Elements

Transposable elements are widely distributed among both eubacteria and archaebacteria and are also found in lower and higher eukaryotes. Many IS elements have been identified in both gram-negative and gram-positive bacteria, and several have been found in Archaebacteria such as *Halobacterium* sp. and *Methanobrevibacter smithii*. Indeed, wherever a serious search has been made for such elements in a bacterial system, they have invariably been found. Failure to detect them is more likely to indicate a deficiency in the detection system rather than an absence of transposable elements, as it seems likely that IS elements are ubiquitous in bacteria.

Transposons, both composite and complex, have been found in many gram-positive and gram-negative bacteria, often in organisms of clinical or veterinary origin. These sources almost certainly account for the preponderance of antibiotic resistance transposons and the relative paucity of other markers. Most bacterial transposons have been found, initially, on plasmids rather than on bacterial chromosomes, although there are a few notable exceptions. This may reflect no more than the fact that once located on a plasmid the ability to spread laterally to other bacteria is markedly enhanced and, hence, the likelihood of detection is also significantly increased. It also seems to be true for some transposons that transposition into a plasmid occurs at a significantly higher frequency than into bacterial chromosomes, despite the greater capacities of the latter. [*See* PLASMIDS.]

F. Nomenclature

The designations given to transposable elements are not usually assigned at random or at the whim of the researchers who discovered them. To avoid accidental assignment of the same code to two different elements, it has been widely accepted that numbers will be allocated from a central directory. This resource is administered by Dr. Ester Lederberg at the Department of Medical Microbiology, Stanford University, California. Periodically, lists of new assignments are published. The convention adopted is simple. Insertion sequences are designated IS followed by one of a set of numbers allocated to the research worker or laboratory from the central directory, upon request. The digits are written in italics. Likewise, transposons are designated Tn followed by an appropriate number (in italics). Regrettably, in my view, deviations from this simple practice have recently appeared in the literature. In these cases additional letters have been added to indicate the bacterial source of the element and the numbering system has been restarted (e.g., ISRm2, an element from *Rhizobium meliloti*, and ISM1, an element from *Methanobrevibacter smithii*, are quite distinct from IS2 and IS1, respectively, which originate from *E. coli*).

II. Insertion Sequence Elements and Composite Transposons

A. History

The first transposable elements to be recognized as such in bacteria were three small cryptic elements, now appropriately designated IS1, IS2, and IS3. They were discovered as the consequence of the analysis of strongly polar mutations in the *gal* and *lac* operons of *E. coli*. Genetically the mutations were found to behave like point mutations but, although they reverted to wild type spontaneously, the frequencies of reversion were unaffected by chemical mutagens known to enhance the generation of base substitution and frameshift mutations. Thus, it was suggested that the damage was caused by the insertion of additional DNA sequences at the

sites of mutation. A series of elegant physical studies using initially λ.gal transducing phages and then electron microscopic analysis of heteroduplex DNA structures provided the experimental evidence for mutation by DNA insertion. These were the first studies to provide physical evidence for DNA transposition, a phenomenon proposed more than two decades previously to explain the switching on and off of pigmentation genes in maize.

B. Structure

IS elements are, structurally, the simplest form of transposable element. Their sizes (Table I), rarely larger than 2 kb, preclude them encoding more than the one or two genes whose products are necessary for transposition. Of those that have been examined in sufficient detail, most display one large open reading frame (ORF) that utilizes the majority of the sequence. In a few cases, the products of these putative genes have been shown to encode proteins; few have been shown to be necessary for transposition, although this is presumed. Some elements

Table I Some Prokaryotic IS Elements

Insertion	Size (bp)	Terminal IRs (bp)	Target (bp)
Elements from gram-negative bacteria			
IS1	768	20/23	9
IS2	1327	32/41	5
IS3	1258	29/40	3
IS4	1426	16/18	11–13
IS5	1195	15/16	4
IS6	820	14/14	8
IS10	1329	17/22	9
IS21	2132	10/11	4
IS50	1534	8/9	9
IS91	~1800	8/9	0
IS150	1443	19/24	3
IS492	1202	0	5
Elements from gram-positive bacteria			
IS110	1550	10/15	NR
IS231	1656	20	11
IS431R	786	14	NR
IS904	1241	32/39	4
ISL1	1256	21/40	3
ISS1	820	18	8
Elements from archaebacteria			
ISH1	1118	8/9	8
ISH2	520	19	10–12
ISH23	~1000	23/29	9
ISH51	1371	15/16	3
ISM1	1381	29	8

NR, not recorded.

also have smaller ORFs, which overlap or are contained within the large ORF. Whether or not these are real genes, the products of which have a role in the mechanism of transposition, remains unknown.

Most IS elements characterized to date to sequence level appear to be unrelated phylogenetically one to the other. Notable exceptions are those that are related to IS3 (e.g., IS2, IS150, IS904) (Table I). Sources of these elements include *Agrobacterium tumefaciens, E. coli, Mycobacterium tuberculosis, Rhizobium* sp., *Shigella dysenteriae, Streptococcus lactis,* and *Xanthomonas campestris.* To date, the family extends to about a dozen members.

An apparent exception to the general arrangement of genes carried by IS elements is IS*1,* one of the smallest transposable elements known. It is 768 bp long, is bounded by near perfect 23-bp IRs, and contains two short, adjacent ORFs, *insA* and *insB,* which have been shown to be necessary for transposition. The two sequences do not overlap and are in different reading frames but are transcribed as an operon. In addition, the sequence accommodates four other ORFs, which overlap *insA* and *insB.* The involvement of these in transposition was conclusively eliminated using site-directed mutagenesis. A series of nonsense mutations were introduced into IS*1* in such a way that only one of the six ORFs was affected by a particular mutation. The consequences for transposition of the element were then determined. Only mutations in the ORFs *insA* and *insB* affected transposition. Similar critical analyses of putative transposition genes have been applied to very few IS elements. However, the demonstration that an ORF is required for a particular function does not mean that the ORF is a gene. This is nicely illustrate by IS*1.* A recent study has demonstrated that while *insA* codes for a protein contiguous with the gene, *insB* does not. Rather, the ORF designated *insB* encodes the second half (C terminus) of a peptide that has the bulk of InsA at its N terminus. This apparent violation of the normal mechanism of protein synthesis, i.e., ignoring the translational stop signal at the end of the *insA* gene, is accomplished by some ribosomes making a −1 frameshift toward the end of the *insA* transcript, before the terminating codon is reached. This puts these ribosomes into the reading frame used by *insB,* and translation can continue until the terminating codon signaling the end of *insB* is encountered. Surprisingly, this frameshifting is not accidental but, rather, is directed by the nucleotide sequence of the messenger RNA.

This InsAB protein is the IS*1* transposase, whereas the InsA peptide functions as a transposition control element. Both proteins compete to bind to the IRs of IS*1;* however, whereas InsAB can mediate transposition, InsA cannot, so, by occluding InsAB from the IRs, InsA can inhibit IS*1* transposition. Interestingly, the transposition frequency is not determined by the absolute amount of either protein but, rather, by their ratio.

Apart from IS*1,* the best studied IS elements are IS*10* and IS*50.* The former constitutes the flanking elements, in inverted repeat, of the composite transposon Tn*10,* which encodes resistance to tetracycline, while IS*50* forms terminal inverted repeats of Tn*5,* a composite element that confers resistance to kanamycin (Km), bleomycin (Bl), and streptomycin (Sm), although the last determinant is silent in some hosts, including *E. coli.* Both transposons have been studied in considerable detail (Figs. 1 and 2). Although these transposons are, in a structural sense, archtype elements, they display several significant differences in their molecular genetics, particularly in the way gene expresion is controlled.

C. Mechanism(s) of Transposition

What is known about the mechanism(s) of transposition of IS elements comes primarily from studies on IS*1,* IS*10,* and IS*50.* One early, important finding was that in most cases IS sequences are flanked by direct repeats (DRs) of short (3–12 bp) sequences that are present only once at the insertion targets. The size of these DRs is element-specific, while the actual sequences are, in general, not specific. It was pointed out that this would be the inevitable result if, in preparation for the insertion, the target site was cut on both strands, with the cleavage sites slightly staggered. If the IS element is then ligated to the target across the short single-strand extensions and these are then filled in (perhaps by DNA repair synthesis), then short DRs will be formed at the junctions of the element and the target. It was found that, although the sequence of the DRs varied from one insertion event to another and the size of the DRs varied from one element to another, for any given element the size of the DRs was constant, suggesting that target site cleavage is element-, rather than host-, directed. These findings have now been extended to the great majority of transposable elements, including phage Mu. Indeed, if during the course of determining a DNA sequence a structure is uncovered that is delineated by IRs and flanked by short DRs, then it is reasonable to conclude that a transposable element has been discovered, even if it

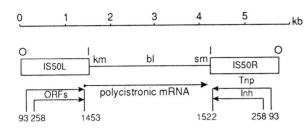

Figure 1 Tn5. Tn5 was detected on the resistance plasmid JR67, whch originated in a strain of *Klebsiella pneumoniae*, when it transposed to phage λ. Tn5 is 5.7 kb, with a central module of 2.6 kb encoding resistance to kanamycin (km), bleomycin (bl), and streptomycin (sm) flanked by inverted copies of IS50, designated IS50L,R. The resistance genes are transcribed as a single operon from a promoter within IS50L. Deletion analysis has indicated that IS50R alone is responsible for Tn5/IS50 transposition. IS50R is 1534 bp, has 19-bp imperfect IRs (7 mismatches), and encodes two peptides designated Inh and Tnp, of 421 and 476 amino acids, respectively. Tnp is the IS50 transposase; Inh is an inhibitor of Tn5/IS50 transposition. (It reduces the frequency of transposition once Tn5/IS50 has been established in a new host. On moving from one cell to another, transposition of IS50/Tn5 is subject to zygotic induction.) The Tnp and Inh proteins are virtually identical, differing only in that the smaller, Inh, lacks the first 55 amino acids of Tnp. The truncation is not produced by posttranslational modification; rather, Inh is produced from its own transcript. IS50R and IS50L differ at only one position, 1453. The change has profound effects. It simultaneously creates the promoter for the resistance operon and premature translational termination signals (ochre codon) for the two peptides encoded by the IS element, both of which are truncated by 23 amino acids at their C termini. The deletions inactivate both proteins. I, junctions of IS50 elements with central resistance module; mRNA, messenger RNA; O, junctions of Tn5 and carrier molecule. Numbers indicate translation initiation and termination points. Arrows indicate direction transcription/translation. [From Bennett, P. M. (1991). Transposons and Transposition. *In* "Modern Microbial Genetics" (U. N. Streips and T. E. Yasbin, eds.). pp. 323–364. Wiley-Liss, New York.]

is no longer able to transpose. With one or two elements, transposition does not generate DRs. In these cases, it is likely that insertion has been into a double-stranded flush cut, rather than that the absence of DRs signals a fundamentally different transposition mechanism.

The transposition mechanism(s) of many IS elements is believed to be conservative; i.e., the element is removed from its donor molecule as a double-stranded structure, and this is inserted into the target. Note that this is not true for all transpositions, as will be seen. The simplest conservative model is that described as the "cut-and-paste" mechanism (Fig. 3). In this process, the two ends of

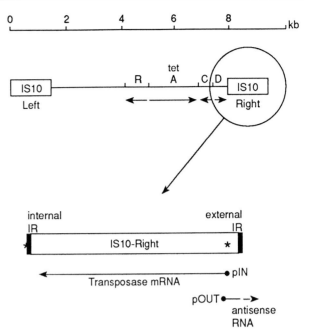

Figure 2 Tn10. Tn10 originated on the resistance plasmid 222 (otherwise called NR1 and R100). It is 6.7 kb with an approximate 4-kb central module encoding inducible tetracycline resistance flanked by inverted copies of IS10, designated IS10R,L. Tetracycline resistance is conferred by the TetA protein, which forms a tetracycline antiport that pumps tetracycline out of the cell. Expression of *tetA* is negatively regulated by TetR and is induced by tetracycline. The divergently transcribed genes *tetC* and *tetD* are also induced by tetracycline, but the functions of their products is not known. The sequences of the two copies of IS10 differ at several positions; of the two, IS10R is predominantly responsible for transposition. The IS10 transposase is encoded by a 1206 open reading frame transcribed toward the *tet* genes from promoter pIN. There is a second promoter, pOUT, located 35 bp downstream from pIN that directs transcription in the opposite direction to pIN and produces a product of 69 nucleotides. This RNA species is remarkably stable with a half-life of approximately 70 min and a level of about five copies per copy of IS10. In contrast, the *tnp* gene transcript is relatively unstable (half-life 40 sec) and of low abundance (approximately 0.25 copy per copy IS10.). These two transcripts have short, complementary sequences at their 5′ ends that can anneal. When this occurs, translation of the transposase messenger RNA is inhibited, an example of antisense RNA control. The transposition activity of IS10 is also regulated by *dam* methylation. Two *dam* sites (*) are involved, one overlapping the −10 box of the *tnp* promoter pIN, the second at the other end of IS10 where tranposase binds. Methylation of the pIN *dam* sequence reduces the efficiency of pIN while methylation of the second site interfers with binding of transposase to that end of the element. Both effects act to damp down transposition activity. However, only the methylation at pIN will affect transposition of Tn10, a differential affect that will tend to enhance the coherence of the composite element. [From Bennett, P. M. (1991). Transposons and Transposition. *In* "Modern Microbial Genetics" (U. N. Streips and T. E. Yasbin, eds.). pp. 323–364. Wiley-Liss, New York.]

the element are brought together and the cognate transposase cuts through both DNA strands at both ends of the element, disconnecting the transposable element from the donor DNA molecule. Although free transposable elements are not normally seen in cells, free, circular forms of IS*10*, complexed with protein, have been seen when the IS*10* transposase is greatly overexpressed, as the result of gene manipulation. Although these DNA–protein structures may be artifacts of the experiment, evidence accumulating with other transposons suggests strongly that the IS*10* circles do indeed represent a transposition intermediate, the existence of which is normally transient. It is believed that the conjoining of the ends of the element prior to cleavage is not directed in any sense but, rather, relies on random collision, probably of IR–transposase complexes, which form at the ends of the element. Once released, the free (circular) form of the element is then inserted into the target site across a staggered cut, probably also made by the transposase. The short single-stranded regions at both ends of the element are then filled in and continuity is restored to both

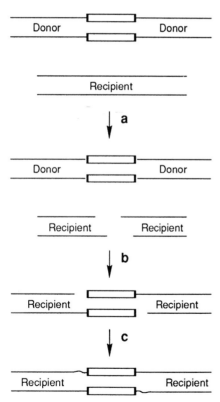

Figure 3 Transposition: cut-and-paste model (see text for details). [From Bennett, P. M. (1991). Transposons and Transposition. *In* "Modern Microbial Genetics" (U. N. Streips and T. E. Yasbin, eds.). pp. 323–364. Wiley-Liss, New York.]

strands by DNA ligase. It seems likely that the deleted donor molecule is then normally lost (degraded).

Some IS elements not only transpose conservatively but can also generate what are called transposition cointegrates (see Fig. 5). These structures result from transposition that involves semiconservative replication of the element. IS*1* is one such. IS-generated cointegrates are relatively stable entities and their formation can be used to isolate previously undetected elements.

D. Transposition of Composite Transposons

Many composite transposons encoding resistance to antibiotics, originating in a broad spectrum of bacteria including both gram-positive and gram-negative organisms have been found and a few such elements encoding functions other than drug resistance have also been identified (Table II). In general, these elements appear not to be phylogenetically related, although different transposons carrying essentially the same drug-resistance gene are known (Table II), as are transposons where completely different central sequences are bracketed by copies of the same IS element (Table II). Each of these structures transposes by a mechanism that treats it simply as an extended version of the element that comprises the terminal repeats. This is possible because each terminal element is delineated by short IRs, so the composite structure is also delineated by the same IRs. An interesting quirk of these systems is that since, apart from the cognate transposase, all that is needed for transposition is a pair of IRs, then a phenomenon called "inside-out transposition" is also possible, providing the donor molecule is circular. In this situation, the terminal elements mediate transposition of a composite structure comprising them, as terminal repeats, and the carrier DNA molecule. This structure is delineated by the pair of IR sequences that are normally situated at the inside junctions of the IS elements and the central module, rather than the outside pair that delineate the original transposon (see Fig. 5).

III. Tn3 and Related Transposons

A. Complex Transposons—General

It is convenient to consider composite transposons as a group because of their structural similarities, although few, if any, are phylogenetically related.

Table II   Some Prokaryotic Composite Transposons

Transposon	Size (kb)	Terminal elements	Target (bp)	Marker(s)[a]
		Elements from gram-negative bacteria		
Tn5	5.7	IS50 (IR)[b]	9	Km Bl Sm
Tn9	2.5	IS1 (DR)	9	Cm
Tn10	9.3	IS10 (IR)	9	Tc
Tn903	3.1	IS903 (IR)	9	Km
Tn1525	4.4	IS15[c] (DR)	(8)[d]	Km
Tn1681	4.7	IS1 (IR)	9	HST
Tn2350	10.4	IS1 (DR)	9	Km
Tn2680	5.0	IS26[c] (DR)	(8)[d]	Km
		Elements from gram-positive bacteria		
Tn3851	5.2	NR	NR	Gm Tb Km
Tn4001	4.7	IS256 (IR)	NR	Gm Tb Km
Tn4003	3.6	IS257 (DR)	NR	Tm

[a] Resistance genes: Bl, bleomycin; Cm, chloramphenicol; Gm, gentamicin; Km, kanamycin; Sm, streptomycin; Tb, tobramycin; Tc, tetracycline. HST, heat-stable enterotoxin.

[b] DR, direct repeat; IR, inverted repeat.

[c] IS15 and IS26 are closely related to IS6.

[d] Figures indicate probable size, as determined for IS6 (Table I).

NR, not recorded.

Similarly, complex transposons show a more complicated genetic arrangement in that genes which do not encode transposition functions have been recruited into and become part of the basic transposable element. Just as the term composite transposon denotes only that such elements have a modular structure with terminal IS repeats, so the implication of the term complex transposon is only that the transposon does not possess a modular structure incorporating terminal IS repeats nor is it a bacteriophage.

B. Tn3

The archtype complex transposon is Tn3, originating on the R plasmid, R1. It is virtually identical to the first transposon to be discovered, Tn1, which originated on the R plasmid RP4. Both of these elements were originally called TnA. Both encode TEM β-lactamases that confer resistance to ampicillin, carbenicillin, and some other β-lactam antibiotics. The two enzymes differ by only one amino acid (a Gln to Lys change at position 37), a change that does not noticably alter the substrate specificity. Both Tn1 and Tn3 are widely distributed among gram-negative bacteria of clinical and veterinary significance, with Tn3 being somewhat more common.

C. Tn3-Related Transposons

Tn3 is the type element for an extended family of phylogenetically related transposons, members of which have been found in gram-negative and gram-positive bacteria. Between them they encode resistance to several antibiotics and to mercuric ions and a couple specify catabolic functions (Table III). The evolutionary relationships among these elements was first revealed when it was discovered that several, apparently unrelated transposons (Tn3, Tn501, Tn551, Tn1721) had different, but clearly similar, short (35–40 bp) IR sequences. Further sequence analysis revealed that the homology was more extensive and encompassed the transposition genes as well. These genes are all of similar size (3 kb) and have clearly evolved from the same ancestral sequence.

D. Transposition Function

Each Tn3-related element encodes a transposase of approximately 1000 amino acids. The genes, designated *tnpA*, are clearly ancestrally related and most terminate within one of their element's IR sequences. The great majority of the family also encode a second recombination enzyme called a

Table III Some Tn3-Like Transposons

Transposon	Size (kb)	Terminal IRs (bp)	Target (bp)	Marker(s)[a]
		Elements from gram-negative bacteria		
Tn1	5	38/38	5	Ap
Tn3	4.957	38/38	5	Ap
Tn21	19.6	35/38	5	Hg Sm Su
Tn501	8.2	35/38	5	Hg
Tn1000	5.8	36/37	5	None
Tn1721	11.4	35/38	5	Tc
Tn1722[b]	5.6	35/38	5	None
Tn2501	6.3	45/48	5	None
Tn2424[c]	25	NR	NR	Hg Cm Am Sm Su
Tn2425[c]	22	NR	NR	Hg Sm Su
Tn2603[c]	22	NR	NR	Hg Ox Sm Su
Tn3926	7.8	36/38	5	Hg
Tn4651	56	32/38	5	xyl
		Elements from gram-positive bacteria		
Tn551	5.3	35	5	Ery
Tn917	5.3	38	5	Ery
Tn4430	4.2	38	5	None
Tn4451	6.2	12	NR	Cm
Tn4556	6.8	38	5	None

[a] Ap, ampicillin; Cm, chloramphenicol; Ery, erythromycin; Hg, mercuric ions; Ox, oxacillin; Sm, streptomycin; Su, sulphonamide; Tc, tetracycline; xyl, xylose catabolism.

[b] A cryptic element that is part of Tn1721.

[c] Elements that are highly homologous to Tn21 (see text).

resolvase. These enzymes, encoded by genes designated *tnpR*, are site-specific recombinases. Each acts at a particular site, designated *res*, which is located adjacent to *tnpR*. Each of these enzymes mediates the second stage of a two-step transposition mechanism (Fig. 5).

E. Family Branches

The Tn3-related transposons split naturally into two main branches of the family. On each branch, the transposition functions of the elements are closely related and may be interchangable. Tn3 is the type element for one branch (which, to date, includes Tn1000 and Tn1331), Tn21 for the other (Table III). Tn21 is larger than Tn3 (20 kb vs. 5 kb) and confers resistance to streptomycin, spectinomycin, and sulphonamide as well as to mercuric (Hg^{2+}) ions. Notwithstanding their different sizes, both elements devote approximately the same genetic capacity, 4 kb, to transposition functions.

The Tn21 branch of the family contains many elements that differ primarily in the number and type of resistance determinants carried (Table III). The ex-

perimental evidence indicates that members of this branch of the family have the same or almost the same transposition functions, which are functionally interchangable. Much of the diversification seen within the subset is likely to be of recent origin. Some elements, such as Tn501 and Tn1721, are less closely related to Tn21 than others such as Tn2424 and Tn2603. Sequence analysis of the transposition functions of Tn501 and Tn1721 indicate significant degrees of divergence from the type element, Tn21, but they are much more closely related to Tn21 than to Tn3. The Tn21-like transposons reflect what is possibly the most successful diversification of a single transposable element discovered to date. Unfortunately, the identity of the parent element, i.e., the one that lacks accessory genes, is unknown.

The structures of Tn3 and some related elements are depicted in Fig. 4, which illustrates the two arrangements of the transposition functions seen in the family. On Tn3 and closely related elements, *tnpR* and *tnpA* are in opposite orientations and are separated by *res*. The two genes are transcribed divergently. On Tn21 and its close relatives, the genes have the same orientation with the arrangement *res tnpR tnpA* IR.

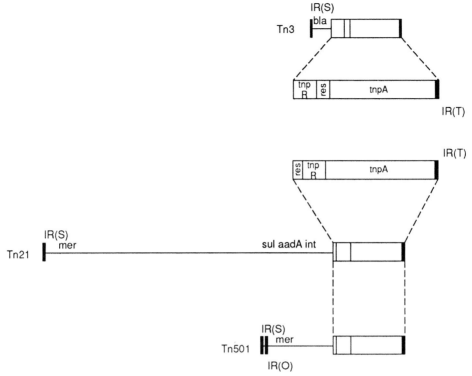

Figure 4 Schematic representation of Tn*3* and related elements. Genes depicted: *tnpA*, transposase; *tnpR*, resolvase; *res*, resolution site; *bla*, TEM β-lactamase; *mer*, resistance to mercuric ions; *sul*, resistance to sulphonamides; *aadA*, aminoglycoside adenylyltransferase A; *int*, integrase. IR(T), inverted repeat adjacent to *tnpA*, and IR(S), second IR, are represented by thick vertical bars. For other details, see text. [From Bennett, P. M. (1991). Transposons and Transposition. *In* "Modern Microbial Genetics" (U. N. Streips and T. E. Yasbin, eds.). pp. 323–364. Wiley-Liss, New York.]

F. Recruitment of Antibiotic Resistance Genes

Molecular analysis of a number of Tn*21*-like elements has revealed that the various antibiotic resistance genes recruited into the structure are not inserted at random. Rather, they are located at only one or two positions, specifically on one side or the other of the *aad* gene if this is carried, or as a replacement for the *aad* gene if it is not. This limited distribution of sites of insertion, given the variety of genes involved and the size of Tn*21*, led to the suggestion that there is a specific mechanism that uses specific sites of insertion to insert into the basic transposon structure. Sequence analysis of the appropriate regions of Tn*21* and related transposons has identified a 59-bp loosely conserved palindromic sequence, analogs of which are found at both ends of the *aad* gene. Further analogs are found as flanking sequences to the other resistance genes that have been recruited, so that the general arrangement of

antibiotic resistance genes (*abr*) and insertion site analogs (ISA) is as follows:

ISA1 *abr1* ISA2 *aad* ISA3 *abr2* ISA4.

In addition, a gene designated *int*, which encodes an enzyme of the integrase family, has been located next to the *aad* gene on Tn*21* (Fig. 4). The product of *int* is necessary for the site-specific integration of resistance casettes. The mechanism of integration has not yet been elucidated.

G. Mechanism of Transposition

Transposition of a member of the Tn*3* family of transposons involves replication of the transposon as well as recombination between the ends of the element and the target site. As is the case for most IS elements, each transposition event also generates a target-site duplication, which is usually one of 5 bp. As in the cut-and-paste mechanism, the duplication is the result of inserting the transposon into a staggered cut at the target site. The mechanisms differ in that the element is not released from the donor molecule. Rather, a single cleavage is made on opposite strands of the DNA at each end of the transposon. The freed ends of the transposon are then ligated to the 5-bp single-strand extensions at the insertion site to generate what is known as a Shapiro intermediate, after J. A. Shapiro, who first proposed its existence. The ligations generate potential origins of replication at both ends of the element that can be used to replicate the transposon. Whether one or both of these is used is not known. When the element has been replicated and double-strand continuity restored by DNA ligase, the product is a cointegrate. This contains the entire sequences of both the transposon donor and target replicons joined by directly repeated copies of the transposon, one at each replicon junction. To release the final product, i.e., a copy of the target replicon with a single transposon insertion, there is a recombination between the two copies of the transposon. Normally, this is mediated by the *tnpR* gene product, resolvase, and the two *res* sites present on the cointegrate. Should either *tnpR* or *res* be damaged, then host-mediated, *recA*-dependent recombination can also resolve the cointegrate to the final transposition product and a molecule indistinguishable from the original transposon donor (Fig. 5).

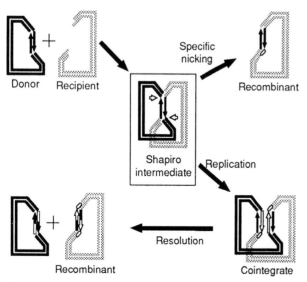

Figure 5 Transposition: replicative model (see text for details). [From Grinsted, J., de la Cruz, F., and Schmitt, R. (1990). *Plasmid* **24**, 163–189, with permission.]

H. One-Ended Transposition

The replicative mechanism of transposition outlined previously indicates explicitly that both ends of the element are involved in the process. This is supported by the observation that removal of an IR sequence seriously impairs the transposition process. Surprisingly, however, TnpA-mediated transposition still occurs, albeit at a much reduced frequency. Such transpositions are referred to as one-ended transpositions and have been reported for Tn3, Tn21, and Tn1721. One-ended transposition is a replicative recombination that utilizes one IR sequence, instead of two, and its cognate transposase. One-ended transposition in these systems generates products that resemble replicon fusions where the donor molecule, rather than a discrete part of it (i.e., the transposon), has been inserted into the recipient molecule. The inserted sequence is flanked by 5-bp DRs, as in normal transposition, consistent with these events being mediated by the cognate transposase. Unlike normal transposition, however, the products are not uniform in size and insertions both less than and greater than the unit size of the donor have been reported, although all form a nested set with the IR sequence defining one end.

These data are not accommodated by the model of replicative transposition (Fig. 5). However, one way that a nested set of insertions, which all start from the same point, may be generated is to use a form of rolling circle replication. In such a system, the transposase makes a single-strand cut on the donor molecule at the solitary IR sequence, precisely as in the two-ended model. The free end of the IR sequence would then be joined to one of the single-strand extensions of the target site, to form a single replication origin. Replication would then start from here, proceed through the IR sequence, and continue round the donor molecule until interrupted. At this point the DNA duplex joined to the recipient molecule would be disconnected completely from the donor and the newly created end would be ligated to the free, single-strand extension of the target to complete the insertion. Repair synthesis, to fill in the short 5-bp gap would complete the process. Whereas the two-ended mechanism precisely determines the sequence to be transposed, rolling circle transposition determines precisely only one end of the insert. How the other end is determined is not known.

I. Transposition Immunity

Tn3 and some of its relatives, but not Tn21, display a characteristic termed transposition immunity. Hence, such elements will not normally transpose on to another DNA molecule that already carries a copy of the transposon. The inhibitory sequence has been identified as the element's IR sequence and a single copy affords significant protection. In the case of Tn3, the immunity can be largely overcome by increasing the level of transposase. The mechanism of transposition immunity is element-specific and poorly understood. The phenomenon has also been demonstrated for Tn7 and bacteriophage Mu.

IV. Conjugative Transposons

A. Origin and Properties

Conjugative transposons have, to date, been found only in gram-positive bacteria. The type element, Tn916, which encodes resistance to tetracycline was discovered in a strain of *Enterococcus faecalis*. As the collective name for these elements implies, they can promote their own transfer from one cell to another, as well as being able to transpose. They differ from plasmids in that they replicate passively as parts of the replicons into which they insert, rather than as autonomous units.

Tn916 is 16.4 kb and is indistinguishable from Tn918, Tn919, and Tn925, which are probably independent isolations of essentially the same element. A related conjugative transposon, Tn1545, is 23.5 kb and encodes resistance to kanamycin and erythromycin in addition to tetracycline. These elements promote their own conjugal transfer at frequencies of 10^{-9} to 10^{-5} per recipient cell and mediate transposition at frequencies of approximately 10^{-5}. [*See* CONJUGATION, GENETICS.]

B. Transposition

Tn916 is typical of the group and is the best studied. Four distinct transposition-related activities have been documented for these elements: (1) conjugal transfer accompanied by transposition into the chromosome of the recipient cell, (2) chromosome to plasmid transposition, (3) zygotic induction of transposition as a result of transformation, and (4) transposon excision. The last of these activities is often tested in *E. coli* rather than in *Enterococcus faecalis* because the frequency of excision is considerably

higher in the former organism. More than 50% of Tn916 encodes conjugation functions. These are located as a block of genes at one end of the element and are not strictly necessary for transposition, although mutants in some interfer with chromosome to plasmid transposition in contrast to zygotic induction of transposition, which is unaffected. Mutations that block transposition map to the other end of the element. Such mutations also inhibit the phenomenon of transposon excision, indicating that this is a transposition-related event. Two transposition genes, designated *xis-Tn* and *int-Tn*, have been identified on Tn1545.

Transposition of Tn916 and related elements is likely to be a two-stage process in which the first step is excision of the element to give a free, circular intermediate. This may then undergo one of two productive fates: (1) insertion into another DNA molecule in the cell (e.g., a plasmid), and (2) conjugal transfer to another cell followed by insertion into the recipient cell's chromosome. The mechanism appears to be very similar to the integration-excision systems of lambdoid phages, both structurally and functionally, involving reciprocal site-specific recombination.

V. Site-Specific Transposons

A. General Comments

Most transposable elements show little or no target site specificity, although a bias toward insertion into certain regions of DNA molecules has been reported for some elements (e.g., IS1 has been reported to favor insertion in AT-rich tracts, as has Tn3). But these are preferences, not requirements. In contrast, a few elements display a marked, if not exclusive, specificity of insertion.

B. Tn7

Tn7 was originally found on the R plasmid R483. It carries two resistance genes, *dfr* and *aadA*, the former encoding a trimethoprim-resistant dihydrofolate reductase while the latter codes for an aminoglycoside adenylyltransferase, which confers resistance to streptomycin and spectinomycin.

Tn7 is 14 kb long, generates 5-bp target site duplications, and inserts in one orientation at high frequency into single sites on the chromosomes of a number of gram-negative bacteria including *E. coli* (and some other enterobacteriaceae), *Klebsiella pneumoniae*, *Pseudomonas* sp., and *Vibrio* sp. At a much lower frequency (10^{-4}), it will also insert, more or less at random, into many different DNA molecules (e.g., plasmids) that lack the specific insertion site.

C. Chromosomal Site of Tn*7* Insertion

The locus of Tn*7* insertion on the *E. coli* chromosome has been located at 84′, between *phoS* (encodes a periplasmic phosphate binding protein) and *glmS* (encodes glucosamine phosphate isomerase) and designated *att*Tn*7*. The 5-bp target site is part of the transcriptional terminator of *glmS*, which is located about 30 bp from the end of the translational reading frame. Unexpectedly, the target site, i.e., the actual site of insertion, is different from the attachment site (i.e., the sequence that determines the locus specificity, namely, *att*Tn*7*). Specifically, *att*Tn*7* is a sequence of approximately 50 bp, which starts 12 bp away from the 5-bp target site and extends into *glmS*. This relationship between *att*Tn*7* and *glmS* is also seen in other bacteria in which there is a specific chromosomal Tn*7* insertion site, such as *Klebsiella pneumoniae* and *Serratia marcescens*. Although the various *att*Tn*7* analogs in the different bacteria are homologous, the target sites differ completely, indicating that the sequence of the actual site of insertion of Tn*7* does not determine specificity. Rather, the site of insertion appears to be determined solely by its distance from *att*Tn*7*.

D. Transposition Functions of Tn*7*

Tn*7* is unusual in the number of transposition functions it requires. Five have been identified, sequenced, and designated *tnsABCDE*. They encode peptides of 31, 78, 63, 59, and 61 kDa, respectively. The first three are required for all transpositions. High-frequency site-specific transposition requires, in addition, the product of *tnsD*, while low frequency random site transposition requires the tnsE product.

E. Mechanism of Transposition of Tn*7*

A Tn*7* *in vitro* transposition system, which uses as a target a plasmid carrying *att*Tn*7*, has been developed. With this system, it has been shown that the transposon is first disconnected from the donor molecule by two double-strand staggered cuts, each of which generates 5′ overhangs of 3 bases at the ends of the element. The DNA of these overhangs derives

from the carrier DNA molecule, not Tn*7*. The free Tn*7* sequence is then joined to the target by ligating the free 3′ ends of the transposon to the 5′ overhangs (5 bp) of the target. Repair processes are then assumed to remove the short ss extensions at each end of the transposon and fill in the single strand gaps that remain, generating the 5-bp DRs that flank Tn*7* insertions.

It has been found, *in vitro*, that the recombination events needed for Tn*7* transposition require a DNA–protein complex containing the transposon donor and target molecules, all four transposition proteins (TnsA, B, C, and D), and adenosine triphosphate (ATP). ATP is thought to be directly involved in the reaction since TnsC has been shown to be an ATP-binding protein, but it may also be needed to ensure that the DNA substrates have the appropriate degrees of supercoiling, maintained by the activity of DNA gyrase.

The cell-free system faithfully reproduces several of the features of Tn*7* transposition that are characteristic of the element. Transposition *in vitro* is site- and orientation-specific with respect to *att*Tn*7*, as it is *in vivo*. Furthermore, transposition *in vitro* displays the phenomenon of transposition immunity: If the target molecule carries a copy of Tn*7*, then transposition of a second copy of Tn*7* into it is blocked.

F. IRs of Tn*7*

Tn*7* is unusual in the length of the terminal sequences needed for transposition; namely, 75 bp is needed at one end of Tn*7* and 150 bp at the other. These sequences form IRs of only 30 bp. Genes flanked by IRs of the shorter sequence will transpose, if transposition functions are provided. In contrast, genes flanked by IRs of the 150-bp sequence do not transpose. The basis of the difference is not known. Each terminal sequence contains several analogs of a 22 bp consensus sequence. These have been shown to be binding sites for TnsB and presumably serve as the basis for the assembly of the DNA-protein complex that is required for transposition.

G. Tn*554*

Tn*554* is a 6.7-kb transposon encoding resistance to erythromycin and spectinomycin that originated in *Staphylococcus aureus*, where it transposes into the chromosome at high frequency, primarily into a single site, designated *att*Tn*554*, in one orientation only. From its sequence, six putative genes were identified on Tn*554*. Of these, five have been confirmed as true genes: *tnsABC* form an operon of transposition functions, which accounts for approximately half of the transposon coding capacity, while *ery* and *spc* code for resistance to erythromycin and spectinomycin, respectively. Tn*554* displays transposition immunity.

The mechanism of transposition of Tn*554* differs from that of most other elements in that not only is the element transposed but so are a few base pairs from the carrier molecule. This short additional sequence is located on one side of the element in the donor but is transferred to the opposite side in the transposition products. The mechanism by which this is achieved is unknown, but the data suggest a variation of the cut-and-paste model that may involve a circular intermediate.

H. Tn*502*

Tn*502* is a poorly characterized gram-negative transposon that encodes resistance to mercuric ions but that appears to be unrelated to Tn*21*, Tn*501*, and other Hg[r] transposons of the Tn*3* family. It is notable because it displays site-specific insertion into a plasmid, rather than into a chromosome. Tn*502* is 9.6 kb and inserts at high frequency in one orientation into a single site on the IncP plasmid RP1 (RP4). When this site is deleted, insertion then occurs at random sites at a much lower frequency. There is insufficient information at present to determine if Tn*502* is related to other site-specific elements.

VI. Transposing Bacteriophages

A. Structure of Mu

Transposing bacteriophages use transposition to replicate. The type element of the group is bacteriophage Mu, which was discovered by Larry Taylor in 1963. Taylor recognized that Mu was a temperate phage and concluded that because lysogenization often created auxotrophs then Mu must be able to integrate at many different sites on the *E. coli* chromosome, some of which cause mutation. It is now well established that Mu replication involves transposition via a Shapiro intermediate (Fig. 5)

The linear genome of Mu is 37 kb, but each phage particle carries a DNA molecule of approximately 39 kb. This is made up of the Mu genome with, on one side, about 150 bp of host DNA and, on the other, 1–2 kb of host DNA. These flanking sequences are different for individual phage genomes

and reflect the last site of insertion of Mu prior to assembly of the phage particle. Genome packaging is by a "headfull" mechanism that disconnects the phage from the carrier molecule by first cutting the DNA approximately 150 bp beyond one end of Mu (the end nearest the replication functions) and then again approximately 39 kb away, on the other side of the phage genome so generating a linear genome flanked by host DNA.

In its genome structure, Mu is a fairly typical phage. It has the usual arrays of genes needed for phage assembly, i.e., head and tail production. It has two replication genes, A and B, which are located close to one end of the genome, conventionally designated the left end, and it has a 3-kb invertible segment, homologous to that found on phage P1, inversion of which changes the host specificity of the phage by altering the type of tail fibers produced.

B. Mu Replication and Integration

It was more than a decade after its discovery that it was realized that Mu replicates by transposition. Since then, study of Mu replication has been invaluable as a model transposition system, particularly since the development of an *in vitro* replication–transposition system for Mu.

Mu uses not one but two forms of transposition as part of its life cycle. Its replication proceeds via the formation of Shapiro intermediates and cointegrates (Fig. 5). These structures are not resolved, as are Tn*3* and related transposon cointegrates, by a site-specific resolution system (*res*/resolvase). If resolution does occur it is the result of host-mediated, *recA*-dependent recombination. Lack of resolution does not prevent further rounds of transposition, i.e., replication, from occurring. Most of these are intramolecular transposition events that automatically result in DNA rearrangements, other than simple element insertion, such as deletions and inversions (see later).

In addition to replicating by transposition, Mu initially is established in a new host by a conservative transposition event. Again, the start point appears to be the formation of a Shapiro intermediate, but one in which the donor molecule is linear instead of circular. Then, instead of replicating the phage DNA, a second set of cleavages at the ends of the genome disconnects it completely from the residual host DNA fragments (Fig. 5). DNA repair synthesis then seals the phage DNA into its new site.

That Mu lysogeny is established by a conservative integration was elegantly demonstrated using phage particles produced in a dam+ host to infect a dam mutant of *E. coli* (which cannot methylate its DNA). The newly integrated phage DNA was found to be fully methylated, not hemimethylated, indicating that no replication occurred prior to or as a consequence of insertion. The conservative transposition used to establish Mu in a new host and phage replication both require the A gene product, which is the Mu transposase.

C. D108 and Other Transposing Bacteriophages

Mu was the first transposing bacteriophage to be discovered. Only one other transposing coliphage is known, D108, which is closely related to Mu. The two phage genomes display 90% sequence homology, with the main divergence being at the ends of the genomes. One such region includes the 5′ ends of the A genes, the consequence of which is that the two A gene products, although they do complement each other, do so only poorly. Transposing phages have also been identified in *Pseudomonas* sp. The sizes of their genomes (approx. 37 kb) and the structures of the DNA packaged to form the phage particles are strikingly similar to those of Mu and D108, but no gross homolgy with Mu has been detected. Putative transposing phages have also been found in *Vibrio* sp., identified on the basis that lysogenization may create an auxotrophic mutation, which was how Mu was originally detected. [*See* BACTERIO-PHAGES.]

VII. Yeast Transposons

A. Types

Finally, mention must be made of transposons found in yeast. Several different types have been identified in *Saccharomyces cerevisiae*. The majority are components of the nuclear DNA, but one, designated Ω, is found in the mitochondrial DNA. No one element is found in both compartments.

Several of the yeast transposons discovered so far, e.g. Ty1, Ty2, Ty3, are what are known as retrotransposons because the way in which they transpose is clearly akin to the mechanism of retroviral replication. Ty1 has been shown to transpose by a process involving a reverse transcriptase step and since Ty2 and Ty3 closely resemble Ty1 in structure it is likely that they also require reverse transcription for transposition.

B. Structure of Ty Elements

The Ty elements are 5–6 kb long and typically have long terminal direct repeats (LTRs), which can themselves transpose (called δ in the case of Ty1 and Ty2, and σ for the LTRs of Ty3). The Ty elements and their LTRs have no IRs as such, although, like many retroviruses, the transposable sequences terminate TG . . . . CA. In keeping with the view that these elements can be regarded as transposons is the finding that intact elements are nearly always flanked by 5-bp duplications of host sequence. Both Ty1 and Ty2 accommodate two ORFs, designated TYA and TYB. The equivalent ORFs of the two elements indicate a considerable degree of similarity between the proteins of the two transposons. Several domains of the Ty1 TYB protein show significant similarities to retroviral *pol* functions. Four in particular have been identified as being related to retroviral protease (pro), integrase (int), reverse transcriptase (rt), and RNaseH (rnh), which occur on the protein in that order. These domains are well conserved between Ty1 and Ty2. There is insufficient sequence data on Ty3 to make reliable comparisons.

C. Transposition of Yeast Transposons

Yeast cells engaged in high-frequency transposition of Ty1 or Ty2 contain large numbers of viruslike particles that contain Ty-RNA, reverse transcriptase, and capsid proteins encoded by the Ty element. It is thought that these particles are transposition intermediates. Transposition would then require the conversion of the RNA to double-stranded DNA and integration of this into the nuclear DNA. That the mechanism of retrotransposon transposition was likely to be analogous to the replication of retroviruses was originally inferred from the structural analysis of the elements and their transcripts, which closely resemble the analogous structures of retroviral proviruses and viral RNA.

In contrast to the retrotransposons, the mitochondrial element, Ω, appears to transpose directly via a DNA form. In this sense, it more closely resembles transposable elements in bacterial cells.

VIII. Consequences of Transposition in Bacteria

A. Genome Rearrangements

The most obvious consequence of transposition is insertion of one DNA sequence into another, an event that may disrupt a gene, so causing a mutation. This property has been widely exploited in genetic analysis and insertional inactivation by transposons is commonly used to locate genes of interest. In addition, transposable elements mediate a variety of DNA rearrangements. The sites of transposon insertion are often hot spots for deletions and inversions. These rearrangements encompass the sequences on one side or the other of the element and not the element itself, which remains in place and intact. Many of these events are now known to be the result of intramolecular transposition events, i.e., where the target site and the transposable element are on the same molecule, by elements that transpose via Shapiro intermediates. Whether a deletion or an inversion results from an intramolecular transposition event depends on how the ends of the element are spliced to the target, i.e., which orientation is "chosen" when the transposon is ligated into the target site. If the orientation is such as would have given direct repeats of the element, then the transposase-mediated reaction automatically fragments the carrier molecule so that the two sections separated by the original insertion and the target site end up on separate circular DNA molecules, each with a copy of the transposon. In general, only one of these molecules will possess an origin of replication. Only this one will be replicated and survive. Conversely, if the transposition generates a duplicate but inverted copy of the transposon, then the section of the carrier molecule separated by the transposon and its new site of insertion is inverted with respect to the rest.

B. Replicon Fusion and Conduction

Replicative transposition via a Shapiro intermediate (Fig. 5) in the absence of resolution functions can result in stable replicon fusions. It is probable that several natural plasmids that appear to be hybrid structures have arisen in this way. The formation of transient cointegrates may result in conduction, i.e., conjugal transfer of a nonconjugative plasmid by a conjugative one. Both plasmids are transferred to the recipient cell as a single DNA molecule where resolution then occurs to release the transposon donor and a derivative of the target molecule that carries a copy of the transposon. The transposon may initially be on either of the participating molecules.

C. Final Thoughts

The ability of transposable elements to insert and to generate deletions and inversions account for much of the macromolecular rearrangement that is observed among related bacterial plasmids. That bacterial chromosomes are subject to the same array of mutational events can readily be demonstrated in the laboratory. Nonetheless, chromosomal rearrangement seems to occur less often than might be expected, given their sizes and the apparent abundance of transposable elements. Whether this reflects the fact that many chromosomal rearrangements prove, immediately or in the longer term, to be deleterious to host survival and so are never established in the population or because chromosomal rearrangements do occur less frequently than we might expect remains to be seen.

Bibliography

Bainton, R., Gamas, P., and Craig, N. (1991). *Cell* **65**, 805–816.

Bennett, P. M., Grinsted, J., and Foster, T. J. (1988). *Methods Microbiol.* **21**, 205–231.

Bennett, P. M., and Hawkey, P. M. (1991). *J. Hos. Infect.* **18**(Suppl. A), 211–221.

Berg, D. E., and Howe, M. M. (eds.) (1989). "Mobile DNA." American Society for Microbiology, Washington, D.C.

Fayet, O., Raymond, P., Polard, P., Prere, M. F., and Chandler, M. (1990). *Molec. Microbiol.* **4**, 1771–1777.

Grinsted, J., de la Cruz, F., and Schmitt, R. (1990). *Plasmid* **24**, 163–189.

Plasterk, R. H. A. (1991). *TIG* **7**, 203–204.

Poyart-Salmeron, C., Trieu-Cuot, P., Carlier, C., and Courvalin, P. (1990). *Molec. Microbiol.* **4**, 1513–1521.

Sakaguchi, K. (1990). *Microbiol. Rev.* **54**, 66–74.

Shapiro, J. A. (ed.) (1983). "Mobile Genetic Elements." Academic Press, New York.

Ureases, Microbial

Harry L. T. Mobley
University of Maryland

I. Enzymology
II. Physiology
III. Genetics
IV. Role in Pathogenesis
V. Summary

Glossary

Accessory polypeptides Nonstructural polypeptides associated with the urease operon that are either essential for activity or play a role in regulation
Pyelonephritis Infection of the kidney resulting in interstitial inflammation and tubule necrosis
Structural subunits Polypeptides that comprise the urease enzyme
Urolithiasis Formation of crystalline stones in the urinary tract

UREASE (classified as urea amidohydrolase; E.C. 3.5.1.5) catalyzes the hydrolysis of urea to yield ammonia and carbamate; the latter spontaneously hydrolyzes to another ammonia and a carbon dioxide molecule. The high-molecular weight, multimeric, nickel-containing enzyme is produced by over 100 bacterial species and is commonly used as a phenotypic marker for speciation. The enzyme plays a role in environmental and rumen nitrogen cycling by allowing urea nitrogen to be incorporated into bacterial protein. Some uropathogenic bacteria produce high levels of urease, which elevates urine pH resulting in formation of kidney and bladder stones. The enzyme is essential for the development of pyelonephritis by *Proteus mirabilis*, a common

uropathogen. In addition, *Helicobacter pylori*, the newly described agent of gastritis and peptic ulceration, apparently requires urease for survival in the human gastric mucosa.

I. Enzymology

A. Kinetics of Urea Hydrolysis

Urease hydrolyzes urea to yield ammonia and carbamate, which spontaneously hydrolyzes to yield another molecule of ammonia and carbonic acid:

$$\underset{\substack{\| \\ H_2N\text{-}C\text{-}NH_2}}{O} + H_2O \xrightarrow{urease} NH_3 + \underset{\substack{\| \\ H_2N\text{-}C\text{-}OH}}{O}$$

$$\underset{\substack{\| \\ H_2N\text{-}C\text{-}OH}}{O} + H_2O \longrightarrow NH_3 + H_2CO_3$$

Subsequently, ammonia equilibrates with water, forming ammonium hydroxide, which results in a rapid increase in pH of the solution:

$$H_2CO_3 \rightarrow H^+ + HCO_3^-$$
$$2\,NH_3 + 2\,H_2O \rightarrow 2\,NH_4^+ + 2OH^-$$

Ureases do not have strong affinities for their substrate. This is reflected by their K_m's for substrate in the millimolar range. K_m's have been reported from 0.1 mM urea to 50 mM urea. Because urea, a principal nitrogenous waste product of mammals, is an abundant compound on earth, substrate limitation is generally not a consideration; therefore, a high-affinity enzyme is not required to scavenge substrate.

B. Structural Properties

Microbial ureases, produced by over 100 bacterial species, are high-molecular weight (200,000–700,000) proteins that are comprised of two or three distinct polypeptide subunits (depending on the species) arranged in a complex stoichiometry. The enzyme apparently is active only when assembled since isolated subunits display no activity. Most microbial ureases purified thus far are made up of three subunits: two in the range of 6000–15,000 daltons and a large subunit in the range of 60,000–75,000 daltons. One genus, *Helicobacter,* appears to produce a fusion of the two small subunits to yield one subunit of about 30,000 daltons and has maintained a large subunit of 66,000 daltons (Fig. 1). In contrast, the well-studied jack bean urease, the first enzyme ever purified (in 1926), synthesizes a single subunit of 93,000 daltons. Six subunits comprise the native enzyme of this plant, *Canavalia ensiformis.*

C. Nickel in Active Site

Ureases are unusual in that they contain nickel ions (Ni^{2+}) in the enzyme-active site. In addition to certain membrane-bound hydrogenases, urease is the only other class of enzyme to contain nickel. Nickel ions are thought to be coordinated in the active site by histidine residues that reside in the large subunit. The putative active site is highly conserved among different species with respect to the amino acid sequence. The most reliable data suggest that there are two Ni^{2+} ions per active site. The ions are thought to be incorporated during assembly of the enzyme and have not been successfully added after assembly of the holoenzyme is complete. No other ion can be substituted to give an active enzyme.

A model for catalysis suggests that two nickel ions complexed by histidines and a sulfhydryl group of cysteine play critical roles in catalysis. The Zerner model proposed for a jack bean urease is applicable also to microbial ureases because the active site is highly conserved among all ureases studied thus far. This scheme is depicted and described in Fig. 2.

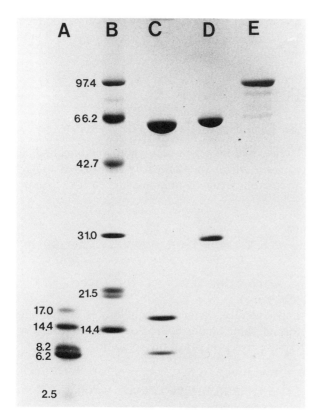

Figure 1 Polyacrylamide gel electrophoresis of purified ureases comprised of one, two, and three distinct subunits. Purified ureases (10 μg protein) were electrophoresed on a 10–20% polyacrylamide gradient gel and stained with Coomassie blue. (A, B) Molecular weight markers in kilodaltons. (C) Purified *Morganella morganii* urease. (D) Purified *Helicoloacter pylori* urease. (E) Jack bean urease purified from a partially purified commercial preparation. [From Mobley, H. L. T., Hu, Li-Tai, and Foxall, P. *Scand. J. gastroenterol.* (Suppl. 1991) (in press).]

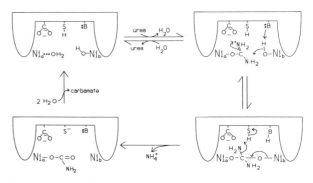

Figure 2 Model for hydrolysis of urea in the urease-active site. One nickel coordinates a water molecule, and the second nickel coordinates a hydroxide. Urea displaces the water molecule, and the positively charged nitrogen is stabilized by a carboxylate group. A general base is proposed to activate the nickel-coordinated hydroxyl group for nucleophilic attack on the urea carbon to form a tetrahedral intermediate. Decomposition of this intermediate and release of ammonia is thought to include general acid catalysis by a nearby thiol group. Finally, carbamate is release with regeneration of the resting enzyme. [Figure and description reprinted with permission from Todd, M. J., and Hausinger, R. P. (1991). *J. Biol. Chem.* **264,** 15835–15842.]

D. Assay of Urease Activity

A number of methods quantitate urease activity using crude lysates or purified protein. Generally, one measures the release of ammonia or carbon dioxide or follows the rise in pH due to ammonium hydroxide formation.

1. pH Indicator Assay

A solution containing urea and phenol red, a pH-indicating dye, is used to detect the change in color of the solution due to a release of ammonia. Starting at pH 6.8, the change in absorbance at 560 nm detected spectrophotometrically is linear between 0.15 and 0.5 absorbance units. This is an excellent assay for routine comparisons of urease activity but is not among the most sensitive assays and is limitated by rising pH during the course of measurement.

2. pH Indicator Assay in Polyacrylamide Gels

This same general principle can be employed to localize native ureases that have been electrophoresed through nondenaturing polyacrylamide gels. After gels are run, they are equilibrated in trays of buffer containing a pH-indicating dye such as cresol red until an acidic pH is maintained. After equilibration, gels are transferred to a urea solution. Urea is hydrolyzed locally at the point of migration of the native enzyme, causing an intense color change, which can be recorded photographically. Images must be recorded quickly because the reaction can occur rapidly and turn a considerable portion of the gel red as ammonia diffuses throughout the gel. The bands can be fixed for photography, however, by addition of lead acetate.

3. Nesslerization Reaction

Unlike the pH indicator assay, this reaction is a fixed time-point assay for which samples must be removed and reacted with Nessler reagent containing KI and HgI_2. An orange color is read spectrophotometrically and compared with a standard curve. Although more sensitive than the pH indicator, it is less sensitive than the indophenol reaction.

4. Indophenol Assay

This is also a fixed time-point, yet sensitive, assay that can detect <0.02 μmol ammonia. Ammonia reacts with phenol hypochlorite at high pH to form indophenol, the blue color of which can be measured spectrophotometrically. Some urease inhibitors interfere with this reaction giving high background absorbance.

5. Coupled Enzyme Assay

Like the pH indicator assay, this is a continuous time-point assay that employs glutamine dehydrogenase to couple ammonia release to the oxidation of nicotinamide adenine dinucleotide, which can be followed at 340 nm. This is a very sensitive reaction that detects 0.003 units of enzyme (1 unit hydrolyzes 1 μmol urea/min) but is inherently more complicated because it depends on the activity of a second enzyme. Also, it is most suitable for purified enzymes because reduced nicotinamide adenine dinucleotide oxidase activity, often present in crude lysates, can interfere with the reaction.

6. Ion-Specific Electrode

Ammonium-specific electrodes can be used to measure the continuous liberation of ammonium ions during urease hydrolysis. However, sensitivity is not high, reaction is not linear, and interference can occur with various monovalent ions such as potassium.

E. Urease Inhibitors

A number of compounds can act as inhibitors of urea hydrolysis by microbial ureases. These reagents have been put to practical use for prevention of urolithiasis (stone formation) in patients infected with urease-positive bacteria and to increase the efficiency of urea nitrogen release in urea-based fertilizers. In addition, inhibitors have been used as molecular probes to study the active site of the enzyme.

The structures of urea and analogs of urea that act as specific urease inhibitors are shown in Fig. 3. The most widely used inhibitor for prevention of human urinary tract stones is acetohydroxamic acid. Besides its stereochemical resemblance to urea, hydroxamates also chelate metal ions (such as Ni^{2+}), which may also play a role in inhibition. Acetohydroxamic acid and hydroxyrea are competitive inhibitors of urease. Some nonspecific inhibitors of urease include phosphate, thiols, and thiol-reactive agents (N-ethylmaleimide and iodoacetimide), boric acid, boronic acid, and fluoride.

Although structural analogs can be potent enzyme inhibitors, the regulatory mechanism of *P. mirabilis* can differentiate urea from these compounds. Urea induces urease synthesis in this species, but no induction is observed with any urease inhibitor shown

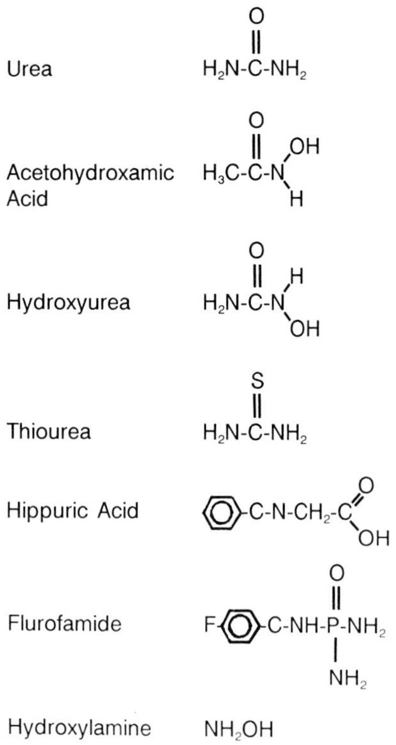

Urea $H_2N\text{-}C\text{-}NH_2$ (O double bond above C)

Acetohydroxamic Acid $H_3C\text{-}C\text{-}N$ with $=O$, OH, H

Hydroxyurea $H_2N\text{-}C\text{-}N$ with $=O$, H, OH

Thiourea $H_2N\text{-}C\text{-}NH_2$ (S double bond above C)

Hippuric Acid   (phenyl)$\text{-}C\text{-}N\text{-}CH_2\text{-}C$ with $=O$, OH

Flurofamide $F\text{-}$(phenyl)$\text{-}C\text{-}NH\text{-}P\text{-}NH_2$ with $=O$, NH_2

Hydroxylamine NH_2OH

Figure 3 Chemical formulas of urea and common urease inhibitors. The formulas for urea, structural analog, and other urease inhibitors. Note similarities of structure as compared to urea. [From Nicholson, E. B., Concaugh, E. A., and Mobley, H. L. T. *Infect. Immun.*, **59**, 3360–3365.]

in Fig. 3. However, urease inhibitors can prolong urease induction in the presence of urea by inhibiting hydrolysis of urea and, thus, maintain a high level of the inducer.

II. Physiology

A. Nitrogen Cycling

Microbial urease plays a central role in the nitrogen cycling of domestic cattle and sheep. These animals, classified as ruminants because they possess a forestomach, are heavily colonized by ureolytic bacterial species. These microorganisms hydrolyze endogenous urea that is either ingested by the animal, diffuses from the bloodstream, or is swallowed in saliva. Liberated ammonia is incorporated into bacterial polymers such as proteins and nucleic acids. This leads to an increase in microbial biomass, which is then used by the ruminant as a rich source of nitrogen. [*See* NITROGEN CYCLE.]

Urea is the principal nitrogenous waste product

for ruminants and other mammals. It is synthesized in the liver where it diffuses into the bloodstream and saliva or is removed by the kidney by excretion. To increase the efficiency of this nitrogen cycling, urea can be added to feedstock.

B. Agricultural Importance of Urease

Numerous soil microorganisms possess urease activity, which plays a central role in agricultural applications. Nitrogen is a key rate-limiting element in the yield of crops and generally must be supplied in the form of fertilizers. To this end, urea is a cheap, convenient, and widely used supplement to commercial fertilizers. Applied urea is hydrolyzed to ammonia by soil bacteria and fungi, thus providing a utilizable source of nitrogen to the growing plant. Urea concentrations must be carefully calculated, however, because excess ammonia is toxic to plant life. The same considerations may cause considerable distress in the suburban environment as well. The death of a favorite bush or tree or patch of lawn may result from excess urea delivered repeatedly in the form of urine from a passing neighborhood dog. Soil bacteria hydrolyze urea delivering a lethal dose of ammonia to the beloved plant.

C. Regulation of Expression

All ureases that have been examined thus far at the nucleotide level appear to display significant homology and may have been derived from common ancestral genes. Although there has been some divergence in the number of distinct subunits (two or three) and native molecular weight, these proteins are clearly a family of closely related enzymes.

The mechanism by which expression of these enzymes is controlled, however, is not so well conserved. Bacterial species, depending on their environment, have evolved a number of regulatory schemes used to control urease expression.

Certain species such as *Morganella morganii*, a human urinary tract pathogen, apparently produces the enzyme constitutively; i.e., environmental factors such as pH, ammonia concentration, presence of glucose, or substrate urea do not influence the amount of urease synthesized.

On the other hand, other species tightly regulate expression. For example, the group of other urease-positive human urinary pathogens belonging to the *Proteus* and *Providencia* genera ordinarily represses urease synthesis unless substrate urea is present. Enzyme synthesis is induced 5–25-fold when urea is added to exponentially growing cells. When substrate is removed by complete hydrolysis, enzyme synthesis is again repressed. The molecular basis for this repression has been worked out only for *P. mirabilis* thus far. A single gene, *ureR,* which lies upstream of the urease structural and accessory genes (see Section III) encodes a repressor protein (33.4 kDa for the single polypeptide) that can repress the urease operon *in trans*. The mechanism of repression is postulated to resemble the classic *lac* operon negative regulatory scheme in which the repressor binds to operator sequences in the absence of substrate, preventing transcription. Substrate presumably binds to the repressor causing an allosteric change in conformation, which releases the repressor from the operator, allowing transcription. For this species, optimal induction of enzyme occurs at concentrations of urea normally found in human urine, i.e., 400–500 m*M*. Thus, synthesis of *P. mirabilis* urease is probably always induced (derepressed) during urinary tract infection. When the organism finds itself, however, in an environment where substrate level is low, the bacterium can conserve energy by shutting off synthesis of an unnecessary protein. The induction process is highly specific for urea and is not observed in the presence of structural analogs of urea such as acetohydroxamic acid, hydroxyurea, thiourea, flurofamide, hippuric acid, hydroxylamine, lysine hydroxamate, or other related compounds.

Other genera such as *Klebsiella* species regulate production of urease in response to the concentration of nitrogenous compounds such as ammonia, glutamine, or lysine in their environment. Urease genes of this genus appear to fall under the control of the complex global nitrogen assimilation by triggering the use of alternate sigma factors for RNA–polymerase-driven transcription. Ntr promoters, recognized by alternate sigma factors, have been found to precede the urease genes of *Klebsiella* species. Phenotypically, urease synthesis is repressed when ammonia concentrations are high and derepressed when the concentration of nitrogenous compounds is low.

Finally, in *Streptococcus salivarius* the urease is regulated by external pH giving optimal expression at 5.5, 100-fold higher than cells grown at pH 7.0. The coupling of pH and gene expression has not been elucidated in this case.

It is interesting that different bacterial species have evolved complex regulatory mechanisms for the control of synthesis of urease. These mechanisms have undoubtedly evolved under the selective pressure of the species' environment. While enzyme structural genes are highly conserved, regulatory mechanisms vary considerably from species to species.

D. Cellular Location

For most species, urease appears to be a soluble cytoplasmic enzyme that is not secreted to the surface of the cell or excreted to the medium. In studies where cells were fractionated into cytoplasm, periplasmic space proteins, and membrane components, urease partitioned overwhelmingly with the cytosolic fraction when compared to enzymes known to partition with each cell compartment. These cell-fractionation experiments are supported genetically in that no signal sequences have been predicted from the nucleotide sequences to precede any urease gene product from any bacterial species. In addition, N-terminal amino acid sequences determined from subunit polypeptides of the native enzyme align with the first amino acids predicted by the nucleotide sequences; i.e., no processing of the urease structural subunits occurs. Because processing is a feature common to most secreted proteins, one would expect the presence of such signal sequences if ureases were indeed secreted.

An exception to these clear-cut fractionation results is found in the genus *Helicobacter,* which includes the species *Helicobacter pylori* and *Helicobacter mustelae* colonizing the human and ferret stomachs, respectively. Urease has been purified from both the cytoplasm and independently from water extracts of whole cells. In addition, monoclonal antibodies directed against *H. pylori* urease bind to the cell surface of exponentially growing cells, suggesting, together with other data, that urease is both a cytoplasmic and cell surface protein in these species.

E. Urea and Ammonium Transport

Most urease assays depend on the measurement of free ammonia liberated by urea hydrolysis. We now know that most ureases reside in the cytoplasm of microorganisms. Therefore, measurement of ammonia depends on entry of urea into the cell, hydrolysis of urea, extrusion of ammonia from the cell, and capture of free ammonia by the assay detection system. Although urease activity can be measured accurately for some species using whole cell preparations, this is not always the case, suggesting that some species possess urea or ammonia transporters. For some species, urease activities of cell lysates exceed activities of whole cells, suggesting that a permeability barrier to urea or ammonia exists. To support this supposition, energy-dependent urea transporters have been described for yeasts, algae, and some bacteria. For other species, urea and ammonia (but not ammonium ions) appear to be freely permeable across membranes. Therefore, when in doubt, it is recommended that for accurate assessment of cellular urease activity, microbial cells should be permeabilized by sonication, French pressure cell lysis, or perhaps toluenization.

III. Genetics

A. Structural Genes

The genes encoding ureases have been cloned from a number of species. The organization of urease gene sequences have been analyzed for several of these including *P. mirabilis*, *Proteus vulgaris*, *Providencia stuartii*, *H. pylori*, *Ureaplasma ureolyticum*, *Klebsiella aerogenes*, *Klebsiella pneumoniae*, *Morganella morganii*, and a rare urease-positive *Escherichia coli*. Perhaps the best-understood operon is that of *P. mirabilis*, for which the complete nucleotide sequence has been determined. Eight genes, designated *ure*, are predicted from the DNA sequence, as shown in Fig. 4. *ureA*, *ureB*, and *ureC* encode the three structural polypeptides that make up the urease enzyme itself. These genes, transcribed on a single messenger RNA, are predicted to assemble into the native enzyme according to the formula $(UreA_2UreB_2UreC_1)_2$. This stoichiometry, estimated from scanning densitometry measured for Coomassie blue-stained subunits on an sodium dodecyl sulfate–polyacrylamide gel, is consistent

with the observed native molecular weight estimated to be between 212,00 and 250,000.

B. Accessory Genes

UreE, UreF, and UreG are the accessory polypeptides that may serve to incorporate Ni^{2+} ions into the active site of the urease enzyme. These genes are homologous to *Klebsiella* genes, which are necessary for production of the active metalloenzyme. UreD's function is unknown.

UreR appears to act as a repressor, shutting off synthesis of the enzyme in the absence of substrate urea (see Section II.C). This polypeptide is transcribed in the opposite direction of the rest of the genes, appears to assemble into a dimer, and displays significant amino acid sequence similarity to several other DNA-binding regulatory proteins of *E. coli* (*appY*, *envY*, and *rhaR*).

Some accessory genes for *H. pylori* appear to be completely unrelated to those of other species. Beside the two structural genes, two genes, *ureC* and *ureD*, of *H. pylori* are required for expression of cloned urease sequences in *Campylobacter jejuni*. These genes share no homology with accessory genes for any species for which the nucleotide sequence has been determined. Mutation of these genes results in loss of urease activity in *C. jejuni*. Interestingly, cloned *H. pylori* urease sequences express activity in *E. coli* only under nitrogen-limiting conditions.

C. Relatedness of All Ureases

DNA sequence analysis has revealed a high degree of homology among the urease genes of all ureases for which the nucleotide sequence has been determined. In addition, the predicted amino acid sequences are highly conserved when compared to that of the jack bean plant urease, suggesting a common evolutionary precursor for ureases of prokaryotes and eukaryotes.

Although urease can have one, two, or three distinct subunits, these enzymes are nevertheless closely related (Fig. 5). For example, in a three-subunit urease (*P. mirabilis*, *M. morganii*), the smallest subunit is homologous with the amino terminal amino acids of the jack bean urease, the medium-size subunit is homologous with internal sequences, and the largest subunit is homologous with the carboxy-terminal sequences of the jack bean enzyme. The same relationship holds true for the two-subunit enzymes of the *Helicobacter* species. Genetic evidence allows us to speculate that,

P mirabilis Urease Operon

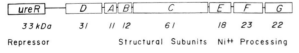

Figure 4 Genetic organization of the urease operon of *Proteus mirabilis*. The organization of genes that comprise the urease operon have been predicted from nucleotide sequencing, transposon mutagenesis, and protein purification. UreA, UreB, and UreC are the structural subunits of urease. UreD's function is unknown. UreE, UreF, and UreG have been suggested to play a role in Ni^{2+} insertion into the holoenzyme. UreR is a repressor that inhibits urease synthesis in the absence of substrate urea.

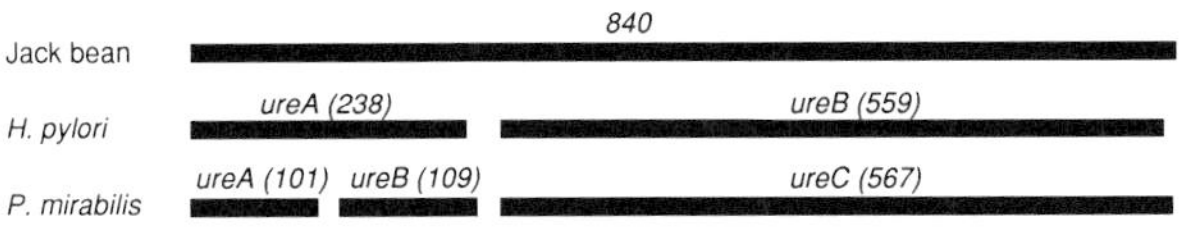

Figure 5 Conservation of amino acid sequence and alignment of subunit genes for urease of jack bean, *Helicobacter pylori*, and *Proteus mirabilis*. The single-subunit polypeptide of jack bean is depicted as a single bar. The predicted amino acid sequence of two subunits of *H. pylori* align with the jack bean subunit as shown. The predicted amino acid sequence of the three *P. mirabilis* subunits align with jack bean subunits and *H. pylori* subunits as shown. *Helicobacter pylori* shares 45.2% exact amino acid matches with the jack bean subunits and 50.2% exact amino acid matches with *P. mirabilis* subunits. [From Mobley, H. L. T., Hu, L. T., and Foxall, P. *Scand J. Gastroenterol.* (Suppl. 1991) (in press).]

during evolution, the two smaller bacterial subunit (see *P. mirabilis* genes in Fig. 3) may have fused to form the 29.5kDa subunit of *H. pylori,* and all three subunit genes may have fused to form the single jack bean subunit.

Ureases from different species appear to have different stoichiometries of subunits in the native enzyme. Jack bean urease is comprised of six identical copies a 93-kDa subunit, whereas *H. pylori urease is apparently comprised of six copies each of the 66-kDa and 29.5-kDa subunits. All other bacterial ureases have been shown to consist of some complex arrangement of three subunits in the range of 63 kDa, 15 kDa, and 6 kDa, as exemplified by purified urease from M. morganii,* That is, the ratios of the subunits do not appear to be 1:1:1 but, rather 1:2:2.

IV. Role in Pathogenesis

A. Infection Stones

Urease is the principal cause of urinary tract infection-induced stones. Enzyme-catalyzed urea hydrolysis results in a rapid rise in urinary pH from 7 to 9, due to the liberation of ammonia. In this environment, polyvalent cations and anions that are normally soluble at neutral pH begin to precipitate in the form of struvite ($MgNH_4PO_4 \cdot 6 H_2O$) and apatite [$Ca_{10}(PO_4)_6 \cdot CO_3$]. These ions complex to form free crystals or use the bacterium itself as a nidus for stone formation, entombing the organism in a mineral casing. Stones can grow to large sizes that can obstruct urine flow by blocking ureters or urethra and have been known to completely fill the renal pelvis, causing so-called staghorn calculi revealed

by X-ray. Previously large stones required surgical removal but can not be pulverized, without surgery, using shock-wave lithotripsy. This procedure focuses sound waves from several directions on the precise coordinates of the stone and disrupts it, producing a fine sand that can be eliminated by urination.

The urinary tract can be infected by a number of urease-positive species (Table I). Among these, *P. mirabilis* is most often implicated in stone formation, perhaps due to urease activity that is significantly higher than that in other uropathogenic species.

Urease-catalyzed stone formation can be a particular problem for patients with indwelling urinary bladder catheters. These foreign bodies predispose patients to polymicrobial infection with a number of urease-positive species, including *P. mirabilis, P. stuartii, M. morganii, Providencia rettgeri,* and *K. pneumoniae.* Crystallization induced by high pH results in encrustation and blockage of urine drainage, requiring catheter removal and replacement. Prior to replacement, infected urine can reflux into the kidney, causing the development of acute pyelonephritis or bactermia. *Proteus mirabilis* has been principally indicated in catheter blockage with stone material.

Stone formation can be retarded or reversed in those individuals with recurrent urinary tract infection by the oral administration of the urease inhibitor and structural analog of urea, acetohydroxamic

Table I Urease-Positive Bacterial Species Causing Human Urinary Tract Infection

Species	Comments
Proteus mirabilis	Species most commonly associated with stones
Klebsiella pneumonia	Urease repressed by high nitrogen
Providencia stuartii	Plasmid-encoded urease; species peculiar to catheter-associated bacteriuria
Morganella morganii	Constitutively produced urease; common in catheter-associated bacteriuria
Proteus vulgaris	Not nearly as common as *P. mirabilis*
Proteus penneri	Newly classified species
Providencia rettgeri	Can have two distinct ureases
Staphylococcus saprophyticus	One of few gram-positive species that cause serious urinary tract infection
Escherichia coli	Very rarely ureolytic (0.1% of isolates)

acid. Concentrated in the urine, this compound inhibits urea hydroysis, thus preventing alkalinization of the urine and subsequent stone formation; however, some side effects of this compound require reduction of dose or cessation of treatment.

B. Pyelonephritis

Pyelonephritis, defined as an acute or chronic inflammation of the kidney and its pelvis, results from bacterial infection and is characterized by interstitial inflammation and tubule necrosis. Although *E. coli,* a primarily nonureolytic species, causes a majority of cases of human acute pyelonephritis, urease elaborated by other infecting species such as *P. mirabilis* appears to contribute significantly to tissue damage, inflammation, and cell invasion.

Proteus mirabilis is the primary urease-producing uropathogen in humans. The role of *Proteus* urease in pyelonephritis has been studied using mouse and rat models of urinary tract infection as well as tissue culture systems. In each case, pyelonephritis or significant cell cytotoxicity was induced by *Proteus* infection. The direct toxicity of urease on renal tissue was demonstrated by using killed *Proteus* suspensions with active and inactivated enzyme; necrosis occured only with the active enzyme. Intracellular tissue culture infection with *P. mirabilis* increased as urea concentration rose. In the mouse model, an ethylmethane sulphonate-generated urease-negative mutant of *P. mirabilis* produced much smaller renal abscesses that the parent strain and lower populations in the kidney.

In contrast to *Proteus,* inoculation with *E. coli* produced almost no intracellular infection in kidney cells, no rise in pH, and no kidney cell injury in tissue culture.

The specific effects of urease deficiency in animal models of pyelonephritis have been examined by treating animals infected with *P. mirabilis* with acetohydroxamic acid. In all studies, kidney abscesses were similar or absent in the acetohydroxamic acid group as compared to animals receiving no inhibitors. In addition, treated animals yielded fewer organisms and suffered few deaths. Finally, a genetically constructed strain of *P. mirabilis* in which the large urease subunit gene was mutated (and, thus, produced an inactive enzyme) was tested in a mouse model of acending urinary tract infection. While the wild-type strain elicited serious acute pyelonephritis ($>10^4$ colony-forming units/g kidney after 48 hr), the mutant strain could rarely successfully colonize the kidney ($<10^2$ colony-forming units/g kidney).

Bovine pyelonephritis is caused by *Corynebacterium renale,* an organsim that possesses a very active urease. Infection with this microorgansim resulted in alkaline urine, growth of *C. renale* from kidney homogenates and urine, and necrosis of the renal tissue. Use of acetohydroxamic acid at high doses reduced the urine pH, decreased the number of colony-forming units of *C. renale* in the kidney, and halted necrosis of the kidney tissue. Infection with a urease-negative mutant of *C. renale* did not result in pyelonephritis. This evidence suggests that the ammonia liberated from ureolysis causes the alkalinization of the urine and may account, in part, for the necrosis of kidney tissue associated with pyelonephritis.

C. Gastritis and Peptic Ulceration

Infection of the human gastric mucosa with *H. pylori* has been unambiguously associated with the development of chronic active gastritis, and a very strong association between the eventual development of peptic ulceration and gastric cancer has been drawn. Urease, produced in extremely high concentrations by this species, is undoubtedly critical for *H. pylori* colonization of the human gastric mucosa. *In vitro,* the bacterium is quite sensitive to the effects of low pH unless urea is present. Initial colonization of the stomach where pH equals 3 or less would be difficult unless the organism can protect itself from exposure to acid. It is postulated that the organism hydrolyzes urea, releasing ammonia, which neutralizes acid allowing survival and initial colonization. Fresh isolates cultured from gastric biopsies are always urease-positive, suggesting the production of urease is an essential phenotype. Furthermore, urease-negative mutants of *H. pylori* have been generated by nonspecific chemical mutagenesis or by selection of naturally occurring mutants and used for colonization experiments in animal models of infection. No urease-negative mutant has colonized the gastric mucosa in these experiments, whereas a high percentage of the wild-type strains established infection. [*See* GASTROINTESTINAL MICROBIOLOGY.]

In addition to the survival benefit of urease, evidence indicates that ammonium hydroxide, generated by urea hydrolysis, contributes significantly to histological damage. It should be emphasized that the ammonium ion per se is not toxic: instead, the hydroxide ion generated by ammonia's equilibration with water. To demonstrate the cytotoxic effect of urease, cell cultures of a human gastric adenocarcinoma cell line were seeded with *H. pylori* and supplemented with various concentrations of urea. Cell viability was found to be inversely proportional to ammonia concentrations generated by urea hydrolysis. Viability was improved when a urease inhibitor, acetohydroxamic acid, was added to the culture prior to exposure to *H. pylori*. Acetohydroxamic acid slowed the liberation of ammonia and reduced the cytotoxic effect.

Similar effects were shown for Vero cells overlaid with filtrates of *H. pylori* Cell rounding and loss of viability were observed in cultures to which 30 mM urea had been added. These changes were associated with a rise in pH. Acetohydroxamic acid reduced this effect significantly. These data suggested that histological damage may result directly from the localized generation of ammonia due to hydrolysis of urea.

Urea hydrolysis has been postulated to have an additional effect. That is, ammonia interferes with normal hydrogen ion back diffusion across gastric mucosa, resulting in cytotoxicity to the underlying epithelium.

V. Summary

In summary, ureases, high-molecular weight, nickel metalloenzymes, are produced by numerous microbial species and play a central role in nitrogen metabolism of the cell. In addition, they are essential for nitrogen cycling in ruminants and provide a mechanism for the efficient delivery of nitrogen to fertilized crops. This enzyme occasionally contributes to the pathogenic process of bacterial infection of mammals, particularly in urinary tract infection and infection of the gastric mucosa. Finally, studies are revealing that all ureases seem to be closely related and may share common ancestral genes.

Bibliography

Clayton, C. L., Pallen, M. J., Kleanthous, H., Wren, B. W., and Tabaqchali, S. (1990). *Nucleic Acid Res.* **18**, 362.

Griffith, D. P., Musher, D. M., and Itin, C. (1976). *Invest Urol.* **13**, 346–350.

Hausinger, R. P. (1986). *J. Biol. Chem.* **261**, 7866–7870.

Jones, B. D., and Mobley, H. L. T. (1987). *Infect. Immun.* **55**, 2198–2203.

Jones, B. D., and Mobley, H. L. T. (1989). *J. Bacteriol.* **171**, 6414–6422.

Jones, B. D., Lockatell, C. V., Johnson, D. E., Warren, J. W., and Mobley, H. L. T. (1990). *Infect. Immun.* **58**, 1281–1289.

Labigne, A., Cussac, V., and Courcoux, P. (1991). *J. Bacteriol.* **173**, 1920–1931.

McLean, R. J. C., Nickel, J. C., Noakes, V. C., and Costerton, J. W. (1985). *Infect Immun.* **49**, 805–811.

Mobley, H. L. T., and Hausinger, R. P. (1989). *Microbiol. Rev.* **53**, 85–108.

Vertebrate Tissues, Serological Specificity

Felix Milgrom
State University of New York at Buffalo

I. Species Specificity
II. Heterophile Specificity
III. Serological Polymorphism within Species
IV. Tissue Specificity
V. Autoantibodies to Tissue Antigens
VI. Altered Autoantigens and Antigens of Pathological Tissues
VII. Conclusions

Glossary

Alloantigen Antigen present in some but not all individuals of a given species

Autoantibody Antibody combining with an autoantigen (i.e., an antigen of the subject producing this antibody)

Heterophile antigen Antigen combining with an antibody produced in response to an apparently unrelated antigen

Oncofetal antigen Antigen that appears in a higher quantity in tumorous and fetal tissues than in normal adult tissues

Species-specific antigen Antigen present in all individuals of a given species and different from an analogous antigen in other species

Tissue-specific antigen Antigen restricted to one tissue, organ, or cell type

Tumor-associated antigen Normal tissue antigen, the concentration of which is increased within a tumor

Tumor-specific transplantation antigen Antigen provoking immune response that leads to rejection of the tumor

THE TERM ANTIBODY was originally introduced by Ehrlich in 1891 to denote proteins appearing in the animal's serum as a result of stimulation exerted by some toxic agents. It was later recognized that all antibodies are serum globulins classified as immunoglobulins. A substance capable of engendering antibody formation was called antigen by Deutsch in 1899, as a shortened version for the original term antisomatogen, meaning a substance generating antibody formation. *In vitro* reactions between antigens and their corresponding antibodies were called serological reactions. Of these, the first to be described were agglutination, meaning clumping by antibodies of an antigen presented in the form of particles, and precipitation, meaning sedimentation by antibodies of a soluble antigen. Subsequently, many other serological reactions were developed.

The most important characteristic of serological reactions is their outstanding specificity. Due to this specificity, antigen preparation can serve as a reagent to detect its corresponding antibodies, a principle on which serological diagnosis of many diseases is based, and antibody-containing serum, referred to as antiserum or immune serum, can serve as a reagent for detection and identification of its corresponding antigen. The latter principle has served in microbiology for identification of microbial species and serotypes within the species. Furthermore, antibody-containing sera have been used as reagents to identify antigens of cells, tissues, organs and body fluids of vertebrate animals. These studies brought important discoveries to be discussed in this article.

I. Species Specificity

Since the very beginning of the twentieth century, immune sera have been employed as reagents identifying antigenic components of vertebrate animals. These studies permitted creation of the term "species-specific antigen," denoting an antigen present in all individuals of a given species but different from an analogous antigen in other species. Species-specific antigens were detected by means of antisera prepared in animals of foreign species, usually rabbits. In early studies, whole sera of various species origin were used most frequently as antigens for preparation of reagent-antisera and for *in vitro* tests. Each reagent-antiserum always gave the strongest reaction with the serum from the same species as the donor of the serum used for immunization. Such reactions were referred to as homologous reactions. The strength of cross-reactions with sera originating from other species, referred to as heterologous reactions, depended on the biological proximity of the species whose serum served for immunization to the species whose serum was tested as *in vitro* antigen. (The adjectives homologous and heterologous are used here as applied originally; regretably, these terms have been applied later in a different connotation.) As could be expected, an antiserum to human serum would give strong cross-reactions with sera of apes, weaker reactions with sera of monkeys, and very weak reactions with sera of oxen or horses. The reactions of this antiserum with sera of vertebrate classes other than mammals (e.g., birds or fish) would most frequently give negative results.

The species origin of the reagent-antiserum determines to a great extent its usefulness in recognizing serological differences among various animals. For example, rabbit antisera, which have served most frequently as reagents in these studies, can clearly distinguish the antigenic differences among sera of mammals; however, they are poor reagents for distinguishing differences among avian sera (e.g., to distinguish chicken serum from pigeon serum). This phenomenon was called "faulty perspective," meaning that, in our example, the rabbit is a species

far distant from birds and consequently is incapable of "seeing the difference" between chicken and pigeon, being "overwhelmed" by strong antigens common to sera of all birds. The procedure recommended for distinguishing the two taxonomically close species is cross-immunization. In our example, chicken antiserum to pigeon serum would react only with pigeon but not with chicken serum, and, conversely, pigeon anti-chicken serum would combine with chicken but not with pigeon serum.

Most of the traditional studies on species specificity employed whole sera as the tested antigens. It was soon realized, however, that serum is composed of several individual proteins. At present, over 30 antigenic serum components can be distinguished. Accordingly, many modern studies on species specificity have been conducted on individual serum proteins. It was shown that the basic principles of species specificity, which were observed in studies on whole sera pertained also to individual serum proteins (e.g., to the serum albumin). Accordingly, albumins of closely related species would give strong cross-reactions and those of remote species would give weak or no cross-reactions.

Application of the procedure of precipitation in gel to studies on species specificity permitted analytical evaluation of interspecies cross-reactions. This is presented schematically in Fig. 1. It may be seen in the first row of this figure that a rabbit antiserum to bovine serum albumin produced very strong cross-reactions with serum albumins of sheep and

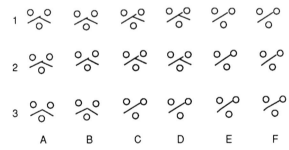

Figure 1 (Row 1) Lower wells: Rabbit antiserum to bovine serum albumin. Upper left wells: Bovine serum albumin. Upper right wells: Serum albumins of (A) sheep, (B) goat, (C) humans, (D) rhesus monkey, (E) chicken and (F) quail. (Row 2) Lower wells: Rabbit antiserum to human serum albumin. Upper left wells: Human serum albumin. Upper right wells: Serum albumins of (A) rhesus monkey, (B) baboon, (C) ox, (D) goat, (E) chicken, and (F) quail. (Row 3) Lower wells: Rabbit antiserum to chicken serum albumin. Upper left wells: Chicken serum albumin. Upper right wells: Serum albumins of (A) quail, (B) duck, (C) ox, (D) goat, (E) humans, and (F) rhesus monkey.

goat, which belong to the same family of Bovidae as ox. The cross-reactions with albumins of primates, humans, and rhesus monkey were weak, and there were no cross-reactions with albumins of birds, chicken, and quail. The spurs extending from the lines of homologous reactions over the lines of heterologous reactions were produced by those antibodies in the reagent antiserum, which reacted only with the homologous albumin but not with heterologous albumins. These spurs were small in tests with sheep and goat albumins, but they were extensive in reactions with albumins of humans and rhesus monkey. In the second row of Fig. 1, reactions are presented that were obtained with an antiserum to human serum albumin. As expected, the cross-reactions produced by this antiserum were strong with albumins of rhesus monkey and baboon but weak with albumins of ox and goat. Here again, no cross-reactions were noted with albumins of birds. Finally, the third row of Fig. 1 presents the reaction with an antiserum to chicken serum albumin. This antiserum gave very strong cross-reactions with albumins of quail and duck origin and no cross-reactions with albumins of mammals. Apparently, this antiserum did not have antibodies that would react with chicken albumin without reacting with quail or duck albumin. Therefore, no spurs were noted that would have extended from the reaction with chicken albumin over the reactions with quail or duck albumin. A quite similar pattern of reactions to the one presented in Fig. 1 would be obtained with other serum proteins and also with antigenic extracts of tissues and their corresponding antisera. For example, an extract of thyroid gland or a purified preparation of thyroglobulin originating from various species would produce similar reactions with antithyroglobulin sera as those presented in Fig. 1 for antialbumin sera.

Demonstration of species specificity does not require application of the intact molecule as the tested antigen. Polypeptide fragments resulting from enzymatic splitting of the original protein molecule frequently retain their species specificity. For several proteins, amino acid sequencing was conducted, which brought some insight into the structures responsible for their species specificity and interspecies cross-reactions. Especially illuminating studies along these lines were conducted on cytochromes. Despite the fact that it is a small molecule (MW 6000) and a rather weak antigen, insulin has also a clearcut species specificity that depends on a recognized chemical structure.

Species specificity is not limited to soluble antigens, but it also can be demonstrated in insoluble antigens on the surface of many cells. The first studies along these lines were conducted in 1898 by Bordet, who could distinguish erythrocytes of various species origin by means of tests with proper antierythrocyte sera. Species specificity of cell-surface antigens could be conveniently shown in tests with cell culture monolayers. Here again, strong cross-reactions were noted when the cultured cells originated from related species and weak cross-reactions were noted when they originated from remote species. Furthermore, cultures of hybrid cells obtained by fusion of cells from two different species (e.g., human, mouse) contained species-specific antigens of both parents (Table I). The expression of human species specificity by human–mouse hybrids was weakened with the deletion of human chromosomes, but it still remained demonstrable in some hybrids that retained only one human chromosome. This and other observations clearly showed that the surface of vertebrate cells contains numerous antigens that carry definite species specificity.

Species specificity has been of considerable theoretical interest, because it reflects evolutionary pathways. These studies also found medico-legal applications. In tests performed in criminal cases, proper antisera may serve for the identification of the species origin of tissue debris and blood stains. Also, such tests may serve to detect adulteration or mislabeling of food products.

II. Heterophile Specificity

Some tissue antigens defy the order of distribution paralleling evolution, which is so characteristic for

Table I Species Specificity of Human–Mouse Hybrid Cells

	Results of mixed agglutination tests with monolayer cultures of:		
	Human cells	Murine cells	Human–mouse hybrid cells
Rabbit antiserum[a] to:			
Human cells	S.P.	Neg.	S.P.
Murine cells	Neg.	S.P.	S.P.

[a] Antisera were absorbed to avoid cross-reactions.
S.P., strongly positive; Neg., negative.

species specificity. The first discovery of such an antigen was made in 1911 by Forssman, who demonstrated that sera of rabbits immunized with guinea-pig kidney suspension gave strong reactions with erythrocytes of sheep but not with those of oxen. The general term heterophile antibodies was coined to denote antibodies combining with antigens that are apparently unrelated to those antigens that stimulated their formation, and the term heterophile reaction has been used for an unexpected serological cross-reaction. It should be stressed that most protein antigens follow the pattern characteristic for species specificity in that they are similar in related and dissimilar in unrelated species. Antigens responsible for heterophile reactions are usually carbohydrates that frequently appear in an unpredictable fashion in remote species and may be responsible for the antigenic similarity between living things as distant as humans of blood group A and pneumococci of type XIV. There are many known heterophile antigen–antibody systems, and of these we will discuss the three most studied because of their importance in medicine.

The Forssman (F) system was originally described by the demonstration that sheep erythrocytes and guinea pig kidney share an antigen. All species could be divided into those that contain the F antigen in their tissues (guinea pig, mouse, sheep, horse, dog, cat, chicken, etc.) and those that do not have this antigen (rabbit, rat, humans, ox, goose, etc.). Only the F-negative species are capable of forming F antibodies. The F antigen is a glycosphingolipid, with a pentasaccharide being the active epitope (i.e., the smallest antigenic configuration recognizable by the antibody). Serologically, this antigen is characterized by its presence on sheep erythrocytes and in guinea pig kidney and by its absence from bovine erythrocytes. Evidence indicates that the F antigen may appear in tumors of F-negative species.

The Hanganutziu–Deicher (H-D) system was first identified in 1924 by means of antibodies formed by patients who received injections of horse sera for therapeutic purposes. Chemically, the H-D antigen is N-glycolylneuraminic acid. Serologically, it is defined by its presence on sheep and bovine erythrocytes as well as in guinea-pig kidney. This antigen was demonstrated in tissues of all mammalian species tested except for humans. H-D antibodies were demonstrated in a small proportion, about 10%, of sera of human patients with various pathological conditions.

The Paul–Bunnell (P-B) system was established in 1932 following the observation that sera of patients with infectious mononucleosis have antibodies combining with sheep and bovine erythrocytes. These antibodies have been found in sera of >90% of patients with infectious mononucleosis, but hardly ever in normal human sera or sera of patients with diseases other than infectious mononucleosis. The P-B antigen has been shown in the glycoprotein fraction of bovine erythrocytes. Two antigenic epitopes could be distinguished, one present on bovine erythrocytes only and another shared by bovine and sheep erythrocytes. Characteristically, the P-B antigen is absent from guinea-pig kidney and from normal human tissues. The detection of P-B antibodies is one of the most valuable serodiagnostic tests available in medicine.

III. Serological Polymorphism within Species

Antigens appearing in some but not all individuals of a given species are called alloantigens. Such antigens were first described in 1900; these were blood group antigens of goats observed by Ehrlich and Morgenroth and human ABO blood group antigens discovered by Landsteiner. The discovery of human blood groups immediately drew considerable attention because of their importance in blood transfusion. In addition to ABO, many other human blood group systems were described, of them the most important in medicine has been the Rh system, because incompatibility in Rh groups between mother and fetus may lead to sensitization of the mother and damage to the fetus by maternal antibodies.

Considerable interest was elicited by the discovery, in the 1950s, of antigens responsible for polymorphism of immunoglobulins (i.e., serum proteins carrying antibody activity). Alloantigens of immunoglobulins, called allotypes, have served as useful markers for tracing the synthesis and distribution of antibodies in humans and animals.

Undoubtedly, the most important alloantigens from a biological point of view are those encoded by the genes of the major histocompatibility complex (MHC). Such antigens have been described in all mammalian species studied; in humans they are called HLA. Two major groups of MHC antigens, Class I and Class II antigens, have been distinguished. MHC antigens act as the strongest transplantation antigens, and the incompatibility in these antigens between donor and recipient is the major obstacle in acceptance of an allograft (i.e., a graft from a donor belonging to the same species but genetically different from the recipient). Also, MHC antigens account for immunological "communication channels," referred to as MHC restriction. The important defense mechanism accomplished by cytotoxicity leading to the elimination of aberrant or infected cells depends on Class I antigens. Specifically, cytotoxic lymphocytes would destroy in a much more efficient way infected or aberrant cells that carry Class I antigens identical with theirs than such cells with nonidentical Class I antigens. On the other hand, Class II antigens on antigen-presenting cells such as macrophages play an important role in stimulation of lymphoid cells (referred to as helper cells) leading to an immune response.

The paramount characteristic of alloantigens is their inheritance. The hereditary character of human ABO blood groups was shown by Dungern and Hirszfeld in 1910, and this initiated the field of human immunogenetics. Determination of alloantigens has served in forensic medicine in cases of disputed parentage. In each case of disputed parentage, alloantigens of the offspring are identified, which could and which could not be inherited from the putative parent. With all alloantigens that can be now defined in specialized laboratories, innumerable human phenotypes can be determined; therefore, it is not surprising that practically in all questionable cases the parentage can be excluded or confirmed. Furthermore, most alloantigens have unequal distribution in various human ethnic groups, which was first shown for the ABO groups by the Hirszfelds in 1919. This led to fruitful anthropological studies.

The chemical structure of many alloantigens has been extensively studied. In some instances, when the basic molecule is a protein (e.g., immunoglobulin), the expression of polymorphism may depend on the structure of one or a few amino acid residues. In other instances, the structure of an alloantigen is determined by a carbohydrate (e.g., specificity of human blood group A antigen depends on the terminal sugar N-acetyl-D-galactosamine, whereas B antigen is characterized by the terminal component D-galactose).

The distribution of alloantigens may be ubiquitous or restricted (see later). For example, ABO antigens appear in many tissues and body fluids, whereas Rh antigens can be found only on erythrocytes. MHC antigens are ubiquitous, but their density differs considerably in individual organs and cells.

Association of alloantigens with species-specific antigens varies for individual antigens. The ABO antigens appear not only in humans, but also in other species and their occurrence seems independent from any species-specific antigens. On the other hand, allotypes are closely associated with the species-specific structures because they appear on immunoglobulin molecules that exhibit strong species specificity. Close association of MHC antigens with species-specific antigens has also been demonstrated.

Detection of alloantigens is usually accomplished by antisera originating from the same species as the tested subject. Because, in most instances, the difference among structures of alloantigens belonging to one system (i.e., encoded by alleleic genes) is subtle, it cannot be readily distinguished by an animal of a foreign species (see the preceding discussion on faulty perspective). On the other hand, within the species, an individual negative for a given alloantigen forms the corresponding antibodies upon exposure to an inadvertent or intentional stimulation by this antigen. Still, formation of such alloantibodies depends on the immunogenic strength of the antigen, and antibodies to very weak alloantigens are formed only exceptionally.

IV. Tissue Specificity

Discussing species specificity, we stressed that phylogenetic differentiation of antigens of vertebrate tissues brought differences among analogous antigens of various species. Still, we emphasized similarities of analogous proteins (e.g., serum albumins) originating from various species of the same vertebrate class (see Fig. 1). In contrast, ontogenetic differentiation of various body components destined to serve different purposes results in the emergence of completely different antigenic configurations. Whereas a cross-reaction between serum albumins from species as far remote as humans and ox may be noted (Fig. 1), there is no cross-reaction at all between albumin and globulin, even those from the serum of the same individual. This brings us to the term tissue specificity. This term has been used by immunologists to denote serological specificity not only of tissues, but also of organs or cells and, therefore, it has been frequently criticized by histologists. Still, we will use this term here in its "traditional" connotation. Those antigens restricted to one tissue, organ, or cell type will be called tissue-specific antigens, and those that can be found throughout the animal body will be called non-tissue-specific or ubiquitous antigens. One could argue that serum proteins are tissue-specific antigens. Albumins are synthesized only by liver cells and globulins by lymphocytes, and accordingly these proteins are "specific" for their synthesizing cells. Serum proteins, however, are shed off to the circulation and may be found throughout the body; therefore, they are classified as ubiquitous antigens. Most investigators consider only those antigens to be tissue specific, which remain confined completely or almost completely to the place of their origin. Only a few examples of tissue-specific antigens will be presented in this article.

The first tissue-specific antigen was described in 1903 by Uhlenhuth, who immunized rabbits with a homogenate of bovine lens and found that antisera produced in this way would react with lens preparations originating from all mammalian species under study, including rabbit, the antibody-producing species, and even with lens preparations from birds, fish, and anphibia. Accordingly, these reactions showed an outstanding tissue specificity and hardly any species specificity. Subsequently, several lens proteins were identified and isolated. All of them exhibited strong tissue specificity and various degrees of species specificity.

Several tissue-specific antigens have been identified in the central nervous system (CNS). In 1928 Witebsky and Steinfeld described an antigen characteristic for brain, which resisted prolonged heating at 100°C. Because this antigen was ethanol-soluble, it was described as a lipid. It had a very strong tissue specificity and hardly any species specificity. Rabbit antiserum to bovine brain reacted with ethanol extracts of brain from all mammals, birds, fish, and anphibia under study. Many years later, this antigen was classified as a galactocerebroside.

Important brain antigens are cerebrosides containing *N*-acetylgalactosamine and *N*-acetylneuraminic acid. There are at least 10 such antigens distinguishable, but 4 of them constitute 90% of the total. One physiologically minor galactoside component increases to 90% of the total in Tay-Sachs disease (amaurotic idiocy), a brain lipoidosis.

Another tissue-specific component of brain, predominantly of gray matter, belongs to a category of tissue antigens referred to as BE antigens for their resistance to boiling and insolubility in ethanol. It has physicochemical properties of a glycoprotein. In addition to a very strong tissue specificity, this antigen also shows a significant degree of species specificity.

Of CNS proteins, myelin basic protein (MBP)

drew most attention because it is the active component in the production of experimental allergic encephalomyelitis (see later). MBP is extracted from delipidated CNS at acidic pH and then separated by column chromatography. It constitutes one-third of the total mass of myelin and 1% of the total weight of CNS. The molecular weight of MBP is about 18,000. It is composed of 170 amino acid residues; their sequence has been well established for the MBP of some species. The peptide fragments of MBP responsible for encephalolitogenicity have been identified and obtained by *in vitro* synthesis. MBP has strong tissue specificity and rather weak species specificity. Open configuration of this protein accounts for its thermal resistance and for the fact that primary structures play a more important role in its antigenicity that secondary and tertiary structures.

Tissue specificity of thyroid gland was studied by Hectoen, Fox, and Shulhof in 1927. They obtained rabbit antisera to partially purified thyroglobulin of several mammalian species and of chicken. Already these studies clearly showed that, in addition to its tissue specificity, thyroglobulin has strong species specificity, because in tests with rabbit antisera, mammalian and chicken thyroglobulins failed to cross-react. Antibodies to thyroglobulin can be detected in virtually all cases of Hashimoto's thyroiditis as well as in many cases of idiopathic myxedema and Graves' disease.

Testis and spermatozoa have been the subject of extensive investigations on tissue specificity. Reactions between spermatozoa and their corresponding immune sera were studied by Henle in 1938, who could distinguish antigens characteristic for the heads and tails of these cells. The unresolved question remains whether the antigens detected on spermatozoa are indeed their integral components or otherwise are absorbed from the seminal plasma. Expression of alloantigens on spermatozoa has been of great interest since the possibility was entertained that spermatozoa of heterozygotes may be separated into two allelic populations. Such separation of, for example, Rh-positive and Rh-negative spermatozoa could be of significant importance in eliminating those Rh-positive that might induce pregnancy with fetomaternal incompatibility and select for insemination those Rh-negative that would be responsible for compatible pregnancy. Thus far, such studies did not result in any reproducible or encouraging results.

Further interest in antigens of testis and spermatozoa stemmed from experiments that showed that immunization of experimental animals with these antigens may result in aspermatogenesis. Several authors reported purification of a tissue-specific glycopeptide antigen from testicular extract. Furthermore, a testis-specific antigen was described in the thermostable ethanol-insoluble (BE) fraction of this organ. Antibodies to testis-specific antigens may be found in the circulation after injury to testis and after vasectomy.

V. Autoantibodies to Tissue Antigens

Autoantibody is an antibody that combines with an autoantigen (i.e., an antigen of the antibody-producing subject). From the very beginning of immunological studies, it was recognized that formation of autoantibodies is a rather exceptional phenomenon. Ehrlich and Morgenroth formulated, in 1900, the famous term *horror autotoxicus* (later referred to as self-tolerance), stressing that normal humans and animals refuse to form autoantibodies because they could be harmful by combining with the body's own components. However, shortly thereafter, observations were presented, which could be considered as exceptions to the horror autotoxicus rule. Of these observations, those pertaining to autoantibodies against tissue-specific antigens are most intriguing, because their formation may be elicited in animal experiments at the will of the investigator employing proper immunization procedures.

It should be stressed that the "auto" nature of an antibody is determined on the basis of the origin of the antigen that reacts with this antibody rather than on the basis of the origin of the antigen that engendered its formation. Accordingly, formation of autoantibodies may be stimulated by an antigen of a foreign species origin and indeed the first autoantibodies to tissue-specific antigens were produced by Uhlenhuth by immunization of rabbits with homogenates of bovine lens. Uhlenhuth himself stressed that these antibodies combined also with the antigenic preparation of rabbit lens and it is obvious that they were autoantibodies. In a similar way, Witebsky and Steinfeld produced brain-specific antibodies by immunization of rabbits with bovine brain preparations. Here again, at least some of these antibodies were autoantibodies that would react with the antigenic preparation of brain of the antibody-producing animal.

Subsequently, autoantibodies to many tissue-specific antigens, including those of thyroid, kidney, adrenal, testis, and pituitary, were produced in animals. To explain the relative ease with which formation of autoantibodies to many tissue-specific antigens was achieved, it was stressed that these antigens are separated from the circulation by their anatomic localization and in many cases by sequestration within the cells. According to most investigators, refusal of autoantibody formation depends on the free contact of the immunological apparatus with the autologous antigens, which establishes self-tolerance. It has been broadly accepted that inaccessibility of many tissue-specific autoantigens accounts for the weak self-tolerance of these antigens, which can be readily overcome by proper immunizing procedures (e.g., by presenting the antigen in a form partially different from the autologous antigen). Application of antigenic preparations from foreign species by Uhlenhuth and by Witebsky and Steinfeld created such conditions unintentionally. Subsequently, in 1965, Weigle altered rabbit thyroglobulin by coupling it with arsenilic or sulfanylic groups and achieved autoantibody formation in rabbits immunized with these preparations. It should be stressed that autoantibodies resulting from such immunizations combined with the native structures of the autoantigens. Accordingly, in the experiments of Uhlenhuth as well as of Witebsky and Steinfeld, they were directed against epitopes shared by autologous and foreign antigens of lens or brain. Those antibodies against epitopes restricted to foreign antigens were obviously not autoantibodies. In a similar way, in Weigle's experiments autoantibodies were only those that were directed against thyroglobulin epitopes that had not been affected by chemical treatment.

In experiments conducted over the last decades, adjuvants have been employed to achieve autoantibody formation upon immunization with tissue antigens. Of these, complete Freund adjuvant, composed of mineral oil, emulsifying agent, and heat-killed mycobacteria, has been used most frequently. This adjuvant promotes response to the immunizing antigen by provoking local granulomatous reaction and most likely also by inducing some molecular transformation of the antigen. Immunization with Freund adjuvant not only conditions formation of autoantibodies to tissue-specific antigens, but it also results in the development by the animal of an experimental autoimmune disease of the organ employed for immunization (e.g., encephalomyelitis, thyroiditis, orchitis).

Autoantibody formation to tissue antigens occurs

also under natural conditions. Important autoantibodies to a tissue antigen were described in 1906 by Wassermann and his colleagues. Wassermann antibodies, characteristic for syphilis, are directed against a ubiquitous tissue antigen, a phospholipid component called cardiolipin, which is devoid of tissue or species specificity. Most likely, these antibodies are formed in response to stimulation exerted by cardiolipin released in immunogenic form from syphilitic lesions. Detection of Wassermann antibodies has played a most important role in the diagnosis of syphilis. Their possible pathogenic role was entertained for decades but has never been substantiated. Autoantibodies against nuclear antigens are characteristic for systemic lupus erythematosus. They are useful for the diagnosis in this disease.

As already mentioned, autoantibodies to thyroid antigens have been regularly detected in several pathological conditions of thyroid gland. Demonstration of these antibodies serves a useful diagnostic purpose and their role in inflicting damage to thyroid gland appears likely. Autoantibodies to other tissue antigens have been demonstrated in various pathological conditions. Some of them are tissue-specific, as those directed against structures of kidney, adrenal, striated muscle, skin, and testicle. In some instances, their pathogenic role may be significant.

In contrast to inaccessible or sequestered antigens, those antigens on the surface of circulating cells elicit strong self-tolerance. Autoantibody formation to such antigens have hardly ever been produced in experiments on normal animals. On the other hand, autoantibody formation to such antigens has been clearly demonstrated under pathological conditions. In autoimmune hemolytic anemia, autoantibodies responsible for destruction of erythrocytes were shown. They are usually directed against an erythrocyte antigen, which can be classified as a "pan" antigen because it occurs on erythrocytes of all humans.

VI. Altered Autoantigens and Antigens of Pathological Tissues

Expression of new antigenic epitopes on a denatured protein (e.g., on serum albumin) has been shown by many investigators. Because these new epitopes are absent from the native protein molecule of the do-

nor, it is not surprising that they elicit, quite readily formation by the donor of antibodies which, however, should not be called autoantibodies. A study along these lines was published in 1960 by Milgrom and Witebsky, who injected rabbits with each animal's own denatured gamma globulin. Antibodies resulting from this immunization produced hardly any reactions with rabbit gamma globulin but reacted very strongly with gamma globulin of foreign species (e.g., human gamma globulin). Apparently, denaturation exposed or created novel antigenic structures on the rabbit gamma globulin, and these were not just random structures but, resembled the gamma globulin structures occurring in species other than rabbit.

Transformation of the configuration of antibody molecules occurs when they combine *in vivo* and *in vitro* with their corresponding antigens. Still, under normal conditions, the immunizing stimulus exerted by such altered antibody molecules does not suffice to elicit formation of anti-antibodies. On the other hand, upon prolonged immunization, anti-antibodies are formed by humans and experimental animals. According to many investigators, rheumatoid factor, characteristic for rheumatoid arthritis, is an anti-antibody produced in response to prolonged formation of antigen–antibody complexes in the patient's body.

The appearance of heterophile antibodies in the course of infectious mononucleosis has been interpreted as the response to a novel antigen. P-B antigen has been demonstrated in organs of patients who died in the early stages of infectious mononucleosis. A hypothesis was advanced that the P-B antigen, appearing under pathologic conditions, probably as a result of a virus infection, is perceived by the patient's immunologic apparatus as a foreign antigen and it engenders formation of antibodies. It is possible that P-B antibodies destroy aberrant cells carrying the P-B antigen, and in this way they contribute to the patient's recovery from infectious mononucleosis. In a similar way, formation of antibodies to the heterophile H-D antigen may be related to the abnormal appearance of this antigen in human tissues from which it is absent under physiological conditions.

Stimulated by studies on tissue specificity, many investigators were interested in detecting hypothetical new antigens elicited by and characteristic for pathological processes.

To detect such antigens, approaches were taken that were very similar to those used for studies on tissue specificity. Accordingly, antisera produced by immunization of animals with preparations of pathological tissues were employed as reagents for studying tissue extracts or body fluids of patients. Following this line of investigation, many investigators reported positive results and described antigens characteristic for amyloidosis, rheumatoid arthritis, rheumatic fever, tuberculosis, etc. Subsequent analysis of these results failed to ascertain disease specificity of such antigens. In most instances, physiological components such as fibrinogen or C-reactive protein appearing in augmented quantities within the pathological tissue and/or throughout the patient's body were taken for novel antigens. To such a category of augmented normal components should be included necrotic antigens described by Hirszfeld in 1937. Hirszfeld noticed correctly that these antigens were shared by many pathological tissues (e.g., tumors, tuberculous caseous tissue, pus). Most likely they reflected the appearance of infiltrating cells in these tissues.

Numerous serological studies on malignant tumors were also conducted by means of antisera raised in foreign species. These studies, initiated at the beginning of the twentieth century, never led to the identification of antigens that would be specific for tumors. On the other hand, in many of these investigations, antigens were identified, which were normal tissue components, increased quantitatively in tumors. Such antigens have been termed tumor-associated antigens (TAA). In the last 30 years, much interest has been devoted to a group of TAA classified as oncofetal antigens because they were demonstrated not only in tumors, but also in fetal tissues. Of these antigens, the most extensively studied were α-fetoprotein and carcinoembryonic antigen.

Alpha-fetoprotein is a serum protein found in a high concentration in fetal blood but in a very low concentration in adult blood. In 1963, Abelev related α-fetoprotein to murine hepatomas and subsequently it was shown that patients with hepatomas and some other tumors have high levels of α-fetoprotein in their sera. Alpha-fetoprotein has a structure similar to serum albumin. It is a glycoprotein with a molecular weight of 70,000 and it contains 4% of carbohydrates. Determination of levels of α-fetoprotein is helpful in the diagnosis of certain malignancies and also of some congenital abnormalities.

Carcinoembryonic antigen was described by Gold and Freedman in 1964 in adenocarcinomas of the human intestinal tract and later it was demonstrated also in other human malignancies. This antigen is a glycoprotein with a molecular weight of approximately 200,000. Demonstration of elevated levels of carcinoembryonic antigen in the circulation of patients is of considerable diagnostic value.

Some tumors have antigens that, in contrast to TAA, are not present in the tumor-bearing animal under physiological conditions. Of these, the most important are tumor-specific transplantation antigens clearly identified in many animal tumors, primarily those induced by carcinogenic agents. Significantly, such antigens are distinguished only by the immunologic apparatus of the host and sometimes of members of the same species, but not by immune sera raised in foreign species. Many tumor-specific transplantation antigens show specificity for an individual tumor. Others are shared by several tumors and some occur as normal alloantigens within the species, but not in the tumor-bearing individual. The presence of tumor-specific transplantation antigens appears to be one of the major factors responsible for homeostatic mechanisms preventing malignant growth in normal individuals and delaying its spread in the tumor-bearing individuals. Accordingly, valid rules proposed are (1) the stronger the antigenicity of the tumor-specific transplantation antigen, the less malignant the tumor, and (2) the weaker the immune responses of the host, the more likely the tumor growth and its spread.

Investigation on antigens characteristic for pathological processes were paralleled by attempts of finding antibodies to such antigens. However, because, as it was shown later, these antigens were normal tissue components, the existing self-tolerance prevented formation of such antibodies in most instances. Claims that patients with colonic carcinomas produce antibodies to carcinoembryonic antigen as well as claims that patients with multiple sclerosis form antibodies against an altered CNS antigen were not confirmed. On the other hand, immune response to heterophile antigens or tumor-specific transplantation antigens does not meet with any refusal on the part of the host, because these antigens do not appear under physiologic conditions.

VII. Conclusions

Studies on serological specificity of vertebrate tissues were most fruitful in identifying many tissue components. Investigations of species specificity brought important information for the science of evolution and found application in forensic medicine. Investigation of polymorphism within the species led to the creation of the field of immunogenetics, contributed greatly to the field of anthropology, and had a decisive impact on the application of blood transfusion and organ transplantation. Studies on tissue specificity permitted the development of several important diagnostic tests and helped greatly in the unravelling of the pathogenesis of diseases elicited by immunological autoaggression. These studies led also to the better understanding of homeostatic mechanisms controlling the tumor growth and opened the possibilities for the immunotherapy of tumors.

Bibliography

Klein, J. (1990). "Immunology," Ch. 2, 7, 18, 24. Blackwell, Boston.

Milgrom, F. (1987). Antibodies in medicine: Past, present and future. In "Cellular, Molecular and Genetic Approaches to Immunodiagnosis and Immunotherapy" (K. Kano, S. Mori, T. Sugisaki, and M. Torisu, eds.), pp. 3–16. University of Tokyo Press, Tokyo.

Milgrom, F. (1988). Scand. J. Rheumatol. **17**(suppl. 75), 2–12.

Milgrom, F. (1989). Immunol. Invest. **18**, XXXI–XLIV.

Milgrom, F., Abeyounis, C. J., and Eaton, R. B. (1984). Immunochemistry of tissue-specific and tumor antigens. In "Molecular Immunology" (M. Z. Atassi, C. J. van Oss, and D. R. Absolom, eds.), pp. 71–89. Marcel Dekker, New York.

Reisfeld, R. A., and Cheresh, D. A. (1987). Human tumor antigens. In "Advances in Immunology," Vol. 40 (F. J. Dixon, ed.), pp. 323–378. Academic Press, Orlando, Florida.

Shulman, S. (1974). "Tissue Specificity and Autoimmunity." Springer-Verlag, New York.

Viruses, Emerging

Stephen S. Morse
The Rockefeller University

Glossary

Arbovirus Arthropod-borne virus; replicates in arthropods (e.g., mosquitoes, biting flies, ticks) and is transmitted by bite to a vertebrate host; arbovirus is an ecological rather than a taxonomic definition: various arboviruses belong to a variety of virus families, including the Flaviridiae, the Bunyaviridae, and the genus *Alphavirus* of the family Togaviridae

Emerging viruses Viruses that have recently increased their incidence and appear likely to continue increasing

Endemic Occurring naturally and constantly in a particular area (as opposed to epidemic)

Epidemic Appearance of a disease in a population at a prevalence greater than expected

Hemorrhagic fever Infection manifested by acute onset of fever and hemorrhagic signs (blood vessel damage as indicated by petechiae on skin, internal bleeding, and, in severe cases, shock); typical of many zoonotic viruses

Incidence Number of people (in a given population) developing a specified disease, or becoming infected, at a specified period of time

Pandemic Epidemic so widespread that it covers virtually the entire world (Gk. *pan*, all)

Prevalence Term in epidemiology, defined as frequency of a disease in a population; an epidemiologic measurement similar to but not technically synonymous with incidence

Vector Agent, usually animate, that serves to transmit an infection (e.g., mosquitoes)

Zoonosis Infection or infectious disease of vertebrate animals that is transmissible under natural conditions to humans

EMERGING VIRUSES are viruses that have recently increased their incidence and appear likely to continue increasing. In less formal terms, they are viruses that have newly appeared in the human population or are rapidly expanding their range, with a corresponding increase in cases of disease. In recent years, a number of viral diseases have been identified for the first time. Some, such as the acquired immune deficiency syndrome, or AIDS, have made their debut alarmingly and dramatically. Other viruses, such as influenza, have long been known for their tendency to reappear periodically to cause major epidemics or pandemics. The reasons for these sudden manifestations of new viral diseases have in general been poorly understood, making it difficult to determine whether or not anything can be done to anticipate and prevent disease emergence. One conclusion to be discussed is that, in many cases, the causes of viral emergence, while complex, may be less random than they seem.

Because of the large number of viruses, this article is selective, focusing on common features shared by emerging viruses; no attempt has been made to include all possible candidates.

I. Defining Emerging Viruses: "New" versus Newly Recognized

A. Categories of Emerging Viruses

The most noticeable category of emerging viruses represents those that seem genuinely "new" in some important respect such as a first appearance in an epidemic of dramatic disease, a sudden increase in distribution, or novel mechanisms of pathogenesis. Many of the emerging viruses that immediately come to mind, such as HIV, fall into this group. Most of this article will consider viruses in this category and what is known about the causes of their emergence.

A second group of potentially emergent viruses consists of viruses that, while not new in the human population, are newly recognized. A recent example is human herpesvirus 6 (HHV-6). Although identified only in the mid-1980s, HHV-6 appears to be almost universal in distribution and has recently been implicated as the cause of roseola (exanthem subitum), a very common childhood disease. Since roseola has been known since at least 1910, HHV-6 is likely to have been widespread for at least decades, if not longer. Suggested roles of HHV-6 in chronic fatigue syndrome or as a cofactor in AIDS are still under scrutiny. Some other viruses discussed later, such as parvovirus B19, are also likely to be in this category.

B. The Newly Recognized

The significance of the newly recognized but common viruses is still debatable. Some have been implicated in various chronic diseases, although much of the evidence is inconclusive. However, as chronic diseases have become increasingly important in industrialized societies, the impact of agents responsible for chronic diseases could be considerable. Otherwise, epidemiologically, newly recognized but common viruses generally do not represent an apparent threat because they are already widespread and may have reached an equilibrium in the population. Recognition of the agent can even be advantageous, offering new promise of more accurate diagnosis and possibly control. Conceivably, with any type of agent, a change in the agent or (more frequently) in host condition might result in a new or more serious disease. This is evident with opportunistic pathogens, agents that generally have limited ability to cause human infections and disease but may do so under suitable circumstances. Immunosuppressed or immunocompromised individuals, such as people with AIDS, are particularly prone to such infections. With many pathogens, expresion of disease may also be altered by such host factors as nutritional or immune status or age at first infection.

C. The Significance of New Technologies in Identifying "New" Viruses

The importance of technological advances in identifying new viruses should be mentioned. Viruses such as HHV-6 and hepatitis C (see later) became apparent because the means were developed to demonstrate their existence. The recognition of HIV depended on the previous development of methods for growing T-lymphocytes in culture, including key methods that were developed in Robert Gallo's own laboratory. In a more general sense, the introduction of tissue culture methods, in the 1940s, was a major breakthrough in the study and characterization of viruses. It can be expected that new tools for detection will uncover new viruses. In particular, many new avenues are opened by the recent development of an exceptionally sensitive technique, the polymerase chain reaction (PCR), which is capable of detecting 1 HIV infected cell in 100,000. Because PCR can detect and amplify DNA in minuscule amounts of sample and is comparatively undemanding, it is rapidly finding favor in many applications. PCR has great potential for disease archeology and the study of evolution. By PCR, many otherwise intractable samples can now be tested, even mummified human bodies 7000 yr old. [See POLYMERASE CHAIN REACTION (PCR).]

II. Examples of Potentially Emerging Viruses

Predicting the greatest threats is a more difficult task, made even more difficult by significant gaps in our knowledge. Although specifics differ, a number of viruses seem likely to merit inclusion on a world list of emerging viruses (Table I). It must be cautioned that nobody can foretell the future. Unanticipated effects of changes in environmental or other conditions can bring an as-yet undescribed or presently obscure zoonotic virus to world prominence, as occurred with HIV a few years ago, or Lassa fever. Conversely, a prominent virus may become submerged again by environmental changes, or even be driven to extinction, as was smallpox. Therefore consideration of emerging viruses should emphasize the principles of viral emergence, and listings of specific viruses are offered here only as examples.

Table I Some Examples of "Emerging" Viruses

Virus	Signs/symptoms	Distribution	Natural host	Transmission
		Orthomyxoviridae (RNA, 8 segments)		
Influenza	Respiratory	Worldwide (?from China)	Fowl, pigs	Respiratory
		Bunyaviridae (RNA, 3 segments)		
Hantaan, Seoul	Hemorrhagic fever + renal syndrome	Asia, Europe, United States	Rodent (e.g., *Apodemus*)	Contact with infected secretions
Rift Valley fever[a]	Fever, ± hemorrhage	Africa	Mosquito, ungulates	Vector: *Aedes* mosquitoes
		Flaviviridae (RNA)		
Yellow fever[a]	Fever, jaundice	Africa, South America	Mosquito, monkey	Vector: *Aedes aegypti* (urban); other *Aedes* species (sylvan)
Dengue[a]	Fever, ±hemorrhage	Asia, Africa, Caribbean	Mosquito, human/monkey	Vector: *A. aegypti* (Asia: also *Aedes albopictus*)
		Arenaviridae (RNA, 2 segments)		
Junin (Argen. HF)	Fever, hemorrhage	South America	Rodent (*Calomys musculinus*)	Contact with infected secretions
Machupo (Boliv.)	Fever, hemorrhage	South America	Rodent (*C. callosus*)	Contact with infected secretions
Lassa fever	Fever, hemorrhage	West Africa	Rodent (*Mastomys natalensis*)	Contact with infected secretions
		Filoviridae (RNA)		
Marburg, Ebola	Fever, hemorrhage	Africa	Unknown	Contact; nosocomial through contaminated needles
		Retroviridae (RNA + reverse transcriptase)		
HIV	AIDS, etc.	Worldwide	?Primate	Blood transfusion; nosocomial through contaminated needles; sexual transmission

[a] Transmitted by arthropod vector.
Underlined bold indicates viruses of special concern for near future.
[Modified from Morse, S. S., and Schluederberg, A. (1990). Emerging viruses: The evolution of viruses and viral diseases. *J. Infect. Dis.* **162**, 1–7. © 1990 by the University of Chicago Press.]

A number of viruses are now prominent as emerging viruses or have been in the recent past. Among these are influenza; several members of the Bunyaviridae, including the hantaviruses (Hantaan, Seoul, and related viruses classified in the genus *Hantavirus*) and Rift Valley Fever (RVF; in a separate genus, *Phlebovirus*, of the Bunyaviridae); yellow fever and dengue (in the Flaviviridae); possibly the arenavirus hemorrhagic fevers, including Junin (Argentine hemorrhagic fever) and Lassa fever; probably the Filoviridae (Marburg and Ebola); human retroviruses, especially HIV but also human T-lymphotropic virus (HTLV); and various arthropod-borne encephalitides, including St. Louis encephalitis and Japanese encephalitis (both in the Flaviviridae), and Venezuelan equine encephalomy- elitis, Eastern equine encephalomyelitis, and (in Australia) Ross River (all three in the *Alphavirus* genus of the Togaviridae). The latter causes a den- guelike disease with arthritis as a frequent compli- cation. These encephalitides are mosquito-borne but have natural (nonhuman) vertebrate hosts as well.

In the United States, some viruses represent a greater potential concern than others. Of the viruses listed, and excluding viruses already established in the United States, influenza, HIV-2, dengue, and possibly some hantaviruses (Seoul) are of greatest immediate potential importance to North America. Proximity makes dengue, which is spreading over the Caribbean basin, a special concern. Some native mosquito-borne encephalitides, such as LaCrosse and California encephalitis (both in the California group of the Bunyaviridae), St. Louis encephalitis, and Eastern equine encephalomyelitis, are generally sporadic but can erupt from time to time. Venezuelan equine encephalomyelitis has been almost entirely eliminated from the United States, but reintroduction is possible.

In considering emerging viruses, it is essential to take a global view. A virus emerging anywhere in the world could, under favorable conditions, reach any other part of the globe within days. Accelerating environmental change and widening immigration also expose new populations to microbes that were once buried in the depths of rainforests or confined to remote villages. For example, in 1989, a man in an Illinois hospital died of Lassa fever. The virus is normally endemic to West Africa; the patient contracted the virus while visiting family in Nigeria but fell ill only after returning home a few days later. The ability of a virus to disseminate throughout the world in a short time is clearly demonstrated by new influenza pandemics, which blanket the globe within a few months of their inception (which is usually in China). Additionally, a number of diseases, such as Korean hemorrhagic fever (with an estimated 100,000 cases a year in China), may be serious causes of illness or death in certain regions of the world, even if they are not an imminent threat to the United States.

Following is a brief description of some of these viruses, arranged alphabetically.

A. Arenaviruses

Members of the Arenaviridae cause hemorrhagic fevers in humans; their natural hosts are rodents. Both Old World and New World representatives are known. New World representatives include Junin (Argentine hemorrhagic fever), whose natural hosts include the rodent *Calomys musculinus,* and Machupo (Bolivian hemorrhagic fever), in the rodent *Calomys callosus.* A new arenavirus, provisionally named Guanarito, was described in 1991 during a dengue epidemic in Venezuela. The major Old World arenavirus, Lassa fever of West Africa, has as its natural host the rodent *Mastomys natalensis.* Lassa became infamous for its high mortality rate in Western medical missionaries, who first came in contact with the virus in the early 1970s. Infected rodent hosts usually shed virus asymptomatically in their urine, and primary infection in humans is gen-

erally by contact with infected secretions from the rodent. Secondary cases of Lassa fever in health care providers or family members can occur through contact with patients' blood or infected secretions.

B. Bovine Spongiform Encephalopathy

See Section II.L, Spongiform Encephalopathies.

C. Dengue and Yellow Fever

See Section II.E, Flaviviruses.

D. Filoviruses (Ebola and Marburg)

The filoviruses (Ebola and Marburg) are among the least understood of all viruses. Their natural hosts are unknown; some believe they originated in primates. Human disease is typically fever with hemorrhage, and mortality can be high. Ebola has caused at least two epidemics in Africa. An epidemic in Zaire in 1976 involved 278 known cases, almost all hospital-acquired through contaminated hypodermic equipment or contact with patients; mortality was about 90%. In the same year, a separate epidemic in the Sudan involved almost 300, with over 50% mortality. Filoviruses in imported Old World monkeys have also been a matter of recent concern. Marburg virus was first identified in 1967 when 25 laboratory technicians in Marburg, Germany, and in Belgrad, Yugoslavia (Serbia), became sick after handling tissues from African green monkeys; seven died. Six medical workers and family contacts subsequently became infected but recovered. Since then, there have been three additional documented primary cases of Marburg, acquired by travelers in Kenya or (one instance) Zimbabwe, all fatal; three individuals who became infected by contact with primary cases all survived. In 1989 and 1990, monkeys in a facility in Reston, Virginia, died suddenly. Although death was due to other causes, a filovirus was isolated as well. The filovirus was originally thought to be Ebola, but additional study determined that the virus originated in Asian macaques and not from Africa. The virus, now variously known as Reston filovirus or Reston strain of Ebola, appears less virulent than classic Ebola. Several animal handlers who apparently became infected did not develop acute disease.

E. Flaviviruses (Dengue and Yellow Fever)

Yellow fever, a mosquito-borne disease characterized by fever and jaundice, remains a significant world health problem. Historically, the impact of yellow fever virus was tremendous. Yellow fever was so devastating to workers in the region that the Panama Canal could be completed only after yellow fever was controlled; every schoolchild has heard the story of Walter Reed and the eradication of yellow fever in the Canal Zone. Historians have documented the high mortality caused by yellow fever in European settlers coming to Africa in the nineteenth century. The development of a yellow fever vaccine was one of the first successes of the Rockefeller Foundation; the same vaccine is still used today. Despite the availability of effective vaccines, yellow fever virus is still unvanquished and is widespread in Africa and South America. Its origins as a human pathogen stretch back several hundred years at least, but human infection is only incidental for yellow fever. The virus is mosquito-borne, but different mosquito species are involved in different settings. The natural cycle of infection is the sylvatic ("jungle") cycle in monkeys, in tropical areas of Africa and South America. In the sylvatic form, the virus is carried by local forest mosquitoes, which can also transmit infection among humans. These human cases occur by incidental infection of people in areas where sylvatic yellow fever is well established. The *Aedes aegypti* mosquito, a species well adapted to living within human habitations (the genus name is Latin for house) is generally the vector in "urban" yellow fever. It is generally believed that transport and movement of people in the slave trade disseminated both the yellow fever virus and the *A. aegypti* mosquito from Africa to other tropical areas. Although *A. aegypti* can be found in some portions of the southeastern United States, the last yellow fever epidemic in the United States was in New Orleans in 1905.

Dengue, another flavivirus borne by some of the same mosquito species as yellow fever, is now in tropical areas worldwide (Africa, Asia, the South Pacific, South America, and the Caribbean) and, if anything, is spreading. Dengue is widespread in the Caribbean basin. Cuba had over 300,000 cases in a 1981 epidemic, Venezuela had an epidemic in the winter of 1990, and Brazil in 1991. Travelers returning to the United States from the tropics occasionally come back with dengue: the federal Centers for Disease Control reported at least 27 confirmed cases of imported dengue (in 17 states) in the United States in 1988. A more severe form, known as dengue hemorrhagic fever, occurs in many areas where dengue is hyperendemic, and has been postulated to result from sequential infection with different dengue viruses that now overlap geographically in many tropical areas. The frequency of dengue hemorrhagic fever is increasing as several types of dengue virus extend their range.

F. Hantaviruses

In the family Bunyaviridae, members of the *Hantavirus* genus (Hantaan, Seoul, and related viruses such as Puumala and others in Europe) cause hemorrhagic fevers with renal syndrome (fever, bleeding, kidney damage, and shock). Various hantaviruses are found in Asia, Europe, and the United States as naturally occurring viruses of rodents. The most prominent member of this family is Hantaan virus, the cause of Korean hemorrhagic fever. Named for a river in Korea, the disease first came to Western attention during the Korean War. At least 3000 U.S. and United Nations troops developed Korean hemorrhagic fever, and over 300 died. The disease has long been known in Asia, and it has been suggested that there is a description of Korean hemorrhagic fever in a Chinese medical text dating to the tenth century. At present, some 100,000 cases of Hantaan are diagnosed annually in China alone, as compared with 471 in 1955 (reasons will be discussed in Section IV. A). The major natural host of Hantaan virus in Asia is the striped field mouse, *Apodemus agrarius*. This rodent is not native to the United States; consequently, Hantaan virus itself is not likely to become established here (of course, the evolution of a host-range variant capable of infecting native rodents cannot be excluded as a theoretical possibility). There are a number of North American hantaviruses with native rodent hosts. Although they have not been as extensively studied, they have not been clearly associated with human disease. Because any new variant either would probably have to compete with these existing viruses or would have to be geographically distinct or have a different host range, introductions of related viruses may be limited. This does not necessarily apply to the rodent hosts, and a new rodent-borne virus could be introduced into the United States if a suitable rodent host established itself here, or if the host were a cosmopolitan rodent such as the domestic rat. For example, another Hantavirus, Seoul virus, is found in rats. The virus was originally identified in rats in Korea and has since been identified in urban rats living in American cities. It has been suggested that rats carried on ships from Asia may have introduced Seoul virus into the United States. In Korea, Seoul virus has caused hemorrhagic fever with renal syndrome similar to Hantaan virus, but it is usually considerably milder. In the United States, acute disease has not been identified, although seropositive individuals have been found in some inner city areas. Some evidence, although inconclusive, suggests a possible association with chronic renal disease, including renal hypertension.

G. Hepatitis

Five viruses, all unrelated to each other, are now known as possible causes of human viral hepatitis. Besides the familiar hepatitis A and B viruses, there are the recently characterized transfusion non-A non-B hepatitis virus (hepatitis C virus); a recently identified water-borne virus from Asia (tentatively named hepatitis E); and delta agent, or hepatitis D.

Hepatitis C is common in the United States and may be responsible for 98% of current posttransfusion hepatitis, especially now that routine testing of blood has greatly reduced hepatitis B in transfusions. Its identification, a tour de force of molecular technology, led rapidly to development of serologic tests for the virus as well as to trials of therapeutic interferon. The recent identification of hepatitis C virus demonstrated the power of molecular biotechnology in isolating and characterizing the viral nucleic acids from infected chimpanzees. The agent is an RNA virus that one researcher calls flaviviruslike (resembling, but not closely related to, yellow fever and dengue viruses). Besides posttransfusion non-A non-B hepatitis, over half of 59 patients with "community acquired non-A non-B hepatitis" (no blood transfusion) in one U.S. county were seropositive, suggesting additional modes of transmission for hepatitis C. From studies in Japan, there is evidence for sexual transmission. Even more recently, hepatitis E has been identified. It is a water-borne disease widespread in Asia and South America and is caused by another RNA virus that has not been fully characterized but appears to belong to a different viral family from the other known hepatitis viruses.

Delta hepatitis, discovered in 1977 in Italy, causes an acute fulminant hepatitis in hepatitis B carriers. Delta is a defective agent, consisting of a small RNA and a distinctive protein known as delta antigen, wrapped in a covering of hepatitis B surface antigen. Delta therefore requires hepatitis B as a helper virus, in essence parasitizing the parasite. Luckily, delta is still not common in the United States, except in groups that are also at high risk for hepatitis B, but is endemic in Italy and parts of South America. Also fortunately, it is comparatively rare in Asia, where hepatitis B is very common. Because it "borrows" the hepatitis B surface antigen and requires co-infection with hepatitis B virus, increasing use of the vaccine for hepatitis B should further reduce the occurrence of disease from delta. In many parts of the world, where hepatitis B immunization may not be widely practiced, delta remains a great potential menace. Delta is unique among known animal viruses due to its small size and the fact that portions of the agent appear to be related to viroids, very small infectious RNA agents of plants. These ultimate parasites are tiny pieces of nucleic acid without a protein coat. Other small RNA agents like delta probably already exist in animals but have not yet been identified. [*See* HEPATITIS.]

H. Influenza

Influenza is one of our most familiar viruses. Annual or biennial epidemics of influenza A are due to antigenic drift, a mutational change in the hemagglutin (H) surface protein of the virus so that the host's immune system no longer recognizes the antigen, and the new virus can reinfect until host immune responses are mounted to the new variant antigen. Pandemics, very large epidemics that occur periodically and involve virtually the entire world, are the result of antigenic shift, a reassortment of viral genes usually involving the acquisition by a mammalian influenza virus of a new hemagglutinin gene from an avian influenza virus. There are some 13 subtypes of the hemagglutin gene, although only a few H subtypes have been associated with human infection. These pandemic strains have generally originated in China. Possible conditions that might bring this about are discussed in Section IV. B.

In June 1991, it was reported that an influenza virus with the H3 hemagglutinin, but distinct from presently circulating varieties, and believed to be of avian origin, had appeared in horses in northeastern China in 1989 and 1990. [*See* INFLUENZA.]

I. Parvovirus B19

Parvoviruses are the smallest DNA viruses of vertebrates (the name *parvo* is from the Latin word for small). The virus particle measures about 20–22 nm in diameter, and the viral DNA genome is about 5000 bases long. Parvoviruses typically replicate in rapidly dividing cells, such as cells lining the intestine or blood cell precursors in bone marrow. Canine parvovirus, discussed elsewhere in this chapter, is another member of this family. Another human parvovirus, adeno-associated virus, has not been associated with human disease. Parvovirus B19 was discovered fortuitously, in sera from healthy blood donors with false-positive reactions for hepatitis B antigen. After the virus was characterized, further studies indicated that approximately 60% of adults are seropositive. The virus is worldwide in distribution. Various evidence implicates B19 as the cause of erythema infectiosum, or fifth disease, a mild, self-limited febrile disease of childhood. B19 has also been associated with aplastic crises in chronic hemolytic anemias (sudden disappearance of blood cell percursors in bone marrow). B19 and other parvoviruses have also been suggested as causes of joint disease, but this is not clearly resolved as of this writing.

J. Retroviruses

Human retroviruses have become the subject of intense scientific interest, largely because of HIV. The HIV pandemic has become one of the defining conditions of the late twentieth Century. HIV-2 has not yet disseminated as far as HIV-1, but the potential exists. While HIV-2 appears to cause less severe disease than HIV-1, AIDS caused by HIV-2 has been documented.

Considerable work has also been done recently on HTLV types I and II. HTLV is widespread in some populations in Asia, the Pacific, and parts of the Caribbean, and HTLV-II has been reported from an isolated aboriginal group in South America, possibly suggesting relative antiquity as a human virus. HTLV-II is also spreading rapidly within certain populations in the United States, such as intravenous drug users. The spectrum of disease due to HTLV is still being defined. HTLV was originally identified in adult T-cell leukemia–lymphoma. In tropical areas, HTLV-I has been associated with a neurological disease, tropical spastic paraparesis, and a related condition, HTLV-I-asociated myelop-

athy, has been defined. It is not known whether HTLV-II is responsible for human disease, although some atypical T-cell leukemias have been advanced as possibilities. Because of serologic cross-reactivity, older studies did not always distinguish the two viruses, and it is possible that some disease attributed to HTLV-I might have been caused by HTLV-II.

More speculatively, various researchers have recently suggested that other neurological diseases or autoimmune diseases may be caused by human retroviruses, either HTLV or presently unknown types. There is only limited evidence in human disease; however, the recent demonstration in mice of viral superantigens, homologs of cellular genes that are carried by a mouse retrovirus and that can induce abnormalities in T-cell development and possibly autoimmunity, suggest that similar roles for human viruses are possible.

Similarly, speculation about endogenous retroviruses is possible, but there are no data to decide the question. Humans, like many other animals, contain endogenous retroviral elements in cellular DNA. Under certain circumstances, an endogenous retrovirus, silent for generations in the host DNA, can regain independent existence and become capable of infecting other hosts. Murine retroviruses have demonstrated this property in the past, and one scientist suggested some years ago that feline leukemia virus, a common and sometimes fatal infection of cats, probably originated in this way from an endogenous rodent retrovirus. Despite this potential, a similar event has never been identified in humans. [*See* RETROVIRUSES; EVOLUTION, VIRAL .]

K. Rift Valley Fever

River Valley fever (RVF) virus, a mosquito-borne virus found in Africa, was first recognized in 1931 as a livestock disease in European breed sheep and cattle introduced into Africa. The apparent recipe, as expressed by Karl M. Johnson, was "foreign animals, local virus, new disease." For unknown reasons, severity of human disease has increased since the virus was first identified. The virus is found naturally in Africa in a number of ungulates (including camels), and RVF caused epizootics in several parts of Africa, with occasional disease in occupationally exposed animal handlers, veterinarians, and butchers, but no human deaths. This changed in 1975 when South Africa experienced an epizootic with some associated human deaths. In 1977, RVF suddenly emerged in a dramatic zoonotic outbreak in Egypt that resulted in thousands of human cases and 598 reported human deaths. Rapid action (quarantine and immunization of livestock in Israel) prevented the virus from spreading over the Mediterranean basin. An outbreak in Mauritania in 1987 followed the Egyptian pattern, but on a smaller scale, with an estimate of 1264 human cases and 224 deaths. The human infection is characterized by a fever, usually with hemorrhaging; retinitis is frequently seen.

L. Spongiform Encephalopathies

Bovine spongiform encephalopathy (BSE), now a cause of great concern in the United Kingdom, is another recently emerged disease. BSE belongs to a family of diseases that also includes scrapie in sheep and whose human counterparts include Creutzfeldt-Jakob disease and kuru. Attempts to implicate conventional viruses or viroids have been unsuccessful, and one researcher has proposed that the spongiform encephalopathy agents are a novel form of infectious material, which he terms prions, infectious self-replicating proteins. Creutzfeldt–Jakob disease is a sporadic presenile dementia (occurring in people below the age expected for senility) with an estimated prevalence of approximately 1 case/1 million population. Familial forms, due to mutation in a gene coding for a specific protein, have recently been identified. Kuru was studied by a researcher who worked out the now well-known epidemiology of kuru, which was sustained within an aboriginal New Guinea population by ritual cannibalism.

At least in crude extracts, strains of the spongiform encephalopathy agents are among the most heat-stable infectious materials known. BSE appears to be an example of interspecies transfer, in this case possibly scrapie from sheep moving to a new host species. Incompletely rendered sheep by-products fed to cattle have been suggested as the vehicle, although the possibility cannot be excluded that BSE existed earlier but was unrecognized. It has been proposed that changes in rendering processes in the late 1970s and early 1980s, using lower temperatures and (perhaps more critically) lesser amounts of solvents, may have permitted some infectious material to survive the process. A similar interspecies transfer seems to have occurred before (in the 1940s), resulting in the disease now known as transmissible mink encephalopathy. The recent identification of spongiform encephalopathy in felines fed BSE-contaminated meat by-products

seems to have a similar history. While risk of BSE to humans is unknown, the relative rarity of Creutzfeldt–Jakob disease, the human equivalent, suggest that risk may be comparatively low. Creutzfeldt–Jakob disease is not noticeably more prevalent among sheep or cattle farmers, even though scrapie has been known in sheep for at least two centuries and there have presumably been many opportunities for human exposure comparable to the acquisition of BSE by cattle. [*See* TRANSMISSIBLE SPONGIFORM ENCEPHALOPATHIES.]

Interestingly, except for the spongiform encephalopathy agents and parvoviruses, most of these emerging viruses contain RNA genomes. While there is no compelling reason for this, one might speculate that this might be related to the diversity and mutability of RNA viruses. At least superficially, RNA viruses represent a great diversity of viruses and replication strategies. This diversity may be due at least in part to the high mutation rates shown by many RNA viruses. This in turn has been attributed to the error-prone nature of RNA replication and especially to the lack of a "proofreading" function in this process.

III. What Are the Origins of "New" Viruses?

A. Possible Sources of New Viruses

While it is not possible to attribute underlying causes or precipitating factors to all episodes of viral emergence, for many at least some causes can be identified. Of course, we cannot be certain that these are the only causes, and the known examples probably show some unintentional selection bias, because the most explainable are the most likely to be included. Nevertheless, it is noteworthy that many episodes are explainable.

An examination of emerging viruses might appropriately ask how a new pathogen might originate. If we assume the constraints of organic evolution, which essentially require that new organisms must descend from an existing ancestor (evolutionary constraints will be briefly considered later), there are fundamentally three sources (which are not necessarily mutually exclusive): (1) the evolution of a new viral variant (*de novo* evolution), (2) the introduction of an existing virus from another species, and (3) dissemination of an agent from a smaller human population in which the agent may have arisen or been introduced originally. The term ''viral traffic'' was recently coined to represent processes involving the access, introduction, or dissemination of viruses to their hosts, as distinct from *de novo* evolution.

B. Evolution of New Viral Variants

Considerable debate has centered around the relative importance of viral evolution versus transfer and dissemination of viruses to new host populations (viral traffic) in the emergence of new viral diseases. Evaluating the significance of *de novo* evolution is complicated by the difficulty of demonstrating that a new isolate is truly newly evolved and not merely a new introduction of an organism that has long existed in nature but was previously unrecognized. There appear to be a few, although relatively few, documented examples in nature; most resemble related viruses in pathogenesis. Antigenic drift in influenza is probably the best-known example of viral emergence due to the evolution of a new variant. Some additional examples of possibly or apparently newly evolved viruses are listed in Table II. The list is not exhaustive, although attempts have been made to make it as inclusive as possible. Most viruses on this list cause diseases typical of their viral families or similar to the parental virus. In both humans and horses, the recombinant Western equine encephalomyelitis, for example, causes a dis-

ease similar to Eastern equine encephalomyelitis, but somewhat milder (its apparent evolutionary advantage is that Western equine encephalomyelitis generally has different insect and bird hosts). A poliolike syndrome recently described in China and South America is not included, because the responsible virus has not yet been characterized, and whether this is newly evolved or the result of viral traffic is not clear.

In addition, there have been a few examples of variants with altered biological properties. One researcher offers the example of an avian retrovirus (reticuloendotheliosis virus strain T, REV-T, which apparently derived from the considerably less virulent REV-A strain) that became more virulent as a result of several accumulated genetic changes. Recent evidence demonstrates that human chronic hepatitis B infection was associated with a mutation in a viral gene for precore protein. Vaccine escape mutants of hepatitis B were also recently described in a few infected individuals. Despite many demonstrations of this phenomenon *in vitro*, this is one of the few documented vaccine escape mutants isolated from natural infection in the field.

C. Role of Viral Traffic

The preceding examples notwithstanding, critical examination of known examples of viral emergence indicates that the overwhelming majority of such

Table II Known or Suggested Newly Evolved Viruses

Virus	Virus family	Remarks	Disease
Rocio encephalitis (Brazil)	Flaviviridae	?Recombinant	H[a]
Western equine encephalomyelitis (United States)	*Alphavirus* genus (Togaviridae)	Recombinant	H
Influenza H5 mutant (chickens, Pennsylvania, 1983)	Myxoviridae	New variant	Severe respiratory infection in chickens
Influenza H7 (seals, United States, 1980)	Myxoviridae		H: Conjunctivitis
Enterovirus 70	Picornaviridae	?New strain	H: Conjunctivitis
REV-T (strain of avian reticuloendotheliosis virus)	Retroviridae	Avian	Fulminant lymphoma in fowl
Friend virus, spleen focus-forming strains	Retroviridae	Mouse	
Canine parvovirus 2	Parvoviridae	Dogs	Enteritis, cardiomyopathy (similar to parvoviruses infecting other species)

[a] H: associated with human disease.
[From Morse, S. S. (1992). Evolving views of viral evolution. *History Phil. Life Sci.* (in press).]

instances can be accounted for by viral traffic. At least over the admittedly limited time span of human history, most emerging pathogens have probably not been newly evolved. Rather, they are existing agents conquering new territory. The overwhelming majority probably already exists in nature and simply gains access to new host populations. The most novel of these emerging pathogens are zoonotic (naturally occurring agents of other animal species); rodents are among the particularly important natural reservoirs. Many instances of emergence can be attributed to precipitating factors that facilitate the introduction of viruses from the environment into human hosts or aid their dissemination or expansion from a smaller human population. The following section will list some of these causes.

IV. Causes Precipitating Viral Emergence

A. Causes of Viral Emergence: The Role of Viral Traffic

The preceding analysis suggests that viral emergence can be viewed as a two-step process, involving the introduction of a virus into a human population followed by dissemination. Emphasis should therefore be placed on understanding the conditions that affect each of these steps.

Many of the known examples of viral emergence share common features. They are usually precipitated by environmental or social changes, often induced by human activities (Table III). The significance of the zoonotic pool is most apparent for viruses, which generally require a host in order to be maintained in nature. Considering that the total number and variety of viruses in animal species is probably very large, this offers a large pool of potential new virus introductions. In such cases, introduction of viruses into the human population is often the result of human activities, such as agriculture, that cause changes in natural environments. Often, these changes place humans in contact with previously inaccessible agents or increase the density of a natural host or vector, thereby increasing the chances of human infection. Examples among the viruses reviewed here include Lassa fever and Argentine hemorrhagic fever, both natural infections of rodents, and probably Marburg, Ebola, and RVF.

For example, Argentine hemorrhagic fever has spread as the pampas have been cleared for maize

planting. The natural host of this virus, the mouse *C. musculinus*, flourishes in this environment, propagating its virus in the process. Numbers of cases of Argentine hemorrhagic fever have increased proportionately. Hantaan, an unrelated virus, is acquired by humans in a similar way: Increased rice planting encouraged the little field mouse *A. agrarius*, the natural host of Hantaan virus. Infected mice shed virus in secretions such as urine. Humans normally become infected during the rice harvest, by contact with infected secretions in the rice fields.

There are numerous additional examples of this situation, in which the emergence of the new virus or pathogen is associated with changing environmental conditions that favor contact of humans with a natural host for an existing virus. Although Lyme disease is bacterial rather than viral, similar environmental conditions are probably responsible for its recent emergence. Another example is monkeypox. The name seems misleading, as various arboreal mammals in the rainforest, mostly squirrels and probably not monkeys, appear to be the natural reservoir hosts. Human monkeypox exposures appear to originate by contact with infected arboreal rodents, as a result of hunting the animals for meat or of exposure while foraging. Incidentally, monkeypox is not an emerging virus, although there had been earlier speculations that it might replace smallpox after the human virus was eradicated. There is no indication that this is occurring, and deforestation in Africa is reducing human exposure to the virus.

Agriculture provides some other unexpected examples. Viroids, infectious agents that consist entirely of small RNA without a protein coat, are spread, as far as we know, entirely by mechanical transmission on agricultural implements such as pruning knives and harvesters. It is speculation, but the evolution of viroids could very likely have been shaped, unbeknownst to its human agents, by these human activities.

B. Pandemic Influenza

More remarkably, the same principles also seem to apply under certain circumstances to viruses in which there is an essential role for viral evolution in the success of new viral variants. Influenza is perhaps the most interesting example. Influenza has been called the oldest emerging virus that is still emerging; influenza A virus is one of the few known examples (aside from some arguable cases, such as

Table III Probable Factors in the Emergence of Some Emerging Viruses

Virus family Virus	Probable factors in emergence
Arenaviridae	
Junin (Argentine HF[a])	Changes in agriculture (maize, changed conditions favoring *Calomys musculinus*, rodent host for virus)
Lassa fever	Human settlement, favoring *Mastomys natalensis*
Bunyaviridae	
Hantaan	Agriculture (Contact with mouse *Apodemus agrarius* during rice harvest)
Seoul	?Increasing population density of urban rats in contact with humans
Rift Valley fever	Dams, irrigation
Oropouche	Agriculture (Cacao hulls encourage breeding of *Culicoides* vector)
Filoviridae	
Marburg, Ebola	Unknown; in Europe and the United States, importation of monkeys
Flaviviridae	
Dengue	Increasing population density in cities and other factors causing increased open water storage, favoring increased population of mosquito vectors
Orthomyxoviridae	
Influenza (pandemic)	?Integrated pig–duck agriculture
Retroviridae	
HIV	Medical technology (transfusion, contaminated hypodermic needles); sexual transmission; other social factors
HTLV	Medical technology (transfusion, contaminated hypodermic needles); sexual transmission; other social factors

[a] HF: hemorrhagic fever.

[Slightly modified from Morse, S. S. (1992). What do we know about the origins of emerging viruses? In "Emerging Viruses" (S. S. Morse, ed.). Oxford University Press, New York.]

HIV, it may be the only example) of an emerging virus whose emergence (actually, for influenza, periodic reemergence) can clearly be ascribed to viral evolution. Although most changes in influenza virus H proteins occur by so-called antigenic drift involving the accumulation of random mutations (this drift can lead to the smaller, but still medically important, influenza epidemics seen every few years), new pandemic influenza viruses arise by a different route, that of major antigenic shifts. Every 20 years or so, influenza A undergoes a major antigenic shift in one key protein, known as the hemagglutinin (H) protein, and a pandemic results. Shifts invariably involve a reassortment of viral genes carried by different influenza strains. Thus, the important event in generating new pandemic influenza strains has, oddly, not been mutational evolution but, rather, reshuffling of existing genes. Where do the genes come from? It has recently been found that most influenza genes are maintained in wild fowl; every known subtype of the H protein can be found in waterfowl, such as ducks. A number of virologists believe that pigs are an important "mixing vessel" allowing influenza virus to make a transition from

birds to humans. Every major influenza epidemic known has originated in south China, which has also long practiced a traditional and unique form of integrated pig–duck farming. Two researchers have suggested that this form of agriculture may facilitate the development of new influenza reassortants by placing ducks, the reservoir of new influenza strains, and pigs, "mixing vessels" for mammalian influenza strains, in close proximity. Agriculture may therefore play the leading role in emergence of this virus as well. Here, too, viral traffic, reassortant viruses from the mixing of animal influenza strains and the transmission of the resulting virus to humans, appears more important than viral evolution for human disease.

C. Arboviruses

Water is an essential factor with mosquito-borne viruses, which include many important diseases worldwide, because many of the insect vectors breed in water. Japanese encephalitis accounts for almost 30,000 human cases annually in Asia, with about 7000 deaths (although immunization programs in several countries promise to control the disease in humans). Incidence of the virus is closely associated with flooding of fields for rice growing. In the outbreaks of RVF in Mauritania, the human cases occurred in villages near dams on the Senegal river.

Rapid urbanization has been blamed for the high prevalence of dengue in Asia. Profusion of water storage containers in cities, necessary to supply the dense and rapidly expanding human population, has caused a mosquito population boom, with a concomitant increase in the transmission of dengue.

Many types of human activities may also disseminate mosquito vectors or reservoir hosts for viruses. As already mentioned, both yellow fever virus and its principal vector, the *A. aegypti* mosquito, are believed to have been spread from Africa via the slave trade. The mosquitoes were apparently stowaways in the large open water containers that were kept on the ships to provide water during the voyage. In a more modern repetition, an aggressive vector of dengue virus, *Aedes albopictus* (the Asiatic tiger mosquito), was recently introduced into the United States in shipments of used tires imported from Asia. From its entry in Houston, Texas, the mosquito has established itself in at least 17 states. In this century, a number of other mosquito species have been introduced to new areas in war material being returned from foreign theaters of action.

The recent rapid spread of raccoon rabies in the United States has a similar cause. In the last few years, rabies in raccoons has moved from the southeast, where it has been localized for some time, to the northeast, and raccoons now seem well on their way to becoming a major wildlife source of rabies in the northeastern United States. After identifying its first rabid raccoon only in October 1989, New Jersey reported 37 rabid raccoons for the first third of 1990 alone. The Centers for Disease Control has implicated sport hunting as the main factor in this explosive spread of raccoon rabies. To ensure an adequate supply of raccoons for hunting, a group of hunters imported Florida raccoons, some of which were apparently rabid, to the area of western Virginia, from whence it spread further north. [*See* RABIES.]

D. Gateways for Viral Traffic: The Expansion of Human Viruses

Highways and human migration to cities, especially in tropical areas, can introduce remote viruses to a larger population. On a global scale, similar op-

portunities are offered by rapid air travel. As another example, HIV probably traveled along the Mombasa–Kinshasa highway and came to the United States presumably through travel. Health officials have also linked the movement of young men from villages to the cities, and resulting freedom from local restraints on behavior, with dissemination of HIV in Africa.

Once introduced into a human host, the success of a newly introduced pathogen then depends on its ability to spread within the human population after introduction. A similar situation would apply to agents already present in a limited or isolated human population, because the agents best adapted to human transmission are likely to be those that already infect people. Here, too, human intervention is providing increasing opportunities for dissemination of previously localized viruses. The example of HIV demonstrates that human activities can be especially important in disseminating newly introduced pathogens that are not yet well adapted to the human host and do not spread efficiently from person to person.

Finally, as a highway for viral traffic, the possibilities for iatrogenic disease should also be mentioned. Cases of Lassa fever, Ebola, and Crimean–Congo hemorrhagic fever, in addition to more familiar viruses, have been acquired by health care workers caring for infected individuals, or spread in hospitals. As demonstrated by well-known examples such as HIV and hepatitis B, many viruses that might not otherwise transmit easily from person to person may be transmitted through transfusions, organ transplants, or contaminated hypodermic needles, allowing the donor's viruses direct access to new hosts. For many viruses that could not spread efficiently from person to person, including HIV, this circumvents their lack of effective means of transmission. As these lifesaving procedures become more widely used, and as the scarcity of donors forces medical centers to look farther afield, it is reasonable to expect more instances.

E. Animal Viruses as Models of Interspecies Transfer

To supplement our knowledge of interspecies transfer of viruses, examples of emerging viruses in animals might also be considered as useful models for interspecies transfer and mechanisms of viral emergence.

Two of the currently best-studied examples are canine parvovirus and seal plague. Canine parvovirus [officially, canine parvovirus type 2 (CPV-2)] first emerged around 1978 as an epidemic disease of dogs. By 1980, the virus spread globally in the dog population, and the virus is now endemic in every country that has been tested. Slight variants, designated as CPV-2a and CPV-2b, appear to have subsequently displaced the original CPV-2 in the dog population. The virus appears closely related to two other parvoviruses, mink enteritis virus and feline panleukopenia virus. The latter is the cause of feline distemper. Typical signs are enteritis and cardiomyopathy, often with leukopenia, in animals infected with these viruses. CPV-2 may have descended from the feline parvovirus; ability to infect the new species may have been conferred by a mutation in the capsid (coat protein) gene. Host range for parvoviruses is at least partly determined by the capsid.

Seal plague, a newly recognized paramyxovirus, is related to measles and canine distemper viruses but is distinct. The virus may have been transferred directly from another species of seal; it has also been speculated that an outbreak of canine distemper might have been related, although this is not clear and is less probable.

In addition to their possible utility as model systems, emerging viruses of other vertebrates are worth further consideration because serious diseases of livestock or of food plants can have great economic impact and, in the worst case, can cause widespread starvation. Occasionally, some might be potential zoonotic viruses, as was RVF.

V. What Restrains Viral Emergence?

Understanding the factors restraining emergence is of great importance, but data are limited. This section, summarizing present knowledge, must necessarily be somewhat speculative in nature.

Although many viruses have high mutation rates and thus potentially may be evolving rapidly, few have shown striking changes in pathogenesis. To survive, viruses must be maintained in nature in a living host. Constraints are imposed by the requirements for a means of transmission and by the relatively few routes by which a virus can infect a host. Such requirements must impose strong selective pressures on a virus. Therefore, while variants are continually being generated, a stabilizing influence presumably is exerted by natural selection as the virus replicates in its natural hosts. Even if viral mutation is unpredictable at present, genotypic variation and phenotypic change are not equivalent, and

evolutionary constraints remain at the phenotypic level at least. This may be the reason that ecological and demographic factors appear to be at least as important in influencing viral emergence as viral mutation or evolution.

Unfortunately, when changes have occurred, they have generally not been predictable. Occasionally, an apparent recombinant virus will emerge, such as Rocio encephalitis or Western equine encephalomyelitis. Why most of them are unsuccessful is not known.

Besides emerging, viruses can also disappear or be displaced by new variants, for reasons that are often poorly understood. Influenza A H7N7, once frequent in horses, has almost disappeared, having apparently been displaced by the H3 subtype. A similar fate may befall H2N2 influenza in humans. The original strain of CPV-2 is being displaced by a variant. However, although smallpox was eradicated, it is not being replaced by monkeypox, despite the latter's ability to cause sporadic human cases. Several recent viral emergences, such as Rocio encephalitis, Lassa fever, and Marburg disease, have fortunately remained limited or have disappeared.

As one scientist has pointed out, many zoonotic introductions fail to become self-sustaining in the human population. The rather restricted range of disease, often severe acute hemorrhagic fevers, and the often limited ability of the virus to spread from person to person, can be offered as evidence that most of these viruses are not well adapted to human infection. Viruses already adapted to humans—a number of which may have been zoonotic introductions in the past that did evolve successfully—and now present in an isolated human population are more likely than newly introduced zoonotic viruses to disseminate if conditions favor their transmission.

Mathematical ecologists have suggested from mathematical analysis that self-sustaining infection involves a trade-off between transmissibility and virulence. As in the case of smallpox, a highly virulent virus can sustain itself if it is also easily transmitted. Many zoonotic introductions cause severe primary disease and, hence, are highly virulent, but they apparently have low transmissibility and, thus, have not yet become established in the human population. This is not an evolutionary imperative for them, because they survive in nature in their natural hosts, to which they are better adapted. Nevertheless, the impact of zoonotic introductions can be severe, as shown by the Ebola outbreaks in Africa. In addition, as already discussed, some can now be transmitted inadvertently by artificial means, such as blood transfusions or contaminated needles, circumventing the requirement to evolve greater transmissibility.

Therefore, understanding restrictions on viral variation and emergence would appear to be of prime importance. Some of this will hinge on improved understanding of viral evolution especially in the ecological context. The relatively low rate of emergence of viral disease may be due in part to the limited entry points whereby viruses gain access to new hosts.

In addition to the considerations already mentioned, successful viruses also need to evade or subvert host defenses. Some disease manifestations may also be the result of or exacerbated by host responses. The macrophage is likely to be one important cell in this process. Other factors restraining the emergence of new viruses, such as inability to spread from person to person, are still not understood at the molecular level. Moreover, an improved understanding of receptors, tissue-specific transcription factors, and tissue-specific mechanisms of cell killing will be essential for defining target cell specificity and host selective pressures restraining viral variation. [*See* EVOLUTION, VIRAL.]

VI. Prospects for Prediction, Control, and Eradication

A. Strategies for Anticipating Viral Emergence

While many of the questions about emerging viruses may have scientific foundations, allocating resources and setting priorities is often more affected by social, economic, and political factors. This is nowhere more apparent than in considering strategies for anticipating and controlling emerging infections. From the discussion in previous sections, it will be apparent that environmental and social factors can be major determinants of viral emergence. In principle, this means that emergence can be better anticipated and potentially controlled. In practice, this requires making political decisions such as defining appropriate and workable precautions for development programs, balancing programs so that both development needs and health protection requirements can be met, and mobilizing resources to accomplish these objectives on a worldwide basis. However, although molecular

technologies such as sensitive immunoassays and the PCR are providing powerful tools for identifying and tracking viruses, resources for global surveillance and control are presently inadequate. Research and clinical facilities are in dangerously short supply, and a critical dearth of trained researchers is expected by the next generation.

From historical experience, new viruses appear most likely to emerge from tropical areas undergoing agricultural and demographic changes and in the periphery of cities in these areas. Surveillance of such areas for the appearance of disease outbreaks or novel diseases would therefore seem advisable. In 1989, an international network of surveillance centers, located in tropical areas, especially on the edges of expanding tropical cities, was proposed. Each center would include clinical facilities, diagnostic and research laboratories, an epidemiological unit that could include disease investigation and local response capability, and a professional training unit, and would be linked to an international network for data analysis and instant response to emergencies.

If we are often the engineers of viral traffic, we need better traffic engineering. At the moment, analytic knowledge of viral traffic is not advanced enough to allow predictions and long-term advance planning based on its principles. However, it is conceivable that such knowledge could be made more systematic in the future, allowing better anticipation of viral and microbial traffic and better predictive ability.

B. Existing Methods for Control of Viral Diseases

Human viruses that have been substantially controlled in industrialized countries by immunization include polio, rubella, and measles. In some populations (primarily American and European travelers from areas without yellow fever, who are immunized before exposure if they are traveling to endemic areas), yellow fever is prevented by immunization. Potential for control by immunization exists for several other viruses, such as hepatitis B, mumps, and perhaps cytomegalovirus; vaccines of reasonable efficacy are available for these viruses. Rabies immunization in domestic animals has greatly reduced the occurrence of human rabies in the United States to one to three cases a year, although wildlife remains a source of potential exposure for both humans and domestic animals.

Public health measures have traditionally been directed to combating transmission or to protecting potential susceptibles through immunization. In addition to immunization, traditional control measures include improved sanitation, mosquito control programs, health certification of travelers, and health inspection of imported livestock. Traditional public health programs have been instrumental in containing many potential threats but also have several drawbacks. Their success with the targeted diseases depends on vigilance and assiduity. Efforts may fall victim to their own success, being prematurely relaxed or abandoned, usually to save money, and allowing the conditions that precipitated the program in the first place to reestablish themselves. Many mosquito control programs have met with this fate after initial partial success.

In many cases, relatively simple solutions are possible, if there is sufficient global resolve to implement them. For example, rebuilding water supply systems in tropical cities to reduce or eliminate open water storage could have a real impact on dengue. In other cases, there may be efficacious vaccines or other preventive measures, but problems in deployment allow old agents, placed under control by improved environmental conditions and public health measures, to regain a foothold or expand. Epidemics of yellow fever in some areas (Nigeria reported 600 deaths in a July 1991 outbreak) are one example. Adding yellow fever immunization to the World Health Organization (WHO) worldwide Expanded Program on Immunization, as has been suggested by some experts, might well prevent further epidemics like the one in Nigeria. Several U.S. cities have experienced epidemics of childhood measles in 1990 and 1991. A recent government report attributes much of the increase in measles cases to cutbacks in childhood vaccination programs, with the result that some children were inadequately immunized or immunized too late. Most programs also cannot contain viruses that can spread efficiently from person to person, such as influenza. Present strategy with influenza is to attempt to track emerging new strains and to immunize when feasible.

C. Prospects and Requirements for Eradication

Given that viruses that had been major scourges in the past have been controlled or (rarely) even eliminated by immunization or public health measures, it is logical to ask the prospects of eradication for some of the diseases discussed here. The greatest victory so far has been smallpox, now officially extinct. One of the most feared of all viral diseases, as well as reputedly one of the most easily transmissible, smallpox has had a long history. One historian suggested that the Spanish conquest of Mexico may have been aided by the effects of smallpox, which the Europeans brought with them. Jenner's work with vaccination makes smallpox one of the first examples of immunization; a form of immunization, using material from smallpox lesions, may also have been practiced earlier in China. In the twentieth century, universal childhood vaccination and health control at borders succeeded in bringing the virus under control in Western industrialized countries, except for occasional imported cases and their contacts. Smallpox was last seen in the United States in 1949. As control measures continued to reduce the geographic distribution of smallpox, eradication was adopted as a feasible goal by WHO. An intensive worldwide campaign of vaccination and surveillance in smallpox-endemic areas eventually succeeded in eliminating the disease. In 1979, with an official certification by WHO that was confirmed in May 1980 by the World Health Assembly, smallpox became the first virus to be declared extinct. [See SMALLPOX.]

This success with a previously feared virus encouraged public health agencies to consider additional candidates for eradication. For successful eradication, self-sustaining infection within the human population must be eliminated, and there must be no natural nonhuman reservoir of infection from which the virus could be reintroduced. In practice, eliminating self-sustaining infection in humans usually involves immunizing a large porportion of the susceptible population and requires effective vaccines. These characteristics—the ability to prevent infection through immunization and the lack of an additional reservoir of the virus—are what made smallpox vulnerable to eradication. On the other hand, yellow fever does not meet these criteria. There is a suitable vaccine, so transmission to people could theoretically be eliminated, but its maintenance in the sylvatic cycle would prevent eradication of yellow fever from the environment. Many of the other viruses discussed here, including by definition all of the zoonotic and arthropod-borne viruses, have natural reservoirs that would render eradication impracticable. Viruses that have no additional reservoirs but that have long periods of infection or transmission, such as human herpesviruses and HIV, would also be difficult to eradicate unless effective life-long immunity could be established in susceptibles. On the other hand, measles and polio meet the criteria and are likely targets for the future. They have no known reservoirs outside humans and effective vaccines are available. Polio eradication, a worldwide goal for the beginning of the twenty-first century, has fallen behind somewhat, but recent efforts in Latin Ameria have been promising. There is no timetable for measles eradication at this time. Thus, even when eradication is possible in principle, the goal may prove elusive or impracticable.

The successes have been encouraging, but the examples of viral emergence and reemergence or resurgence should caution us against complacency. Continuing improvement in nutrition and sanitation has been responsible for reducing the overall impact of infectious disease in industrialized countries. However, most of the world has still to benefit from these simple improvements, with predictable consequences. With respect to viral emergence, human intervention is providing increasing opportunities for dissemination of previously localized viruses, as in the case of HIV and dengue. Because human activities are a key factor in emergence, anticipating and limiting viral emergence is more feasible than previously believed but requires mobilizing effort and funds, especially on behalf of the third world. One speculation is that episodes of disease emergence may become more frequent as environmental and demographic change accelerate. The evidence suggests that both the scientific and the social challenges of viral emergence are likely to continue confronting us for the foreseeable future.

Acknowledgments

I thank Joshua Lederberg, Mirko Grmek, Howard Temin, John Holland, Frank Fenner, Walter Fitch, Gerald Myers, Edwin D. Kilbourne, Robert E. Shope, Thomas P. Monath, Karl M. Johnson, Peter Palese, Hugh Robertson, Baruch Blumberg, and other colleagues for invaluable comments and discussions. In modified form, some of the views and interpretations in portions of this chapter are patterned on those expressed in my earlier articles in other publications including "The Origins of 'New' Viral Diseases," *Environmental Carcinogenesis and Ecotoxicology Reviews* C9[2] (1991).

I am supported by the National Institutes of Health, U.S. Department of Health and Human Services (RR 03121 and RR 01180). Work on emerging viruses was additionally supported by the Division of Microbiology and Infectious Diseases (DMID), National Institute of Allergy and Infectious Diseases, National Institutes of Health (NIH), and the Fogarty International Center of NIH. I am especially grateful to Dr. John R. La Montagne, Director, DMID, and to Dr. Ann Schluederberg, Virology Branch Chief, for their support and encouragement.

Bibliography

Benenson, A. S. (ed). (1990). "Control of Communicable Diseases in Man," 15th ed. Amrican Public Health Association, Washington, D.C.
Fields, B. N., Knipe, D. M., *et al.* (eds.) (1991). "Fields Virology," 2nd ed. Raven, New York.
LeDuc, J. W. (1989). *Rev. Infect. Dis.* **11**(suppl. 4), S730–S735.
May, R. M., and Anderson, R. M. (1983). *Proc. R. Soc. (London)* *B* **219,** 281–313.
McNeill, W. (1976). "Plagues and Peoples." Doubleday, New York.
Miller, J. A. (1989). *BioScience* **39,** (September), 509–517.
Morse, S. S. (1991). *Perspect. Biol. Med.* **34,** 387–409.
Morse, S. S. (ed.) (1992). "Emerging Viruses." Oxford University Press, New York.
Morse, S. S., and Schluederberg, A. (1990). *J. Infect. Dis.* **162,** 1–7.
Wilson, M. E. (1991). "A World Guide to Infections. Diseases, Distribution, Diagnosis." Oxford University Press, New York.

Wastewater Treatment, Municipal

Ross E. McKinney
University of Kansas

I. Wastewater Characteristics
II. Microbes of Importance
III. Synthesis-Endogenous Metabolism
IV. Energy Reactions
V. Aerobic Treatment Systems
VI. Facultative Lagoons
VII. Anaerobic Treatment Systems
VIII. Pathogenic Organisms
IX. Future Biological Treatment Systems

Glossary

Activated sludge Aerobic-dispersed growth microbial bioreactor

Anaerobic digester Anaerobic dispersed growth microbial

BOD5 5-Day, 20°C biochemical oxygen demand

Denitrification Reduction of NO_3-N and/or NO_2-N to N_2

Endogenous metabolism Biological metabolism in the absence of organic substrate

Facultative lagoons Wastewater lagoons with aeration equipment that exhibits both aerobic and anaerobic metabolism

Nitrification Oxidation of NH_3-N to NO_2-N and/or NO_3-N

Trickling filters Attached-growth microbial bioreactor using stacks of rocks or plastic media

MUNICIPAL BIOLOGICAL WASTEWATER TREATMENT is of vital importance to the future of society. It is concerned with the removal of contaminants from the wastewaters generated and collected in all cities and communities from the small to the very large. Without proper treatment of municipal wastewaters, the rivers, lakes, and coastal waters would become nothing more than cesspools with destruction of aquatic life and transmission of enteric diseases. While the United States has demonstrated the value of municipal wastewater treatment in preserving our water resources for aquatic recreation and economic development, as well as in the reduction of enteric diseases, many areas of the world standout as examples of destruction that have been created by the lack of adequate municipal wastewater treatment systems. The key to modern municipal wastewater treatment systems lies in the application of the natural biological stabilization reactions in a controlled engineering environment. The self-purification reactions that have existed since the beginning of time have been increased a thousand times over to provide stabilization of the waste organics in a few hours instead of weeks and months. By creating the proper environment in concrete and steel tanks along the banks of the rivers and lakes, it is possible to use the natural bacteria and protozoa to provide for a high degree of treatment. The success of municipal wastewater treatment lies with the engineers who design and construct the treatment plants and with the operators who make certain that the treatment systems operate as they were designed to operate. Over the years, municipal wastewater treatment systems have been called upon to provide greater degrees of treatment, pushing the current systems to their maximum capabilities.

I. Wastewater Characteristics

The design of municipal wastewater treatment plants should be based on the influent wastewater and the desired effluent characteristics. The required treatment in any wastewater plant will be the influent characteristics minus the effluent characteristics. It is important to recognize that wastewater characteristics are measured as to both concentration (mg/liter) and quantity (lb/day or kg/day).

A. Oxygen Demand

Oxygen demand of municipal wastewaters has been one of the major reasons why municipal wastewater treatment is required for all municipalities in the United States. Excessive discharge of oxygen-demanding materials in municipal wastewaters reduced the oxygen resources in the streams, rivers, and lakes, killing fish and destroying the value of these water resources. The BOD5 test was developed as a reasonable measure of the biological oxygen demand (BOD). The fact that it required 5 days to determine the BOD5 data led to the development of a rapid chemical oxygen demand (COD) test. The COD data could be obtained in a few hours, but it measured both the biochemical oxygen demand and the nonbiological chemical oxygen demand. Studies on pure organic compounds indicated that the COD data were close to the theoretical oxygen demand with the BOD5 data about 0.58 times the theoretical oxygen demand. The 0.58 value was related to the growth rate of microbes in the BOD bottle. With normal microbial seed, the conversion factor was about 0.58, with a higher-than-normal microbial seed, the conversion factor was >0.58, and with insufficient microbial seed, the conversion factor was <0.58. For domestic municipal wastes, the microbial seed is close to normal. Industrial wastes can produce variations in the conversion factor unless an acclimated microbial seed is used. Thus, by using both BOD5 data and COD data, it is possible to estimate the biodegradable COD (BCOD) and the nonbiodegradable COD. Normal domestic municipal wastewaters will contain about 200 mg/liter BOD5, 345 mg/liter BCOD, and 450 mg/liter COD. The data indicates that about 105 mg/liter COD is nonbiodegradable.

B. Suspended Solids

Suspended solids are also important contaminants. Normal domestic municipal wastewaters contain 200 mg/liter total suspended solids (TSS) with 80% volatile suspended solids (VSS; 160 mg/liter). Suspended solids create turbidity in water and make it look "dirty." Suspended solids adversely affect small fish, which are essential in the biological chain. Approximately 65% of the VSS in raw municipal wastewaters are biodegradable and exert about 100 mg/liter BOD5. It is important to realize that some of the VSS are biodegradable and some are not. BOD5 is not a separate parameter from suspended solids and soluble organics. BOD5 is the oxygen utilized by the metabolism of the biodegradable suspended solids or the biodegradable soluble organics. Suspended solids are the mass of materials in suspension. The suspended solids that settle readily are termed settleable solids, whereas the tiny particles remaining in suspension are largely colloidal.

C. Nutrients and Trace Metals

Microbial metabolism requires nitrogen (N) and phosphorus (P) for cell mass synthesis. Cell mass starts with 11% N and 2.2% P. Microbes normally use ammonia nitrogen (NH_3-N) and phosphate phosphorus (PO_4-P). Without adequate N and P, the microbes cannot produce normal protoplasm or metabolize organics. In addition to N and P, the microbes need various trace metals such as iron, cobalt, and zinc. The trace metals are part of the metallic activators for enzymes.

D. Alkalinity and pH

The microbes need an environment with a pH between 6 and 9. The alkalinity keeps the pH between the normal levels. Alkalinity is generated by the carriage water and by proteins and urea in the wastewaters. The alkalinity in the carriage water is related to sodium and calcium bicarbonates. The protein and urea alkalinity is created as the proteins and urea are metabolized and ammonia nitrogen is released. The ammonia reacts with the carbon dioxide to produce ammonium bicarbonate alkalinity. The alkalinity keeps the pH from shifting very quickly when acids are produced by metabolism. Alkalinity is one of the simplest parameters to measure and yields more information than most realize.

E. Temperature

The rate of metabolism is a chemical reaction that is affected by temperature as well as by other parameters. The rate of metabolism changes by a factor of 2 with each 10°C temperature change between 5° and 40°C. The temperature between 5° and 40°C is the mesophilic temperature range. Increasing the temperature kills the mesophilic microbes. The death of the mesophilic microbes allows the thermophilic microbes to grow. At 40°C, the thermophilic bacteria begin to exert their ability to survive and predominate. The thermophilic microbes double their rate of metabolism until the temperature exceeds 65°C. The thermophiles may survive to even higher temperatures, but the rate of endogenous respiration as well as synthesis creates a real problem of survival.

II. Microbes of Importance

A. Bacteria

The primary microbes of importance in wastewater treatment systems are bacteria, which metabolize the organics in wastewaters with the production of new microbial cell mass. The bacteria that can metabolize the maximum amount of the different organics predominate. Bacteria are single-cell plants that must metabolize soluble organics. The soluble organics are taken into the bacteria and metabolized. Surface enzymes convert suspended organics to soluble organics. Because bacteria weigh approximately 10^{-12} g each, a very large population of bacteria is necessary to stabilize the organics in municipal wastewaters. While most bacteria in wastewater treatment systems utilize organics for their metabolism, there is an important group of bacteria that utilize inorganic compounds for their metabolism. As a net result, the two groups of bacteria do not compete with each other for their nutrients and both grow in the same environment. Normal municipal wastewaters contain between 10^5 and 10^7 bacteria/ml, depending upon the time of travel in the sewer collection system.

B. Fungi

The fungi are similar to the bacteria but are multicellular organisms rather than single-cell organisms. The fungi are larger than the bacteria and cannot compete with the bacteria for organics under normal environmental conditions. The fungi tend to be filamentous and present too much mass per surface area. Fungi are strict aerobes and cannot grow in the absence of oxygen. Municipal wastewaters contain fungi spores, primarily from the soil.

C. Algae

The algae are the third group of plant cells. Algae are true photosynthetic microbes, requiring light for energy while using inorganics for cell protoplasm. As a net result, the algae do not compete with the bacteria and the fungi for nutrients. Like fungi spores, the algae enter municipal wastewaters from the soil.

D. Protozoa and Rotifers

Protozoa are single-cell animals that live on bacteria and small algae, helping to remove the dispersed bacteria and algae from the system. Flagellated protozoa are not very efficient energy gatherers and cannot compete with the higher forms of protozoa. The free-swimming ciliated protozoa are the most efficient protozoa and metabolize tremendous quantities of bacteria. When the energy level of the system decreases, the free-swimming ciliated protozoa give way to stalked ciliated protozoa, which are attached to floc particles and can metabolize bacteria in the nearby vicinity with a lower expenditure of energy than the free-swimming ciliated protozoa. Suctoria are a special group of stalked protozoa that eat free-swimming ciliated protozoa rather than bacteria. It is important not to confuse suctoria with stalked ciliated protozoa. Rotifers are multicellular animals that can eat small particulates as well as bacteria and algae. The rotifers can attach themselves to floc particles and graze on the bacteria on the floc surface. Because of their large size, protozoa and rotifers are easily recognized under the microscope and are often used as indicators of the biochemical characteristics of wastewater treatment systems.

E. Crustaceans and Higher Animals

Crustaceans and higher animals are only found in wastewater treatment systems that are highly stable. The crustaceans and higher animals can consume bacteria, algae, and protozoa as well as small particles of organics. Because of their large size and complexity, the crustaceans and higher animals play a very limited role in wastewater treatment systems.

III. Synthesis-Endogenous Metabolism

A. Energy-Synthesis Metabolism

The basis of metabolism is the production of new cell mass. The bacteria process nutrients to obtain energy, which is used for the synthesis of cell mass. Nutrient metabolism is brought about by a series of enzyme reactions within the bacteria. Small, soluble molecules easily move through the cell wall and are metabolized first, whereas suspended organics must be hydrolyzed to small, soluble organics by enzymes on the cell surface. Rapid metabolism of suspended organics requires a very large bacteria population with good mixing to provide contact between the suspended solids and the bacteria.

Because of limited amounts of enzymes within the bacteria, the energy obtained from the metabolism of the nutrients is immediately transferred to the reactions creating synthesis of cell components. Some of the enzymes that are used to degrade the nutrients for energy can be used for synthesis as well. Examination of the energy-synthesis reactions indicates that the same amount of energy is required to produce a unit of cell mass regardless of the substrate being metabolized. The bacteria require about one-third of the energy in the nutrients while converting two-thirds of the energy to cell mass. The one-third–two-third relationship is a rather rough measure that results with mixed microbial populations in wastewater treatment systems.

B. Endogenous Respiration

The cell mass formed by the energy-synthesis reactions is not a stable product. In the absence of an external substrate, the bacteria utilize some of their cell mass for energy to remain alive. Under aerobic conditions, researchers found that the mixed microbial populations undergo endogenous respiration at the rate of 2% of the active mass per hour at 20°C. At one time, it was thought that if the rate of synthesis was balanced by the rate of endogenous respiration, no excess microbes would exist to be removed from the system. Studies on endogenous respiration indicated that there was a residual mass of dead cell material that remained as VSS, which could not be further degraded in a reasonable time period. The dead cell mass accumulated at the rate of 20% of the rate of endogenous metabolism. Thus, endogenous respiration could reduce the microbial mass 80%.

IV. Energy Reactions

A. Aerobic

The ability of the microbes to obtain energy from metabolism of nutrients is essential for biological wastewater treatment. Aerobic metabolism requires that oxygen be used as the ultimate electron acceptor and yields the maximum amount of energy to the microbes for metabolism. For this reason, aerobic metabolism yields the most cell mass production per unit of substrate metabolized. Unfortunately, the solubility of oxygen is limited in water, only 9.1 mg/liter at 20°C and 1 atmosphere air pressure. Aerobic metabolism produces carbon dioxide and water as end-products in addition to the cell mass. These two end-products are the most oxidized forms of carbon and hydrogen and can be discharged to the environment without creating a further oxygen demand. The bacteria can reduce the nutrients in solution to a very low level under aerobic conditions, although they cannot remove 100% of the nutrients. The rate of aerobic metabolism is normally controlled by the rate of oxygen transfer. Mixing is the most important factor affecting the rate of aerobic metabolism because it determines the rate of dispersion of nutrients, microbes, and dissolved oxygen (DO).

Ammonia nitrogen is an end-product of protein metabolism and is not the most stable form of nitrogen. Ammonia nitrogen can be metabolized by the *Nitroso*-bacteria to nitrite nitrogen and then by the *Nitro*-bacteria to nitrate nitrogen. Nitrate nitrogen is the most stable form of nitrogen. Hydrogen sulfide can also be metabolized by a special group of bacteria, the *Thiobacillus*, to sulfate, similar to the nitrifying bacteria. Both the nitrifying bacteria and the sulfide-oxidizing bacteria utilize carbon dioxide and water for nutrients to form cell components.

B. Anaerobic

In the absence of DO, the bacteria metabolize nutrients anaerobically. The ultimate electron acceptor is shifted from DO to other materials with a lower energy yield and a lower cell mass production per unit of nutrients metabolized.

1. Denitrification

Nitrates are the preferred electron acceptor for metabolism in the absence of DO because nitrates produce almost as much energy to the bacteria as DO.

Nitrates are reduced to nitrites, then to nitrous oxides, and finally to nitrogen gas. Because nitrogen gas is relatively insoluble in water, it is lost to the atmosphere as tiny bubbles. The process of nitrate reduction to nitrogen gas is known as denitrification. It should be noted that if the bacteria need nitrogen for cell protoplasm and no ammonia nitrogen is available, the bacteria will reduce nitrates to ammonia for cell synthesis. Nitrates will not be reduced to ammonia for the energy reaction. This difference is important.

2. Sulfate Reduction

One group of bacteria, the *Desulfo-*, can use sulfates as their electron acceptor for cell synthesis reactions. The sulfates can be reduced to thiosulfates, sulfites, sulfur oxides, free sulfur, or hydrogen sulfide. The available sulfates and substrate energy determine which bacteria predominate and how far the metabolic reaction proceeds. In excess nutrients and limited sulfates, the normal condition for domestic wastewaters, sulfate reduction proceeds to hydrogen sulfide. Hydrogen sulfide has greater solubility in water than oxygen, but it forms a gas that diffuses from the water because air does not contain hydrogen sulfide to keep it in solution. The sulfate-reducing bacteria do not obtain as much energy from their metabolism as aerobic bacteria and will not compete against them. The hydrogen sulfide produced by the bacteria is a toxic gas that is not stable when discharged back into the environment.

3. Organic Acid Production

In the absence of DO, nitrates, or sulfates, the bacteria often use the nutrients themselves as electron acceptors. In effect, the bacteria oxidize a portion of the nutrients to obtain energy while reducing a portion of the nutrients. The bacteria lose some of their energy in the ultimate electron transfer and have very little energy for cell synthesis. The net result is limited cell mass production and considerable amounts of reduced nutrients unmetabolized. From a practical point of view, the oxygen demand potential of the end-products is almost equal to that of the initial nutrients. The metabolism only changed the form of the nutrients. The primary metabolic end-products are organic acids such as lactic acid, acetic acid, propionic acid, and butyric acid. The organic acids can also be reduced to aldehydes and alcohols to a limited extent. One of the major problems is the need to neutralize the organic acids to keep the pH >6 or further metabolism would be reduced.

4. Methane Production

If not for the methane bacteria, the organic acids would accumulate in the anaerobic environment. The methane bacteria metabolize acetate to form methane and carbon dioxide. A second group of methane bacteria can reduce carbon dioxide to form methane and water. Hydrogen is transferred from the bacteria metabolizing the organic acids to acetic acid. The methane contains most of the energy from the original substrate and is relatively insoluble in water. Methane gas will be lost to the atmosphere, where it will exert an environmental impact unless it is collected and used as an energy source. Carbon dioxide will also exceed solubility and will be lost to the atmosphere, which normally only has about 0.04% carbon dioxide. The methane bacteria can reduce the organics to a relatively low level, producing a low organic effluent.

V. Aerobic Treatment Systems

Municipal wastewaters were initially treated by simple sedimentation to remove the settleable solids and the floating matter. Primary sedimentation normally reduced 60–70% suspended solids and 30–40% BOD5. The aerobic biotreatment systems were developed to provide further removal of the remaining suspended solids and BOD5.

A. Trickling Filters

Trickling filters are beds of rock media, 2–3 in. in size, stacked to a depth of 6–10 ft over a clay tile underdrain system (Fig. 1). A rotary distributor ap-

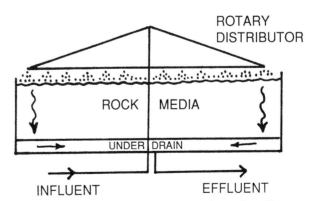

Figure 1 Schematic cross-section through a rock media trickling filter showing the rotary distributor applying the wastewaters over the rock and collection through the underdrain system at the bottom of the trickling filter.

plies the settled sewage over the top of the rock bed. The bacteria begin to grow where the rocks touch and provide a quiescent place for growth. Because of the rock surface, new bacteria growth move outward and along the rock surface. Air in the void space between the rocks supplies the oxygen for aerobic metabolism. Temperature differential through the rock bed causes a draft that permits air to move through the void space in the trickling filter to maintain the availability of oxygen. The underdrain channels are designed to permit air movement through the trickling filter.

The growth of bacteria stimulates protozoa and higher animals to grow on the slime surfaces. With time, microbial mass would fill the void spaces and clog the filter if hydraulic flow did not wash the excess microbes from the rock surfaces. At low organic and hydraulic application rates, the hydraulic flow is not adequate to strip the growth from the rocks. The microbial mass actually dies and is washed from the filter at periodic intervals, even at the low hydraulic loading rates. Examination of the microbial growth indicates that only the microbes at the surface between the liquid and the microbial mass are aerobic. The bacteria below the surface shift from aerobic to anaerobic metabolism. The net effect is that endogenous respiration under anaerobic conditions slowly destroys the microbial adhesive layer and allows the growth to drop off.

The rotary distributor sprinkles the wastes over the rocks as uniformly as possible along the diameter of the filter. The wastes produce a hydraulic wave that moves down across the surface of the microbial growths to the bottom of the filter and into the underdrain system for collection and removal from the filter. While it would appear that the liquid only takes 30 sec to move through the filter, the traveling wave actually mixes with the adhered liquid layer on the top of the microbial growth. The nutrients from the applied wastes are mixed with the surface water layer and diluted as the traveling wave moves down through the filter. End-products of surface metabolism are picked up and nutrients are left behind. The microbes on the surface quickly metabolize the organics before the next surge of liquid passes by. The net effect is that the liquid applied to the top of the filter remains in the filter for much longer than 30 sec. The nutrient concentration is greatest at the top of the filter and decreases with depth through the filter. Unfortunately, it is not possible to remove all the nutrients from the applied wastewaters. It would require a filter of infinite depth for complete metabolism.

At low organic loading rates, the nitrifying bacteria will be found in the bottom of the filter. As the organic loading rates are increased, the normal bacteria overgrow the nitrifying bacteria and prevent nitrification. Filamentous bacteria and fungi are found in trickling filters. The filaments help hold the normal bacteria on the rock media and break apart as the microbial growths drop off the rock media. Nematodes and worms live in the microbial layers on the rock. Larvae of the psychoda fly also find the microbial growths at the edge of the trickling filter a suitable environment for growth. Final sedimentation tanks are used after trickling filters to remove the settleable suspended solids. The treated effluent from trickling filters can meet the Environmental Protection Agency (EPA) 30 mg/liter TSS and 30 mg/liter BOD5 effluent criteria only at very low loading rates, requiring large areas of trickling filters. Because trickling filters have anaerobic metabolism, odors and nuisances are often produced, making trickling filters of limited use in the United States at the present time.

A shortage of good rock media resulted in the development of plastic media trickling filters after World War II. Plastic media has much more void space per unit of volume and is not easily clogged. Microbial growth on plastic sheets tends to be very thin, limiting the microbial populations and efficiency of metabolism. One advantage of the plastic media was the ability to create filters 14–21 ft tall.

B. Activated Sludge

Activated sludge is a dispersed microbial growth system that depends on the bacteria flocculating and separating by gravity sedimentation. Activated sludge is also an aerobic system utilizing diffused aeration, mechanical aeration, or a combination of both for oxygen transfer and mixing. Settled microbial floc is continuously recycled back to the aeration tank to provide an adequate microbial population for rapid stabilization of the nutrients. Excess microbial growth must be either wasted from the system or lost in the effluent as excess suspended solids are produced continuously in the activated sludge system. Examination of the bacteria in activated sludges indicates that the important bacteria are soil bacteria such as *Pseudomonas, Alcaligenes, Achromobacter, Flavobacterium,* and *Bacterium.* The flagellated protozoa include *Peranema, Bodo, Oikomonas,* and *Monas,* whereas the ciliated protozoa include *Lionotus, Paramecium, Colpidium, Euplotes, Aspidiscus,* and *Stylonychia,* to name a few of the major groups. The stalked ciliated protozoa include *Vorticella, Epistylis, Opercularia,* and *Carchesium.*

1. Conventional Systems

Conventional activated sludge systems are either plug flow systems with long, narrow aeration tanks or completely mixed systems using round or square aeration tanks. The activated sludge system includes sedimentation tanks after the aeration tanks to allow the activated sludge to separate from the treated waste waters. The settled activated sludge must be collected and returned to the aeration tanks with the excess sludge wasted from the system. The return activated sludge (RAS) is added to the incoming wastewaters at the head end of the aeration tank and removed at the opposite end of the aeration tank. The net result is a high nutrient concentration at the head end of the aeration tank with a high oxygen demand rate. Use of spiral flow aeration systems results in a greater oxygen demand at the head end of the aeration tank than that met with normal aeration but has sufficient oxygen at the discharge end of the aeration tank. The diffused aeration system has been modified by removing diffusers from the far end of the aeration tank and placing them at the head end of the aeration tank, creating tapered aeration down the tank. The high oxygen demand at the head end of the aeration tank creates an environment that stimulates filamentous bacteria to predominate in the mixed liquor. The activated sludge with a large fraction of filamentous bacteria settles poorly and creates problems in the sedimentation tank. The filaments kept the floc particles apart and required a larger fraction of the settling tank volume. Hydraulic variations that normally occurred results in loss of excessive amounts of suspended solids in the treated effluent, creating violations of the effluent criteria.

The complete mixing activated sludge system eliminates the high initial oxygen demand as well as providing more efficient utilization of the oxygen added to the system. Complete mixing activated sludge has proven its ability to toxic industrial wastes, more efficiently than plug flow activated sludge systems (Fig. 2). The bacteria are able to metabolize the organics as quickly as they are added to the aeration tank. It has been found that by proper design of the aeration system and tank size, wastewaters of any influent concentration could be treated to any desired concentration. A typical example of the efficiency of complete mixing was the treatment of an industrial wastewater having a mixed phenolic

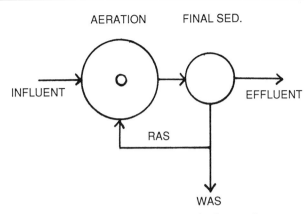

Figure 2 Schematic diagram of a completely mixed activated sludge system utilizing a single mechanical aerator in the center of the aeration tank and the final sedimentation tank for activated sludge separation and collection for return to the aeration tank (RAS) and wasting from the system (WAS).

concentration of 3000 mg/liter with the production of an effluent containing only 0.04 mg/liter phenol in a single-stage aeration system. Because of more efficient oxygen utilization, complete mixing systems can be made smaller or produce less waste sludge. Unfortunately, some complete mixing systems have been designed with less aeration, creating less mixing and allowing filamentous bacteria to grow and adversely affect the floc characteristics.

Originally, researchers believed that activated sludge floc was created by *Zooglea ramigera,* but research at the Massachusetts Institute of Technology (MIT) in 1950 demonstrated that floc formation could be created by other bacteria when all of the nutrients had been metabolized. It was also noted that *Z. ramigera* growth was stimulated by aromatic compounds but did not grow to a significant extent in domestic wastewaters. Floc formation is an energy relationship that requires a balance between growing bacteria and inert suspended solids. Very long aeration periods with limited microbial growth have been shown to cause floc to break apart and not reform. Protozoa are important organisms to remove the dispersed bacteria, but they cannot metabolize dead cells or tiny inert particles.

2. Contact Stabilization

Contact stabilization is a modification of activated sludge that utilize the ability of activated sludge to adsorb suspended solids very quickly and to stabilize the concentrated sludge mixture rather than the dilute mixture. A short mixing time (10–15 min) provide sufficient time for adsorption of the suspended solids. The activated sludge is then concentrated by gravity settling. The return activated sludge is aerated directly for 2 hr instead of the normal 6–8 hr. The contact stabilization process works best with raw municipal wastewaters rather than primary settled effluent, since the activated sludge cannot adsorb soluble organics. While some soluble organics can be absorbed into the bacteria cells, best removal occurs when sufficient aeration is used for mixing to allow complete metabolism of the soluble nutrients. The adsorbed nutrients are allowed to be stabilized as before in a concentrated return activated sludge aeration tank. With proper design, contact stabilization results in greater treatment in the same treatment tanks.

3. Extended Aeration

Extended aeration activated sludge grew out of the desire to completely oxidize the nutrients in the aeration tank. The extended aeration tank is designed on the basis of 24 hr aeration of raw wastewaters, eliminating the need for primary sedimentation tanks. The mixed liquor suspended solids (MLSS) are allowed to build up to 30 or 40 days solids retention time (SRT_a) under aeration. Because the MLSS are largely dead cells and inert suspended solids, with little active bacteria and protozoa, rotifers are the predominant animals that keep the effluent clarified. The long SRT_a period allows the activated sludge to undergo aerobic digestion. Most extended aeration plants are small systems, providing complete mixing in the aeration tank.

4. Oxidation Ditches

One of the more interesting forms of extended aeration activated sludge is the oxidation ditch where the aeration tank configuration has been modified as a shallow oval. A mechanical brush aerator is placed across the oval channel to provide both aeration and mixing. Initially, the tanks were 3–4 ft deep and provided 24–48 hr aeration. Separate sedimentation tanks provide the separation of the suspended solids and their return to the oxidation ditch. Recently, oxidation ditches have been modified to provide for depths of 10–12 ft and shorter activated sludge retention periods, 6–8 hr.

5. Sequencing Batch Reactors

Sequencing batch reactors (SBRs) provide batch-fed operations on a continuous flow system and the elimination of separate final sedimentation tanks with sludge return equipment. The aeration tanks

are used for both aeration and sedimentation. The system is allowed to accumulate some liquid with aeration. The tank fills and aerates until the wastes are stabilized. The aeration is stopped and the MLSS are allowed to settle in the aeration tank. The supernatant is decanted after settling, and the fill process is started again. Two aeration tanks are used so that one can be accepting wastes while the other tank is settling solids and removing supernatant. The small size of SBRs makes them completely mixed with a varying volume. The efficiency of oxygen transfer as the tank fills is lower than when the tank is full and can result in the growth of filamentous bacteria, the same as in conventional activated sludge systems with long, narrow aeration tanks. SBR plants are used primarily in small communities.

6. Pure Oxygen Systems

The solubility of oxygen in water increases as the partial pressure of oxygen in the gas increases. Air contains 23% oxygen by weight. Pure oxygen can approach 98–99% oxygen, giving a greater driving force. To take advantage of the greater oxygen driving force, pure oxygen systems are divided into several compartments in series. While each compartment is completely mixed, the series operation creates a decreasing oxygen demand rate from the first compartment to the last compartment. So as not to lose the oxygen into the atmosphere, the tanks are covered and connected with both the gaseous phase and the liquid phase operating in series. The aeration retention times have been reduced to match the increased oxygen transfer, 1–3 hr. The net effect has been overloading the first aeration compartment, creating oxygen-deficient conditions and growth of filamentous bacteria. The production of carbon dioxide results in an increase in carbon dioxide in the overhead gas and a lowering of the pH to 6 or less. The net effect has been to bleed off excess oxygen and carbon dioxide to keep the pH from adversely affecting the bacteria metabolism. While a complete mixing pure oxygen system was developed, it has not been promoted to date.

7. Biological Indicators

Microscopic examination of activated sludge on a routine basis can provide considerable information about the chemical characteristics of the various systems. Normally, the bacteria will form dense floc particles with few dispersed bacteria. Under oxygen-deficient conditions filamentous bacteria will grow rapidly and predominate over the normal floc-producing bacteria. Very old activated sludge floc will show heavy filamentous bacteria growth even though there is more than adequate DO in the mixed liquor. Proper organic loading and adequate DO is essential for good bacterial development. Protozoa have been used as indicators of the biological activity of the different activated sludge systems. Adequate DO is required for development of protozoa. Because the protozoa metabolize the dispersed bacteria, the protozoa reflect the bacteria response in the system. The flagellated protozoa cannot compete with the ciliated protozoa but can grow to a limited extent. Predomination of flagellated protozoa indicates a very young activated sludge system that has yet to produce the normal bacteria. As the bacteria population increases, the free-swimming ciliated protozoa also increase. The free-swimming ciliated protozoa require very large populations of bacteria for nutrients. As the free-swimming bacteria population is reduced with limited organics remaining in solution, the free-swimming ciliated protozoa give way to stalked ciliated protozoa. The stalked ciliated protozoa are good indicators of a high-quality effluent as far as soluble biodegradable organics are concerned. As the SRT_a increases, the inert suspended solids and dead cell mass accumulate and rotifers predominate. The simplicity of microscopic evaluation of activated sludge makes it the best tool for operators.

C. Rotating Biological Contactors

RBCs consist of large plastic discs with ridges to allow microbial growth to be retained and to allow the excess liquid to move off of the growth. The discs rotate slowly through a shallow tank of wastewaters, collecting fresh wastes and then moving into the air for oxidation. The excess wastewaters drain back into the shallow tank. Little power is required to turn the discs, making the operation inexpensive. Unfortunately, poor design resulted in plastic that failed, shafts that broke, and bearings that failed. Operation of the discs in series resulted in overloading the initial set of discs. Filamentous bacteria grew so quickly that the growth bridged between the discs. Anaerobic conditions resulted with the production of strong odors, making it necessary to cover the discs with a plastic cover. Corrosion of metal products quickly occurred. With proper loading, the growth can be controlled and a high-quality effluent can be produced. Producing nitrification is possible if desired. The RBC is an activated sludge

system with growth on discs. The hydraulic flow between the discs is responsible for shearing the biological growth off the discs. A final sedimentation tank is needed to remove the excess suspended solids produced by the RBC system.

D. Activated Biofiltration Towers

Activated biofiltration (ABF) towers is an interesting development to utilize waste redwood scraps. The redwood media are fabricated in strips that are stapled to cross beams and stacked on top of each other and placed in towers. The incoming wastes are mixed with return activated sludge from a final sedimentation tank that follows the aeration tank after the ABF tower. Tower underflow can also be recycled back to the feed sump to maintain a fixed hydraulic flow rate to the top of the ABF tower. Microbial growths cover the rough redwood planks and allow rapid flow of wastewaters with good oxygen transfer to produce a high rate of microbial synthesis. The flow of wastewaters over the ABF tower washes the excess growth from the system. The soluble organics are partially metabolized while the suspended solids simply pass through the tower to the aeration tank. The aeration tank completes the metabolism of the nutrients left. It is even possible to nitrify in the aeration tank. Misunderstanding about the basic biochemistry of the ABF tower operations created design problems initially, but it has been demonstrated that ABF towers can produce good results when properly applied.

E. Aerated Lagoons

Aerated lagoons were developed as a result of wastewater lagoons being overloaded. Aeration equipment was added to the lagoons, producing aerated lagoons. It was quickly recognized that domestic wastewater aerated lagoons should be designed with a 24-hr aeration period and complete mixing. A normal facultative lagoon should follow an aerated lagoon to permit stabilization of the microbes generated in the aerated lagoon. The aerated lagoons are designed with either mechanical surface aeration equipment or diffused aeration. Adequate power for mixing should be provided to prevent the suspended solids from settling in the aerated lagoon. Because no suspended solids are settled and returned to the aerated lagoon, the microbial growth are basically a dispersed growth system. The bacteria are normal but remain dispersed with limited protozoa growth.

VI. Facultative Lagoons

Facultative lagoons are simply shallow ponds, 4–5 ft deep, that retain the wastewaters for 90–120 days. The incoming wastewater flow are discharged into a large cell with a sharp drop in velocity. The suspended solids quickly settle out around the influent pipe. The colloidal and soluble organics move toward the effluent pipe. The settled solids undergo anaerobic degradation with the production of organic acids and a drop in pH. The bacteria slowly begin to metabolize the organic acids and produce methane. The anaerobically stabilized organics are lost to the atmosphere as methane. The shallow liquid layer allows oxygen to be transferred to the water from the air above the lagoon. Wind action over the lagoon surface allows even more oxygen transfer and moves the wastewaters around the lagoon.

The release of ammonia nitrogen and phosphorus in the wastewaters allows algae to grow at the surface of the lagoon where adequate light is available. The algae utilize the carbon dioxide produced by the bacteria and bicarbonates from the carriage water to grow and produce oxygen as an end-product. Green algae grow at the lagoon surface with *Chlorella* predominating during cold weather, when very few rotifers and crustaceans are available to metabolize the *Chlorella*. In warm weather, the protozoa, rotifers, and crustaceans grow on the *Chlorella,* allowing other algae to grow and predominate. In the midwestern part of the United States, *Scenedesmus* and *Ankistrodesmus* can survive because of sharp-pointed cell structures. The growth of *Daphnia* in the spring has often completely eliminated the algae for a few days until the resistant algae could grow on the available nutrients. Needless to say, the *Daphnia* began to starve and soon reached normal population levels.

The algae in facultative lagoons respond to sunlight variations by moving to the optimum depth for efficient metabolism. During the course of a normal day, the sunlight increases from zero just before dawn to a maximum and then back to zero at night. Because the sunlight prevents the algae from processing the light energy, the algae move deeper below the water surface. The water absorbs the light energy. Many of the algae are motile with flagella, but *Chlorella* is a nonmotile algae that utilizes gas pressure to adjust for light conditions. The buoyancy of the cells determines the proper light level for the nonmotile algae. In the dark, algae demand oxygen for endogenous respiration the same as bacteria and fungi. Cell mass is degraded with the release of carbon dioxide, water, and ammonia nitrogen. The algae also produce dead cell mass that is not readily biodegraded in the lagoon. The dead cell mass suspended solids slowly settle by gravity or remain in suspension by wind mixing of the lagoon contents. The next morning, the endogenous endproducts from dark metabolism are metabolized back to new cell mass by the algae. Under very heavy metabolism, the algae growth can become so thick that further growth is limited by light penetration. The net effect is that the effluent contains large quantities of algae, raising the suspended solids and BOD5 above EPA effluent criteria. Because the effluent suspended solids and BOD5 are related to algae rather than to wastewater characteristics, the EPA has allowed small treatment plants to have greater effluent suspended solids in the summer months, when algae growth is a maximum.

In an effort to improve effluent quality from facultative lagoons, the system is normally divided into two or three ponds in series. The first pond normally contains 50% of the total volume and stabilizes the nutrients. A submerged effluent structure is designed to allow drawoff from below the lagoon surface to minimize carryover of algae from the first pond. The next two ponds normally are of equal size and have similar submerged drawoffs. In this way, the best possible effluent quality is produced. Studies on fecal coliform bacteria have indicated that these indicator bacteria die off by competition with the other bacteria for nutrients and by protozoa and rotifers as predators. The reduction in coliforms and expected survival of pathogens is a function of liquid retention time in the facultative lagoon system. The three ponds in series provides a greater retention time than a single lagoon of the same volume. Short-circuiting of fluid flow through lagoons tends to produce actual retention times much less than the theoretical displacement time. Plastic baffles placed across the lagoons can control short-circuiting of flow and can improve the hydraulic retention characteristics of lagoons.

As facultative lagoons become overloaded, sulfate-reducing bacteria will grow in the sludge solids on the bottom of the lagoon. The hydrogen sulfide will diffuse from the sludge and rise as tiny bubbles to the surface. Blue-green algae such as *Oscillatoria* and *Anacystis* will begin to grow because they can metabolize the hydrogen sulfide with light energy. The green algae are usually more efficient than the blue-green algae and will predominate until the lagoon is overloaded. The blue-green algae tend to form mats at the lagoon surface, blocking light penetration at the surface and dampening wind mixing. Nuisance conditions soon follow. The blue-green algae have often been blamed for producing the hydrogen sulfide and nuisance conditions, but it should be recognized that the blue-green algae are simply responding to the environmental conditions and are good organisms for indicating that the facultative lagoon is being overloaded and help is needed. Facultative lagoons represent the simplest effective method to treat municipal wastewaters from small communities and will continue to be used extensively throughout the world.

VII. Anaerobic Treatment Systems

Anaerobic treatment has been used for many years to treat concentrated waste streams. Economics has dictated that primary sludge be treated by anaerobic systems rather than by aerobic systems in large plants. While secondary sludges have also been treated by anaerobic systems, engineers have only recently recognized the problems created by secondary biological sludges. The pretreatment regulations of the EPA have forced many industries to look at anaerobic pretreatment prior to discharge into municipal sewers. The microbiology and biochemistry of anaerobic systems has been much more difficult to study than for aerobic systems.

A. Sludge Digestion

Anaerobic digesters are designed to handle the primary sludge and the waste activated sludge (WAS). In order to insure a good anaerobic environment, anaerobic digesters are covered and operated in such a manner as to prevent air from entering the digester. The primary sludge can be concentrated to 4.0–6.0% total solids (TS) prior to being pumped to the anaerobic digester. A slow speed positive displacement pump with a sludge density meter on the discharge line is used to pump the sludge. It is important to control the sludge pumping to minimize the addition of extraneous water. Sludge is pumped at intervals to keep biological decomposition in the primary sedimentation tank to a minimum. Even with regular pumping the primary sludge will have

an acid pH, around 4 to 5, and several hundred mg/L ammonia nitrogen, confirming bacterial metabolism within the settled sludge. WAS will normally be between 0.8 to 1.0% concentration. The high population of bacteria, living and dead, keeps the WAS from concentrating as much as primary sludge. Many treatment plants utilize air flotation thickening to concentrate the WAS to 3.0–4.0%. Even after flotation thickening WAS occupies a large part of the volume of the anaerobic digester.

The primary sludge and the WAS are passed through an external heat exchanger to raise the temperature to about 37°C as they are added to the digester (Fig. 3). Digesting sludge is recycled continuously through the external heat exchange to maintain the digester temperature at the maximum temperature for rapid digestion. Various types of mixers have been used to provide good mixing in anaerobic digesters. Mechanical mixers and gas mixers have been used with varying degrees of success. The viscosity of the sludge makes mixing in the digester difficult. In single tank digesters the mixing is limited to the upper half of the digester as the lower part of the digester is used for concentrating the digested sludge. With two stage digesters the first stage is well mixed and heated; while the second stage is used as a digested sludge thickener with supernatant being returned back to the primary sedimentation tank for further treatment.

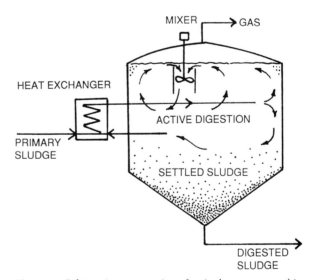

Figure 3 Schematic cross-section of a single stage anaerobic digester employing a small mechanical mixer in a short draft tube and an external heat exchanger to keep the temperature at 37°C. The digested sludge settles and concentrates at the bottom of the digester prior to removal.

Anaerobic digesters have from 20 to 60 d total hydraulic retention time, more than adequate for complete metabolism of the biodegradable organics. As previously indicated, the primary sludge contains 60–70% biodegradable VS, giving a potential digestion of 50% TS. Half of the primary sludge solids will pass through the digester untouched. If WAS is added to the digester, the biodegradable VS fraction of the WAS could vary from 10 to 60% VS, giving limited reduction in VS in the digester.

The digesting sludge contains four major groups of bacteria that are critical to the digestion process. The first group of bacteria are the acid forming bacteria, which hydrolyze the complex organics to soluble compounds that can be absorbed within the cell. *Clostridium* is a key obligate anaerobe; but most of the bacteria are facultative bacteria growing in an anaerobic environment. The organic acids produced by metabolism must be neutralized by ammonium bicarbonate if the pH is to be maintained at a suitable level for good metabolism, above pH 6.5. The organic acids are metabolized by the second group of bacteria using beta oxidation to produce acetic acid. The hydrogen removed from the organic acids has been postulated to be transferred to a group of methane bacteria which reduce carbon dioxide to form methane and water. Using energy relationships from Battley (1987) the energetics of the beta oxidation reactions produce a positive energy requirement for the reaction to occur. Hamilton (1988) and others have pointed out that the energy reactions produce an energy release if the hydrogen gas is kept at a low concentration. The third group of bacteria are the methane bacteria that reduce carbon dioxide or split acetate. The acetate utilizing bacteria are different from the carbon dioxide reducing methane bacteria. The acetate methane utilizing bacteria split acetate to form methane and carbon dioxide. Because the energy yield for acetate metabolism is low, the acetate utilizing methane bacteria are the most sensitive group of bacteria in the anaerobic digester and can be used as indicators for digester operations. The fourth group of bacteria are the sulfate reducing bacteria. By reducing the sulfates in the digester to sulfides, the sulfate reducing bacteria create a strongly reduced environment, required by the methane bacteria.

Anaerobic metabolism does not yield much energy for bacteria synthesis, limiting the growth of the different groups of bacteria. Good digestion requires a balanced bacteria population for sustained operations. The very nature of digesting sludge makes it difficult to carryout quantitative bacteriological studies. As a net result, most studies have dealt with the overall biochemical relationships in the digesters. R.E. Speece (1988) examined 30 municipal plants with anaerobic digesters to determine their operational characteristics. The raw sludge fed to the digesters averaged 4.7% TS with 3.3% VS. The VS destruction in the anaerobic digesters averaged 54% VS. The VS destroyed were converted to digester gas, methane and carbon dioxide. Approximately 15.0 cf gas was produced for each lb VS destroyed. The gas contained 67% methane and 0.22% hydrogen sulfide. The digesters had a median pH of 7.1 with 2,500 mg/liter alkalinity and 200 mg/liter volatile acids. These data are typical for municipal wastewater treatment plants.

VIII. Pathogenic Organisms

Even though municipal wastewater treatment plants have demonstrated their ability to help reduce the spread of various enteric diseases, questions concerning the fate of the pathogens are raised from time to time. The environments in the different biological treatment processes have been found to be unsuitable for growth of pathogenic organisms. Pathogenic bacteria tend to die off by predation from protozoa and higher animals as well as by starvation. Only a few pathogenic bacteria form spores that permit them to survive in an unfavorable environment. Over the years a number of studies have been made on the survival of viruses. Viruses are tiny parasitic organisms that are highly specific. Large-scale immunization for polio provided the opportunity to examine the efficiency of wastewater treatment for reducing the viruses in the treated effluents. It was observed that activated sludge systems were very effective in reducing the attenuated polio viruses. Recently, Rao *et al.* (1987) examined the survival of rotaviruses which cause infant diarrhea. Rotaviruses are found extensively in municipal wastewaters. An activated sludge plant was found to remove 94–99% of the rotaviruses from the municipal wastewater. It appeared that the removal of rotaviruses was related to the efficiency of the activated sludge process in capturing the mixed liquor suspended solids in the final sedimentation tanks. *Endamoeba* and *Giardia* are two protozoa pathogens found in municipal wastewaters from time to time. These pathogenic protozoa survive by producing cysts; but the cysts are easily removed in the wastewater treatment processes. The data on enteric diseases in the United States indicate that municipal wastewater treatment plants are reasonably effective in reducing the pathogens and protecting the health of the public.

IX. Future Biological Treatment Systems

As more research is carried out on the microbiology and biochemistry of municipal wastewater treatment systems, a better understanding will be developed on more efficient designs and operational procedures. New treatment systems will be created for both aerobic and anaerobic treatment. Biological nitrogen and phosphorus removal systems are currently being constructed and tested. In a few years some of the designs may become standards such as the activated sludge process. High-rate anaerobic systems should be possible to treat WAS in smaller tanks so that primary sludge can be treated more efficiently in its own system. High-rate algae–bacteria systems should be able to move from the research phase to the field phase as engineers de-velop a better appreciation for the microbes and how they respond. In the past, the engineers have constructed the systems and expected the microbes to fit into those systems. More often than not the biological systems failed to provide the desired results. By understanding the microbiology and their biochemistry, the engineers should be able to develop the optimum designs for the microbes as well as providing the operators with the knowledge necessary to obtain the optimum effluent quality without any environmental damage. The future for biological treatment of municipal wastewaters never looked brighter than it does at the present time.

Bibliography

Battley, E. H. (1987). *"Energetics of Microbial Growth."* Wiley-Interscience, New York.

Hamilton, W. A. (1988). 2 Energy transduction in anaerobic bacteria. *In* "Bacterial Energy Transduction" (C. Anthony, ed.) pp 83–149, Academic Press, San Diego.

Ingraham, J. L., Maaloe, O., and Neidhardt, F. C. (1983). "Growth of the Bacterial Cell." Sinauer Associates, Sunderland, Massachusetts.

Rao, V. C., Metcalf, T. G., and Melnick, J. L. (1987). *Wat.Res.,* **21,** 171–177.

Speece, R. E. (1988). *Wat.Res.,* **22,** 365–372.

Waterborne Diseases

Anita K. Highsmith
Centers for Disease Control

Sidney A. Crow
Georgia State University

Glossary

Endemic Prevalent in a certain locality or population; said of a disease or agent

Epidemic Outbreak of disease that spreads rapidly and extensively among many individuals in an area and is widely prevalent

Etiologic agent Cause of disease as determined by medical, diagnostic, and epidemiologic evidence

Heterotrophic bacteria Group of microorganisms that are ubiquitous in the environment and use organic substances as nutrient and energy sources

Indicator organism Organism whose presence in water may alert investigators to the source of waterborne disease

Waterborne disease Incident of illness in which two or more persons experience a similar illness after consumption or use of water, or epidemiologic evidence implicates the water as the source of illness

WATERBORNE INFECTIOUS DISEASES are widespread and transmittable through three primary modes (ingestion, contact, and inhalation) and are caused by a variety of microorganisms (bacteria, viruses, fungi, and protozoa). Over the last 10 years, cases of waterborne illness caused by *Giardia lamblia*, a parasitic agent, have increased due to the ingestion of chlorinated but unfiltered surface water. *Shigella sonnei* was the most commonly implicated

bacterial pathogen associated with human illness following consumption of water contaminated with human waste. Waterborne diseases are underreported; in >50% of the reported cases, the cause is unknown. Additional waterborne illnesses can be attributed to biological toxins and chemical contaminants. This article discusses the basic nature of waterborne infectious diseases with consideration of their mode of transmission, the etiologic agent, and the application of water.

I. Background

Water is one of the most basic requirements for life. Its presence is so important to life that its absence was often used as an indicator of lack of life in early extraterrestrial explorations. In the United States, based on an average personal use of water (50–150 gal/day), it is estimated that 39% of a potable water supply is used for domestic purposes (bathing, laundry, and dishwashing), 30% for watering lawns and filling swimming pools, 29% for toilets, and 1% for drinking. In the United States, the average per capita consumption is 48 oz of water per day.

Over time, methods of water treatment have been developed to provide fresh, aesthetically clean water. However, global changes have occurred and threaten the environment with both pathogenic microorganisms and toxic chemicals. Water supplies are frequently overused and contaminated. Providing safe, clean water to the world's population is an issue facing every country in the 1990s. Developing water resources programs is vital to economic progress and to disease control. One method for addressing these needs will be to consider implementing the water treatment technology currently used in developed countries throughout the world.

Waterborne disease may vary from simple allergic reactions such as swimmers' itch, to contact dermatitis, to intestinal disorders (diarrhea, which can become life threatening), to severe meningoencephalitis caused by a variety of organisms. All types of microorganisms may live in grossly contaminated water and can potentially trigger a waterborne outbreak. The scope of waterborne illnesses includes several major classes of disorders: infectious diseases, intoxications, and poisoning by nonbiological agents. The emphasis of this article is on infectious disease caused by bacteria, viruses, fungi, and protozoa. Although cases of waterborne illness resulting from poisoning by nonbiogenic agents and biotoxins in water are recognized, incidence of these is not extensively documented.

In examining the nature of waterborne diseases, several perspectives are possible: (1) classification by usage (drinking, recreation, and agriculture/ industry), (2) classification by mode of transmission (ingestion, inhalation, and contact), and (3) classification by etiologic agent (bacteria, virus, fungi, and protozoa). This article examines waterborne diseases based on primary modes of transmission: ingestion, contact, and inhalation. Further emphasis will be placed on the practical usage of water and the etiologic agents (bacteria, virus, fungi, protozoa, or chemical) associated with illness, recognizing that all subcategories may not be relevant to each classification.

II. Occurrence

It has been estimated that 80% of all disease worldwide is due to improperly sanitized food and water, simply because most people in the world do not have access to clean water for drinking. Contaminated water causes diarrhea and as many as 25,000 deaths in infants and young children daily. Outbreaks of water-related illnesses are underreported in the United States and in other countries throughout the world. Endemic cases of illness occur with minor discomfort to individuals, whereas some epidemics may be severe enough to cause death. An example is the ongoing pandemic caused by *Vibrio cholerae* in South America.

Water systems are classified by the United States Environmental Protection Agency (EPA) as follows: community, noncommunity, and individual. Community public water systems (municipal systems) are public or investor-owned water systems that serve large or small communities, subdivisions, or trailerparks with at least 15 service connections

or 25 year-round residents. Noncommunity public water systems (semipublic water systems) are those of institutions, industries, camps, parks, hotels, or service stations that may be used by the general public. Individual systems (private water systems), which are generally wells and springs, are those used by one or several residences or by persons traveling outside populated areas. In the United States, over 200,000 public utilities treat and convert wastewater to potable water.

Several different types of deficiencies (484) in water supply systems led to waterborne disease outbreaks from 1971 to 1985. The number and percentage of deficiencies by source water included untreated surface water (33, 7%), treated groundwater (153, 32%), treatment deficiencies (188, 39%) distribution system deficiencies (81, 17%), and miscellaneous (29, 6%).

Based on estimates that community water systems serve 180 million people, noncommunity water systems serve 20 million people, and individual water systems serve 30 million people, one researcher reported that the incidence of waterborne disease during 1983–1986 was 243 cases of illness per 1 million persons served per year by noncommunity water systems, 26 cases per 1 million persons per year by community systems, and 7 cases per 1 million persons per year by individual systems.

In the United States, waterborne disease records have been tabulated since 1920. In 1971, the EPA and Centers for Disease Control published the first annual summary of waterborne diseases. It has been documented that in the United States the most common source of waterborne disease by far is drinking water. Over 1405 outbreaks were recorded in the United States between 1920 and 1980. These were associated with either municipal or private water systems. Of these cases (386,144), >75% were not attributed to a specific etiologic agent. Under the worst possible sanitary conditions, it is conceivable that almost any disease organism could survive in water and could thus be transmitted through ingestion of contaminated water, inhalation of aerosols, or contact with contaminated water in recreational or agricultural/industrial practices. Those diseases commonly associated with the varied uses of water are more restricted and are listed in Table I.

Between 1981 and 1983, 112 outbreaks associated with contaminated potable water were reported in the United States, causing illness in 28,791 persons; 14 outbreaks and 394 cases of illness were caused by consumption of contaminated water from nonpo-

Table I Selected Etiological Agents of Major
Waterborne Diseases

	Etiological agent	Illness/disease
Viral		
	Hepatitis A and B viruses	Infectious hepatitis
	Poliovirus	Poliomylitis
	Norwalk virus	Gastroenteritis
	Coxsackie A and B viruses	Respiratory and cardiovascular disease
	Rotavirus	Gastroenteritis
Bacterial		
	Toxigenic *Escherichia coli*	Gastroenteritis
	Salmonella typhi	Typhoid fever
	Vibrio cholerae	Cholera
	Pseudomonas sp.	Gastroenteritis
	Campylobacter sp.	Gastroenteritis
	Legionella sp.	Pneumonia
	Leptosprira sp.	Leptospirosis
Fungal		
	Aspergillus sp.	Allergic and respiratory disease, toxicosis (via mycotoxins)
	Cryptococcus neoformans	Respiratory disease
	Histoplasmasis capsulatum	Respiratory disease
	Candida albicans	Candidiasis
	Various dermatophytes	Athlete's foot, etc.
Protozoa		
	Entamoeba histolytica	Dysentery
	Giardia lamblia	Gastroenteritis
	Cryptospiridium sp.	Gastroenteritis
	Acanthamoeba sp.	Corneal lesions
	Naegleria sp.	Meningoencephalitis

table sources. During 1984 and 1985, 41 outbreaks of waterborne illness occurred. For the 3-yr period 1986–1988, 50 outbreaks were reported due to ingested potable water, causing 25,846 cases of illness. Twenty-four states and Puerto Rico reported at least one outbreak during this period; 34% of the outbreaks were reported from three states: Pennsylvania (7), Colorado (6), and Vermont (4). Outbreaks occurred in all months except December; 26% occurred in July. Table II lists waterborne disease outbreaks by year and type of water supply system for 1971–1988. From 1989 to 1990, >50% of reported outbreaks of acute gastroenteritis were attributed to unknown etiologic agents, while 26 outbreaks were documented as caused by water intended for drinking and 30 outbreaks were associated with contaminated recreational water. *Giardia lambia* and *Escherichia coli*, OH157, were the most commonly

associated microorganisms causing illness from water intended for drinking. As mentioned earlier, up to one-half of the waterborne illnesses in the United States are reported with unknown etiology due to an inability to recover the pathogen at the index time of exposure or to inadequate laboratory methods for isolating the agent.

III. Routes of Transmission

A. Ingestion

By far the most common source of water-associated illness is the direct consumption of water; however, the common usage of drinking water for other applications occasionally results in the development of disorders unrelated to the consumption of the water.

Furthermore, standards include regulations for water and the standards by which a water supply are judged. Standards for water quality are designed to ensure that a measure comparison is possible for qualitative and quantitative values. Guidelines are recognized statements of procedure. Outbreaks of waterborne diseases attributable to drinking water seem to be cyclical in nature. The outbreaks thus do not reflect the technological advances in water treatment. For instance, the number of outbreaks in 1980 was higher than any year except 1941 and 1942 in the 61-yr period examined (1920–1980). Whether this reflects a genuine increase in diseases or is a reflection of better reporting and epidemiologic investigations is not clear. Factors influencing infection include the number of microorganisms, pathogenic virulence, host susceptibility, and mode of entrance.

During the period 1975–1985, common infectious diseases caused by ingestion of drinking water contaminated by bacteria included campylobacteriosis, cholera, enterotoxigenic *E. coli*, gastroenteritis, salmonellosis, typhoid fever, and shigellosis. Viral diseases included hepatitis A, Norwalk gastroenteritis, and rotavirus gastroenteritis. Fungal diseases are apparently less often associated with ingestion of contaminated water. Although amebiasis due to consumption of contaminated water is rare in the United States, the incidence of giardiasis and cryptosporidiosis is increasing. [*See* Water, Drinking.]

Waterborne campylobacteriosis has been an important cause of epidemics and sporadic outbreaks in backpackers. Enterotoxigenic *E. coli* gastroenteritis is rare in the United States, but several outbreaks have been documented.

Table II   Waterborne Disease Outbreaks, 1971–1988

Year	Water supply system			Total	Total cases
	Community	Noncommunity	Individual		
1971	8	8	4	20	5184
1972	9	19	2	30	1650
1973	6	16	3	25	1762
1974	11	9	5	25	8356
1975	6	16	2	24	10,879
1976	9	23	3	35	5068
1977	14	18	2	34	3860
1978	10	19	3	32	11,435
1979	24	13	8	45	9841
1980	26	20	7	53	20,045
1981	14	18	4	36	4537
1982	26	15	3	44	3588
1983	30	9	4	43	21,036
1984	12	5	10	27	1800
1985	7	14	1	22	1946
1986	10	10	2	22	1946
1987	8	6	1	15	22,149
1988	4	8	1	13	2128
Total (%)	234 (43)	246 (45)	65 (12)	545	136,833

[Courtesy of the Centers for Disease Control.]

Annually, over 2 million cases of nontyphoid salmonellosis are documented in the United States. Most of the rare cases of typhoid fever (500/yr) reported in the United States appear to be acquired during foreign travel. Although outbreaks of shigellosis more commonly are due to person-to-person transmission, common source outbreaks are less frequent but identifiable. Norwalk viruses have also been increasingly indicated as sources of waterborne and foodborne outbreaks. The occurrence of rotavirus in waterborne diseases has been recognized since an outbreak in a Colorado ski resort in 1981.

B. Contact

Disease outbreaks associated with water used for recreational purposes meet the same criteria used for waterborne outbreaks associated with drinking water. However, outbreaks associated with recreational water include illnesses due to exposure to or unintentional ingestion of freshwater or marine water, but exclude wound infections caused by water-related organisms.

Recreational waters are generally considered to be freshwater swimming pools, whirlpools, and naturally occurring fresh and marine surface waters. Fresh and marine recreational waters are often polluted with animal and human waste and industrial and agricultural products. This is illustrated by the increased number of outbreaks associated with recreational water, illness resulting from occupational exposure to environmental water conditions, and illness resulting from ingestion, inhalation, or skin contact with contaminated water.

The risk of acquiring a waterborne illness depends on a number of factors including quality of water and nature of activity as well as the health condition of the individual (e.g., *Vibrio parahaemolyticus* in a compromised host). Numerous reports of the development of typhoid fever after swimming in contaminated waters illustrate the significance of this mode of infection. Leptospirosis is chiefly transmitted through water contact and is invariably fatal. Primary amebic meningoencephalitis results from the traumatic introduction of amebas from water into nasal passages.

Indicators of recreational water quality include the coliform group, species of *Pseudomonas, Streptococcus, Staphylococcus,* and, in some rare cases, *Legionella.* Coxsackie viruses A and B, adenovirus types 3 and 4, hepatitis A, and a variety of viruses that cause gastroenteritis have been isolated from recreational water and associated with illness. Other microorganisms such as *Mycobacterium, Candida albicans,* and species of *Naegleria* and *Acanthamoeba* have also been isolated from recreational water and found to cause waterborne disease. Contact-associated waterborne illnesses can be divided into two broad categories: (1) those resulting from introduced (allochthohaus) species, usually arising from anthropogenic activities such as sewage treatment, and (2) those resulting from native (autochthonuous) bacteria. The instances of the former can be reduced or eliminated by sewage treatment and disinfection procedures. Incidents of the latter are not influenced by these factors.

Some researchers suggest that separate criteria be used for each type of bathing water. Routine examination of recreational water for the presence of microorganisms is not recommended, except in the case of special studies or an investigation of a potential waterborne disease outbreak, in which case the microbiologic assays should be focused on the known or suspected pathogen.

The presence of microorganisms in water is not the only factor involved in infectious disease transmission. Other factors in disease may be the individual host susceptibility, immersion time, bather load, gender, and use of clothing. For example, a bather's risk for *Pseudomonas aeruginosa* dermatitis or folliculitis apparently depends on immersion in water colonized by *P. aeruginosa,* skin hydration with altered skin flora, and toxic reactions to extracellular enzymes or exotoxins produced by *P. aeruginosa.*

A variety of heterotrophic bacteria have been isolated following contact with contaminated water. These bacteria survive in water for different periods of time and under different conditions. For example, *Campylobacter* sp. survive in stream water for 2–18 days, whereas *Legionella* sp. can be isolated from pond water for 250 days or longer. *Shigella* sp. survive in filtered river water for 28 days; however, *Salmonella* die off in <1 day in well water. [*See* HETEROTROPHIC MICROORGANISMS.]

C. Inhalation

Microbiologically contaminated water droplets have resulted in waterborne diseases. Generally, the water particles are aerosolyzed into the respiratory tract, resulting in pneumonia. One example of such transmission would be outbreaks caused by *Legionella* sp. Other documented cases of illness have been related to contaminated water in showers, pool water, hot tub tanks, cooling towers, spray mists, and respiratory therapy equipment. The use of contaminated water for agricultural and industrial purposes, such as irrigation, has resulted in outbreaks in a number of foreign countries. The increased demand placed on irrigation sources in the United States, as well as the deteriorating conditions of many surface water resources, may lead to a serious hazard in the United States as well.

IV. Etiologic Agents

As already mentioned, four types of microorganisms serve as the etiologic agents causing waterborne disease. The clinical syndromes and incubation periods of those most common in the United States are summarized in Table III.

A. Bacteria

Bacteria have diverse metabolic and environmental tolerances in the natural aquatic environment as well as on artificial laboratory media. Although studies have shown that certain gram-negative bacteria frequently are isolated under similar habitat conditions, these conditions extend over a wide range. For example, gram-negative bacteria such as *P. aeruginosa* can grow in water with trace organics, whereas other gram-negative bacteria such as *Legionella* sp. may be more nutrient-dependent and need aquatic environments rich in organics or minerals for survival. Gram-negative bacteria such as *Klebsiella, Enterobacter,* and *Psuedomonas* sp. produce slime or a biofilm that aids in protecting the bacteria against chemical disinfectants and may greatly contribute to their persistence in water. The growth rate for gram-negative bacteria varies: Coliform bacteria such as *Flavobacterium* may take up to 7 days.

Microorganisms also produce various extracellular materials, such as toxins and enzymes. Although

Table III Clinical Syndromes and Incubation Periods of Infectious and Chemical Agents Causing Acute Waterborne Disease in the United States

Agent	Incubation period	Clinical syndrome
Bacteria		
Campylobacter jejuni	2–5 days	Gastroenteritis, often with fever
Enterotoxigenic *Escherichia coli*	6–36 hr	Gastroenteritis
Salmonella	6–48 hr	Either gastroenteritis, often with fever; enteric fever; or extraintestinal infection
Salmonella typhi	10–14 days	Enteric fever—fever, anorexia, malaise, transient rash, splenomegaly, and leukoplenia
Shigella	12–48 hr	Gastroenteritis, often with fever and bloody diarrhea
Vibrio cholerae 01	1–5 days	Gastroenteritis, often with significant dehydration
Yersinia enterocolitica	3–7 days	Either gastroenteritis, mesenteric lymphadenitis, or acute terminal ileitis; may mimic appendicitis
Viruses		
Hepatitis A	2–6 wk	Hepatitis—nausea, anorexia, jaundice, and dark urine
Norwalk virus	24–48 hr	Gastroenteritis—of short duration
Rotavirus	24–72 hr	Gastroenteritis, often with significant dehydration
Parasites		
Entamoeba histolytica	2–4 wk	Varies from mild gastroenteritis to acute fulminating dysentery with fever and bloody diarrhea
Giardia lamblia	1–4 wk	Chronic diarrhea, epigastric pain, bloating, malabsorption, and weight loss
Chemicals		
Fluoride	<1 hr	Nausea, vomiting, and abdominal cramps
Heavy Metals		
Antimony		
Cadmium		
Copper		
Lead		
Tin		
Zinc, etc.	<1 hr	Nausea, vomiting, and abdominal cramps, often accompanied by a metalic taste
Other		
Pesticides		
Petroleum products	Varies	Variable

[Courtesy of the Centers for Disease Control.]

the mechanisms of pathogenicity for these organisms are complicated and are not fully understood, some evidence indicates that these products may play a role in the organism's pathogenicity. For instance, studies with *Klebsiella* and *Psuedomonas* have demonstrated intraspecies differences in endotoxin production, which may be related to virulence. For example, *Klebsiella* produces little of this toxin, whereas *E. coli* produces large amounts that may be highly virulent. In addition to endotoxin production, *Pseudomonas* sp. produce another toxin, exotoxin A, that is thought to be the most toxic product elaborated by this organism. Although *P. aeruginosa* produces a variety of extracellular enzymes, recent studies on *P. aeruginosa* strains isolated from whirlpool water and bathers demonstrated significant dif-

ferences in extracellular enzyme production compared with control strains. The effects of cholera toxin following ingestion of *V. cholerae* from contaminated drinking water are well established. [*See* ENZYMES, EXTRACELLULAR.]

Bacteria associated with contact transmission react with a wide range of biochemical substrates in water; they can also demonstrate antimicrobial resistance and the potential for plasmid transfer. Examples are the antimicrobial-resistant coliforms in drinking water, the resistant *Klebsiella* found in rivers, and the multiresistant strains of *Pseudomonas* isolated from whirlpool water.

In addition, temperature affects the persistence of gram-negative bacteria in water: *Pseudomonas* and *Legionella* are frequently found in water with higher temperatures compared with organisms such as *Yersinia* that persist in natural water at lower temperatures. Seasonal variations in gram-negative bacteria populations in drinking and recreational water have been documented. Although increased bacterial loads in distribution systems have been observed during the spring and summer, the exception has been shown with *Yersinia enterocolitica*. This bacterium persists in water at cold temperatures for longer periods of time.

Diverse metabolic activities and environmental tolerances are also seen in laboratory analyses of water samples. All viable bacteria from mixed populations in any given water sample cannot be cultivated on a single medium, at any one temperature, or at one specific length of incubation. Enough emphasis cannot be placed on the selection of media for isolating and recovering gram-negative bacteria from water samples. The absence of a pathogen does not indicate that it is not present if adequate media, appropriate temperature, and periods of incubation are not included in the protocol design.

B. Viruses

A range of viral agents have been documented in water and waterborne diseases. Improperly treated sewage allows viruses to enter drinking water supplies. Some viruses such as the picornaviruses are hardy agents and are capable of surviving for periods of time in the free state in water. Unlike most bacterial pathogens, viruses cannot multiply outside their host. Coliform counts do not correlate with viral counts in water. Recognition of viral waterborne disease is difficult due in part to vaiable incubation period, inconsistent clinical signs or symptoms, and widespread lack of methodology for detection. Viruses that have easily recognized routes of transmission and short, uniform incubation periods produce characteristic, easily recognizable diseases. A multistate outbreak of gastroenteritis among an estimated 5000 persons was caused by commercially produced ice made from well water contaminated with Norwalk-like virus among an estimated 5000 persons. Hepatitis viruses have been implicated in 60 outbreaks between 1946 and 1980.

C. Fungi

Over 984 fungal species have been isolated from unchlorinated groundwater, chlorinated surface water systems, and service mains.One study reported that three-fourths of all pipe surfaces examined contained some level of filamentous fungi, including *Penicillium*, *Nomurae*, *Sporocybe*, and *Acremonium*, and that one-half contained yeast forms, such as *Cryptococcus* and *Rhodotorula*. Presumptive evidence indicates that *Geotrichum candidum* is a causal agent of one form of chromomycosis. *Aspergillus fumigatus* has been shown to cause pulmonary aspergillosis. A common fungal disease associated with water is candidiasis, caused by *C. albicans*. While waterborne diseases have rarely been directly associated with fungi, illness associated with fungal infections are more likely observed in allergic reactions due to water and air dispersements.

D. Protozoa

In the United States in recent years, illness has occurred following ingestion of drinking water contaminated by protozoa and, in some cases, after contact with bathing water, including recreational water. These organisms cause diarrhea or gastroenteritis. Species of protozoa, *Giardia* sp., *Entamoeba histolytica*, and, rarely, *Balantidium coli* have caused illness following ingestion of water contaminated with these microorganisms. Coliform bacteria are poor indicators for the presence of *Giardia* or *E. histolytica* in treated water because of the increased resistance of the protozoan to inactivation by disinfection. Boiling water is an effective means of control in outbreaks associated with drinking water contaminated with protozoa. The infective stages of many parasitic roundworms and flatworms can be transmitted to humans through drinking water.

Giardia lamblia was responsible for 23,000 cases of giardiasis in 90 outbreaks of illness between 1965 and 1984. *Giardia lamblia* continues to be the most commonly implicated parasitic agent causing waterborne disease. The primary route of transmission is usually ingestion of chlorinated but unfiltered surface water. Although positive water samples have had low concentrations of *Giardia*, only a few cysts (>10) are required to cause human infection.

A number of parasitic infections caused by multicellular entities are also acquired by ingestion. The nature of these diseases and their characteristics are beyond the scope of a discussion of disease microbiology. Disease caused by guinea worm (*Dracunculus medinensis*) and the human schistosomes are particularly important; however, dracunculiasis, or guinea worm disease, is the only infectious disease that is solely transmitted by contaminated water. It is estimated that between 5 and 15 million persons may be infected yearly.

V. Conclusion

Although the incidence of waterborne diseases cannot be directly correlated with any meaningful parameters because of the myriad of factors associated with the development of illness, clearly a better understanding of natural microbial communities and the survival of introduced species would clarify many issues. Better methods to monitor indicator species as well as the indigenous opportunistic pathogens associated with water would allow more complete epidemiologic surveillance and the rapid recognition of agents associated with illness. This recognition and the timely response of public health agencies could lead to an overall reduction in the cases of waterborne disease resulting from infection with common microbiologic agents.

Bibliography

Cabelli (1983). Health effects criteria for marine recreational water. EPA publication 600/1-80.031.

Centers for Disease Control (1988). MMWR: Water-related disease outbreaks, 1985. 6/88.
Centers for Disease Control (1990). MMWR: Guidelines for investigating clusters of health events. 7/27/90.
Centers for Disease Control (1990). MMWR: Viral agents of gastroenteritis, public health importance and outbreak management. 4/27/90.
Centers for Disease Control (1990). MMWR: Water-related disease outbreaks, 1986–1988. 3/90.
Craun, G. F. (1986). "Waterborne Diseases in the United States." CRC Press, Boca Raton, Florida.
Dufour, A. P. (1984). Health effects criteria for fresh recreational water. EPA publication 600/1-84.004.
Environmental Protection Agency (1974). Safe Drinking Water Act (Public Act 93-523) of 1974.
Highsmith, A. K. (1988). Water in health care facilities. In "Architectural Design and Microbial Contamination" (R. B. Kundsin, ed.), pp. 81–102. Oxford University Press, New York.
Hopkins, D. R. (1984). Eradication of dracunculiasis. In "Water and Sanitation: Economic and Sociological Perspectives" (P. G. Bourne, ed.), pp. 93–114. Academic Press, New York.
Nagy, L. A., and Olson, B. H. (1985). Occurrence and significance of bacteria, fungi, and yeasts associated with distribution pipe surfaces. In "Proceedings of the Water Quality Technology Conference" (Denver), pp. 213–238. American Water Works Association.
Rosenzweig, W. D., Minnigh, J., and Pipes, W. O. (1986). *J. Am. Water Works Assoc.* **78,** 53–55.

Water, Drinking

Paul S. Berger, Robert M. Clark, and Donald J. Reasoner
U.S. Environmental Protection Agency

Glossary

Biofilm Biologically active layer on surfaces exposed to water consisting of adsorbed inorganic and organic chemicals and microorganisms held together by a matrix of organic polymers produced and excreted by the microorganisms

Disinfection Chemical or physical process used to destroy or otherwise inactivate pathogenic microorganisms

Indicator organism Organism that is used to monitor the microbial quality of drinking water

THE PRIMARY OBJECT of the microbiology of drinking water is to prevent waterborne disease outbreaks. A drinking water system can minimize the likelihood of such an outbreak by employing proper treatment and control practices, and by monitoring the effectiveness of these practices. This article addresses these issues, and briefly describes how regulations in the United States attempt to protect the public from pathogens in the drinking water.

I. Introduction

Safe drinking water is one of the oldest public health concerns known. Ancient civilizations practiced water treatment, as evidenced by Egyptian inscriptions and Sanskrit writings. Sand filtration was in use by some cities in the 16th century, but chlorination was not introduced until the first decade of the 20th century. These two treatment practices dramatically decreased the incidence of waterborne disease, although waterborne disease is still a problem in the United States and elsewhere. [See WATERBORNE DISEASES.]

Control of waterborne disease depends on the presence of natural and artificial barriers. Natural barriers may include soil and bedrock that inactivate, remove, or otherwise prevent passage of fecal pathogens into well water. Systems using surface water may control human activities on the watershed supplying the system, thereby creating another barrier. Artificial barriers include various practices for treatment of the raw source water and programs for protecting the integrity of the underground pipe network that transports the treated water to consumers. A water system must employ at least one, and usually several, of these barriers to control waterborne disease. Regardless of the nature of the barrier system used, it must be adequately monitored to insure its continued effectiveness.

This article discusses how water is treated and monitored to insure its microbiological safety for human consumption. The primary focus is on practices in the United States, although those of other countries will be mentioned. The article also includes a brief description of the drinking water regulations for microbiology in the United States.

II. Indicator Organisms in Microbial Monitoring

Ideally, specific detection of the various pathogenic agents of waterborne disease would be the most direct approach in reducing their prevalence. Unfortunately, in actual practice, this is quite impractical because of a number of limitations. The variety of potential waterborne pathogens—including various species of bacteria, viruses, and protozoa—makes a search for all, especially on a routine basis, extremely time-consuming, expensive, and difficult.

Moreover, the efficiency of current techniques for recovering and detecting known waterborne pathogens in drinking water is often very low. Many pathogens are so fastidious in their environmental and nutritional requirements that only highly specialized laboratory techniques can be used to detect them. Some pathogens cannot yet be cultivated in the laboratory; some can only be cultivated with difficulty. Moreover, if pathogens were present in drinking water, their concentrations would usually be sufficiently small to require analysis of large-volume samples. Finally, extended delays would usually be involved in carrying out identification procedures, under the extremely risky assumption that laboratories involved in water testing would have such resources.

Because of the problems with trying to detect specific enteric pathogens, sanitary indicator organisms are used as surrogates. Such indicator organisms are used to assess the microbiological quality of drinking water. The ideal drinking water indicator has the following attributes.

• It is suitable for all types of drinking water.
• It is present in sewage and polluted waters at much higher densities than fecal pathogens.
• Its survival time in water is at least that of waterborne pathogens.
• It is at least as resistant to disinfection as the waterborne pathogens.
• It is easily detected by simple, inexpensive, laboratory tests in the shortest time with accurate results.
• It is stable and nonpathogenic.
• It is generally not present in waters uncontaminated by mammalian feces.

The sanitary indicators generally employed worldwide are total coliforms, fecal coliforms, or both. Some countries also use *Escherichia coli* and other organisms. Total coliforms constitute a group of closely related bacteria in the family *Enterobacteriaceae* that are usually not pathogenic and are widespread in ambient water. They are usually, but not necessarily, associated with sewage. Total coliforms are not defined in precise taxonomic terms, but by whether they can produce acid and gas from lactose on selective culture media designed to recover enteric bacteria. These media vary by country. Total coliforms include most species of the genera *Enterobacter*, *Klebsiella*, *Citrobacter*, and *Escherichia*, although some species of *Serratia* and

other genera are also often included. Total coliforms are used to determine the effectiveness of water treatment, to monitor the integrity of the underground pipe network called the distribution system, and as a screen for fresh fecal contamination. Treatment and other water system practices that provide coliform-free water should also reduce pathogens to minimal levels. A major shortcoming of the use of total coliforms as an indicator is that they are only marginally adequate for predicting the potential presence of some pathogenic protozoa and viruses, since total coliforms are less resistant to disinfection than are these other organisms.

Fecal coliforms are a subset of the total coliform group, primarily including *E. coli* and a few thermotolerant strains of *Klebsiella*. Fecal coliforms and *E. coli* are more suitable indicators than total coliforms of fresh fecal contamination, and most waterborne pathogens are associated with fecal contamination. However, total coliforms are a more suitable indicator than fecal coliforms or *E. coli* for determining the vulnerability of a system to fecal contamination, especially in the absence of fecally contaminated samples at the times and locations of sample collection. Total coliforms are more sensitive than fecal coliforms or *E. coli* in assessing system vulnerability because total coliforms are usually present at much higher concentrations in source waters than these other indicators, and they are relatively more resistant to chlorine and other environmental stresses. Fecal coliform or *E. coli* monitoring may be preferable in countries where monitoring for total coliforms is impractical due to consistently high densities of these organisms in the drinking water. Most countries that have formal drinking water rules, including the United States, monitor for total coliforms. Many of these countries, including the United States, also require fecal coliform or *E. coli* monitoring.

In addition to the indicators just described, other drinking water indicators also have been used or suggested, usually as a supplement to the initial indicators. These supplemental indicators include fecal streptococci, enterococci, coliphage, *Clostridium perfringens*, *Bacteroides*, heterotrophic bacteria (as measured by plate counting techniques such as the Heterotrophic Plate Count, HPC), and other organisms. Several chemicals have also been mentioned, for example, fecal sterols, assimilable organic carbon, and disinfectant residuals. For monitoring ambient waters, the U.S. Environmental Protection Agency (EPA) suggests the use of enterococci or

E. coli as indicators. [*See* BIOMONITORS OF ENVIRONMENTAL CONTAMINATION.]

III. United States Drinking Water Regulations

A. General

The EPA publishes enforceable regulations under the Safe Drinking Water Act, which was passed by the United States Congress in 1974 and subsequently amended several times. Regulations under this Act apply to all public water systems, that is, to those systems that regularly serve 25 or more people at least 60 days out of the year, or that have 15 or more service connections. The number of public water systems in the United States is about 200,000.

In 1986, amendments to the Safe Drinking Water Act required the EPA to regulate total coliforms, turbidity, viruses, *Giardia, Legionella,* and heterotrophic bacteria. The amendments also required the EPA to publish regulations specifying criteria under which filtration was needed for systems supplied by surface water. In addition, the amendments specified that the EPA was to publish regulations requiring that all water systems disinfect, with appropriate criteria for waivers (called variances). In response, in 1989, the EPA revised the existing regulation on total coliforms and developed the Surface Water Treatment Requirements. In the near future, the Agency will publish a regulation on ground-water disinfection. The Total Coliform Rule and the Surface Water Treatment Requirements are briefly described here, as are several anticipated provisions of the Ground-Water Disinfection Rule. The Code of Federal Regulations (40 CFR 141 and 142) provides detailed requirements about the two existing rules. The *Federal Register* (FR) also provides detailed requirements, along with the rationale for these requirements.

B. Total Coliform Rule

The Total Coliform Rule (54 FR 27544; June 29, 1989) was revised in June 1989, and became effective on December 31, 1990. This regulation sets a maximum contaminant level (MCL) for total coliforms as follows. For systems that collect 40 or more samples per month, no more than 5.0% may have any coliforms; for systems that collect fewer than 40 samples per month, no more than one sample may

be total coliform-positive. If a system exceeds the MCL for a month, it must notify the public using mandatory language developed by the EPA. The required monitoring frequency for a system depends on the population served. This frequency ranges from 480 samples per month for the largest systems to once annually for certain of the smallest systems. The regulation also requires all systems to have a written plan identifying where samples are to be collected.

If a system has a total coliform-positive sample, it must take three (for small systems, four) repeat samples within 24 hours of being notified of the positive sample. In addition, systems that collect fewer than five samples per month must, with some exceptions, collect at least five routine samples the next month of operation. Both routine and repeat samples count toward calculating compliance with the MCL.

If any sample is total coliform-positive, the system must also test the positive culture for the presence of either fecal coliforms or *E. coli.* Any positive fecal coliform or *E. coli* test must be reported to the state. If two consecutive samples at a site are total coliform-positive and one is also fecal coliform- or *E. coli*-positive, the system is in violation of the MCL and must notify the public using more urgent mandatory language than used for the presence of total coliforms alone.

The Total Coliform Rule also requires each system that collects fewer than five samples per month to have the system inspected every 5 years (10 years for certain types of systems using only protected and disinfected ground water). This on-site inspection (referred to as a sanitary survey) must be performed by the state or by an agent approved by the state.

The regulation also defines when a state may invalidate a total coliform-positive sample, and when a laboratory must invalidate a total coliform-negative sample. It also specifies which analytical methods are approved for compliance samples. The sample volume must be 100 ml, regardless of the method used.

C. Surface Water Treatment Requirements

The Surface Water Treatment Requirements (SWTR) (54 FR 27486; June 29, 1989) became effective on December 31, 1990. This rule requires all systems using surface water, or ground water under the direct influence of surface water, to disinfect. It also requires all such systems to filter their water, unless the system can demonstrate that it (1) has an effective watershed control program, as determined by the state, (2) complies with the Total Coliform Rule and the MCL for trihalomethanes (40 CFR 141.12), (3) has not had a waterborne disease outbreak in its current configuration, and (4) uses source water that contains no more than 20 fecal coliforms per 100 ml or no more than 100 total coliforms per 100 ml and has a turbidity that does not exceed 5 Nephelometric turbidity units (NTU), and (5) meets stringent disinfection conditions.

The SWTR also controls the presence of *Giardia lamblia,* viruses, and *Legionella* in surface water by specifying that water treatment meet certain removal/inactivation levels. Specifically, the SWTR requires systems to insure that water treatment removes/inactivates at least 99.9% of the *Giardia lamblia* cysts and at least 99.99% of the enteric viruses. The regulation, and associated EPA guidance, assists the system in meeting these removal/inactivation levels by identifying pertinent CT values (C, disinfectant concentration in mg/liter; T, time of disinfectant contact with the water in min). CT values are provided for *Giardia* and enteric viruses by disinfectant type (e.g., chlorine, chloramines, ozone, chlorine dioxide), water pH, and temperature.

The regulation also requires a system using surface water to have at least a 0.2 mg/liter disinfectant residual continually entering the distribution system, at least a detectable disinfectant residual in the distribution system, and an annual on-site inspection of the treatment facility. The regulation specifies the monitoring frequency for determining the disinfection residuals just mentioned and (for unfiltered systems) for testing source water quality.

D. Ground-Water Disinfection Rule

The Ground-Water Disinfection Rule (GWDR) will not be published and implemented for several years (from 1991). The regulation will apply to all 185,000 public water systems that use ground water; exceptions will be those systems that are under the direct influence of surface water and, consequently, must comply with the SWTR. Under the GWDR, every ground-water system will be required to practice continuous disinfection, unless the system can obtain a waiver on the grounds that it is not vulnerable to fecal contamination. An evaluation of vulnerabil-

ity will be based on whether the well is subject to viral contamination from external sources.

The GWDR will also control the presence of viruses, heterotrophic bacteria, and perhaps *Legionella* in ground water by specifying that a combination of natural treatment (i.e., by the soil or bedrock) and disinfection meet certain removal/inactivation levels. If a system needs to practice disinfection, the regulation will specify the minimum level of disinfectant entering the distribution system and within the distribution system, as well as monitoring requirements. The rule may also set maximum disinfectant levels to prevent the formation of excessive levels of toxic disinfection by-products.

In developing the GWDR and other future regulations, EPA will assess the risk of waterborne disease associated with specific pathogens. This assessment will facilitate the development of a risk model(s) that would specify, for example, the level of pathogen inactivation/removal needed by a ground-water system and, thus, whether that system would be required to disinfect. A satisfactory assessment of risk depends on the evaluation of a number of factors, including infectious dose, virulence, ratio of infection to disease, identification of susceptible populations, the extent to which a pathogen is present in water and the percentage that are viable, and level of water treatment. The development of risk assessment and modeling for waterborne pathogens is presently in its infancy.

IV. Treatment and Control

A. Sources of Water Supply

Drinking water sources can be divided into two categories, surface water and ground water; treatment varies with the source. Microbiologically, ground waters tend to be of better quality than surface waters and, consequently, require less intensive treatment. Most ground-water supplies used for drinking water are pumped wells (dug, drilled, or driven), artesian wells, or springs.

Surface water sources include streams and rivers, natural ponds and lakes, and manmade impoundments and reservoirs. These sources represent precipitation runoff that is not lost to the atmosphere by evaporation and does not enter the ground via infiltration and percolation. Surface waters become contaminated by human pathogens through direct and indirect inputs of municipal sewage and other

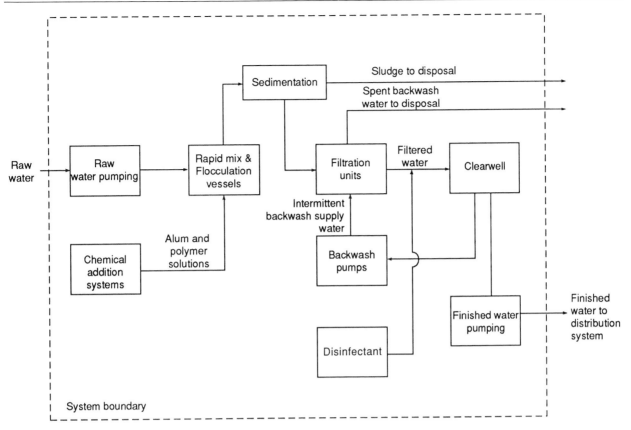

Figure 1 Conventional filtration system for drinking water treatment.

sources of human and animal excreta. Removal or inactivation of these microorganisms to provide safe drinking water involves a multibarrier concept, using a variety of treatment processes in addition to measures aimed at controlling or reducing source water pollution (source protection).

Bacterial concentrations in ground water, as measured by the HPC, are usually below 100 colony-forming units (CFU)/ml; coliforms are normally absent. Bacterial concentrations in surface waters depend on the degree of contamination with human and animal fecal material. Pristine and relatively un-contaminated surface waters commonly contain bacterial concentrations of 10^1–10^4 CFU/ml, whereas contaminated surface waters contain more than 10^6 CFU/ml. Coliform bacteria concentrations in surface waters range from $< 1/100$ ml to more than $10^6/100$ ml, and fecal coliforms range from $< 1/100$ ml to more than $10^5/100$ ml. Raw water containing more than 2000 fecal coliforms/100 ml should not be used as a drinking water source if at all possible.

B. Water Treatment

The major drinking water treatment processes for controlling microbiological occurrence in surface waters include coagulation and flocculation, sedimentation, filtration, and disinfection. Most ground waters are only disinfected, if they are treated at all. Figure 1 is a schematic of the unit processes in a conventional drinking water treatment system used for surface waters. Several additional treatment processes are widely used, for example, softening, fluoridation, and iron removal, but these will not be addressed in this article.

1. Coagulation, Flocculation, and Sedimentation

In the first step, raw water is pumped into a rapid mix unit, where chemicals are added to destabilize particles in the water electrostatically (i.e., make them "sticky"). This step is referred to as coagulation. Then water enters a flocculation basin, which is often a series of chambers with slow moving paddles. During flocculation, the destabilized particles are brought into contact with each other so that aggregation into larger particles, or flocs, can occur. Microorganisms become trapped by or attached to the flocs. The optimal coagulation practice varies with water pH and temperature, raw water turbidity, and type of coagulant(s) used. Commonly used coagulants include aluminum sulfate (alum), calcium oxide (lime), ferrous sulfate, and ferric chloride. Often, coagulant aids such as activated silica, bentonite clay, and polyelectrolytes (synthetic polymers of varying charge) are also used to reduce the concentration of primary coagulants employed.

After flocculation, water enters the sedimentation basin(s), where flocs are given time to settle by gravity. Some systems omit this step and feed the flocculated water directly to filters (direct filtration). If the source waters are highly turbid, systems may have an additional sedimentation step before coagulation (presedimentation). Sludge in the sedimentation basins is removed and discharged to a municipal sewer, lagooned, or dewatered and hauled to a landfill. The sedimentation basins are generally equipped with sludge removal mechanisms; if not, they usually have a sloping bottom so that most of the sludge flows out with the water when the basin is drained for cleaning.

2. Filtration

The next treatment step is filtration, which removes suspended and colloidal material that has not settled. The filtration unit consists of steel or concrete vessels containing granular materials such as graded sand, anthracite, and/or gravel. Three types of filters are commonly used: rapid granular filters, slow sand filters, and diatomaceous earth (DE) filters.

The most commonly used filter in the United States is the rapid granular filter. These filters may consist of silica sand, anthracite coal, and/or other materials. Filters are usually set up to provide gradation in filter media; the largest particles are on top and the smallest particles are on the bottom. In a rapid granular filter, the rate at which water is applied is at least 2 gal/min per square foot of surface area.

Rapid granular filters gradually accumulate a large amount of particles that impedes water flow; thus, the filters must be periodically cleaned. In this process, water flow through the filter is reversed in a process termed "backwashing." Backwashing expands the filter media, thereby releasing trapped particles into the water. Jets of water, air injection, or mechanical agitation at the surface may improve

the process. The filter backwash water should be disposed of in an environmentally acceptable manner.

Slow sand filters are similar to the rapid granular filters, except that water flows through the filter at a much slower rate, 0.05–0.15 gal/min per square foot of surface area, and pretreatment (coagulation, flocculation, and sedimentation) is often omitted. Removal occurs primarily in the upper portion of the sand, by straining, sedimentation, adsorption, and chemical and microbiological action. As contaminants are removed, a layer of deposited material called a "schmutzdecke" forms. This schmutzdecke contains a large number of bacteria that break down organic material in the water. When the schmutzdecke clogs the filter, it must be removed by scraping the top layer of sand to improve water flow. Normally, slow sand filters are used by systems with relatively clean source water (turbidity of 10 NTU or less and no undesirable chemical contaminants). They require far less maintenance than rapid granular filters.

For diatomaceous earth filtration, solids are removed by passing water through a thin filter consisting of a layer of diatomaceous earth supported on a rigid base (septum). Diatomaceous earth is composed of crushed siliceous shells of diatoms (microscopic algae). To maintain adequate water flow, additional diatomaceous earth called body feed is continually added to the raw water during operation. Like slow sand filters, systems that use diatomaceous earth are usually small with relatively clean source water, and often pretreatment is omitted.

Granular activated carbon (GAC) filters are sometimes used in conjunction with rapid granular filters to control taste and odor problems. In Europe, many systems ozonate the water before applying it to the GAC filters. This promotes the growth of a high density of heterotrophic bacteria in the filter; the resulting high metabolic rate significantly reduces the level of organic substances in the water as it passes through the filter.

3. Disinfection

Disinfection is the primary means for inactivating pathogenic microorganisms in water. For systems using ground water, disinfection is generally the only treatment practiced. Several disinfectants are available, including chlorine, chloramines (chlorine combined with ammonia or organic amines), ozone, chloride dioxide, and ultraviolet light. All are oxidizing agents.

Chlorination is, by far, the most common disinfec-

Table I Microorganism Removal Efficiency by Water Treatment Unit Processes

Unit process	Removal (%)			
	Bacteria	Viruses	Protozoa	Helminths
Storage reservoirs	80–90	80–90	—[a]	—
Aeration	—	—	—	—
Pretreatment[b]	90–99	90–99	>90	>90
Hardness reduction				
high lime	90–99.9	99–99.9	—	—
low lime	90–99	90–99	—	—
Slow sand filtration				
without pretreatment	35–99.5	10–99.9	59–94	—
with pretreatment	90–99.9	90–99.9	59–99.98	—
Rapid granular filtration				
without pretreatment[b]	0–90	0–50	0–90	—
with pretreatment except sedimentation[b]	90–99	90–99	90–99	—
with pretreatment[b]	90–99.9	90–99	90–99.9	—
Diatomaceous earth filtration with pretreatment and precoating of filter	90–99.9	99–99.96	99–99.999	—
Activated carbon	—	10–99	—	—
Disinfection	99–99.99	99	27–78	—
Full conventional treatment (pretreatment, filtration, and disinfection)	99–99.9999	99.9–>99.99	99.9–99.98	—

[a] Not known.
[b] Pretreatment includes coagulation, flocculation, and sedimentation.
[Reprinted in part from Amirtharajah, A. *AWWA Journal*, Vol. 78, No. 3 (March 1986), by permission. Copyright © 1986, American Water Works Association.]

tion technique practiced in the United States. The dose of chlorine usually applied is sufficiently high to meet chlorine demand (i.e., the tendency of organic substances and ammonia to react with chlorine) and still leave a sufficient concentration (chlorine residual) to inactivate microorganisms throughout the distribution system. A major shortcoming is that chlorine combines with organic substances in the water to produce chlorinated by-products, some of which are toxic (e.g., trihalomethanes, such as chloroform). Unlike chlorine, chloramines do not result in any significant trihalomethane formation, but their microbial inactivation rates are significantly less than that of chlorine. Chlorine dioxide also does not generate significant by-product formation, but its intermediates (chlorite and chlorate) are toxic. Like chlorine, chloramines and chlorine dioxide are useful because they provide disinfectant residuals in the water throughout the distribution system.

Ozone is the most effective disinfectant generally available, and is widely used in water treatment in Europe. Fewer toxic by-products have been identified with ozone than chlorine, but ozone is more expensive and does not leave a significant residual. There is increased interest in the United States in using ozone as a predisinfectant, followed by chlorine or chloramine. This process would allow a system to control the formation of toxic by-products, yet maintain a disinfectant residual in the distribution system.

Ultraviolet light is sometimes used by smaller systems and individual domestic systems, especially those using ground water. The optimum wavelength for biocidal effectiveness is 254 nm. Dosage is expressed as the product of radiation intensity (μW) and the time (sec) per unit area (cm^2). Ultraviolet light does not produce disinfectant by-products in water; however, it leaves no disinfectant residual. Its effectiveness is reduced by high turbidity, air bubbles, some dissolved chemicals that block light penetration, lack of reliable methods or meters to measure dosage, and equipment maintenance and reliability considerations.

C. Microbiology of Treatment Processes

The removal of microorganisms by the treatment processes through the filtration step is variable due to a variety of factors, including fluctuating source water quality, coagulant dose, pH, water temperature, depth of filter, filter medium particle size, and filtration rate.

Table I shows drinking water treatment processes

and the approximate percentage removal/inactivation of microorganisms achieved by each. Removal percentages represent only removal percentages for that stage, not the cumulative percentage. It is difficult to establish the cumulative removal percentages from one step to the next because of considerable variation in the data reported from different studies. These differences reflect different source waters and qualities, different treatment processes, and different operating conditions. Thus, direct comparison of study results is difficult, but sufficient studies have been conducted to permit general characterization of the effectiveness of water treatment processes for removing microorganisms.

1. Slow Sand Filters

The efficiency of microorganism removal by slow sand filters is influenced by several factors, including the particle size of the filter medium and the extent to which the scum layer (schmutzdecke) has developed. A filter with a smaller particle size is more efficient in microbe removal, but results in shortened filter runs, that is, more frequent filter cleaning is necessary to maintain a suitable water flow rate. In addition, new or cleaned slow sand filters require some conditioning (ripening) before they are effective in removing microorganisms. The ripening period allows a biologically active schmutzdecke to build up on the particles and in the filter bed. This layer then assists in the filtration of other particles and colloids from the water. The schmutzdecke is important in microbe removal, particularly for bacteria and viruses. Virus removal by sterile sand is negligible, and removal by clean nonsterile sand is variable but poor.

Removal of coliform bacteria by slow sand filters ranges from about 83% with new filter sand to nearly 100% for sand with an established schmutzdecke. Removal of poliovirus Type 1 by slow sand filters ranges from 22 to 96% with clean sand to ≥ 99.9% for sand with an established biological population. Removal of *Giardia* cysts (protozoan) by slow sand filtration ranges from 59 to 99.98%.

Cold water temperatures generally decrease microorganism removals by slow sand filtration. At water temperatures ≤ 5°C, removal of heterotrophic bacteria (measured by heterotrophic plate count) and coliforms decreases by about 2% and 2–10%, respectively, compared with removals at temperatures greater than 15°C. Virus removals decrease by about 0.5% and protozoan cyst removals decline by 0–6.2%. Increased water flow through the filter (filtration rate) also results in decreased microorganism removals. [*See* HETEROTROPHIC MICROORGANISMS.]

2. Rapid Granular Filters

Systems that use rapid granular filtration must first pretreat the raw source water (coagulation, flocculation, and sedimentation) for effective microbial removal, unless the water is clear (turbidity less than 10 NTU). Effective pretreatment should reduce the turbidity of muddy surface water to a level well below 1 NTU. Effective pretreatment and filtration collectively should reduce the turbidity to 0.1 NTU or less.

Microorganism removals achieved by pretreatment and rapid granular filtration are high. Bacterial removals (heterotrophic bacteria and coliforms) range from 86 to 98.8%, and virus removals range from 90 to >99.99%. The coagulation and filtration processes generally can achieve removals of protozoan cysts ranging from 83 to 99.99%. Efficient removal of *Giardia* and other protozoan cysts by filtration is dependent on an adequate coagulant dose. A change in alum dose from 5 mg/liter to 10 mg/liter can increase the removal of *Giardia* cysts from 96 to > 99%.

Factors that adversely affect removal efficiency include interruption of chemical feed (coagulants and polymers), poor filter efficiency at the beginning of a filter run, sudden increases in water flow, and turbidity breakthrough that can occur with higher filter head loss (resistance to water passage through the filter) at the end of a filter run. Any of these factors can seriously degrade the microbiological quality of the filtered water. In addition, some source waters are so highly polluted that full treatment (pretreatment, filtration, and disinfection) may not achieve a suitable level of microbiological reduction.

3. Diatomaceous Earth

Bacterial removals by diatomaceous earth (DE) filtration are affected by the grade of DE used. When fine DE is used that yields a median pore size of 1.5 μm in the filter cake, bacterial removals of nearly 100% can be achieved. Lower percentage removals occur when coarser DE is used that provides increased median pore size. However, by chemical conditioning of coarser grades of DE, good bacterial removals can be obtained.

Diatomaceous earth filtration can satisfactorily remove viruses from water but, to be most effective, the raw water must be pretreated with a coagulant aid (polymer) or the DE filter cake must be chemically conditioned to enhance virus attachment to the filter material during filtration. Overall, DE filtration is most effective for removal of microorganisms in the size range of *Giardia* or *Entamoeba histolytica* cysts. *Giardia* cyst removals of 99% or greater can be achieved by proper DE filter operation. The use of DE to remove smaller organisms (bacteria and viruses) can be improved by coating it with aluminum hydroxide precipitate. This coating gives the DE a positive surface electrical charge. Bacteria and viruses with a negative electrical charge are then removed by surface attachment.

4. Disinfection

a. General Considerations

The final treatment process for drinking water is chemical or physical disinfection intended to inactivate any coliforms and pathogenic microorganisms that penetrate the filter. The effectiveness of disinfection is a function of the types of organisms to be inactivated, the quality of the water, the type and concentration of the disinfectant, the exposure or contact time, and the temperature of the water.

As stated previously, CT values are used to iden-

tify the level of removal/inactivation provided by a given disinfectant for an organism under a specific environmental condition. These values are useful for comparing biocidal efficiency. Table II provides CT values for several organisms. Most of the available CT data for microorganisms of health concern were developed from laboratory studies that might not be indicative of field operations.

Water temperature can affect disinfection rates (and, thus, CT values). Microorganism inactivation rates decrease as water temperature decreases. Low water temperature therefore represents a worst case condition for chemical disinfection. Water pH can also affect disinfection rates. In most water systems, the pH is kept in the range of 7–9. Water pH, for example, determines the proportions of the most important chlorine species, hypochlorous acid (HOCl) and hypochlorite ion (OCl^-). Lower pH values (pH 6–7) result in the formation of HOCl which is favorable for rapid inactivation, whereas higher pH values (pH 8–10) result in formation of OCl^-, which results in slower inactivation. For chlorine dioxide (ClO_2), which does not dissociate, inactivation is more rapid at higher pH values (pH 9) than at lower pH values (pH 7). Ozone disinfection efficacy does not appear to be affected by pH.

Table II   Inactivation of Microorganisms (99%) by Chemical Disinfectants

Microorganism	Disinfectant	pH	Temperature (°C)	Concentration (mg/liter)	Contact time (min)	CT (mg · min/liter)
Escherichia coli	HOCl	6.0	5	0.1	0.4	0.04
	OCl^-	10.0	5	1.0	0.92	0.92
	NH_2Cl	9.0	5	1.0	175	175
		9.0	15	1.0	64	64
	$NHCl_2$	4.5	15	1.0	5.5	5.5
	ClO_2	7.0	5	0.3	1.8	0.54
	O_3	7.2	1	0.07	0.083	0.006
	O_3	7.2	1	0.065	0.33	0.022
Poliovirus Type 1	HOCl	6.0	5	0.5	2.1	1.05
	OCl^-	10.0	5	0.5	21	10.5
	NH_2Cl	9.0	15	10	90	900
	$NHCl_2$	4.5	5	100	140	14,000
		4.5	15	100	50	5000
	ClO_2	7.0	5	0.5	12.0	6.0
		7.0	21	0.3	5.0	1.5
		9.0	21	0.4	1.0	0.4
	O_3	7.2	5	0.15	1.47	0.22
Giardia lamblia cysts	HOCl	6.0	5	2.0	40	80
	$NH_2Cl/$ $NHCl_2$	7.5	3	2.4	220	528
G. muris cysts	O_3	7.0	5	0.15	12.9	1.94
Entamoeba histolytica cysts	HOCl	6.0	5	5.0	18	90

An important factor in bacterial inactivation is the phenomenon of cell injury. Disinfection and other environmental stresses may cause nonlethal physiological injury to waterborne bacteria. This phenomenon causes problems for monitoring water quality and calculating CT values, because injured bacteria may not grow on selective media normally used to detect or enumerate the bacteria. Thus, the actual number of viable cells may be underestimated. In some cases, injured pathogens remain infective. Problems with detecting injured cells can be mitigated by the use of media and procedures that remain selective, yet permit the injured cells to repair metabolic damage.

b. Microorganism Inactivation

i. Chlorine Table II shows that enteric viruses (represented by poliovirus Type 1) are more resistant to inactivation by chlorine than are bacteria (represented by *E. coli.*), and protozoan cysts are nearly two orders of magnitude more resistant than the enteric viruses. Differences in effectiveness of HOCl and OCl$^-$ against the viruses and bacteria are also shown. [*See* ENTEROVIRUSES.]

ii. Chloramines Comparison of chloramines with chlorine for disinfection of microorganisms (Table II) shows that, in general, for all types of microorganisms, CT values for chloramines are higher than CT values for free chlorine species. However, CT values for *Giardia lamblia* cysts are lower, in contrast to the results for free chlorine.

iii. Chlorine Dioxide Chlorine dioxide CT values in Table II show that, at pH 7.0, ClO_2 is not as strong a bactericide and virucide as HOCl. However, as the pH is increased, the efficiency of ClO_2 for inactivation of viruses increases. CT data for protozoan cyst inactivation is not available.

iv. Ozone Overall, comparison of CT values for ozone with those for chlorine and ClO_2 indicates that ozone is a much more effective biocide than the other disinfectants. *Escherichia coli* is about 10-fold (1 $\log_{10}$) more sensitive to ozone (Table II) than poliovirus Type 1. *Giardia muris* cysts are about 10-fold more resistant to ozone than poliovirus Type 1. Since ozone is a powerful oxidant, it reacts rapidly with both microorganisms and organic solutes and is very useful as a primary disinfectant.

The order of microbial disinfectant efficiency is $O_3 > ClO_2 > HOCl > OCl^- > NH_2Cl > NHCl_2 >$ RNHCl (organic chloramines). However, for technical reasons, practical handling considerations, cost and effectiveness, the frequency of use of disinfectants by utilities in the United States is generally chlorine $>>$ chloramines $> O_3 > ClO_2$.

v. Ultraviolet Light Sensitivity of the various microbial groups to ultraviolet light is similar to that for chemical disinfectants. Enteric bacteria are most sensitive, followed by enteric viruses; protozoan cysts are least sensitive. Organisms that are sublethally injured by UV light exposure may, under appropriate conditions, be able to repair the damage (i.e., photoreactivation or dark repair). Ranges of UV dosages required for 99.9% inactivation of microorganisms of concern in drinking water are: bacteria, 1400–12,000 $\mu W \cdot sec/cm^2$; viruses, 21,000–46,800 $\mu W \cdot sec/cm^2$; and protozoan cysts, 105,000–300,000 $\mu W \cdot sec/cm^2$. The UV disinfection values given for protozoan cysts are not practical with current UV technology used for water treatment.

D. Distribution Systems

1. Description

Water transmission and distribution systems are needed to deliver water to the consumers. Distribution systems represent the major investment of a municipal water works and consist of large mains that carry water from the source or treatment plant, service lines that carry water from the mains to the buildings or properties being served, and storage reservoirs that provide water storage to meet demand fluctuations, for fire fighting use, and to stabilize water pressure. The branch and loop (or grid) are the two basic configurations for most water distribution systems.

The layout of a branch system is similar to a tree branch, with smaller pipes branching off from larger pipes throughout the area served. This system, or a derivative of it, is normally used to supply rural areas where water demand is relatively low and long distances must be covered. Disadvantages of this configuration are the possibility that a large number of customers will be without service should a main break occur, and the potential water quality problems in parts of the system resulting from the presence of stagnant water. System flushing should be accomplished at regular intervals to reduce the possibility of water quality problems.

The loop configuration currently is the most widely used distribution system design. This configuration consists of connected pipe loops throughout the area to be served. Good design practices for smaller systems call for feeder mains to form a loop approximately 1 mile (1600 m) in radius around the center of the town, with additional feeder loops according to the particular layout and geography of the area to be served. The area inside and immediately surrounding the feeder loops should be gridded with connecting water mains on every street.

The most commonly used pipes for water mains are ductile iron, prestressed concrete cylinders, polyvinyl chloride (PVC), reinforced plastic, steel, and asbestos cement.

2. Microbiology

Microbiologically, water distribution systems are interesting bacterial ecosystems that present a real challenge to the water utilities in terms of maintaining good quality water with low bacterial densities. The construction characteristics, operation, and maintenance of a water distribution system provide ample opportunities for microbial recontamination of the treated water during distribution. Pipe joints, valves, elbows, tees, and other fittings as well as the vast amount of pipe surface provide both changing water movement and stagnant areas where bacteria can attach and colonize.

a. Biofilms in Water Distribution Systems

Bacteria found in water distribution systems can be classified into indigenous (autochthonous) and exogenous (allochthonous) populations. The indigenous organisms are well-adapted biofilm-forming bacteria that represent a stable ecosystem that is difficult to eradicate. The exogenous bacteria are contaminants that are transported into the system by a variety of mechanisms. Exogenous bacteria include organisms that survived the treatment and disinfection processes or were present in inadequately treated water, contaminants introduced through water line breaks and repairs, and contaminants introduced through cross-connections and back-siphonage events. A cross-connection is any direct connection between a drinking water distribution system and any nondrinkable fluid or substance. Backsiphonage occurs when the atmospheric pressure exceeds the water supply pressure or when the pressure at a point of use exceeds the supply pressure.

The development of a permanent biofilm in the distribution system occurs because the bacteria find physical and chemical conditions conducive to colonization and growth at the solid surface/water interface. These conditions include an ample supply of nutrients assimilabile organic carbon for growth, a relatively stable temperature, and some degree of protection from exposure to harmful chemicals such as the disinfectant(s) used to treat the water. [See BIOFILMS AND BIOFOULING.]

When an adequate disinfectant residual is maintained in the water throughout a distribution system, growth of bacteria is usually well controlled and the density of bacteria in the water will remain low—in the range of < 10 to several hundred CFU/ml (HPC). The choice of the medium and method used to determine the bacterial density may result in low or high counts of heterotrophic bacteria. In general, rich culture media and incubation at 35°C will yield lower counts than dilute nutrient media and incubation at 20–28°C. The disinfectant residual concentration needed to control the growth of the bacteria varies from one water system to another. Factors that appear to be critical are pH, temperature, dissolved organic carbon (DOC) concentration, AOC concentration, and type of disinfectant used. The disinfectant residual will also reduce or suppress extensive biofilm growth if the residual is maintained throughout the system.

It is generally recommended that, for chlorine, a free residual of at least 0.2 mg/liter be maintained. For monochloramine (NH_2Cl), a residual of 0.4 mg/liter is usually the goal. Higher concentrations of disinfectant residual may be applied and maintained if necessary. However, if the water contains high levels of DOC, it may be difficult to maintain an adequate disinfectant residual to control bacterial growth and still have water that is aesthetically acceptable to the consumers.

Bacterial concentrations in distribution water vary from < 1 CFU/ml in the water leaving the treatment plant to as high as 10^5–10^6 CFU/ml in water from slow flow or stagnant areas of the distribution system. The concentrations of bacteria in the water and on the pipe surfaces vary spatially and temporally in the distribution system. Bacterial densities in the pipe wall biofilm and in sediments may reach 10^7 CFU/cm^2. The biofilm contributes bacteria to the flowing water through shear loss (erosion) and by migration of actively motile bacterial cells into the water. Table III lists some of the bacteria commonly found in drinking water, sediments,

Table III Bacteria Found in Treated Distribution Water, Sediment, and Distribution System Biofilm

Microorganism	Distribution water	Sediment	Biofilm
Pseudomonas vesicularis	X	X	X
diminuta	X		X
cepacia			X
pickettii			X
fluorescens		X	
stutzeri		X	X
paucimobilis	X		
maltophilia	X		
Alcaligenes spp.			X
Acinetobacter spp.	X		X
Moraxella spp.	X	X	X
Arthrobacter spp.	X	X	X
Agrobacterium radiobacter			X
Corynebacterium spp.			X
Bacillus spp.			X
Yeasts			X
Enterobacter agglomerans			X
cloacae	X		
Micrococcus spp.	X		
Flavobacterium spp.	X	X	X
CDC Group II J	X		
Klebsiella oxytoca	X		
Mycobacterium spp.	X		X

and biofilm. Many of these bacteria are found in both the water and the biofilm, indicating the influence of the biofilm on bacterial quality of distribution water.

Information about the development and control of biofilms in drinking water is sparse. In addition, there is little data on other issues related to biofilms, including the role of iron and sulfur bacteria in biofilm development, the influence of pipe material on biofilms, the role of biofilms in nitrification for systems that use chloramines as a secondary disinfectant, and the effect of added corrosion inhibitors on bacterial populations in biofilms.

b. Biofilm Control by Disinfectants

Although a free chlorine level of 0.2 mg/liter may control bacteria in the water, extensive biofilm can develop on the pipe surfaces. At higher free chlorine residuals (0.8 mg/liter) a patchy biofilm may develop on the pipe surfaces, but a longer period of time is required. The presence of chlorine retards the development and affects the spatial distribution of biofilm. The type of disinfectant residual provided selects for bacteria that are more tolerant to the specific disinfectant used. The biofilm environment provides protection for the cells by diffusional resistance and neutralization of the disinfectant. There-

fore, biofilm organisms are less inhibited by the disinfectant residual than are planktonic cells. Differential effectiveness of chlorine and monochloramine for control of biofilm growth has been shown. Monochloramine, because it is less reactive and therefore more persistent, apparently can penetrate the biofilm and is more effective than chlorine in controlling biofilm growth.

c. Coliform Biofilm Problems

In some cases, coliforms that gain entry into the distribution system may attach to pipes or pipe sediments and proliferate, thus becoming a biofilm constituent. The intermittent, sporadic, or persistent sloughing of coliform bacteria from biofilms into the water of the distribution system may cause systems to repeatedly violate standards for total coliforms. This problem is most frequently associated with utilities that use a surface water supply where the water temperature is 15°C or greater. The presence of a coliform biofilm does not signify that the system is vulnerable to outside contamination. However, no definitive process currently exists to differentiate between coliforms associated with biofilms and the penetration of coliforms—especially injured coliforms—from outside sources. Therefore, public safety dictates that all coliforms detected by testing be regarded as representing system vulnerability, unless strong evidence suggests otherwise.

Remediation measures may include one or more of the following: raising or lowering the water pH, changing the concentration or type of disinfectant residual applied to the water, applying corrosion control chemicals to the water, and pursuing an aggressive water main flushing program. Corrosion control chemicals (orthophosphates, polyphosphates, silicates) may inhibit biofilm formation or may coat existing biofilm so bacteria are prevented from sloughing off into the water.

The utility should also review its treatment operations and increase monitoring of the quality of water entering the distribution system, to be sure that inadequate or failed treatment is not responsible for the total coliform occurrences and that *E. coli* is not present in the water. In addition, the utility should review the operation and management of the water distribution system to insure that no *E. coli* are present, a disinfectant residual is present, and an adequate cross connection control program is in effect. The system should also monitor water quality using an alternative indicator (e.g., heterotrophic bacteria in addition to total coliforms).

Finally, the water utility should insure that large

volumes of water held in storage in reservoirs, standpipes, or above-ground tanks are not the source of the total coliform problem. Often a large volume of stored stagnant water or an open reservoir without any chlorine residual has been found to be the source of total coliforms in the water system because it permitted growth of the organisms. The organisms then were fed into the distribution system during warm-water periods when consumer demands were increased, or when events such as a large fire resulted in increased water demand that drew the stored water into the system.

Another factor in coliform occurrences is that, during warm-water periods, it is more difficult to maintain a disinfectant residual in the water in all parts of the distribution system. Since chemical reaction rates increase with increasing temperature, the disinfectant reacts more rapidly with dissolved organic chemicals in the water and in the biofilm. This increased reaction rate is often compounded by a lack of knowledge of system hydraulics and of the actual water movement throughout the system. In many cases, there are large areas where minimal movement of the water occurs. It is becoming increasingly likely that many water utilities may create microbial or chemical contaminant problems by their failure to understand how their systems work, and that better system management can reduce those problems.

V. Alternative Water Sources

In areas in which both surface and ground-water sources may not be suitable as potable source water, other sources of drinking water may be necessary. Bottled water is one alternative source, but is very expensive and is therefore not viewed as a long-term solution.

A. Bottled Water

The bacterial quality of noncarbonated bottled water varies both among brands and from lot to lot within a brand. Chemical quality varies as well. Bacterial densities in noncarbonated bottled water, determined by plate count methods, vary from 0 to $> 1.0 \times 10^5$ CFU/ml. Fresh noncarbonated bottled waters often have low bacterial densities, but during storage prior to sale the bacteria multiply, often by several orders of magnitude. Carbonated (sparkling) bottled waters generally have low or no detectable bacteria because of the low pH. In most

bottled waters examined, coliform bacteria have been absent.

Bottled water quality is regulated in the United States by the U.S. Food and Drug Administration, but must comply with EPA drinking water regulations. A significant difference between bottled waters and treated municipal waters is that chemical disinfection (other than ozonation) is not used in bottled water production. The primary means of disinfecting bottled water, if any is used at all, is ultraviolet irradiation, which leaves no residual. However, because there is no disinfectant residual, bacteria in the bottled water that survive UV disinfection will grow during storage of the bottled water.

B. Emergency Drinking Water

Treatment of water for drinking in an emergency generally involves either boiling the water or chemically disinfecting the water using liquid chlorine laundry bleach, tincture of iodine, or iodine or chlorine tablets. Water to be disinfected should be strained through clean cloth to remove particles and floating matter.

For heat disinfection, the strained water should be boiled vigorously for at least one full minute and allowed to cool before use. For chemical disinfection, add liquid chlorine bleach (4–6% chlorine) at a rate of 2 drops (clean clear water) or 4 drops (cloudy water) per quart, mix thoroughly, and allow to stand for 30 min. A slight chlorine odor should be detectable in the water; if not, repeat the dosage and let stand for an additional 15 min before using. If tincture of iodine is used, add 5 drops (clean clear water) or 10 drops (cloudy water) of 2% tincture of iodine solution per quart of water. Allow the water to stand for 30 min before use. For chlorine or iodine tablets (obtained from drug or sporting goods stores), follow the instructions on the package. Keep all purified water in clean closed containers.

C. Point-of-Use/ Point-of-Entry Treatment

Point-of-use/point-of-entry (POU/POE) treatment is the application of technology commonly used for industrial and drinking water treatment scaled down for individual home use. In the past, POU/POE treatment devices were only used for nonhealth-related water conditions such as taste, odor, color, turbidity, iron, and hardness. POU/POE treatment has taken on an expanded role because of the detection of toxics and carcinogens in drinking water. A variety of treatment technologies may be incorporated into POU/POE treatment systems, including adsorption [granular activated carbon (GAC) or powdered activated carbon], reverse osmosis, filtration, distillation, ultraviolet irradiation, ozonation, and air stripping. One or more of these technologies may be combined in the design of a POU or POE treatment device.

Activated carbon filtration is the most widely used point-of-use system for home treatment of water. Carbon units are normally the easiest to install and maintain. Operating costs are usually limited to periodic filter replacement, and performance in removing organic contaminants is good. The performance of an individual unit depends on a combination of factors, such as unit design, type and amount of activated carbon, and contact time of the water with the carbon. Most units use GAC in their design, although other types of carbon such as pressed block, briquettes of powdered carbon, and powdered carbon are available.

Bacterial growth in activated carbon units may be a problem. The organic material adsorbed onto the carbon provides nutrients for bacterial growth. Stagnation periods between operations allow time for bacteria to grow. To help prevent this problem, some units contain GAC impregnated with silver or some other bacteriostatic agent. Silver, although useful against some bacteria (e.g., coliforms and related enteric bacteria) as a bacteriostatic agent, does not effectively inhibit the growth of all bacteria in GAC POU/POE filters. Bacterial levels in water leaving GAC filters, with or without silver, range from 10^2 to 10^5 CFU/ml, but it may take longer for bacterial levels in a GAC filter with silver to reach the higher concentrations. When a GAC filter is used as the key treatment technique and the incoming water is not microbiologically safe, water from the unit should be disinfected either chemically or by UV irradiation to reduce bacterial levels in the water and to inactivate opportunistic pathogens that may colonize the carbon. Some carbon block units do not contain silver but claim bactericidal effects, probably due to straining through small carbon pore size (0.5 μm). Some POU/POE systems use a disinfectant such as ozone or UV irradiation to control bacteria.

VI. Conclusions

The technology available today is capable of providing microbiologically safe drinking water to all consumers. In practice, however, many systems are still vulnerable to waterborne disease. Reasons include (1) inability to fund the installation and use of adequate technology, (2) ignorance of the need for adequate technology, (3) inadequate or improper monitoring to insure that barriers against waterborne disease remain intact, (4) lack of real-time monitoring data, and (5) lack of adequate data on the effectiveness of various treatment practices against certain pathogens. In addition, with increased population pressure, source water quality may decline over time, especially in areas where water is scarce; thus, the technology used becomes inadequate. These problems are common on a global basis. The challenge globally is to highlight the imperative of safe drinking water and to find solutions to these problems.

Bibliography

American Public Health Association (1989). "Standard Methods for the Examination of Water and Wastewater," 17th ed. Washington, D.C.

Amirtharajah, A. (1986). *J. Amer. Water Works Assoc.* **78,** 34–49.

Clark, R. M., and Tippen, D. L. (1990). Water supply In "Standard Handbook of Environmental Engineering" (R. Corbitt, ed.), pp. 5.1–5.225. McGraw-Hill, New York.

Committee on the Challenges of Modern Society (1984). "Drinking Water Microbiology." EPA 570/9-84-006. U.S. Environmental Protection Agency, Office of Drinking Water, Washington, D.C.

Craun, G. F. (ed.) (1986). "Waterborne Diseases in the United States." 295 pp. CRC Press, Boca Raton, Florida.

McFeters, G. A. (ed.) (1989). "Drinking Water Microbiology." 502 pp. Springer-Verlag, New York.

Montgomery, J. M. (1985). "Water Treatment Principles and Design." 696 pp. John Wiley and Sons, New York.

O'Melia, C. R. (1985). Coagulation In "Water Treatment Plant Design" (R. L. Sanks, ed.), pp. 65–81. Ann Arbor Science Publishers, Ann Arbor, Michigan.

U.S. Enviromental Protection Agency (1991). "Manual of Small Public Water Supply Systems." EPA 570/9-91-003. Washington, D.C.

Wine

Keith H. Steinkraus
Cornell University

Glossary

Dry Absence of sweetness in wine
Enology Science of wine-making
Must Expressed grape juice or the mashed grape before and during fermentation
Vinification Technology of wine-making
Viticulture Growing of grapes

WINE is an alcoholic beverage made by fermentation of grape juice or other sugar-containing substrates including honey, sugar cane, fruit juices, and other plant juices containing sugars such as palm tree sap, floral extracts, and *Agave,* the century cactus plant. Wine is distinguished from beer in which the fermentable sugars are derived principally from starches by means of amylases produced in cereal grains, particularly barley, by germination or malting. The term ''wine'' probably should not be applied to beverages that involve the hydrolysis of starch in cereals through ptyalin in saliva introduced by chewing, the hydrolysis of starch to sugars via fungal or bacterial amylases, or the application of microorganisms that hydrolyze the starch to sugars and then ferment the sugars to ethanol. Thus, South American chicha made by chewing maize and subse-

quently fermenting it should probably be called a maize beer. Rice wine, such as Japanese sake, made by overgrowing boiled rice with *Aspergillus oryzae* as a source of amylases and then fermenting with yeast should probably more correctly be termed rice beer; however, accepted usage may preclude this.

This article will restrict the term ''wine'' to products made by the fermentation of sugar-rich substrates in which starch hydrolysis is not required.

I. Antiquity of Wine Fermentation

Wine fermentation is one of the oldest and perhaps the oldest fermentation known to humans. Honey diluted with rain water ferments spontaneously with yeasts in the environment. Fruit juices ferment spontaneously with yeasts on the fruit. Mexican pulque, one of the most ancient beverages of the Mayans and Aztecs, required that the core of the *Agave* cactus plant be harvested and mashed, and then it underwent spontaneous fermentation by yeasts and other microorganisms in the environment.

Wine was truly a gift of God (or nature) to primitive humans. Fruit juices ferment spontaneously, yielding a product that preserves and, in fact, enhances the nutritive value and retains essential flavors of the fruit. Conversion of the sugar to ethanol and ethanol's striking effect on the psyche of the consumer led naturally to the conclusion that the ''spirit'' of the fruit had been extracted for the benefit of the consumer.

Wine or alcoholic fermentation was present on earth many millions of years before humans evolved or were created. Fossil microorganisms have been found in rocks more than 3 billion years old. Plants, the basis for human's food, likely evolved about a billion years later. Thus microorganisms required for fermentation and plants providing the substrates were well established on earth at least a billion years before humans.

Microorganisms have the primary task of recycling organic matter including sugars; yeasts in the environment ferment the sugars to alcohol as fruits become overripe and fall to the ground. Alcohol is one of the products of recycling reactions. It, in turn, is further fermented to acetic acid or respired to CO_2 and H_2O in continuing recycling reactions.

There are many references to wine in the Bible. The book of Genesis, Chapter IX, verses 20 and 21, states: "And Noah began to be a husbandman, and he planted a vineyard. And he drank of the wine and was drunken." Some anthropologists have suggested that it was the alcoholic fermentation that caused humankind to change from hunter/gatherer to agriculture to satisfy human's desire for alcohol.

From the beginnings of civilization, wines have been associated with culture, art, and religion as early humans considered wines a miraculous conversion. Alcoholic beverages have always been closely related to politics, and even today, governments use alcoholic fermentations as major methods of raising taxes. In very primitive times in South America, emperors could hold office only as long as they provided the people with a sufficient supply of chicha. From ancient times, wines have played a vital role in religious ceremonies, and they continue to do so today. From the Middle Ages on, religious orders and monasteries had a major role in the manufacture and supply of wines for religious and also other cultural activities.

II. Alcoholic Fermentation

The biochemistry of the alcoholic fermentation is among the most thoroughly studied enzyme–substrate sequences ever investigated. Black, in the eighteenth century, reported that ethyl alcohol and carbon dioxide were the products of sugar fermentation. Lavoisier, in the same century, reported that 95.9 parts of sugar yielded 57.7% ethyl alcohol, 33.3% carbon dioxide, and 2.5% acetic acid. Gay-Lussac was the first to show that glucose fermentation yielded approximately equal weights of ethanol and carbon dioxide (45 parts glucose yielded 23 parts alcohol and 22 parts carbon dioxide). Pasteur found that 100 parts sucrose yielded 105.4 parts of invert sugar, which was fermented to 51.1 parts of ethyl alcohol, 49.4 parts of carbon dioxide, 3.2 parts of glycerol, and 1.7 parts of succinic acid. In Neuberg's normal fermentation (Scheme 1), 1 mole of glucose yields 2 moles of pyruvic acid, which, in turn, yield 2 moles of carbon dioxide and 2 moles of acetaldehyde, which, in turn, are reduced to 2 moles of

ethanol. Detailed research has shown that there are actually 12 intermediate steps between glucose and ethyl alcohol (the Embden-Meyerhof scheme), which account for the production of 2 moles of carbon dioxide, 2 moles of ethanol, and smaller quantities of glycerol from 1 mole of glucose.

Buchner in the nineteenth century produced ethyl alcohol from sugar using cell-free juice, made by grinding yeast cells with sand under pressure and filtering to remove whole cells. This demonstrated that it was enzymes that were required for fermentation and not the whole cells themselves.

III. Yeasts

The principal microorganisms in the alcoholic fermentation of wines are yeasts belonging to genus *Saccharomyces*, in particular *S. cerevisiae*, *S. cerevisiae* var. *ellipsoideus*, *S. bayanus*, and *S. oviformis*. Although yeasts belonging to genus *Saccharomyces* are generally present on the grapes or other fruits and they are capable of fermenting the juice to wine, it is generally accepted procedure, particularly in wine companies, to inoculate pure cultures of selected wine yeasts. These may be selected for flavor, production of little or no foam, ability to ferment at low temperatures, or flocculation/sedimentation behavior. They are more often selected for absence of any undesirable flavors in wines.

To ensure that the selected wine yeast predominates over wild yeasts and other microorganisms that may be present, it is also common practice to add 20 to 50 mg sulfur dioxide per liter to the juice, usually in the form of bisulfite, before inoculating the selected yeast, which is generally tolerant or acclimated to these concentrations of sulfite.

Saccharomyces cerevisiae and other wines yeasts have the ability to grow and metabolize sugars such as glucose, fructose, and sucrose aerobically or anaerobically. Aerobic growth yields far more energy and the yeasts multiply vigorously, but little, if any, ethanol is produced. Anaerobically, yeasts dissimilate glucose according to the Emden-Meyerhof scheme. Theoretically 1 mole of glucose (180 g) yields 2 moles (92 g) of ethanol and 2 moles (88 g) of carbon dioxide. Actually, yields are slightly less because of the production of glycerol and other minor compounds; nevertheless, it is practical to consider that 2 g of sugar yields slightly less than a gram of ethanol. Anaerobic dissimilation of 1 mole of glucose yields 56 kcal of energy to the yeast, part of which is used in the fermentation itself and a portion in multiplication of the yeast.

Yeasts multiply by budding, and the percentage of yeast cells showing buds when examined microscopically in culture or in a fermenting wine showing buds is a measure of the vigor of the culture.

Calculated on an overall fermentation basis, an individual wine yeast cell produces about 30 million molecules of ethanol/cell/sec. In the earlier stages of fermentation, a yeast cell may produce 100 million ethanol molecules/cell/sec. As ethanol content reaches 12% v/v, the rate of ethanol production may fall to 10 million molecules/cell/sec. To maintain high levels of ethanol production, it is necessary to introduce small amounts of oxygen (e.g., 13% of oxygen saturation) into the must. Some wine producers wanting to achieve high concentrations of ethanol expose the must to air during the early stages of fermentation.

IV. Malo-Lactic Fermentation

Another group of microorganisms accompanying the yeast in wine fermentations is the lactic acid bacteria. Lactic acid bacteria–yeast interactions are rather common in food/beverage fermentations. Grape juice contains mainly tartaric acid and malic acid. If the acidity is too high for best flavor, it may be desirable for lactic acid bacteria to ferment the malic to lactic acid, which decreases the total acidity and raises the pH of the wine. It also modifies the aroma, body, and aftertaste (Henick-Kling). If the grape juice is low-acid, a malo-lactic fermentation may be undesirable as it will decrease the acidity still further making the wine taste flat. Malo-lactic acid fermentation should occur during the alcoholic fermentation or during holding before final clarification and fining as it increases the microbiological stability, decreases fermentable carbohydrates, and produces inhibitory bacteriocins in the wine (Henick-Kling). If malo-lactic fermentation occurs in the bottle, it may result in cloudy wine or carbonation of a still wine. [*See* Bacteriocins: Activities and Applications.]

V. Nobel Mold

Under high-humidity conditions, unripened grapes may become infected with *Botrytis cinerea*, the no-

bel rot. If the humidity decreases, the grapes lose moisture and, in fact, become shriveled. Such grapes yield a juice with a higher sugar content, the grape flavor may also be concentrated, and the mold may introduce a delicate flavor. Under ideal conditions, very fine flavors can be produced from grapes infected with the nobel mold.

VI. Microorganisms Causing Spoilage in Wines

The major organisms causing spoilage in wines are those belonging to genus *Gluconobacter* and genus *Acetobacter*. *Acetobacter* includes the organisms producing acetic acid/vinegar from ethyl alcohol. *Acetobacter* is highly aerobic and, in fact, grows on the surface of the must or wine if oxygen is present. The acetic acid bacteria are not generally a problem as long as the fermenting must and the wines are kept totally anaerobic. During active fermentation, carbon dioxide is being rapidly produced, and it blankets the must, keeping it anaerobic.

Other wine spoilage organisms include film-forming yeasts such as genera *Hansenula* and *Pichia*. *Hansenula* is an esterifying yeast that produces fruity aromas such as ethyl acetate. Osmotolerant yeasts such as *Schizosaccharomyces pombe* may produce haze in stored wines and bottles. *Brettanomyces* produces off-flavors if it develops during storage.

VII. Microbiology of Sherry Wines

Sherry wines have an oxidized flavor/color/aroma and are highly acceptable among many wine consumers. The oxidized flavor arises by conversion of a portion of the ethanol to acetaldehyde and other complex reactions. A type of sherry flavor can be introduced by baking a still wine. The still wine is "baked" (i.e., held at elevated temperatures with air being injected into the wine) to produce the sherry flavor/aroma. In a true fermented sherry, the still wine is filled into half-filled barrels that are incubated in the sun often on the roof of the winery. A type of baking does occur over a long period of time, and the surface of the wine is exposed to air. The sherry flavor is produced by film-forming yeasts such as *Saccharomyces oviformis* and *S. bayanus* or related yeasts. In the absence of these yeasts, *Acetobacter* could invade and spoil the wine.

VIII. Wine Grapes

Grape wines are made mainly from species *Vitis vinifera*. This is particularly true in Europe, where the major producers are France and Italy. The native grapes in the United States belong to the species *Vitis labrusca* and several other species that have a characteristic fruity, floral ("foxy") aroma, distinctly different from *vinifera* grapes. Hybrids of *V. vinifera and V. labrusca* have been developed to yield grape varieties able to withstand a colder climate and yield wines more similar to the European types. *Vitis rotundifolia* is an important species in the southeastern United States. Most American wine is produced in California, based on cultivation of *V. vinifera*. *Vitis vinifera* wines are becoming increasingly more important in the northeast United States.

IX. Wine Classification

Wines are classified according to color [white, red, or rosè (pink)] and how much ethanol they contain. Wines with 7% to 14% ethanol are described as table wines; more than 14% ethanol designates fortified wines such as port and sherry. Wines are dry (nonsweet), semi-dry, semi-sweet, or sweet. Even a dry (nonsweet) wine contains small quantities of sugar.

Still wines given a secondary fermentation in the bottle or in a tank to produce carbon dioxide under pressure are called "sparkling." Outside the United States, the term "champagne" is legally reserved for use only with sparkling wines made in the Champagne region of France. Wines are also designated, at times, by the type of grape from which they have been made.

There are thousands of cultivars of *Vitis* species. Many yield wines with flavors/aromas reflecting their distinct type. The reader is referred to the references at the end of this article for further information. A very wide variety of flavors/aromas/colors of wines can be produced. The climate and the soil on which the grape variety grows and the particular growing year can have a marked effect on the flavor of the resulting wine. This leads to "vintage" years in which a particular variety yields unusually high-quality wine.

X. Steps in Manufacture of Wine

1. *Harvesting:* Grapes should be harvested when they are at their peak of maturity with desired flavor with a high sugar content and desirable acidity. They should be as intact as possible and be processed quickly.
2. For white wines, the stems are removed and the grapes are crushed. The skins are retained for red wines. Sulfite may be added to inhibit oxidation and growth of the natural microbial flora.
3. Pressing is separation of the juice from the skins, seeds, and pulp—the pomace. Pectinase may be added to facilitate juice extraction. Free-run juice, the juice running from the press before pressure is applied, is considered to be best for the highest-quality wines.
4. *Fermentation:* Fermentation can proceed either through the action of the naturally present yeast or by inoculation of a selected pure yeast culture. In industry the yeast inoculum may be produced in the company's own laboratory or it may be purchased as an active dry yeast from companies specializing in production of industrial cultures. Fermentation temperature is generally 10° to 15°C for white wines and somewhat higher for red wines. The higher the temperature, up to about 30°C, the more rapid the fermentation; however, the slower, lower temperature fermentations are considered to lead to higher-quality wines. Less volatile aroma is lost. As the fermentation proceeds, the actively metabolizing yeasts produce heat and the temperature of the must tends to rise. If not controlled, the temperature can reach a level where yeast cells are killed or the ethanol becomes so inhibitory that the maximum ethanol content is decreased.

Sugar content of the juice must be sufficient to yield a wine with the desired alcohol content. This requires a sugar content of about 13% w/v to yield a 7% v/v alcohol wine or about 22% to yield a 12% v/v alcohol wine. The sugar content of the juice may be raised before or during fermentation to adjust the final ethanol content. Water may be added to lower the acidity where the law allows amelioration.

A weight of carbon dioxide gas almost equivalent to the weight of ethanol produced is released during fermentation. This causes the fermentation to "boil" during its most active phases.

The fermentation gradually ends as the fermentable sugar is consumed. With red wines, the pomace, skins, seeds, and other insoluble matter must be removed. This is usually performed when color and tannin extraction have reached the maximum toward the end of fermentation or several days after completion of the fermentation. Separation of the pomace is accomplished by draining and pressing the must. With white wines, the pomace has already been removed during the pressing before fermentation.

5. *Racking:* As the fermentation nears completion, the yeast cells tend to settle, and after fermentation has been completed, the wine can be decanted or siphoned (racked) as the first step in clarification of the wine. Racking is repeated periodically until most or all the yeast cells (the lees, tartrate crystals, and other insolubles) have been separated from the wine. This may be combined with cold stabilization, which is accomplished by storing the wine at about −2°C. Cold stabilization removes excess tartrates and other materials that might cause cloudiness in the bottle later.

6. *Clarification:* Most wine consumers like a crystal clear wine with no haze or sediment. To accomplish this objective, gelatin, isinglass, or egg white may be added to the wine. These ingredients combine with tannin and form a fine precipitate that removes some of the materials that might cause turbidity. White wines may not contain sufficient tannin. In that case, some tannin sufficient to react with the added protein may be necessary to produce the precipitate and stabilize the wine. Bentonite, a clay, is also widely used for clarification. Ion exchange may be used to change potassium to sodium tartrate, thus increasing the solubility and reducing the likelihood of its precipitation later.

7. *Bottling:* Wines are generally bottled in dark green or brown bottles to decrease deleterious effects on quality that can be caused by sunlight. Small amounts of sulfite may be added to inhibit oxidation. Sweet wines may be fortified with ethanol (18–20%) or the wines may be pasteurized or sterile filtered (the choice for high quality wines) to prevent growth of contaminants in the bottle.

8. *Aging:* Aging occurs after fermentation. It can occur in tanks, in barrels, or in the bottle. The flavor may be improved through esterification reactions in which small portions of ethanol combine with organic acids in the wine. Aging improves some wines but has little effect on others.

The great enemy of stability in wine is oxygen and oxidation. Barrels, tanks, or bottles holding the wine must be kept free of oxygen. This is done principally by keeping the containers full or replacing the atmosphere in the container with nitrogen or carbon dioxide. In the case of sherries, however, there is a deliberate oxidation to modify the flavor, as discussed earlier.

XI. Sparkling Wines

Sparkling wines require unique processing, at least in the secondary fermentation that occurs in the bottles or in tanks under pressure (Charmat process). The initial sugar content of the grapes should yield a dry, still wine with about 10% v/v ethanol. The base wine, cuvee, is stabilized, fined, and filtered. It is often a blend of several wines to obtain the acidity and alcohol content desired with a clean flavor—no off-flavors or off-odors. Special wine yeasts that tend to agglomerate and thus can be more easily removed after the secondary fermentation are generally used. These same yeasts may also be used for the primary fermentation if desired. It is particularly important that the yeasts do not produce any off-odors or off-flavors, as part of the sparkling wine flavor is derived from the yeast cells as they autolyze after the secondary fermentation. Sugar addition for the secondary fermentation is carefully controlled, as too little added will lead to a flat sparkling wine and too much sugar can lead to dangerous pressures and even explosions of the bottles. A sugar concentration of 2.4% w/v is enough to produce 7 atmospheres pressure in the sparkling wine bottles, which generally will withstand 8 atmospheres pressure.

The secondary fermentation can occur either in the bottle or in special pressurized tanks designed for the purpose (Charmat process). The tanks offer the most efficient, lowest cost method of manufacturing sparkling wines; however, fermentation in the bottle with the longer contact of the wine with the yeast, particularly during autolysis, even though less efficient and more costly, generally yields the highest-quality sparkling wines. If fermentation occurs in the bottle, the yeast cells and any precipitates must be removed after the secondary fermentation. This involves the riddling process whereby bottles are placed neck down in racks and turned a quarter turn every day or so until the yeast and all sediment in the bottle is in the neck against the cork or often now a crown seal. The bottles are then cooled to close to freezing, and the neck of the bottle containing the yeasts and precipitates is frozen. The bottle is then opened and held at such angle that the ice plug is blown out of the bottle. The volume lost is immediately replaced with a small quantity of sugar in wine (sometimes brandy) called "dosage." A dry sparkling wine is called brut and contains from 0.5% to 1.5% sugar; sec indicates 2.5% to 4.5% sugar; doux contains 10% sugar. The bottles are sealed with a final cork and stored on their sides to age or labeled and sent to the distributor.

XII. Fortified Wines

Although it is possible to ferment wines to ethanol levels of 18% v/v or higher if sufficient sugar is present and a lengthy fermentation is acceptable, wines containing higher concentrations of ethanol (i.e., 17% to 21%) have been generally fortified to the higher ethanol concentration by the addition of wine spirits produced by distillation of wine. Wine spirits contain about 95% ethanol v/v and are generally manufactured by distillation of lower-quality wines.

Sweet dessert wines such as port, muscatel, or sherry with sugar contents of 12% generally start with grapes containing more than 25% sugar. Such wines are fortified by the addition of wine spirits to an ethanol content of 17% to 21% v/v. *Acetobacter* is inhibited at this ethanol concentration, and such wines are quite stable at room temperature even in partially filled bottles.

XIII. Flavor and Acceptability

Wine preferences are strictly personal. No other fermented beverages offer so much variety in flavor, aroma, color, sweetness, and carbonation. Consumers should drink what they like within their eco-nomic means. There are many excellent values at reasonable cost on the market. Many new wine consumers prefer the sweet wines initially but, over time, come to prefer dry wines, which provide more of the subtle flavor/aromas characteristic of individual grape varieties and which vary from different countries and parts of the world.

Bibliography

Amerine, M. A. (1980). "The Technology of Wine Making," 4th ed. Avi, Westport, Connecticut.
Amerine, M. A., and Roessler, E. B. (1983). "Wines. Their Sensory Evaluation." W. H. Freeman & Co., San Francisco.
Davis, C. R., Wibowa, D., Eschenbruch, R., Lee, T. H., and Fleet, G. H. (1985). *Am. J. Enol. Vitic.* **36**, 290–301.
Henick-Kling, T. (1988). Yeast and bacteria control in winemaking. *In* "Modern methods of Plant Analysis," new series, vol. 6 (H. F. Linskens, and J. F. Jackson, eds.). Springer Verlag, Berlin.
Henick-Kling, T. (1991). Malolactic fermentation. *In* "Wine Microbiology and Biotechnology." (G. H. Fleet, ed.). Harwood Academic Publisher (Chur).
Johnson, H. (1977/1985). "The World Atlas of Wine." Simon and Schuster, New York.
Lafon-Lafourcade, S. (1983). Wine and brandy. *In* "Biotechnology, vol. 5. Food and Feed Production with Microorganisms" (H. J. Rehm and G. Reed, eds.). VCH, New York.
Morse, R. A., and Steinkraus, K. H. (1975). Wines from the fermentation of honey. *In* "Honey, A Comprehensive Survey" (E. Crane, ed.). Heinemann, London.
Peynaud, E. (1984). "Knowing and Making Wine." John Wiley and Sons, New York.
Pool, R. and Henick-Kling, T. (1991). "Production Methods in Champagne." New York State Agricultural Experiment Station. Cornell University, Geneva, New York.
Reed, G., and Nagodawithana, T. W. (1991). "Yeast Technology," 2nd ed. Van Nostrand Reinhold, New York.
Subden, R. E. (1984). *Dev. Ind. Micro.* **25**, 221–229.
Troost, R. (1988). "Technologie des Weines." Handbuch der Lebensmitteltechnologie. Ulmer (Stuttgart).
Vine, R. P. (1981). "Commercial Winemaking." Avi, Westport, Connecticut.

Linkage Maps

Linkage map of *Bacillus subtilis*   Genetic map of *B. subtilis* 168. The single, circular chromosome is divided into 360° and presented in linear segments for convenience. [Reproduced, with permission, from O'Brien, S. J. (ed.) (1990). "Genetics Maps: Locus Maps of Complex Genomes." 5th ed. pp. 2.28–2.53. Cold Spring Harbor Laboratory Press, Cold Spring Harbor, New York.] Map begins on page 407.

Linkage map of *Escherichia coli*   Linear scale drawings representing the circular linkage map of *E. coli* K-12. The time scale of 100 min, beginning arbitrarily with zero at the *thr* locus, is based on the results of interrupted conjugation experiments. Parentheses around a gene symbol indicate that the position of that marker is not well known and may have been determined only within 5 to 10 min. An asterisk indicates that a marker has been mapped more precisely but that its position with respect to nearby markers is not known. Arrows above genes and operons indicate the direction of transcription of these loci. [Reproduced, with permission, from Bachmann, B. (1990). *Microbiol. Rev.* **54**, 132–135.] Map begins on page 410.

Linkage map of *Salmonella typhimurium*   Linkage map of *S. typhimurium*, represented as 10 segments. The scale of 100 min begins at zero for the *thr* loci. The segmented line to the right of the gene symbol indicates that the genes are jointly transduced; the numbers to the right of the segmented line indicate the linear distance between genes. This linear distance was determined from the fragment of joint transduction and was calculated by assuming that the length of P22, KB1, and ES18 transducing fragments is 1 min, whereas that of P1 is 2 min, and applying the formula developed by Wu to convert the percentage of joint transduction to map distance. Parentheses around a gene symbol indicate that the location of the gene is known only approximately, usually from conjugation studies. An asterisk indicates that a gene has been mapped more precisely, usually by phage-mediated transduction, but that its position with respect to adjacent markers is not known. Arrows to the extreme right of genes and operons indicate the direction of mRNA transcription by these loci. Daggers are shown to the right of a few genes; these genes of *S. typhimurium* are carried on an F-prime factor, and this plasmid was shown to complement *E. coli* K-12 mutations of that gene; mutant alleles of *S. typhimurium* have not been tested directly. [Reproduced, with permission, from Sanderson, K. E. (1988). *Microbiol. Rev.* **52**, 486–487.] Map begins on page 414.

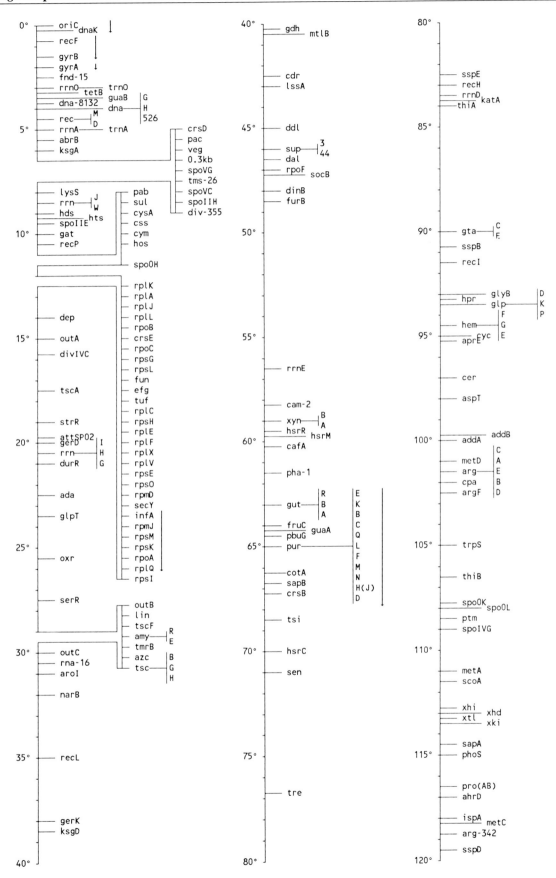

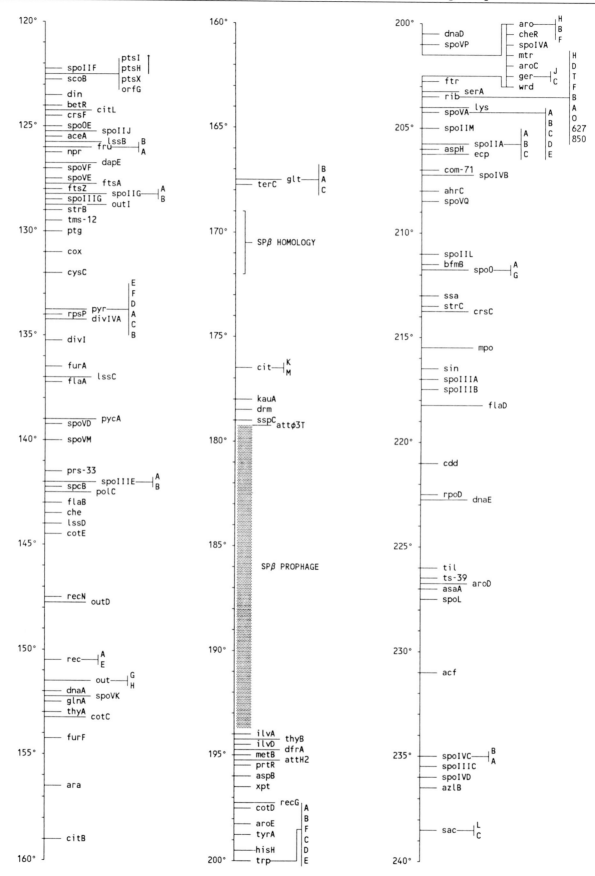

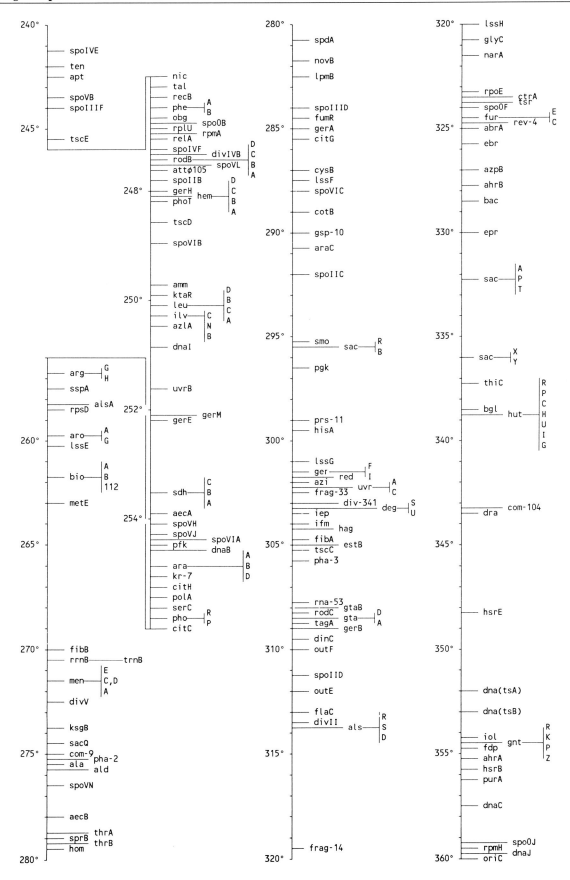

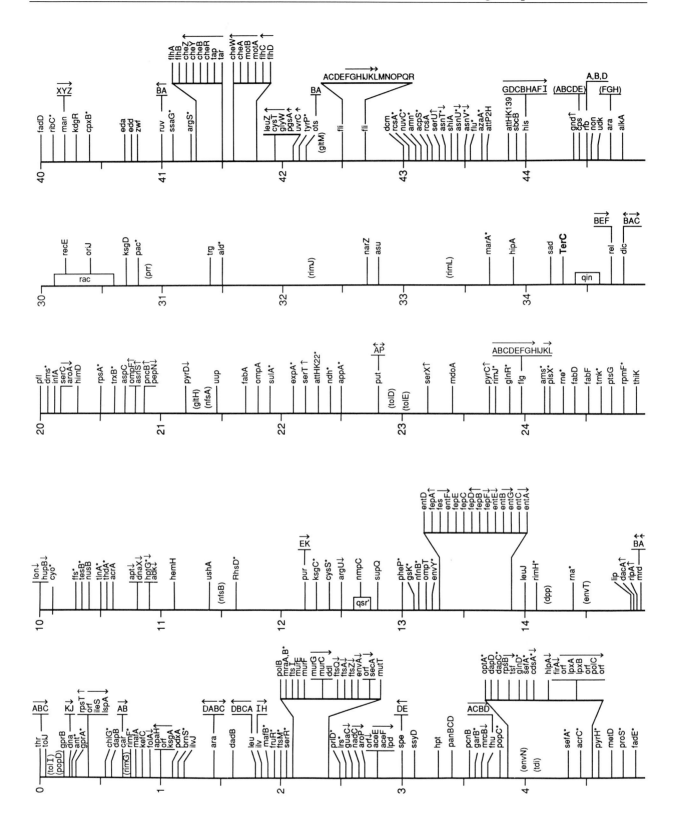

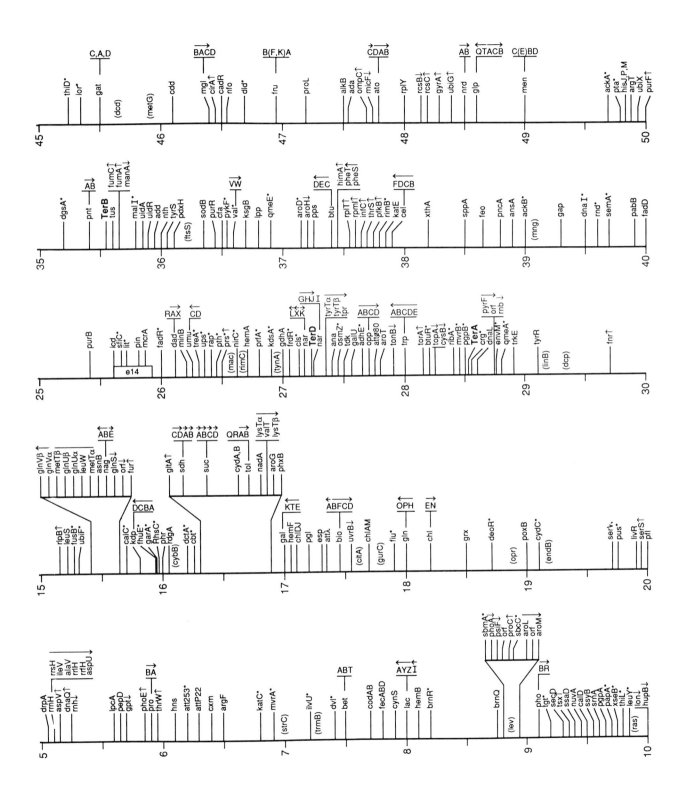

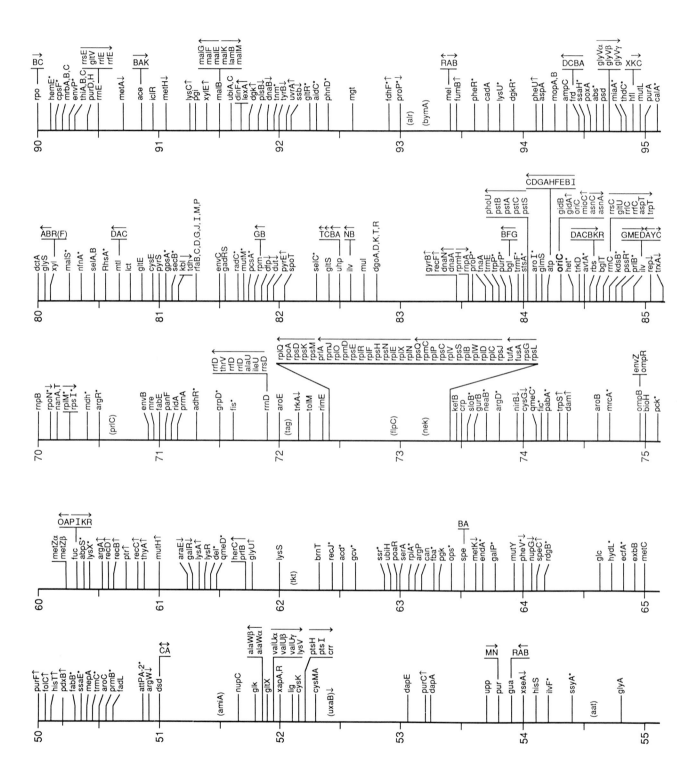

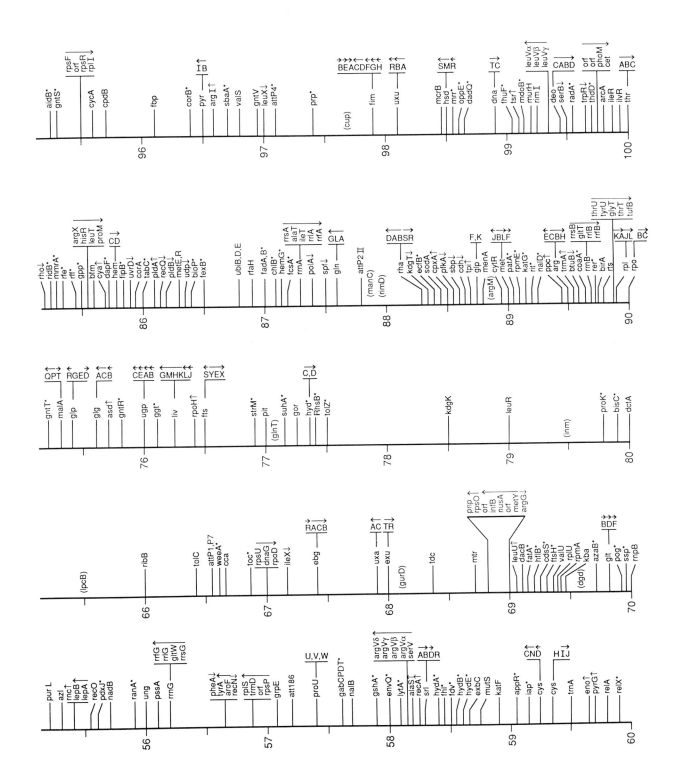

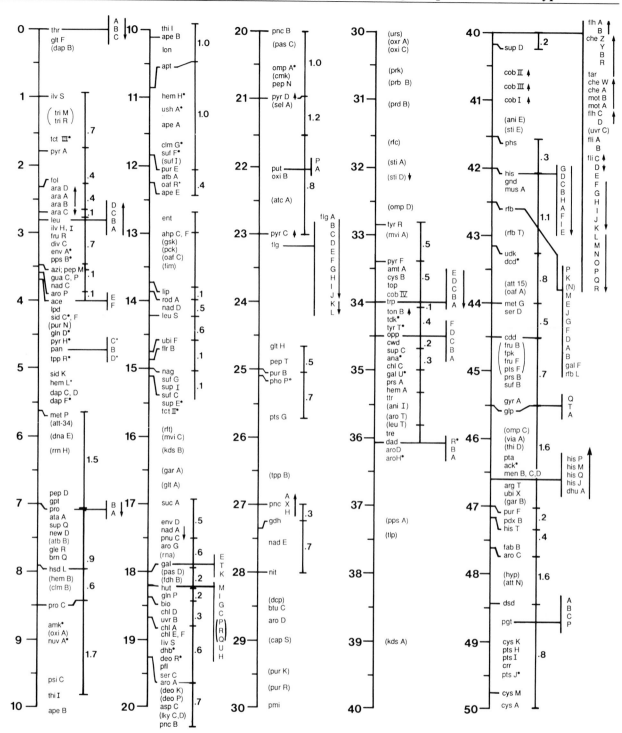

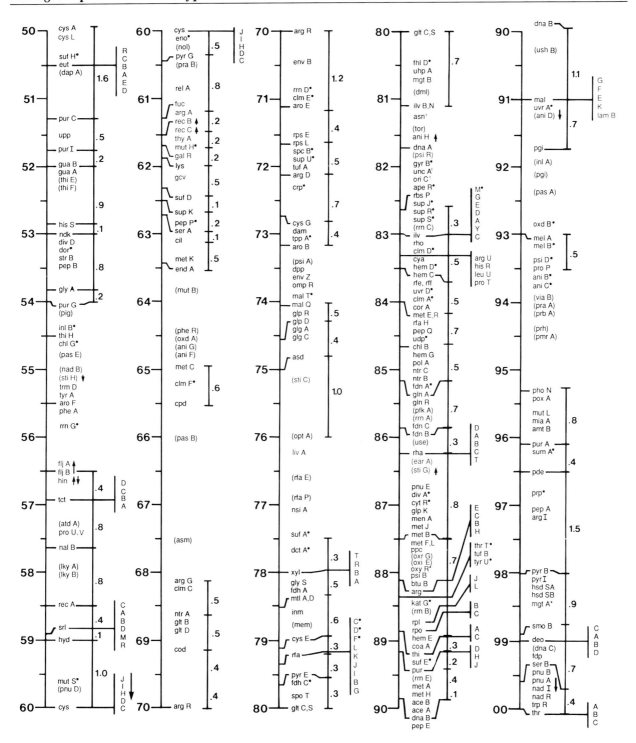

Contributors

Numbers in parentheses indicate the volumes and pages on which the authors' contributions begin.

I. Jerome Abramson (4:61)
U.S. Food and Drug Administration
Center for Drug Evaluation and Research
Reisterstown, Maryland 21136-5609

Hans-Wolfgang Ackermann (1:203)
Faculty of Medicine, Université Laval
Cité Universitaaire
Québec, Canada G1K 7P4

Louis A. Actis (3:431)
Oregon Health Sciences University
Portland, Oregon 97201

Edwin W. Ades (2:467)
Centers for Disease Control
Atlanta, Georgia 30333

Daniel Amsterdam (4:157)
School of Medicine
SUNY at Buffalo, and Erie County
    Medical Center
Buffalo, New York 14215

Eurico Arruda (3:555)
Department of Internal Medicine
University of Virginia School of Medicine
Charlottesville, Virginia 22903

John P. Atkinson (1:539)
Howard Hughes Medical Institute Research
    Laboratories
Department of Medicine
Division of Rheumatology
Washington University School of Medicine
St. Louis, Missouri 63110

Ronald M. Atlas (3:363)
Department of Biology
University of Louisville
Louisville, Kentucky 40292

Jean-Paul Aubert (1:467)
Unite de Physiologie Cellulaire
Institute Pasteur
Paris 15EME, France

S. F. Barefoot (1:191)
Department of Food Science and
    Microbiology
A-203 J Poole Agricultural Center

Clemson University
Clemson, North Carolina 29634

Tamar Barkay (3:65)
U.S. Environmental Protection Agency
Environmental Research Laboratory
Sabine Island, Gulf Breeze, Florida 32561

Larry L. Barton (4:135)
Department of Biology
University of New Mexico
Albuquerque, New Mexico 87131

Pierre Béguin (1:467)
Unite de Physiologie Cellulaire
Institute Pasteur
Paris 15EME, France

Abbas M. Behbehani (4:33)
The University of Kansas School of
    Medicine
Kansas City, Kansas 66160

Angela Belt (2:561)
The Biotic Network
Sonora, California 95370

Peter M. Bennett (4:311)
Department of Pathology and Microbiology
University of Bristol
Bristol BS8 1TD, United Kingdom

Paul S. Berger (4:377)
U.S. Environmental Protection Agency
Office of Groundwater and Drinking Water
Washington, D.C. 20460

Mathias Bernhardt (1:495)
Wisconsin State Laboratory of Hygiene
University of Wisconsin
Madison, Wisconsin 53706

Andrew N. Binns (1:37)
Department of Biology
University of Pennsylvania
Leidy Laboratory of Biology
Philadelphia, Pennsylvania 19104-6018

Carolyn M. Black (1:123)
Centers for Disease Control
Atlanta, Georgia 30333

R. Blaszczyk (2:473)
Department of Chemical Engineering
University of Western Ontario
London, Ontario, Canada N6A 5C1

Seymour S. Block   (4:87)
Department of Chemical Engineering
College of Engineering
University of Florida
Gainesville, Florida 32611

Jean-Marc Bollag   (1:269)
Environmental Resources Institute
The Pennsylvania State University
University Park, Pennsylvania 16802

Wendy B. Bollag   (1:269)
Environmental Resources Institute
The Pennsylvania State University
University Park, Pennsylvania 16802

Claude Bollet   (4:179)
Hospital Salvator
Bd De Sainte-Marquerite-BP51
13274 Marseille Cedex 9, France

Diane C. Bosse   (2:467)
Centers for Disease Control
Atlanta, Georgia 30333

George H. Bowden   (3:269)
Department of Oral Microbiology
Faculty of Dentistry
University of Manitoba
Winnipeg, Canada R3T 2N2

Daniel K. Brannan   (1:593)
Abilene Christian University
Abilene, Texas 79699

Bonnie Bratina   (3:121)
Gray Freshwater Biological Institute
University of Minnesota
Navarre, Minnesota 55392

William F. Braun, Jr.   (3:217)
University of Missouri
College of Veterinary Medicine
Veterinary Medical Diagnostic Laboratory
St. Louis, Missouri 65205

N. J. Brewin   (3:239)
John Innes Institute
Norwich NR4 7UH
United Kingdom

Stephanie K. Brodine   (2:95)
Epidemiology Department
Naval Health Research Center
San Diego, California 92134

Marvin P. Bryant   (3:97)
Department of Animal Sciences
University of Illinois at Urbana-Champaign
Urbana, Illinois 61801

Jerry M. Buysse   (2:29)
Department of Enteric Diseases
Walter Reed Army Institute of Research

Washington, D. C. 20012

Gerhard C. Cadée   (1:67)
The Netherlands Institute for Sea Research
1790 AB Denburg
POB 59 Texel, The Netherlands

Stan W. Casteel   (3:217)
University of Missouri
College of Veterinary Medicine
Veterinary Medical Diagnostic Laboratory
St. Louis, Missouri 65205

K. -J. Cheng   (2:311)
Agriculture Canada Research Station
Lethridge, Canada

Chee-Kok Chin   (3:409)
Department of Horticulture
Rutgers, The State University of New Jersey
New Brunswick, New Jersey 08903

Robert M. Clark   (4:377)
U.S. Environmental Protection Agency
Office of Groundwater and Drinking Water
Washington, D.C. 20460

Jill E. Clarridge III   (2:331)
Baylor College of Medicine
Baylor University
Waco, Texas 76703

Douglas J. Cork   (3:357; 4:151)
Institute of Gas Technology
Illinois Institute of Technology
Chicago, Illinois 60616

J. William Costerton   (1:277; 2:311)
Faculty of Science
Department of Biological Sciences
University of Calgary
Calgary, Alberta, Canada T2N 1N4

Jorge H. Crosa   (3:431)
Oregon Health Sciences University
Portland, Oregon 97201-3098

Sidney A. Crow   (4:391)
Georgia State University
Atlanta, Georgia 30303

Stephanie E. Curtis   (1:627)
Department of Genetics
North Carolina State University
Raleigh, North Carolina 27695

Richard D'Ari   (2:611)
Institut Jacques Monod and Centre National
   de la Recherche Scientifique
Universite Paris 7
75221 Paris Cedex, France

Albert E. Dahlberg   (3:573)
Division of Biology and Medical
   Science

Brown University
Providence, Rhode Island 02912

Stephen F. Dealler   (4:299)
Department of Microbiology
University of Leeds
Leeds LS2 9JT, United Kingdom

Ph. de Mico   (4:179)
Hospital Salvator
Bd De Saite-Marquerite-BP51
13274 Marseille Cedex 9, France

Rutger de Wit   (4:105)
Centro de Investigacion y Desarrollo
Barcelona 08034, Spain

Edward F. Delong   (2:405)
University of California, Santa Barbara
Santa Barbara, California 93106

Gerald A. Denys   (2:493)
Department of Pathology and Laboratory
   Medicine
Methodist Hospital of Indiana
Indiapolis, Indiana 46202

Warren A. Dick   (4:123)
Agronomy Department
Ohio State University
Wooster, Ohio 44691

C. H. Dickinson   (2:449)
Ridley Building, University of Newcastle
Newcastle Upon Tyne
NE1 7RU, United Kingdom

David J. Dickinson   (4:231)
Department of Biology
Imperial College of Science, Technology, and
   Medicine
Prince Consort Road
London SW7 2BB, United Kingdom

Roy H. Doi   (2:259)
Department of Biochemistry and Biophysics
University of California, Davis
Davis, California 95616

A. E. Douglas   (4:165)
Department of Zoology
University of Oxford
Oxford OX1 3PS, United Kingdom

J. A. Downie   (3:239)
John Innes Institute
Norwich NR4 7UH, United Kingdom

R. J. Doyle   (1:479)
Microbiology Health Science
Department of Microbiology and Immunology
University of Louisville School of
   Medicine
Louisville, Kentucky 40292

Harold L. Drake   (1:1)
Universitat Bayreuth
Bayreuther Institut fur Terrestrische
Okosystemforschung (BITOK)
Dr. -Hans-Frisch-Str. 1-3
Postfach 10 12 51
D-8580 Bayreuth, Germany

Gordon R. Dreesman   (2:371)
BioTech Resources, Inc.
San Antonio, Texas 78249

Karl Drlica   (1:517)
Public Health Research Institute, and New
   York University
New York, New York 10016

Herbert L. DuPont   (2:63)
Center for Infectious Diseases
The University of Texas Medical School
School of Public Health
Houston, Texas 77030

Daniel Dykhuizen   (3:351)
Division of Biological Sciences
Department of Ecology and Evolution
SUNY at Stony Brook
Stony Brook, New York 11794-5245

Henry L. Ehrlich   (3:75, 283)
Department of Biology
Rennsselaer Polytechnic Institute
Troy, New York 12180

Gerald H. Elkan   (1:285)
Department of Microbiology
North Carolina State University
Raleigh, North Carolina 27695

Jeanne M. Erickson   (3:371)
Department of Biology and The Molecular
   Biology Institute
University of California, Los Angeles
Los Angeles, California 90024

Larry E. Erickson   (1:363)
Center for Hazardous Substance Research
Department of Chemical Engineering
Kansas State University
Manhattan, Kansas 66506

Patrizia Fabrizio   (3:611)
The Agouron Institute
La Jolla, California 92037

Matthew J. Fagan   (1:431)
Department of Biology
University of California, San Diego
La Jolla, California 92093

Brendlyn D. Faison   (2:335)
Oak Ridge National Laboratory
Oak Ridge, Tennessee 37831

Spencer B. Farr   (3:315)
   Department of Molecular and Cellular
      Toxicology
   Harvard School of Public Health
   Boston, Massachusetts 02115
Brian A. Federici   (2:521)
   Department of Entomology
   University of California, Riverside
   Riverside, California 92521
James L. Fishback   (4:255)
   University of Kansas School of
      Medicine
   Kansas City, Kansas 66160
Samuel B. Formal   (2:29)
   Department of Enteric Diseases
   Walter Reed Army Institute of Research
   Washington, D.C. 20012
Patricia L. Foster   (2:107)
   Boston University School of Public
      Health
   Boston, Massachusetts 02118
Andrea L. Friedman   (3:481)
   University of Miami
   Coral Gables, Florida 33124
Herman Friedman   (3:481)
   Department of Medical Microbiology and
      Immunology
   University of South Florida College of
      Medicine
   Tampa, Florida 33612
S. Marvin Friedman   (4:217)
   Department of Biological Sciences
   Hunter College of the City University
      of New York
   New York, New York 10021
Herbert C. Friedmann   (3:1)
   Department of Biochemistry
   University of Chicago
   Chicago, Illinois 60637
Daniel Y. C. Fung   (2:209)
   Department of Animal Science
   Kansas State University
   Manhattan, Kansas 66506
Geoffrey M. Gadd   (2:351)
   Department of Biological Sciences
   University of Dundee
   Dundee DDA 4HN, United Kingdom
George J. Galasso   (1:137)
   National Institutes of Health
   Bethesda, Maryland 20892
Edward L. Gershey   (2:571)
   Department of Laboratory Safety

Rockefeller University
New York, New York 10021
Xavier Gidrol   (3:315)
   Department of Molecular and Cellular
      Toxicology
   Harvard School of Public Health
   Boston, Massachusetts 02115
Thomas P. Gillis   (2:601)
   Gillis W. Long Hansen's Disease
      Center
   Carville, Louisiana 70721
Jan C. Gottschal   (1:559)
   Department of Microbiology
   University of Groningen
   Kerklaan 30, The Netherlands
W. D. Grant   (1:73)
   Department of Microbiology
   University of Leicester
   Leicester LEI 9HN, United Kingdom
Douglas R. Green   (4:243)
   La Jolla Institute for Allergy and
      Immunology
   La Jolla, California 92037
D. A. Grinstead   (1:191)
   Department of Food Science and
      Microbiology
   Clemson University
   Clemson, North Carolina 29634
Lawrence Grossman   (2:9)
   School of Hygiene and Public Health
   The Johns Hopkins University
   Baltimore, Maryland 21205
Tetsuo Hamamoto   (1:81)
   The RIKEN Institute
   2-1 Hirosawa
   Wako, Saitama 351-01, Japan
Ian R. Hamilton   (3:269)
   Department of Oral Microbiology
   Faculty of Dentistry
   University of Manitoba
   Winnipeg R3T 2N2, Canada
Martin Alva Hamilton   (4:75)
   Center for Interfacial Microbial Process
      Engineering and Department of
      Mathematical Statistics
   Montana State University
   Bozeman, Montana 59717
Peter M. Hammond   (3:451)
   Center for Applied Microbiology and
      Research
   Division of Biotechnology
   Proton Down

Salisbury, Wiltshire SP4 OJG, United
Kingdom
Richard S. Hanson (3:121)
Gray Freshwater Biological Institute
University of Minnesota
Minneapolis, Minnesota 55455
Susan Harlander (2:191)
Department of Food Science and Nutrition
The University of Minnesota
St. Paul, Minnesota 55108
K. M. Harmon (1:191)
Department of Food Science and
Microbiology
Clemson University
Clemson, North Carolina 29634
Mark A. Harrison (1:247)
Department of Food Science and
Technology
University of Georgia
Athens, Georgia 30602
Robert C. Hastings (2:601)
Gillis W. Long Hansen's Disease Center
Carville, Louisiana 70721
Harold Hatt (1:615)
American Type Culture Collection
Rockville, Maryland 20852
Frederick G. Hayden (3:555)
Department of Internal Medicine and
Pathology
University of Virginia School of Medicine
Charlottesville, Virginia 22903
Jack A. Heinemann (1:547)
N.I.H., NIAID, LMSF
Rocky Mountain Laboratories
Hamilton, Montana 59840
Charles E. Helmstetter (1:529)
Florida Institute of Technology
Melbourne, Florida 32901
Anita K. Highsmith (4:391)
Centers for Disease Control
Atlanta, Georgia 30333
Paul D. Hoeprich (1:107)
Department of Internal Medicine
School of Medicine
University of California, Davis
Davis, California 95616
Dallas G. Hoover (1:181)
Department of Food Science
University of Delaware
Newark, Delaware 19716
Koki Horikoshi (1:81)
The RIKEN Institute

2-1 Hirosawa
Wako, Saitama 351-01, Japan
Thomas W. Huber (4:67)
College of Medicine
Texas A&M University
Temple, Texas 76502
Rick E. Hudson (2:289)
Department of Ecology and Evolutionary
Biology
University of Arizona
Tucson, Arizona 85721
Scott Hultgren (3:143)
Department of Molecular Microbiology
Washington University School of Medicine
St. Louis, Missouri 63110-1092
Jennie C. Hunter-Cevera (2:561)
The Biotic Network
Sonora, California 95370
Abdallah M. Isa (1:513; 2:59)
School of Arts and Sciences
Tennessee State University
Nashville, Tennessee 37209
William D. James (4:23)
Department of Medicine
Walter Reed Army Medical Center
Dermatology Services
Washington, D. C. 20307
Gilbert Jay (1:17)
Department of Virology
Jerome H. Holland Laboratory
American Red Cross
Rockville, Maryland 20885
Kaj Frank Jensen (3:249)
Institute of Biological Chemistry
University of Copenhagen
1168 Copenhagen K, Denmark
Rolf D. Joerger (1:37)
U. S. Department of Agriculture,
and University School of Medicine
Department of Biology
Schools of Arts and Sciences
Leidy Laboratory of Biology
Philadelphia, Pennsylvania 19104-6018
Douglas H. Jones (3:433)
Department of Pediatrics
Division of Infectious Diseases
University of Iowa College of Medicine
Iowa City, Iowa 52242
Shahid Khan (3:193)
Department of Anatomy and Structural
Biology
Department of Physiology and Biophysics

Albert Einstein College of Medicine
Bronx, New York 10461

John Kidwell   (3:159)
Department of Microbiology and
Immunology
Stanford University
Stanford, California 94305

Edwin D. Kilbourne   (2:505)
Mount Sinai School of Medicine
New York, New York 10029

Robert D. Kiss   (1:351)
Redwood City, California 94065-1630

Donald A. Klein   (3:565)
University of Colorado
Fort Collins, Colorado 80524

Robert C. Klein[1]   (2:571)
Department of Cell Biology
Rockefeller University
New York, New York 10021

Thomas W. Klein   (3:481)
Department of Medical Microbiology and
Immunology
College of Medicine
University of South Florida
Tampa, Florida 33612

Michael J. Klug   (2:45)
Kellogg Biological Station
Natural Science Office
Michigan State University
Hickory Corners, Michigan 49060

Tokio Kogoma   (3:315)
University of New Mexico
Albuquerque, New Mexico 87106

N. Kosaric   (2:473)
Department of Chemical Engineering
University of Western Ontario
London, Ontario N6A 5C1

James P. Krueger   (3:357)
Micro Nutrient Technologies
Chicago, Illinois 60603

C. P. Kurtzman   (1:621)
U.S. Department of Agriculture
Northern Regional Research Laboratory
Peoria, Illinois 61604

Richard W. Lacey   (4:299)
Department of Microbiology
University of Leeds
Leeds LS2 9JT, United Kingdom

---

[1] *Present Address:* Doncet and Mainka, P.C., Peekskill,
New York

Michael M. C. Lai   (1:573)
Department of Microbiology, and
Howard Hughes Medical Institute
University of Southern California
School of Medicine
Los Angeles, California 90033

James A. Lake   (3:403)
Molecular Biology Institute and Department
of Biology
University of California, Los Angeles
Los Angeles, California 90024

Hilary M. Lappin-Scott   (1:277; 2:311)
University of Exeter, EX4 4QJ Exeter
United Kingdom, and Faculty of Science
Department of Biological Sciences
University of Calgary
Calgary, Alberta, Canada T2N 1N4

David S. Latchman   (4:291)
University College and Middlesex School of
Medicine
The Windeyer Building
University College, London
London W1P 6DB, United Kingdom

Robert Latham   (2:439)
Saint Thomas Hospital
Division of Infectious Diseases
Nashville, Tennessee 37202

Joshua Lederberg   (2:419)
Rockefeller University
New York, New York 10021

Martha W. Ledford   (1:401)
New York, New York 10021

Richard A. Ledford   (1:401)
Department of Food Science
Cornell University
Ithaca, New York 14853

Richard E. Lenski   (2:125)
Center for Microbial Ecology
Michigan State University
East Lansing, Michigan 48824

John F. Levy   (4:231)
Department of Biology
Imperial College of Science, Technology, and
Medicine
London SW7 2BB, United Kingdom

Eric P. Lillehoj   (1:387)
Cambridge Biotech Corporation
Rockville, Maryland 20850

Rongtuan Lin   (2:611)
Department of Biology
Concordia University
Montreal, Quebec, Canada H3G 1M8

M. Kathryn Liszewski (1:539)
Howard Hughes Medical Institute
Research Laboratories
Washington University School of Medicine
St. Louis, Missouri 63110

John H. Litchfield (4:11)
Battelle Memorial Institute
Columbus, Ohio 43201

E. Littler (2:381)
Department of Molecular Sciences
Wellcome Research Laboratories
Beckenham, Kent BR3 3BS, United
Kingdom

R. C. Loehr (1:369)
Department of Civil Engineering
University of Texas, Austin
Austin, Texas 78712

William F. Loomis (3:305)
Department of Biology
University of California, San Diego
La Jolla, California 92093

Roderick I. Mackie (3:97)
Department of Animal Sciences
University of Illinois at Urbana-Champaign
Urbana, Illinois 61801

Robert M. Macnab (2:165)
Department of Molecular Biophysics and
Biochemistry
Yale University
New Haven, Connecticut 06511

Andrea Maka (4:151)
Institute of Gas Technology
Illinois Institute of Technology
Chicago, Illinois 60616

Vedpal S. Malik (1:387)
Philip Morris Research Center
Richmond, Virginia 23251

Rocco L. Mancinelli (3:229)
NASA-Ames Research Center
Moffett Field, California 94305

Michael D. Manson (1:501)
Department of Biology
Texas A&M University
College Station, Texas 77843

Karl Maramorosch (3:417)
Department of Entomology
Cooks College
Rutgers, The State University of New Jersey
New Brunswick, New Jersey 08903

Robert E. Marquis (1:161)
University of Rochester
Rochester, New York 14642

Thomas J. Marrie (1:277)
Dalhousie University
Department of Medicine, and Division of
Infectious Diseases
Victoria General Hospital
Halifax, Nova Scotia, Canada B3H 2Y9

Abdul Matin (3:159)
Department of Microbiology and
Immunology
Stanford University
Stanford, California 94305

Ann G. Matthysse (1:29)
Department of Biology
The University of North Carolina at
Chapel Hill
Chapel Hill, North Carolina 27599

Alison C. Mawle (1:123)
Centers for Disease Control
Atlanta, Georgia 30333

Michael P. McCann (3:159)
Department of Microbiology and
Immunology
Stanford University
Stanford, California 94305

Kenneth D. McClatchey (1:171)
Department of Pathology
University of Michigan
Ann Arbor, Michigan 48109

Johnjoe McFadden (3:203)
Department of Biological Sciences
University of Surrey
GU125XH Guildford Surrey
United Kingdom

Martina McGloughlin (2:259)
Department of Biochemistry and
Biophysics
University of California, Davis
Davis, California 95616

Ross E. McKinney (4:363)
Department of Civil Engineering
University of Kansas
Lawrence, Kansas 66045

Dorothy McMeekin (2:157)
Department of Botany and Plant
Pathology
Michigan State University
Kedzie Laboratory
East Lansing, Michigan 48824

Edward A. Meighen (1:309)
Department of Biochemistry
McGill University
Montreal, Quebec, Canada H3G 1Y6

Joseph L. Melnick (2:69)
Division of Molecular Virology
Baylor College of Medicine
Houston, Texas 77030

Paul Messner (1:605)
Center for Ultrastructure Research, and
Ludwig Boltzmann Institute of Molecular
Nanotechnology
Vienna A1060, Austria

Richard E. Michod (2:289)
Department of Ecology and Evolutionary
Biology
University of Arizona
Tucson, Arizona 85721

Felix Milgrom (4:337)
Department of Microbiology
SUNY at Buffalo
Buffalo, New York 14214

Linda Ann Miller (1:297)
Department of Microbiology/
Immunology
Holy Redeemer Medical Center
Meadowbrook, Pennsylvania 19406

Robert V. Miller (3:509)
Department of Microbiology
Oklahoma State University
Stillwater, Oklahoma 74078

Michael L. Misfeldt (2:539)
Department of Molecular Microbiology and
Immunology
University of Missouri
Columbia, Missouri 65212

S. K. Mishra (4:53)
KRUG Life Sciences
Houston, Texas 77058

Tapan K. Misra (2:361)
Department of Microbiology and
Immunology
University of Illinois College of
Medicine
Chicago, Illinois 60680

Harry L. T. Mobley (4:327)
University of Maryland School of
Medicine
Department of Medicine
Division of Infectious Diseases
Baltimore, Maryland 21201

Craig A. Molgaard (2:95)
Division of Epidemiology and Biostatistics
Graduate School of Public Health
San Diego State University
San Diego, California 92182

Richard Y. Morita (2:617, 625)
Department of Microbiology
Oregon State University
Corvallis, Oregon 97331-3804

Stephen A. Morse (4:1)
National Center for Infectious Diseases
Department of Health and Human Services
Centers for Disease Control
Atlanta, Georgia 30333

Stephen S. Morse (2:141; 4:347)
The Rockefeller University
New York, New York 10021

Robb E. Moses (2:17)
Department of Molecular and Medical
Genetics
The Oregon Health Sciences University
Portland, Orgeon 97201

Roy H. Mosher (1:97)
Department of Biology
Dalhousie University
Halifax, Nova Scotia, Canada B3J 4J1

C. G. Nettles (1:191)
Department of Food Science and
Microbiology
A-203 J Poole Agricultural Center
Clemson University
Clemson, North Carolina 29634

Elaine Newman (2:611)
Department of Biology
Concordia University
Montreal, Quebec, Canada H3G 1M8

Austin Newton (1:443)
Department of Molecular Biology
Princeton University
Princeton, New Jersey 08544-1014

Jerry Nielsen (3:59)
Calumet, Oklahoma 73014

Kenneth M. Noll (1:149)
Department of Molecular and
Cell Biology
University of Connecticut
Storrs, Connecticut 06269

Staffan Normark (3:143)
Department of Molecular Microbiology
Washington University School of
Medicine
St. Louis, Missouri 63110-1093

David A. Odelson (2:45)
Central Michigan University
Mount Pleasant, Michigan 48858

Noriko Ohta (1:443)
Department of Molecular Biology

Princeton University
Princeton, New Jersey 08544-1014

Thomas J. Palker   (3:545)
Division of Rheumatology and
   Immunology
Department of Medicine
Duke University Medical Center
Durham, North Carolina 27710

J. Todd Parker   (2:467)
Centers for Disease Control
Atlanta, Georgia 30333

Julie Parsonnet   (2:245)
Division of Geographic Medicine
Stanford University School of Medicine
Stanford, California 94305

John A. Patterson   (3:623)
Department of Animal Sciences
Purdue University
West Lafayette, Indiana 47907

Ronald J. Patterson   (3:257)
Department of Microbiology
Michigan State University
East Lansing, Michigan 48824

Eldor A. Paul   (3:289)
Department of Crop and Soil Science
Michigan State University
East Lansing, Michigan 48824

Henrik Pedersen   (3:409)
Department of Chemical and Biochemical
   Engineering
College of Engineering
Rutgers, The State University of New Jersey
Piscataway, New Jersey 08855

Carl L. Pierson   (1:171)
Department of Pathology
University of Michigan
Ann Arbor, Michigan 48109

D. L. Pierson   (4:53)
NASA/Johnson Space Center
Houston, Texas 77058

Olga Pierucci   (1:529)
Roswell Park Cancer Institute
Buffalo, New York 14263

Pieter Postma   (3:339)
E. C. Slater Institute for Biological
   Research
University of Amsterdam
Spui 21, 1012 WX, The Netherlands

James A. Poupard   (1:297)
SmithKline Beecham Pharmaceuticals
King of Prussia, Pennsylvania 19406

K. L. Powell   (2:381)
Department of Cell Biology
Wellcome Research Laboratories
Beckenham, Kent BR3 3BS, United
   Kingdom

Fergus G. Priest   (2:81)
Department of Biological Sciences
Heriot Watt University
Riccarton, Edinburgh EH14 4AS,
   Scotland

Anthony P. Pugsley   (3:461)
Unité de Génétique Moléculaire
Institut Pasteur
25 rue du Dr. Roux
Paris 75015 Cedex 15, France

Reinhard Rauhut   (3:611)
Division of Biology
California Institute of Technology
Pasadena, California 91125

R. L. Raymond   (1:369)
Du Pont Environmental Remediation
   Services
Newark, Delaware 19702

Donald J. Reasoner   (4:377)
U.S. Environmental Protection Agency
Office of Groundwater and Drinking
   Water
Washington, D.C. 20460

Alice G. Reinarz   (1:409)
Department of Microbiology
University of Texas at Austin
Austin, Texas 78712

Gregory R. Reyes[2]   (2:371)
Gene Labs, Inc.
Redwood City, California 94063

Emily Riggs   (3:545)
Division of Rheumatology and Immunology
Department of Medicine
Duke University Medical Center
Durham, North Carolina 27710

Monica Riley   (3:21)
Marine Biological Laboratory
Woods Hole, Massachusetts 02543

Maria C. Rivera   (3:403)
Molecular Biology Institute and Department
   of Biology
University of California, Los Angeles
Los Angeles, California 90024

[2] *Present Address:* Triplex Pharmaceuticals, The Woodlands, Texas 77380

Donald W. Roberts (2:521)
Boyce-Thompson Institute for Plant
Research
Cornell University
Ithaca, New York 14853

Rudolf R. Roth (4:23)
Travis Air Force Base
Washington, D.C. 94535

David G. Russell (3:143)
Department of Molecular Microbiology
Washington University School of
Medicine
St. Louis, Missouri 63110-1093

Milton H. Saier, Jr. (1:431)
Department of Biology
University of California, San Diego
La Jolla, California 92093

Charles E. Samuel (2:533)
Department of Biological Sciences
Division of Molecular and Cellular
Biology
University of California, Santa Barbara
Santa Barbara, California 93106-9610

Mary Ellen Sanders (2:1)
Consultant, Dairy and Food Culture
Technologies
Littleton, Colorado 80122

Gary S. Sayler (1:417)
Department of Biology
University of Tennessee
Knoxville, Tennessee 37996

Moselio Schaechter (2:115)
Department of Molecular Biology and
Microbiology
Tufts University School of Medicine
Boston, Massachusetts 02111

Kenneth F. Schaffner (3:111)
University of Pittsburgh
Pittsburgh, Pennsylvania 15206, and George
Washington University
Washington, D.C. 20052

William Schaffner (2:439)
Department of Medicine
Vanderbilt University Hospital
Nashville, Tennessee 37232

Ronald F. Schell (1:495)
Wisconsin State Laboratory of Hygiene, and
Department of Medical Microbiology and
Immunology
University of Wisconsin
Madison, Wisconsin 53706

Robert H. Schiestl (3:45)
Department of Molecular and Cellular
Toxicology
Harvard School of Public Health
Harvard University
Boston, Massachusetts 02115

Lawrence R. Schreiber (2:23)
U.S. Department of Agriculture
Agriculture Research Service
Delaware, Ohio 43015

Jeffrey E. Segall (1:501)
Department of Anatomy and Structural
Biology
Albert Einstein College of Medicine at
Yeshiva University
Bronx, New York 10461

Howard M. Shapiro (2:177)
West Newton, Massachusetts 20165

Leora A. Shelef (3:529)
Department of Nutrition and Food
Science
Wayne State University
Detroit, Michigan 48202

Leonard H. Sigal (2:639)
University of Medicine and Dentistry of New
Jersey
Robert Wood Johnson Medical School
New Brunswick, New Jersey 08903-0019

Simon Silver (2:549)
Department of Microbiology and
Immunology
College of Medicine
The University of Illinois at Chicago
Chicago, Illinois 60680

Clifford Siporin (3:489)
G. D. Searle
Skokie, Illinois 60077

Uwe B. Sleytr (1:605)
Center for Ultrastructure Research, and
Ludwig Boltzmann Institute of Molecular
Nanotechnology
Vienna A1060, Austria

David L. Smalley (4:213)
The Health Science Department
College of Medicine
Department of Pathology
University of Tennessee
Memphis, Tennessee 38163

Richard B. Smittle (2:219)
Silliker Laboratories Group, Inc.
Stouchsburg, Pennsylvania 19567

M. D. Socolofsky (2:51)
Louisiana State University, and Louisiana
    State University Agricultural Center
Baton Rouge, Louisiana 70803
John N. Sofos (4:43)
Department of Animal Sciences, and
    Department of Food Science and Human
    Nutrition
Colorado State University
Fort Collins, Colorado 80523
Terje Sørhaug (4:201)
Department of Dairy and Food Industries
Agricultural University of Norway
1432 Ås, Norway
Hiroshi Souzu (2:231)
The Institute of Low Temperature
    Science
Hokkaido University
Sapporo 060, Japan
David T. Spurr (1:257)
Agriculture Canada Research Station
Saskatoon, Saskatchewan, Canada S7N OX2
Erko Stackebrandt (3:171)
Department of Microbiology
University of Queensland
Queensland 4072, Australia
James T. Staley (2:393)
School of Medicine
Department of Microbiology
University of Washington
Seattle, Washington 98195
Keith H. Steinkraus (4:399)
Cornell University
Ithaca, New York 14850
Gregory Stephanopoulos (1:351)
Department of Chemical Engineering
Massachusetts Institute of Technology
Cambridge, Massachusetts 02139
Linda D. Stetzenbach (1:53)
Harry Reid Center for Environmental
    Studies
University of Las Vegas
Las Vegas, Nevada 89154
Dennis L. Stevens (1:641)
Infectious Disease Section
Veterans Affairs Medical Center
Boise, Idaho 83702, and
University of Washington
Seattle, Washington 98195
Robert E. Stevenson (1:615)
American Type Culture Collection
Rockville, Maryland 20852

Valley Stewart (1:89)
Division of Biological Sciences
Section of Microbiology
Cornell University
Ithaca, New York 14853
Nicklas Strömberg (3:143)
Faculty of Onocology
Götteburg Universitet
Guadrers, Medicinarengaten 12
Gotebory, Sweden S-400 33
Fred Stutzenberger (3:327)
Department of Microbiology
Clemson University
Clemson, South Carolina 29634
Ian W. Sutherland (1:339)
Department of Cell and Molecular
    Biology
University of Edinburgh
Edinburgh EH9 3JG, Scotland
Michael I. Tait (1:339)
Aberdeen University
Aberdeen AB9 1AS, Scotland
Barrie F. Taylor (3:29)
Rosensteil School of Marine and
    Atmospheric Chemistry
University of Miami
Miami, Florida 33149-1098
David N. Taylor (2:29)
Department of Enteric Diseases
Walter Reed Army Institute of Research
Washington, D.C. 20012
Rudolf K. Thauer (3:1)
Philipps University
FB Biology
DW-3550 Marburg/Lahn
Germany
J. M. Thomas (1:369)
Rice University and National Center for
    Ground-Water Research
Department of Environmental Science and
    Engineering
Rice University
Houston, Texas 77251
Jan S. Tkacz (1:331)
Merck Sharp & Dohme Research
    Laboratories
Rahway, New Jersey 07065
William J. Todd (2:51)
Louisiana State University, and
    Louisiana State University Agricultural
    Center
Baton Rouge, Louisiana 708033

Sue Tolin (2:281)
Plant Pathology-Physiology-Weed
Science
Virginia Polytechnic Institute and State
University
Blacksburg, Virginia 24061-0330
Marcelo E. Tolmasky (3:431)
Oregon Health Sciences University
Portland, Orgeon 97201
Robert K. Trench (3:129)
Department of Biology
University of California, Santa Barbara
Santa Barbara, California 93106
Joseph G. Tully (3:181)
National Institute of Allergy and Infectious
Diseases
National Institute of Health
Frederick Cancer Research Faculty
Frederick, Maryland 21701
Anne Vidaver (2:281)
Department of Plant Pathology
University of Nebraska
Lincoln, Nebraska 68583
Leo C. Vining (1:97)
Department of Biology
Dalhousie University
Halifax, Nova Scotia, Canada B3J 4J1
Jonathan Vogel (1:17)
Jerome H. Holland Laboratory
American Red Cross
Rockville, Maryland 20855
Milton Wainwright (2:419)
Department of Molecular Biology and
Biotechnology
University of Sheffield
Sheffield S10 2TN, United Kingdom
Marianne Walch (1:585)
Department of the Navy
Naval Surface Warfare Center
White Oak Laboratory
Silver Spring, Maryland 20903
Mark Walderhaug (2:549)
U.S. Food and Drug Administration
Washington, D.C. 20204
William H. Wallace (1:417)
Department of Biology
University of Tennessee
Knoxville, Tennessee 37996
John L. Wang (3:257)
Department of Biochemistry
Michigan State University
East Lansing, Michigan 48824

C. H. Ward (1:369)
Rice University and National Center for
Ground-Water Research
Department of Environmental Science and
Engineering
Houston, Texas 77251
John E. Ward, Jr. (1:37)
Department of Biology
Southern Methodist University
Dallas, Texas 75222
Emilio Weiss (3:585)
Naval Medical Research Institute
Chevy Chase, Maryland 20815
Sybil Wellstood (2:319)
U.S. Food and Drug Administration
Kensington, Maryland 20857
Elizabeth A. Werner (3:257)
Department of Microbiology
Michigan State University
East Lansing, Michigan 48824
Bryan A. White (3:97)
Department of Animal Sciences
University of Illinois at
Urbana-Champaign
Urbana, Illinois 61801
Stanley T. Williams (2:457)
Department of Genetics and Microbiology
The University of Liverpool
Liverpool L69 3BX, United Kingdom
J. T. Wilson (1:369)
R. S. Kerr Environmental Research
Laboratory
U.S. Environmental Protection Agency
Ada, Oklahoma 74820
T. Michael A. Wilson (3:417)
Agbiotech Center, Cooks College
Rutgers, The State University of New Jersey
New Brunswick, New Jersey 08903
Ulrike Wintersberger (3:45)
Institute of Tumor Biology and Cancer
Research
University of Vienna
A-1090 Wien, Vienna, Austria
Martin F. Wojciechowski (2:299)
Department of Ecology and Evolutionary
Biology
Biological Sciences
University of Arizona
Tucson, Arizona 85721
William H. Wunner (3:497)
The Wistar Institute
Philadelphia, Pennsylvania 19104-4268

Charles Yanofsky (4:281)
Department of Biological Sciences
Stanford University
Stanford, California 94305

Marylynn V. Yates (1:321)
Department of Soil and Environmental
   Sciences
University of California, Riverside
Riverside, California 92521

Philip L. Yeagle (1:455)
University of Buffalo School of
   Medicine
Buffalo, New York 14214

J. P. W. Young (3:239)
John Innes Institute
Norwich NR4 7UH, United Kingdom

Alan A. Yousten (2:521)
Department of Biology
Virginia Polytechnic Institute, and State
   University of Virginia
Blacksburg, Virginia 24061-0406

Robert A. Zimmermann (3:573)
Lederle Graduate Research Center
Department of Biochemistry and Molecular
   Biology
University of Massachusetts
Amherst, Massachusetts 01003

Stephen H. Zinder (3:81)
Department of Microbiology
Cornell University
Cornell, New York 14853

Ian M. Zitron (1:217)
Wayne State University School of Medicine
Detroit, Michigan 48201

Jordanka Zlatanova (4:265)
Institute of Genetics
Bulgarian Academy of Sciences
1113 Sofia, Bulgaria

Judith W. Zyskind (3:519)
Department of Biology
San Diego State University
San Diego, California 92182

Subject Index

A

Abelson murine leukemia virus, **1:** 242
ABO blood group system, **4:** 340–341
Abomasum, defined, **3:** 623
Abscesses, oral, pathogenesis, **3:** 279
Accessory polypeptides, defined, **4:** 327
Acclimation, defined, **3:** 129
Acemannan (antiviral agent), **1:** 145
Acetate
    methanogenesis, **3:** 105–106
    methanogenesis substrate, **3:** 84
    rumen fermentation product, **3:** 633
Acetate-oxidizing rod, **3:** 87
Acetic acid bacteria, and spoilage of beer, **1:** 253
Acetobacter, and wine spoilage, **4:** 401
Acetogenesis
    defined, **1:** 1
    processes causing, **1:** 1
Acetogenic bacteria, **1: 1–15.** See also Acetyl-CoA
    pathway; Methanogenesis
    ATP synthesis via oxidative phosphorylation, **1:**
        13–14
    bioenergetics, **1:** 13–14
    characteristics, **1:** 1
    classification, **1:** 5
    defined, **1:** 1
    distinction from other acetate-forming bacteria,
        1: 2–3
    environmental roles, **1:** 14–15
    growth, **1:** 3, 5
    growth-supportive substrates and substrate-
        product stoichiometries, **1:** 12
    isolation, **1:** 3, 5
    list of isolates, **1:** 4
    origins, **1:** 3
    physiologic potentials and regulation, **1:** 12–13
    terminology and definition of, **1:** 2–3
    vitamins required for growth, **1:** 3, 5
Acetotrophic methanogens. See under
    Methanogens
Acetyl-CoA pathway, **1: 5–12**
    acetyl-CoA synthase and autotrophic formation
        of acetyl-CoA, **1:** 8–11

carbonyl reaction path, **1:** 10
detailed description, **1:** 7–8
distinguishing characteristic of acetogenic
    bacteria, **1:** 3
enzymes of autotrophic portion of the pathway,
    1: 10
general overview, **1:** 6–7
heterotrophic and autotrophic processes, **1:** 7
history, **1:** 5–6
methyl reaction path, **1:** 10
poly-β-hydroxybutyric acid synthesis and, **1:**
    347, 348
sorbic acid and, **4:** 46
variations used by other bacterial groups, **1:**
    11–12
Acetyl-CoA synthase
    and autotrophic formation of acetyl-CoA, **1:**
        8–11
    binding sites, **1:** 8–10
    potential functions in obligate anaerobe
        bacteria, **1:** 11
    role in acetyl-CoA pathway, **1:** 7
Acetylcholine, botulinum toxin and, **4:** 215
Acetylcholine receptor, **1:** 464
Acetylene reduction technique, nitrogen fixation,
    3: 232–233
Acetyltransferases, **1:** 102
Acholeplasma distribution
    in arthropods, **3:** 190
    in plants, **3:** 190–191
Acid-fast bacteria, defined, **3:** 203
Acid proteases, **2:** 85
Acid rain
    environmental effects, **4:** 126, 132
    microbial desulfurization, **4:** 151
Acidianus, **1:** 156
Acidolysis, and microbial metal extraction from
    minerals, **3:** 78
Acidotherumus, **1:** 156
Acinetobacter
    hydrocarbon metabolism, **3:** 365
    skin, **4:** 25
    taxonomy, **4:** 180
Acini, defined, **3:** 497

DNA transfer, **1:** 553–554
  plasmid and host, **1:** 553–555
  surface exclusion, **1:** 554
cyanobacteria, **1:** 631
defined, **1:** 37, 181, 547, **2:** 299, **3:** 509
evolutionary impact, **1:** 558
gene nomenclature, **1:** 548
gene transmission, **1:** 548, 549
genetic transformation compared with, **2:** 299–300
history, **1:** 547–548, **2:** 431
host-range determinants, **1:** 555
other DNA transmission systems, **1:** 555–557
parallels in DNA transmission processes, **1:** 557
plant cell transformation by Agrobacterium, **1:** 44–46
plasmid and host, **1:** 549–555
  conjugal recipient, **1:** 553–555
plasmid families, **1:** 548
plasmid incompatibility (Inc) groups, **1:** 548
plasmids, **1:** 427
role of pili, **1:** 549–551
transposition and phage infection compared with, **1:** 557
Conjugative transposons, **4:** 321
Conjunctivitis
  acute hemorrhagic, **2:** 73, 74
  defined, **4:** 33
  inclusion, **1:** 513–514
  in smallpox, **4:** 37
Conserved, defined, **1:** 37
Consortia of bacteria, defined, **1:** 277. *See also* Biofilms and biofouling
Constitutive, defined, **1:** 181, 191
Constitutive expression, defined, **1:** 417
Constitutive gene, defined, **2:** 361
Containment of genetically modified organisms, defined, **2:** 281
Contingency table analysis, defined, **4:** 75
Continuous culture, **1: 559–572**
  basic equipment, **1:** 565–566
  defined, **1:** 431, **1:** 559
  derepression of enzymes, **1:** 567
  design characteristics, **1:** 565–566
  ecological applications, **1:** 566–571
  mixed cultures, **1:** 563–564, **1:** 569–570
  mixotrophic metabolism, **1:** 568
  multiple-substrate-limited growth, **1:** 562–563
  nonsubstrate-limited growth, **1:** 564–565
  physiological applications, **1:** 566–571
  principles, **1:** 559–565
  pure culture studies, **1:** 566–569

selective enrichment cultures, **1:** 570–571
single-substrate-limited growth, **1:** 559–562
summary, **1:** 571–572
turbidostats, **1:**564, **1:** 565
Contrast (microscopy), defined, **2:** 51
Convenience foods, refrigerated, **3:** 541–542
Convention on the Prohibition of the Development, Production and Stockpiling of Bacteriological (Biological) and Toxin Weapons and on their Destruction (1972), **1:** 4, 5–6
Convoluta roscoffensis, **3:** 132, 134
Coordinated Framework for the Regulation of Biotechnology, **2:** 261–262
Copiotrophic bacteria, defined, **2:** 393, 619
Copiotrophic habitat, defined, **3:** 29
Copolymers. See Polymers
Copper
  bioleaching, **3:** 284–285
  plasmid-governed transport, **2:** 559
Copper chrome arsenate, wood preservation, **4:** 241
Coprococcus, **2:** 329
Coprotease
  defined, **3:** 509
  RecA protein activity, **3:** 511–512, 513
Corals, and microalgal-invertebrate symbiosis, **3:** 131, 141
Corculum cardissa, **3:** 133
Coronaviruses, **1: 573–583**
  classification, **1:** 573–574
  cytopathic effects, **1:** 581
  defined, **1:** 573
  diseases in humans and animals, **1:** 574, 582–583
  establishment of Coronaviridae family, **1:** 573
  feline infectious peritonitis virus, **1:** 583
  genetics
    defective-interfering particles, **1:** 582
    RNA recombination, **1:** 581–582
    temperature-sensitive mutants and complementation, **1:** 581
  growth and host range, **1:** 574
  members of Coronaviridae and their diseases, **1:** 574
  murine coronaviruses, **1:** 582–583
  nonstructural proteins
    defined, **1:** 573
    post-translational modification and processing, **1:** 580–581
  replication cycle, **1:** 578–581
    adsorption to cell receptors, **1:** 577

probability of infection, due to environmental
contamination, **1:** 325
rotaviruses, **2:** 67
Salmonella, **2:** 64–65
Shigella, **2:** 64
vibrios, **2:** 66
viral agents, **2:** 67
Yersinia enterocolitica, **2:** 66
Enterotoxigenic *Escherichia coli* (ETEC)
defined, **2:** 115
foodborne illness, **2:** 215
pathogenesis, **2:** 120, 122, 251
Enterotoxins, defined, **2:** 63, 268, **3:** 529
Enterovirus type 70, **2:** 74
Enterovirus type 71, **2:** 74
Enterovirus type 72, taxonomy, **2:** 70
Enterovirus types 68–71, taxonomy, **2:** 69–70
Enteroviruses, **2: 69–80.** *See also* Hepatitis A virus;
Hepatitis E virus
diagnosis of infections, **2:** 74–75
diseases, **2:** 72–75
Coxsackievirus diseases, **2:** 72–74
echovirus diseases, **2:** 74
other enterovirus diseases, **2:** 74
poliomyelitis, **2:** 72
environmental contamination caused by, **1:** 323
epidemiology, **2:** 75–77
foodborne infection, **2:** 217
immunity, **2:** 77–78
intestinal infections, **2:** 249
pathogenesis, **2:** 71–72
prevention, **2:** 78–80
properties, **2:** 70–71
replication, **2:** 71
resistance to chlorine water treatment, **4:** 394
taxonomy, **2:** 69–70
Entner-Doudoroff glycolytic pathway, **1:** 153, 154,
155, 156
Entomophaga maimaiga, as insecticide, **2:** 530
Entomophthorales, as insecticides, **2:** 529
Entropy
defined, **1:** 455
and hydrophobic effect in cell membranes, **1:**
456
env gene
human immunodeficiency virus, **1:** 22
retroviruses, **3:** 546, 547–548
Env protein
destruction of CD4$^+$ target cells, **1:** 25
as target for immune system in HIV infections,
1: 22
Envelope, defined, **1:** 203

Enveloped virus fusion, **1:** 464–465
Environmental contamination
biomonitors for. *See* Biomonitors of
environmental contamination
bioremediation. *See* Bioremediation
Environmental Defense Fund, **2:** 277
Environmental introduction of genetically
engineered organisms. *See under* Genetically
engineered organisms
Environmental manipulation, fungal insecticides,
2: 531
Environmental Protection Agency, U.S.
bioremediation technology regulation, **1:** 371
chemical waste and emissions, **2:** 594–595
hazard communication guideline, **2:** 584–585
hazardous waste treatment regulation, **2:** 336
medical waste regulations, **2:** 596
oversight of genetically modified organisms, **2:**
263–266, 284
Biotechnology Advisory Committee (BAC), **2:**
264
Biotechnology Science Advisory Committee
(BSCA), **2:** 265
Clean Water Act, **2:** 265
Environmental Response, Compensation, and
Liability Act (CERCLA), **2:** 265
examples of applications for field tests, **2:** 267,
268
Federal Insecticide, Fungicide, and
Rodenticide Act (FIFRA), **2:** 264
Resource Conservation and Recovery Act
(RCRA), **2:** 265
risk assessment, **2:** 271
Risk Assessment research program, **2:** 265
Toxic Substances Control Act (TSCA), **2:** 264,
265
proposed Maximum Contaminant Level Goals
for viruses and *Giardia,* **1:** 330
Enzyme engineering, defined, **2:** 191
Enzyme-linked immunosorbent assay. *See* ELISA
technology
Enzymes. *See also* specific enzymes
activities decreased by L-leucine, **2:** 613
activities increased by L-leucine, **2:** 613
antibacterial, **3:** 428
antifungal, **3:** 428
antimicrobial agent inactivation, **4:** 163
bacteriolytic, **1:** 184
biodegradative enzymes, **1:** 271–272, 275
biotechnology, **1:** 392–394
cell disruption, **3:** 452
cell-free enzyme systems, **2:** 206

hospital. *See* Epidemiology, hospital

HTLV-I, **2:** 104

incidence rate, defined, **2:** 96–97

incubation period, defined, **2:** 97

infection, defined, **2:** 97

influenza, **2:** 516–518

 genetic reassortment, **2:** 103–104, 513

 influenza A, **2:** 517–518, 519

 influenza B, **2:** 518

 influenza C, **2:** 518

 interpandemic (epidemic) influenza, **2:** 517

 pandemic influenza, **2:** 517

 seasonal factors, **2:** 517

Koch's postulates, **3:** 116–118

models for biological warfare, **1:** 305

modern era, **2:** 102–103

molecular, **3:** 176–177

 low molecular weight RNA profiles, **3:** 177

 plasmid profiles and fingerprints, **3:** 176–177

 restriction endonuclease digestion of DNA, **3:** 176

 restriction fragment length polymorphism, **3:** 176

poliomyelitis, **2:** 75–77

quarantine, defined, **2:** 97

rhinovirus infections, **3:** 559–560

seroepidemiology, defined, **2:** 97

theoretical controversies, **2:** 101–102

toxoplasmosis, **4:** 258–260

transmission of infection, defined, **2:** 97–98

vector-borne infection, defined, **2:** 98–99

Epidemiology, hospital, **2: 439–447.** *See also* Sterilization

 antimicrobial resistance, **2:** 445

 bacteremias, **2:** 444–445

 Clostridia difficile colitis, **2:** 445

 control measures, **2:** 445–446

 employee health programs, **2:** 446–447

 infections in immunocompromised patients, **2:** 445

 pneumonia, nosocomial, **2:** 443–444

 surgical wound infections, **2:** 442–443

 surveillance methods, **2:** 439–441

 urinary tract infections, **2:** 441–442

Epidermis, defined, **4:** 33

Episomes

 coined by Jacob and Wollman, **2:** 432

 compared with bacteriophages, **1:** 211

Epistasis, defined, **3:** 351

Epitopes

 categories of, **1:** 221–222

 conserved epitopes and the immune response, **2:** 468

 defined, **1:** 17, 217, 221, **2:** 505, **3:** 497

 epitope-specific antibodies, **1:** 221

 in rabies virus identification, **3:** 500, 502

Epstein-Barr virus. *See also* Herpesviruses

 associated with Burkitt's lymphoma and other neoplasms, **1:** 243, **2:** 390

 binding with complement receptors, **1:** 546

 gene expression, **2:** 385

 induction of neoplasms in HIV infection, **1:** 27

 vaccine for, **2:** 389

Equine encephalitis, Western, **2:** 104

Equine leukoencephalomalacia, fumonisin, **3:** 226

Equipartition, chromatid, defined, **1:** 529

Ergonovine, uterine contraction, **1:** 332

Ergosterol, pathways for synthesis in fungi, **1:** 110

Ergot alkaloids

 in animals, **3:** 224–225

 biopharmaceutins, **1:** 332–333

 intoxication, **1:** 332–333

Ergotamine, in migraine headaches, **1:** 332

Ergotism. *See* Mycotoxicoses

Erlich, Paul, **2:** 424, 425

Erwinia

 oligogalacturonate lyases, **3:** 334

 pectinase biosynthesis regulation, **3:** 336

Erwinia amylovora

 characteristics, **2:** 159

 fire blight symptoms, **2:** 158

 hypersensitive reaction genes (hrp), **2:** 162

 prevention and control of blight, **2:** 161

 recombinant DNA technology, **2:** 161–163

 transmission, penetration, and infection of plants, **2:** 160

Erwinia carotovora, **3:** 332

Erwinia chrysanthemi

 exopolygalacturonase, **3:** 332–333

 metalloprotease secretion pathway, **3:** 476

Erynia, as insecticide, **2:** 531

Erysipeloid, **2:** 332

Erysipelothrix rhusiopathiae, **2:** 332

Erythema chronicum migrans, defined, **2:** 639

Erythrocyte membrane, and antigenic variation in malaria, **1:** 128–129

Erythrocyte processing of immune complexes (EPIC), **1:** 545–546

Erythrocytes, attachment to cell receptors, **3:** 144

Erythromycin

 antibiotic resistance, **1:** 98

 produced by *Saccharopolyspora erythraea*, **1:** 99

G

Gymnosperms, **4:** 231–232
Gypsy moth, fungal insecticides against, **2:** 530
Gyrase. *See* DNA gyrase

H

Habitats
  defined, **2:** 561, **4:** 231
  marine. *See* Marine habitats, bacterial
  methylotrophs, **3:** 123–124
HACCP (hazard analysis and critical control point). *See under* Food quality control
Hadley, Phillip, **2:** 432
Haemophilus influenzae
  antigenic variation, **1:** 132
  DNA binding, **2:** 305–306
  genetic transformation, **2:** 303, 305–306
Hafnia alvei, evolution, **4:** 198
Hairy roots. *See* Agrobacterium
Half-life, defined, **3:** 489
Half-saturation constant for growth (continuous culture), defined, **1:** 559
Haloalkaliphiles, defined, **1:** 73, 81
Halobacteria
  defined, **1:** 73
  Natronobacterium, **1:** 79
  photosynthesis, **3:** 404
Halogenated organic wastes, biodegradation, **2:** 342–343
Halophiles, marine bacteria, **3:** 31
Halophilic Euryarchaeotes
  heterotrophic characteristics, **2:** 399
  hypersaline environments, **1:** 151, 153
  intermediary metabolism, **1:** 153
  photometabolism, **1:** 153
Hamycin, as antifungal agent, **1:** 115
Hand, foot, and mouth disease, **2:** 73
Hanganutziu-Deicher (H-D) antigen-antibody system, **4:** 340
Hansen, Emil Christian, **1:** 248
Hansen's disease. *See* Leprosy
Hantaan virus, **4:** 351, 356
Hantaviruses, **4:** 351–352
Haplodiscus, microalgal-invertebrate symbiosis, **3:** 134
Haploid, defined, **2:** 289
Hapten, defined, **1:** 217–218, 231
Hardwoods, defined, **4:** 231, 232
Hastings, J. W., **1:** 315

HAV. *See* Hepatitis A virus
Hayes, William, **2:** 431
Hazard analysis and critical control point (HACCP). *See under* Food quality control
Hazard communication standard
  community right-to-know, **2:** 592–593
  defined, **2:** 571
  OSHA, **2:** 584–585
Hazardous materials. *See also* Chemicals, hazardous; Hazardous waste treatment
  defined, **2:** 571
  OSHA standard, **2:** 578
  shipment, **2:** 594, 597–598
Hazardous Substances Act, **2:** 593
Hazardous waste, defined, **2:** 335, 493
Hazardous waste treatment, **2: 335–349.** *See also* Bioremediation; Infectious waste management; Laboratory safety and regulations
  biochemical basis for microbial technologies, **2:** 342–345
  biodegradation, **2:** 342–343
  biosorption, **2:** 344–345
  biotransformation, **2:** 343–344
  conventional wastes, **2:** 338
  current waste-management technologies, **2:** 338–342
  development of microbial processes, **2:** 345–348
    application of candidate organisms, **2:** 347
    competitiveness with nonbiological technologies, **2:** 348
    genetically engineered organisms, **2:** 348
    naturally occurring organisms, **2:** 345–347
    psychological issues, **2:** 348
    sources of cultures, **2:** 346
    technical issues, **2:** 347–348
  disposal, **2:** 339
  examples, **2:** 337
  future directions, **2:** 349
  magnitude of the problem, **2:** 338
  microbial technologies, **2:** 340–342
    activated sludge, **2:** 341
    aerated lagoons, **2:** 341
    anaerobic digestors, **2:** 341
    anaerobic filters, **2:** 341
    composting, **2:** 341
    infectious waste, **2:** 497
    land treatment, **2:** 341–342
    landfilling, **2:** 342
    rotating biological contactors, **2:** 341
    trickling filters, **2:** 340
    waste stabilization ponds, **2:** 341

expansion by cytokines, **4:** 245
functions, **3:** 482, **4:** 244–245
in rabies immunity, **3:** 506, 507
rabies virus G protein and, **3:** 503
rabies virus N protein and, **3:** 502
Hemagglutininesterase (HE) proteins, coronaviral, **1:** 576, 580
Hemagglutinins, influenza virus, **2:** 508–509
Heme
defined, **3:** 1
formation, **3:** 2, 8–9
Hemicellulases, in the rumen, **3:** 630
Hemicellulose
decomposition, **3:** 294
structure, **1:** 467
Hemin-binding protein gene, *Shigella* species, **2:** 37–38
Hemolymph test, rickettsial identification, **3:** 606, 609
α-Hemolysin, **3:** 474–475
α-Hemolytic streptococci, **2:** 325–326
Group D, **2:** 325–326
nutritionally variant, **2:** 326
oral group, **2:** 326
Streptococcus pneumoniae, **2:** 326
β-Hemolytic streptococci, **2:** 323, 325
Hemolytic uremic syndrome, **2:** 32
Hemorrhagic colitis. *See* Colitis
Hemorrhagic conjunctivitis, acute, **2:** 73, 74
Hemorrhagic fever
antiviral agents, **1:** 141
defined, **4:** 347
epidemiology, **2:** 103–104
viral agents causing
arenaviruses, **4:** 350
filoviruses, **4:** 350
hantaviruses, **4:** 351–352
Henle, Jakob, **2:** 99
Henle-Koch postulates, **2:** 99
HEPA filters. *See* High-Efficiency Particulate Air filters (HEPA)
Hepadnaviruses, **2:** 376
Heparin, produced as an exopolysaccharide, **1:** 343
Hepatitis, **2: 371–380**
antiviral agents, **1:** 141
as a cause of detrimental immune complexes, **1:** 540
characteristics of viral agents, **2:** 371–372
emerging viruses, **4:** 352
enteric viruses, epidemiology, **2:** 372–375
hepatitis A virus, **2:** 372–374
hepatitis B virus, **2:** 374–375
introduction, **2:** 371–372

parenteral viruses, epidemiology, **2:** 375–379
hepatitis B virus, **2:** 376–377
hepatitis C virus, **2:** 378–379
hepatitis Delta virus, **2:** 377–378
Hepatitis A virus, **2:** 372–374
endemic regions, **2:** 373
incidence cycles, **2:** 373–374
laboratory diagnosis, **2:** 372–373
mode of transmission, **2:** 374
persistent infection, **2:** 374
principle reservoir, **2:** 373
Hepatitis B virus, **2:** 376–377
e antigen, defined, **2:** 371
endemic regions, **2:** 377
hepatocellular carcinoma and, **2:** 377
induction of neoplasms in HIV infection, **1:** 27
infectious waste management, **2:** 494, 495
OSHA bloodborne pathogens standard, **2:** 586–587
laboratory diagnosis, **2:** 376
mode of transmission, **2:** 377
persistent infections, **2:** 377
principal reservoir, **2:** 376
surface antigen, defined, **2:** 371
Hepatitis C virus, **2:** 378–379, **4:** 352
core antigen, defined, **2:** 371
endemic regions, **2:** 379
hepatocellular carcinoma and, **2:** 379
laboratory diagnosis, **2:** 378
mode of transmission, **2:** 379
persistent infections, **2:** 379
principal reservoir, **2:** 378–379
recent identification of, **4:** 352
Hepatitis Delta virus, **2:** 377–378
endemic regions, **2:** 378
hepatocellular carcinoma and, **2:** 378
laboratory diagnosis, **2:** 377–378
mode of transmission, **2:** 378
persistent infections, **2:** 378
principal reservoir, **2:** 378
recent identification of, **4:** 352
Hepatitis E virus, **2:** 374–375, **4:** 352
endemic regions, **2:** 374–375
incidence cycles, **2:** 375
laboratory diagnosis, **2:** 374
mode of transmission, **2:** 375
persistent infection, **2:** 375
principal reservoir, **2:** 374
recent identification of, **4:** 352
Hepatocellular carcinoma
hepatitis B virus and, **2:** 377
hepatitis C virus and, **2:** 379

Hyphal stage, defined, **4:** 23
Hyphomycetes, as insecticides, **2:** 529
Hypovirulent pathogen, defined, **3:** 417

I

Ice cream. See Dairy products
Ice inoculation (freeze-drying), defined, **2:** 231
Icosahedral symmetry, defined, **3:** 555
Identification
    defined, **2:** 457
    probabilistic, **2:** 457, 462–463
Identification coefficients, defined, **2:** 457
Identification of bacteria, computerized, **2:**
        457–466
    applications, **2:** 465–466
    data sources, **2:** 458
    identification matrix applications, **2:** 461–463
        numerical codes, **2:** 462
        probabilistic identification, **2:** 462–463
    identification matrix construction, **2:** 459–461
        selection of characters for, **2:** 460–461
        selection of data base, **2:** 459–460
    identification matrix evaluation, **2:** 463–465
        practical, **2:** 463–465
        source of computer programs, **2:** 465
        theoretical, **2:** 463
    principles, **2:** 457–458
    rapid identification, defined, **1:** 171
    selection of diagnostic characters, **2:** 458–459
Idiopathic, defined, **3:** 203
Idoxuridine
    first practical antiviral agent, **1:** 138
    treatment of herpes keratitis, **1:** 142
IFNs. See Interferons
IgA, in gonococcus infection, **4:** 7
IgA deficiency, selective, **1:** 241
IgG, in gonococcus infection, **4:** 7
IgM antibody, defined, **2:** 371
IgM antibody capture assay, **2:** 59–60
Immediate hypersensitivity reactions, **1:** 241
Immobilization of hazardous wastes, defined, **1:** 369
Immobilization of sulfur
    defined, **4:** 123
    mineralization/immobilization, **4:** 130–131
Immune complexes
    autoimmune diseases and, **1:** 540
    extrinsic antigen exposure and, **1:** 540–541
    inflammation caused by, **1:** 540–541
    persistent infection and, **1:** 540

Immune response
    feedback inhibition, **2:** 468
    immune suppression and, **2:** 467
    normal response to antigens, **2:** 468
    primary and secondary response, **1:** 221
    primary response to immunogens, **1:** 228
    provoked by cell wall materials, **1:** 485
Immune sera. See Serological specificity
Immune suppression, **2: 467–472**
    defined, **2:** 467
    enteroviral infections and, **2:** 77
    hospital infection and, **2:** 445
    immune response and, **2:** 467
    immune response to antigens, **2:** 468
    induced causes, **2:** 468–469
        nonpathogenic suppression, **2:** 469
        pathogenic suppression, **2:** 468–469
    manipulation
        biological response modifiers, **2:** 472
        cytokines, **2:** 470–472
        immunosuppressive drugs, **2:** 472
        modulatory drugs, **2:** 470
    natural causes, **2:** 469–470
        aging, **2:** 470
        congenital and acquired immunodeficiencies,
            2: 469–470
        pregnancy, **2:** 469
    summary, **2:** 472
    toxoplasmosis and, **4:** 255–256
    viral, **2:** 469
Immunity. See also B cells;
        Psychoneuroimmunology; T cells
    anamnestic response to immunogens, **1:** 228
    cell-mediated immunity, **1:** 220
    clonal selection hypothesis, **1:** 220
    defined, **3:** 417, 481
    enterovirus diseases, **2:** 77–78
    humoral immunity
        defined, **1:** 220, **3:** 482
        evolution of, **1:** 222
        humoral response of B cells, **1:** 227–230
        regulation of humoral response, **1:** 239–240
    leprosy, **2:** 608–610
    mucosal immune system, **1:** 221
    natural
        gonorrhea, **4:** 7
        syphilis, **4:** 10
    passive immunity of fetus and newborn, **1:**
        221–222
    poliomyelitis, **2:** 77–78
    skin microbiology, **4:** 31
    smallpox, **4:** 38

arctic and antarctic environments, **2:** 625
high mountains, **2:** 626
microbial activities in
　cold caves, **2:** 633
　polar oceans, **2:** 633
　polar regions, **2:** 632–633
　trophosphere, **2:** 633
microorganisms in
　adaptation, **3:** 530
　incidence
　　antarctic soil samples, **2:** 630–631
　　arctic and other cold northern soil samples,
　　　2: 629
　　nonpolar soil samples, **2:** 628
　other thermo-groups, **2:** 627
　psychrophiles, **2:** 626–627
　"psychrophilic" enzymes, **2:** 631–632
　psychrotrophs, **2:** 626
　temperature-growth profiles, **2:** 627–628
　thermosensitivity, **2:** 628, 631
oceans, **2:** 625–626, **3:** 36
summary, **2:** 633, 637
upper atmosphere, **2:** 626
Lrp, defined, **2:** 611. *See also* Leucine/Lrp regulon
LTR. *See* Long terminal repeat region
Lucanidae, **4:** 177
Luciferases
defined, **1:** 309, 311
expression in prokaryotes and eukaryotes, **1:**
　318–319
in vitro applications, **1:** 317–318
primary and quaternary structures, **1:** 313
purification and stability, **1:** 313
Luciferin, defined, **1:** 171, 311
Luminescence. *See* Bioluminescence, bacterial
Lupus erythematosus, systemic, **1:** 242
Luria, Salvador Edward, **2:** 430, 433
Lux, defined, **1:** 309
Lux genes
cloning of *lux* genes, **1:** 316
common *lux* structural genes, **1:** 316
other *lux* genes, **1:** 316
Lwoff, André Michael, **2:** 427, 434
Lyme disease, **2: 639–646.** *See also Borrelia
　burgdorferi; Rickettsias*
clinical description, **2:** 640–641
arthritis, **2:** 641
cardiac features, **2:** 640–641
early, disseminated disease, **2:** 640–641
early, localized disease, **2:** 640
late disease, **2:** 641
neurological features, **2:** 640

opthalmologic features, **2:** 641
　tertiary neuroborreliosis, **2:** 641
defined, **2:** 639
diagnosis, **2:** 641–644
antigen detection assays, **2:** 644
ELISA technology, **2:** 642–643
immunoblot (Western blot), **2:** 643
indirect immunofluorescence, **2:** 642
polymerase chain reaction (PCR), **2:** 643–644
T-cell proliferative response, **2:** 643
environmental conditions related to, **4:** 356
history, **2:** 639–640
immunologic changes caused by *Borrelia
　burgdorferi,* **2:** 645–646
pregnancy and, **2:** 641, 645
therapy, **2:** 644–645
early, disseminated disease, **2:** 644
early, localized disease, **2:** 644
late disease, **2:** 644–645
vaccine development, **2:** 646
Lymphadenitis, defined, **3:** 203
Lymphocytes. *See* B cells; T cells
Lymphokines, **1:** 233–235. *See also* Cytokines
cytokines, **1:** 234
defined, **1:** 641, **2:** 539, **3:** 481
effect of cytokines on B cells, **1:** 235
monokines, **1:** 234
produced by macrophages, **3:** 482
Lymphotoxin, **1:** 643–644, **2:** 546
Lyophilization
defined, **1:** 615, 621
method for preserving culture collections, **1:** 622
Lysis, for detecting bacteria in blood, **1:** 496
Lysogenic bacteria, in space flight conditions, **4:**
　58
Lysogeny
defined, **1:** 181, **3:** 509
history, **2:** 434
Lysozymes
defined, **1:** 479
digestion of cell walls, **1:** 482
LysR, defined, **2:** 611

M

M proteins. *See* Membrane proteins
Machupo (Bolivian hemorrhagic fever), **4:** 350
Macrocyclic tetrapyrrole biosynthesis in bacteria,
　3: 1–19

δ-aminolevulinic acid formation, **3:** 5–6
  conversion to porphobilinogen, **3:** 6
  genetics, **3:** 6
  pathways, distribution of, **3:** 5–6
chlorophyll and bacteriochlorophyll
    biosynthesis, **3:** 9–10
coenzyme F$_{430}$ formation, **3:** 11
heme formation, **3:** 8–9
  genetics, **3:** 8–9
  insertion of iron into protoporphyrin IX, **3:** 8
siroheme formation, **3:** 10–11
structure, **3:** 3–4
uroporphyrinogen III conversion
  to protoporphyrin IX, **3:** 7–8
  to siroheme, coenzyme F$_{430}$, and corrinoids,
    3: 10
uroporphyrinogen synthesis, **3:** 6–7
  deaminating polymerization of
    porphobilinogen, **3:** 6–7
  hydroxymethylbilane formation, **3:** 7
  uroporphyrinogen III formation, **3:** 7
vitamin B$_{12}$ formation, **3:** 11–17
  C-methylation and associated processes, **3:**
    12–13
  chemistry, **3:** 11–12
  corrin ring side-chain amidations, **3:** 15
  corrinoid α, β-ligand attachment, **3:** 14–15
  descobaltocorrinoids, **3:** 13–14
  genetics, **3:** 16–17
  nucleotide loop formation, **3:** 15–16
  side-chain amide group attachment, **3:** 14–15
Macrolide antibiotics, **1:** 98
Macrophages
  defined, **2:** 467
  factors produced by, **3:** 482–483
  HIV infection and, **1:** 26–27
  immunity and, **3:** 482–483, 484, **4:** 250
  leprosy, **2:** 610
  replication of human immunodeficiency virus,
    1: 27–28
  survival of pathogens inside, **3:** 152
Macules, defined, **4:** 33
Magnesium, chromosomal-governed transport, **2:**
  552
Magnification, defined, **2:** 51
Major histocompatibility complex
  class I, defined, **4:** 243
  class II, defined, **2:** 467, **4:** 243
  defined, **4:** 244
  direction of immune response, **4:** 248, 249
  HLA antigens, **4:** 341

interferon-induced expression of HLA antigens,
  2: 537
MHC antigens, **2:** 470–471
T-cell receptor ligand recognition and, **4:** 247
Malaria. *See Plasmodium*, antigenic variation
Malic dehydrogenase, **2:** 631–632
Malpighian epithelium, defined, **4:** 33
Malt
  in beer brewing, **1:** 248–249
  defined, **1:** 247
Maltose
  improvement for breadmaking, **1:** 404
  produced by α-amylase, **2:** 83, 84
Maltose binding protein, bacterial chemoreceptor,
  1: 503
Manganese
  chromosomal-governed transport, **2:** 554
  respiration, organic matter decomposition, **3:**
    297
Mantoux test, tuberculosis control, **3:** 212
''Many-many modeling,'' scientific method, **3:**
  112–113
Map, defined, **3:** 21
Map locations of bacterial genes, **3: 21–27**
  conservation of gene order among related
    bacteria, **3:** 21–22
  cyanobacteria, **1:** 630
  functional consequences of gene location, **3:**
    23–24
    genes of related function, **3:** 23–24, 25
    supercoiled domains, **3:** 23
  gene location and orientation, **3:** 24–26
    gene dosage, **3:** 24
    replicating arm length, **3:** 25
    transcription direction, **3:** 24, 26
  linkage maps
    Bacillus subtilis, **4:** 407
    Escherichia coli, **4:** 410
    Salmonella typhimurium, **4:** 414
  unusual properties of chromosome termination
    region, **3:** 26–27
  vitamin B$_{12}$ formation, **3:** 16–17
Marboran. *See* Methisazone
Marburg virus, **4:** 350
Marek's disease virus, **2:** 389
Marine ecosystems
  bioremediation of oil spills, **3:** 367–368
  flow cytometry applications, **2:** 187
  sulfide-containing, **4:** 111
Marine habitats, bacterial, **3: 29–44**. *See also* Deep
  sea habitats
  benthic, **3:** 29

Penciclovir, **2:** 388
Penicillin. *See also* β-Lactams
  antibiotic resistance, **1:** 97, 99
  discovery of, **1:** 97, 99
  "ditch plate" demonstration, **4:** 158
  semi-synthesis of, **1:** -100
  structure of, **1:** 100
  sulfur metabolism, **4:** 146–147
  syphilis treatment, **4:** 8
Penicillin binding proteins (PBPs)
  antibiotic resistance and, **1:** 100–101
  interference with cell wall biosynthesis, **1:** 488
  number in eubacteria, **1:** 100
  relationship with serine-based β-lactamase, **1:** 102
  used as markers, **4:** 199
Penicillium, sorbates and, **4:** 49
Penicillium notatum, **4:** 158
1,3-Pentadiene, from sorbate decarboxylation, **4:** 49
PEP: carbohydrate phosphotransferase system, **3:** **339–349.** *See also* Phosphoenolpyruvate (PEP)
  chemotaxis, **3:** 348–349
  genetics, **3:** 344–345
  levansucrase synthesis, **2:** 88
  phosphotransferase system
    components, **3:** 339–341
      interactions with nonphosphotransferase system, **3:** 348
    protein structure and properties, **3:** 341–343
    regulation
      gram-negative organisms, **3:** 345–347
      gram-positive organisms, **3:** 347–348
    transport, **3:** 343–344
Peplomers
  in coronaviruses, **1:** 574
  defined, **1:** 573, **3:** 497
Peptic ulceration, **4:** 334–335
Peptides
  coding for, **3:** 308–309
  plant disease resistance, **3:** 428
  primitive metabolism, **3:** 310
  protocells, **3:** 311
  synthetic
    in rabies immunization, **3:** 503
    rabies virus N protein and, **3:** 502
Peptides, antifungal
  cilofungin, **1:** 120–121
  saramycetin, **1:** 121
Peptidogens, eubacteria, **2:** 397
Peptidoglycans
  alkaliphiles, **1:** 84–85

antibiotics interfering with biosynthesis, **1:** 487
biosynthesis, **1:** 486–487
chemotaxonomic analysis, **4:** 191
defined, **1:** 97, 479, **2:** 311, **3:** 75, 97
enzymic cell disruption, **3:** 452
in gonoccocal pathogenesis, **4:** 7
required for flagellar motion, **1:** 483
structure, **1:** 479–480
Peptidyl transferase, defined, **3:** 573
Peptococcus, **2:** 327
Peptostreptococcus, **2:** 327, 329
Peracetic acid
  gaseous sterilant, **4:** 95–96
  liquid sterilant, **4:** 97–98
Percent relative likelihood, taxonomy, **4:** 196, 197
Perialgal vacuole, **3:** 129
Peridinin-chlorophyll α-protein (PCP) complexes, **3:** 132, 138
Perinuclear space, defined, **3:** 257
Periodic selection, **3:** **351–355**
  defined, **3:** 351
  dynamics, **2:** 130–131, **3:** 351–352
  experimental test, **3:** 353–354
  Fisher model, **3:** 353–354
  models of evolution, **3:** 352–353
  role of mixis, **3:** 354–355
  Wright model, **3:** 353–354
Periodontal disease, **3:** 269
  oral microbial ecology, **3:** 271
  pathogenesis, **3:** 278–279
  spirochetes, **4:** 71
Periodontal pocket, defined, **3:** 269
Peripheral protein, defined, **1:** 455
Periplasmic flagella, defined, **4:** 67
Periseptal annuli, gram-negative bacteria, **1:** 531–532
Peristalsis, defined, **2:** 268
Peritrichous, defined, **2:** 157
Permease proteins, **1:** 434
Permissible exposure limits
  air contaminants, **2:** 578
  defined, **2:** 571
Peroxide. *See* Hydrogen peroxide
Pesticide biodegradation, **3:** **357–361**
  biochemistry, **3:** 358–359
  chemistry, **3:** 357–358
    benzene nucleus, **3:** 357
    chloroaromatics, **3:** 357–358
  molecular biology, **3:** 359–360
  recalcitrance, **1:** 273
  taxonomy of microorganisms for, **3:** 358

Rhodospirillaceae, **4:** 128
Ribavirin
  antiviral effectiveness
    arenaviruses, **1:** 141
    human immunodeficiency virus, **1:** 145
    respiratory syncytial virus, **1:** 141
  combined with other drugs, **1:** 146
Ribonuclease
  defined, **1:** 181
  pancreatic, **4:** 222
Ribonucleoside diphosphate reductase, **3:** 323–324
Ribonucleotides, reduction, **3:** 254, 256
Ribose binding protein, bacterial chemoreceptor, **1:** 503
Ribosomal proteins
  evolutionary conservation, **3:** 582–583
  primary structure, **3:** 573–574
  secondary and tertiary structure, **3:** 574
Ribosomal RNA. *See* RNA, ribosomal
Ribosomes, **3: 573–583**
  antibiotic resistance, **1:** 99
  biosynthesis and assembly, **3:** 582–583
    eukaryotes, **3:** 583
    prokaryotes, **3:** 582–583
  defined, **1:** 97, **3:** 573
  electron microscopy, **2:** 58
  function, **3:** 579–582
    antibiotics and, **3:** 581–582
    elongation, **3:** 581
    initiation, **3:** 580–581
    termination, **3:** 581
  microbial starvation, **3:** 161
  protection from tetracyclines, **1:** 101
  protocells, **3:** 311
  structure
    archaebacteria, **1:** 158, **3:** 574
    arrangement of proteins, **3:** 577–579
    eukaryotes, **3:** 575
    models of ribosomes, **3:** 578–579
    primary r-protein structure, **3:** 573–574
    primary ribosomal RNA structure, **3:** 574
    prokaryotes, **3:** 573–575
    secondary and tertiary r-protein structure, **3:** 574
    secondary ribosomal RNA structure, **3:** 574–575, 576
    structure-function relationships, **3:** 579
    subunits and ribosomes, **3:** 575, 577–579
    tertiary ribosomal RNA structure, **3:** 575
  synthesis, **3:** 263–265

thermostability, **4:** 221–222
α-Ribosylthymine triphosphate, *de novo* synthesis, **3:** 254–255, 256
Ribozymes
  defined, **2:** 143, **3:** 417
  plant disease resistance, **3:** 425
Rice paddies, methanogenesis, **3:** 92, 95
Rickettsia akari
  habitat, **3:** 594
  human disease, **3:** 603
Rickettsia australis, **3:** 594–595
Rickettsia canada, **3:** 593
Rickettsia conorii
  habitat, **3:** 594
  human disease, **3:** 603
Rickettsia diaporica, **3:** 587
Rickettsia japonica, **3:** 595
Rickettsia prowazekii
  habitat, **3:** 591, 593, 595–596
  human disease, **3:** 586–587, 602
  isolation, **3:** 607
Rickettsia quintana, **3:** 595–596
Rickettsia rickettsii
  habitat, **3:** 593–594
  human disease, **3:** 587, 602–603
  isolation, **3:** 607
Rickettsia sibirica, habitat, **3:** 594
Rickettsia tsutsugamushi
  habitat, **3:** 595
  human disease, **3:** 603
  isolation, **3:** 607
  vaccine, **3:** 607–608
Rickettsia typhi
  habitat, **3:** 593
  human disease, **3:** 587
  isolation, **3:** 607
Rickettsias, **3: 585–610**
  animal diseases
    pathogenesis, **3:** 603–604
    pathogens, **3:** 588
  cultivation of established strains, **3:** 604–607, 609
  defined, **3:** 585
  ecophysiology, **3:** 597–601
    host cells and defense mechanisms, **3:** 598
    survival and multiplication in host cells, **3:** 598–601
  habitat, **3:** 591, 593–597
    Coxiella burnetii, **3:** 596, 609
    Ehrlichieae, **3:** 596–597
    Rochalimaea, **3:** 595–596
    scrub typhus rickettsia, **3:** 595

Standards
  defined, **2:** 493
  infectious waste management, **2:** 502
  laboratory safety. *See* Laboratory safety and
    regulations
Staphylococcus
  coagulase-negative staphylococci, **2:** 322–323
  numerical taxonomy, **4:** 184
  skin microbiology, **4:** 23–24, 26–27, 29–31, 32
  subspecies, **2:** 320, 322–323
Staphylococcus aureus
  food poisoning, **2:** 211
  gastroenteritis, **3:** 61
  hospital infections, **2:** 443, 444
  milk contamination, **2:** 4
  pathogenicity, **2:** 320, 322
Staphylococcus epidermidis, **2:** 322, 444
Staphylococcus haemolyticus, **2:** 322
Staphylococcus lugdunensis, **2:** 322
Staphylococcus saprophyticus, **2:** 322
Staphylococcus schleiferi, **2:** 322
Starch degradation, in the rumen, **3:** 631
Starch-hydrolyzing enzymes, **2:** 82–84
  α-amylase, **2:** 82–83
  β-amylase, **2:** 83
  applications, **2:** 83–84
  debranching enzymes, **2:** 83
  glucoamylase, **2:** 83
Starter cultures
  defined, **2:** 1
  fermented dairy products, **2:** 6
  genetic improvement, **2:** 198–199
Starvation-survival
  defined, **2:** 617, 620
  oligotrophic environments, **2:** 619–624
    degrees of starvation-survival, **2:** 621–622
    patterns of, **2:** 620
    physiological processes, **2:** 622–624
  rumen microorganisms, **3:** 638
Starvation-survival, genetics, **3: 159–170**
  energy generation, **3:** 160–162
    endogenous respiration, **3:** 162
    protein degradation, **3:** 161
    reserve polymers, **3:** 160–161
    RNA, **3:** 161
    stringent response, **3:** 161–162
  scavenging capacity enhancement, **3:** 162–166
    carbon starvation-escape response, **3:**
      165–166
    iron starvation-escape response, **3:** 165
    nitrogen starvation-escape response, **3:**
      162–164

phosphorous starvation-escape response, **3:**
  164–165
starvation gene regulation, **3:** 168–169
  Cst versus Pex genes, **3:** 168
  KatF gene, **3:** 168
  σ^{32}, **3:** 168–169
starvation proteins, **3:** 166–168
  defined, **3:** 159
  stress resistance, **3:** 169
  stress resistance proteins, **3:** 169
Statistical methods for microbiology, **4: 75–85.** *See
  also* Bioassays in microbiology
  descriptive, **4:** 77–81
    graphics, **4:** 77–79
    measures, **4:** 79–81
  epidemiological use, **2:** 102–103
  measurement, **4:** 76–77
  probability models, **4:** 81–83
    continuous random variables, **4:** 81–82
    discrete random variables, **4:** 82
    fundamental assumptions, **4:** 81
    goodness-of-fit, **4:** 82–83
  statistical inference, **4:** 83–85
    point estimation and associated uncertainty,
      4: 83–84
    significance testing, **4:** 84–85
Steady state (continuous culture), defined, **1:** 559
Steam/compaction treatment, infectious waste, **2:**
  497–499
Steam sterilization, infectious waste, **2:** 496
Stem cell, defined, **3:** 45
Stemphylium
  degradation of cyanide, **2:** 343
  potential use for waste treatment, **2:** 346
Stephenson, Marjorie, **2:** 427
Sterigmatocystin
  in animals, **3:** 222
  food poisoning, **2:** 212
  in humans, **3:** 220
Sterilization, **4: 87–103**
  chemical sterilization, **4:** 95–98
    chlorine dioxide, **4:** 96, 98
    ethylene oxide, **4:** 95–96
    formaldehyde, **4:** 96, 98
    gaseous sterilants, **4:** 95–96
    gluteraldehyde, **4:** 97
    hydrogen peroxide, **4:** 96, 97
    liquid sterilants, **4:** 97–98
    peracetic acid, **4:** 96, 97–98
    β-propiolactone, **4:** 96
  commercial sterility, defined, **2:** 1
  defined, **4:** 87

T

T–2 toxin, **3:** 220–221, 223
T-cell anergy
  defined, **2:** 601
  lepromatous leprosy, **2:** 609
T-cell growth factor (interleukin-2). *See under*
  Interleukins
T-cell receptors
  composition, **4:** 247
  ligand recognition, **4:** 247
  rabies virus epitopes and, **3:** 502
  TCR complex, defined, **4:** 244
  TCR ligand, defined, **4:** 244
T-cell replacing factor (interleukin-5). *See under*
  Interleukins
T-cell testing for Lyme disease, **2:** 643
T cells, **4: 243–253.** *See also* B cells; Immunity
  antigen-directed immune response, **4:** 248, 249
  antigen presentation, **4:** 247–252
  antigen processing, **4:** 248, 249
  antigen stimulation, **4:** 247–248
  cell-mediated immunity in leprosy, **2:** 602,
    608–610
    genetics, **2:** 609–610
    macrophages, **2:** 610
    T-cell anergy, **2:** 609
    T-cell responses, **2:** 609
    tuberculoid leprosy, **2:** 602–603
    vaccines, **2:** 610
  changes induced by *Borrelia burgdorferi*, **2:**
    645–646
  cyclosporin A, **1:** 333–334
  cytokines
    feedback in immune response, **4:** 252
    interleukin-2 activation of T cells, **1:** 644
  defined, **2:** 467, **4:** 244
  functions, **4:** 244–247
  HIV vaccine design and, **1:** 134–135
  immune response mechanisms, **4:** 248–252
  normal functions, **3:** 482
  pathologic response to infection or parasites, **4:**
    252–253
    autoimmune disease, **4:** 253
    granulomatous reaction, **4:** 253
    Leishmania infections, **4:** 252
    superantigens, **4:** 253
  replication of human immunodeficiency virus,
    1: 27–28
T cells, cytotoxic. *See* Cytotoxic T cells
T cells, helper. *See* Helper T cells

T-DNA, Agrobacterium
  basic infection cycle, **1:** 38
  defined, **1:** 37
  DNA delivery process, **1:** 44–46
    movement of transferred intermediate, **1:**
      45–46
    production of transferred intermediate, **1:**
      44–45
  genes involved in tumorigenesis, **1:** 38–39
  genetic engineering with Agrobacterium, **1:**
    49–50
  movement and integration inside plant cells, **1:**
    46–47
  opine production, **1:** 47–48
  signaling involved in DNA transfer, **1:** 39, 41–43
  summary of *vir* gene products, **1:** 40
  tumorigenesis and hairy root formation, **1:**
    48–49
  vir gene induction, **1:** 41–43
T lymphocytes. *See* T cells
T-suppressor factors. *See* Suppressor factors,
    immunologic
Tachycardia
  defined, **1:** 495
  detection of bacteria in blood, **1:** 495
Tachyzoite, defined, **4:** 255
Tailed bacteriophages. *See also* Bacteriophages
  Myoviridae, **1:** 205
  Podoviridae, **1:** 205
  Siphoviridae, **1:** 205
Tannins, inhibition of enzymes, **2:** 618
TAR region, human immunodeficiency virus, **1:**
    21–22, 23
Target site, defined, **4:** 311
tat gene, human immunodeficiency virus, **1:** 23
Tat protein-TAR region, human
    immunodeficiency virus, **1:** 21–22
Tatum, E. L., **2:** 427, 430
Tatum, E. M., **1:** 547
Taxis, defined, **2:** 165. *See also* Chemotaxis
Taxon
  bacterial identification, **2:** 458
  defined, **2:** 259, 457, **3:** 171
  homogeneity, **2:** 460
  identification of natural populations, **3:** 172
  objective criteria to define, **3:** 172–173
Taxonomic methods, **4: 179–200**
  applications, **4:** 196–200
  bacterial identification, **2:** 458
  chemical, **4:** 189–192
    cytoplasmic membrane, **4:** 191–192
    end-products of metabolism, **4:** 191–192

V

X

Index of Related Titles

ISBN 0-12-226894-6

9 780122 268946

90018